Lecture Notes in Computer Science 11080

Commenced Publication in 1973
Founding and Former Series Editors:
Gerhard Goos, Juris Hartmanis, and Jan van Leeuwen

More information about this series at http://www.springer.com/series/7409

Mathias Weske · Marco Montali
Ingo Weber · Jan vom Brocke (Eds.)

Business Process Management

16th International Conference, BPM 2018
Sydney, NSW, Australia, September 9–14, 2018
Proceedings

 Springer

Editors
Mathias Weske
Hasso-Plattner-Institute
University of Potsdam
Potsdam
Germany

Ingo Weber (iD)
Data61
CSIRO
Eveleigh, NSW
Australia

Marco Montali (iD)
Free University of Bozen Bolzano
Bolzano
Italy

Jan vom Brocke
University of Liechtenstein
Vaduz
Liechtenstein

ISSN 0302-9743   ISSN 1611-3349   (electronic)
Lecture Notes in Computer Science
ISBN 978-3-319-98647-0   ISBN 978-3-319-98648-7   (eBook)
https://doi.org/10.1007/978-3-319-98648-7

Library of Congress Control Number: 2018950653

LNCS Sublibrary: SL3 – Information Systems and Applications, incl. Internet/Web, and HCI

This Springer imprint is published by the registered company Springer Nature Switzerland AG
The registered company address is: Gewerbestrasse 11, 6330 Cham, Switzerland

Preface

The 16th International Conference on Business Process Management provided a forum for researchers and practitioners in the broad and diverse field of business process management. To accommodate for the diversity of the field, this year the BPM conference introduced a track structure, with tracks for foundations, engineering, and management. These tracks cover not only different phenomena of interest and research methods, but, consequently, also ask for different evaluation criteria. Each track had a dedicated track chair and a dedicated Program Committee. The track chairs, together with a consolidation chair, were responsible for the scientific program.

BPM 2018 was organized by the Service Oriented Computing Research Group, School of Computer Science and Engineering, University of New South Wales, in collaboration with research groups at Macquarie University, the University of Technology Sydney, and the Service Science Society. The conference was held in Sydney, Australia, during September 9–14, 2018.

The conference received 140 full paper submissions, well distributed over the tracks. Each paper was reviewed by at least three Program Committee members, and a Senior Program Committee member who triggered and moderated scientific discussions and reflected these in an additional meta-review. We accepted 27 excellent papers in the main conference (acceptance rate 19%), nine in each track. 14 submissions appeared in the BPM Forum, published in a separate volume of the Springer LNBIP series.

Implementing the track system, we can report that (i) BPM continued to attract excellent papers from the core BPM community and (ii) BPM 2018 managed to attract excellent papers from the management discipline. In the words of BPM conference founder and long-time chair of the BPM Steering Committee Wil van der Aalst: "The track system works". There were also lessons learned, which we expose in a short paper that can be found in these proceedings.

In the foundations track led by Marco Montali, core BPM topics including process discovery and performance analysis were represented. There were also papers on conceptual modeling aspects including domain-specific process modeling, process collaborations, and aspects related to time in business processes. The engineering track was led by Ingo Weber. Structurally quite similar to the foundations track, there were papers related to phenomena that have been discussed at BPM in recent years, but there were also papers that open the conference to new topics, for instance machine learning aspects in BPM. The extension in breadth of the BPM conference can be found mainly in the management track, led by Jan vom Brocke. It is interesting to see that several papers concentrate on challenges related to method analysis and method selection. The value of technical solutions is analyzed with respect to their impact and usability in a concrete business context. This track also features papers about digital innovation and the role of business process management in this context.

In his keynote, Manfred Reichert provided an engineering perspective on business process management, by investigating the relationship of BPM technology and

Internet-of-Things scenarios. Brian Pentland took a management perspective on business processes by looking at patterns of actions in organizations and by proposing a novel role of BPM techniques. Sanjiva Weerawarana introduced Ballerina, a middleware platform that can play an important role in future BPM integration scenarios.

Organizing a scientific conference is a complex process, involving many roles and many more interactions. We thank all our colleagues involved for their excellent work. The workshop chairs attracted 11 innovative workshops, the industry chairs organized a top-level industry program, and the demo chairs attracted many excellent demos. The panel chairs compiled an exciting panel, which opened doors to future research challenges. Without the publicity chairs, we could not have attracted such an excellent number of submissions. Younger researchers benefited from excellent tutorials; doctoral students received feedback about their work from experts in the field at the Doctoral Consortium. The mini-sabbatical program helped to bring additional colleagues to Australia. The proceedings chair professionally interacted with Springer and with the authors to prepare excellent volumes of LNCS and LNBIP.

The members of the tracks' Program Committees and of the Senior Program Committees deserve particular acknowledgment for their dedication and commitment. We are grateful for the help and expertise of sub-reviewers, who provided valuable feedback during the reviewing process and engaged in deep discussions at times. BPM 2018 had a dedicated process to consolidate paper acceptance across tracks. During the very intensive weeks of this phase, many Senior Program Committee members evaluated additional papers and were engaged in additional discussions. Special thanks goes to these colleagues, who were instrumental during this decisive phase of the reviewing process.

Finally, we thank the Organizing Committee and the Local Arrangements Committee, led by Boualem Benatallah and Jian Yang. The development of the program structure was particularly challenging, because with the new track structure this year many more papers were accepted than traditionally at BPM. Still, we managed to avoid concurrency between main conference papers, while proving a packed, exciting program. Through their generous support, the sponsors of the conference had a great share in its success. We thank the conference partner Data61, the Platinum sponsor Signavio, the Gold sponsors Celonis and IBM Research, and the Bronze sponsors Bizagi and Springer for their support. We also thank the University of New South Wales and Macquarie University for their enormous and high-quality support.

September 2018

Mathias Weske
Marco Montali
Ingo Weber
Jan vom Brocke

Organization

BPM 2018 was organized by the University of New South Wales, in collaboration with Macquarie University, the University of Technology Sydney, and the Service Science Society, and took place in Sydney, Australia.

Steering Committee

Mathias Weske (Chair)	University of Potsdam, Germany
Boualem Benatallah	University of New South Wales, Australia
Jörg Desel	University of Hagen, Germany
Schahram Dustdar	Vienna University of Technology, Austria
Marlon Dumas	University of Tartu, Estonia
Wil van der Aalst	RWTH Aachen University, Germany
Michael zur Muehlen	Stevens Institute of Technology, USA
Stefanie Rinderle-Ma	University of Vienna, Austria
Barbara Weber	Technical University of Denmark, Denmark
Manfred Reichert	Ulm University, Germany
Jan Mendling	Vienna University of Economics and Business, Austria

Executive Committee

General Chairs

Boualem Benatallah	University of New South Wales, Australia
Jian Yang	Macquarie University, Australia

Program Chairs

Mathias Weske (Consolidation Chair)	University of Potsdam, Germany
Marco Montali (Chair Track I)	University of Bolzano, Italy
Ingo Weber (Chair Track II)	Data61|CSIRO, Australia
Jan vom Brocke (Chair Track III)	University of Liechtenstein, Liechtenstein

Industry Chairs

Fabio Casati	University of Trento, Italy
Gero Decker	Signavio, Germany
Surya Nepal	Data61|CSIRO, Australia

Workshops

Florian Daniel	Politecnico di Milano, Italy
Hamid Motahari	IBM, Almaden Research Center, San Jose, USA
Michael Sheng	Macquarie University, Australia

Demo Chairs

Raffaele Conforti	The University of Melbourne, Australia
Massimiliano de Leoni	Eindhoven University of Technology, The Netherlands
Barbara Weber	Technical University of Denmark, Denmark

Publicity Chairs

Cinzia Cappiello	Politecnico di Milano, Italy
Daniela Grigori	Université Paris Dauphine, France
Oktay Türetken	Eindhoven University of Technology, The Netherlands
Lijie Wen	Tsinghua University, China

Sponsorship and Community Liaison Chairs

François Charoy	University of Lorraine, France
Onur Demirors	Izmir Institute of Technology, Turkey and UNSW Sydney, Australia
Fethi Rabhi	UNSW Sydney, Australia
Daniel Schlagwein	UNSW Sydney, Australia

Panel Chairs

Athman Bouguettaya	The University of Sydney, Australia
Mohand-Saïd Hacid	Université Claude Bernard Lyon 1, France
Manfred Reichert	Ulm University, Germany

Tutorial Chairs

Marcello La Rosa	The University of Melbourne, Australia
Stefanie Rinderle-Ma	University of Vienna, Austria
Farouk Toumani	Blaise Pascale University, France

Doctoral Consortium Chairs

Yan Wang	Macquarie University, Australia
Josep Carmona	Universitat Politècnica de Catalunya, Spain

Mini-Sabbatical Program Chairs

Shazia Sadiq	The University of Queensland, Australia
Moe Thandar Wynn	Queensland University of Technology, Australia

Local Organization Liaison and Networking Chair

Ghassan Beydoun   University of Technology Sydney, Australia

Local Arrangements Committee

Olivera Marjanovic   University of Technology Sydney, Australia
  (Co-chair)
Lina Yao (Co-chair)   UNSW Sydney, Australia
Kyeong Kang   University of Technology Sydney, Australia
Wei Zhang   Macquarie University, Australia

Proceedings Chair

Luise Pufahl   University of Potsdam, Germany

Web and Social Media Chair

Amin Beheshti   Macquarie University, Australia

Track I (Foundations)

Senior Program Committee

Florian Daniel   Politecnico di Milano, Italy
Dirk Fahland   Eindhoven University of Technology, The Netherlands
Giancarlo Guizzardi   Free University of Bozen-Bolzano, Italy
Thomas Hildebrandt   IT University of Copenhagen, Denmark
Marcello La Rosa   The University of Melbourne, Australia
John Mylopoulos   University of Toronto, Canada
Manfred Reichert   Ulm University, Germany
Jianwen Su   University of California at Santa Barbara, USA
Hagen Völzer   IBM Research - Zurich, Switzerland
Matthias Weidlich   Humboldt-Universität zu Berlin, Germany

Program Committee

Ahmed Awad   Cairo University, Egypt
Giuseppe De Giacomo   Sapienza University of Rome, Italy
Jörg Desel   Fernuniversität in Hagen, Germany
Claudio di Ciccio   Vienna University of Economics and Business, Austria
Chiara Di Francescomarino   Fondazione Bruno Kessler-IRST, Italy
Rik Eshuis   Eindhoven University of Technology, The Netherlands
Hans-Georg Fill   University of Bamberg, Germany
Guido Governatori   Data61|CSIRO, Australia
Gianluigi Greco   University of Calabria, Italy
Richard Hull   IBM, USA
Irina Lomazova   National Research University Higher School
                              of Economics, Russian Federation
Alessio Lomuscio   Imperial College London, UK

Fabrizio Maria Maggi	University of Tartu, Estonia
Andrea Marrella	Sapienza University of Rome, Italy
Heinrich C. Mayr	Alpen-Adria-Universität Klagenfurt, Austria
Oscar Pastor	Universitat Politècnica de València, Spain
Geert Poels	Ghent University, Belgium
Artem Polyvyanyy	Queensland University of Technology, Australia
Wolfgang Reisig	Humboldt Universität zu Berlin, Germany
Arik Senderovich	University of Toronto, Canada
Andreas Solti	Vienna University of Economics and Business, Austria
Ernest Teniente	Universitat Politècnica de Catalunya, Spain
Daniele Theseider Dupré	Università del Piemonte Orientale, Italy
Victor Vianu	University of California San Diego, USA
Lijie Wen	Tsinghua University, China

Track II (Engineering)

Senior Program Committee

Jan Mendling	Vienna University of Economics and Business, Austria
Cesare Pautasso	University of Lugano, Switzerland
Hajo A. Reijers	Vrije Universiteit Amsterdam, The Netherlands
Stefanie Rinderle-Ma	University of Vienna, Austria
Pnina Soffer	University of Haifa, Israel
Wil van der Aalst	RWTH Aachen University, Germany
Boudewijn van Dongen	Eindhoven University of Technology, The Netherlands
Jianmin Wang	Tsinghua University, China
Barbara Weber	Technical University of Denmark, Denmark

Program Committee

Marco Aiello	University of Stuttgart, Germany
Amin Beheshti	Macquarie University, Australia
Andrea Burattin	Technical University of Denmark, Denmark
Cristina Cabanillas	Vienna University of Economics and Business, Austria
Josep Carmona	Universitat Politècnica de Catalunya, Spain
Fabio Casati	University of Trento, Italy
Jan Claes	Ghent University, Belgium
Francisco Curbera	IBM, USA
Massimiliano de Leoni	Eindhoven University of Technology, The Netherlands
Jochen De Weerdt	Katholieke Universiteit Leuven, Belgium
Remco Dijkman	Eindhoven University of Technology, The Netherlands
Marlon Dumas	University of Tartu, Estonia
Schahram Dustdar	Vienna University of Technology, Austria
Gregor Engels	University of Paderborn, Germany
Joerg Evermann	Memorial University of Newfoundland, Canada
Walid Gaaloul	Télécom SudParis, France
Avigdor Gal	Technion, Israel

Luciano García-Bañuelos   University of Tartu, Estonia
Chiara Ghidini Fondazione   Fondazione Bruno Kessler-IRST, Italy
Daniela Grigori   University of Paris-Dauphine, France
Dimka Karastoyanova   University of Groningen, The Netherlands
Christopher Klinkmüller   Data61|CSIRO, Australia
Agnes Koschmider   Karlsruhe Institute of Technology, Germany
Jochen Kuester   Bielefeld University of Applied Sciences, Germany
Henrik Leopold   Vrije Universiteit Amsterdam, The Netherlands
Raimundas Matulevicius   University of Tartu, Estonia
Massimo Mecella   Sapienza University of Rome, Italy
Hamid Motahari   IBM, USA
Jorge Munoz-Gama   Pontificia Universidad Católica de Chile, Chile
Hye-Young Paik   The University of New South Wales, Australia
Luise Pufahl   University of Potsdam, Germany
Manuel Resinas   University of Seville, Spain
Shazia Sadiq   The University of Queensland, Australia
Minseok Song   Pohang University of Science and Technology,
                                 South Korea
Stefan Tai   Technical University of Berlin, Germany
Samir Tata   IBM, USA
Arthur Ter Hofstede   Queensland University of Technology, Australia
Farouk Toumani   Blaise Pascal University, France
Moe Wynn   Queensland University of Technology, Australia

Track III (Management)

Senior Program Committee

Joerg Becker   European Research Center for Information Systems,
                                 Germany
Alan Brown   University of Surrey, UK
Mikael Lind   University of Borås, Sweden
Peter Loos   Saarland University, Germany
Amy Looy   Ghent University, Belgium
Olivera Marjanovic   University of Technology, Sydney, Australia
Jan Recker   University of Cologne, Germany
Maximilian Roeglinger   University of Bayreuth, Germany
Michael Rosemann   Queensland University of Technology, Australia
Schmiedel Theresa   University of Liechtenstein, Liechtenstein
Peter Trkman   University of Ljubljana, Slovenia

Program Committee

Peyman Badakhshan   University of Liechtenstein, Liechtenstein
Alessio Braccini   University of Tuscia, Italy
Patrick Delfmann   University of Koblenz-Landau, Germany

Peter Fettke	German Research Center for Artificial Intelligence (DFKI) and Saarland University, Germany
Kathrin Figl	Vienna University of Economics and Business, Austria
Thomas Grisold	Vienna University of Economics and Business, Austria
Marta Indulska	The University of Queensland, Australia
Mieke Jans	Hasselt University, Belgium
Janina Kettenbohrer	University of Bamberg, Germany
John Krogstie	Norwegian University of Science and Technology, Norway
Xin Li	City University of Hong Kong, Hong Kong, SAR China
Alexander Maedche	Karlsruhe Institute of Technology, Germany
Willem Mertens	Queensland University of Technology, Australia
Charles Moeller	Aalborg University, Denmark
Oliver Mueller	IT University of Copenhagen, Denmark
Markus Nuettgens	University of Hamburg, Germany
Ferdinando Pennarola	Università L. Bocconi, Italy
Flavia Santoro	Universidade Federal do Estado do Rio de Janeiro, Brazil
Anna Sidorova	University of North Texas, USA
Silvia Inês Dallavalle de Pádua	University of São Paulo, Brazil
Vijayan Sugumaran	Oakland University, USA
Oliver Thomas	University of Osnabrück, Germany
Harry Wang	University of Delaware, USA
Charlotte Wehking	University of Liechtenstein, Liechtenstein
Axel Winkelmann	University of Würzburg, Germany
Dongming Xu	The University of Queensland, Australia
Weithoo Yue	City University of Hong Kong, Hong Kong, SAR China
Sarah Zelt	University of Liechtenstein, Liechtenstein
Michael Zur Muehlen	Stevens Institute of Technology, USA

Additional Reviewers

Alessio Cecconi
Alexey A. Mitsyuk
Alin Deutsch
Anna Kalenkova
Anton Yeshchenko
Armin Stein
Azkario Rizky Pratama
Bastian Wurm
Benjamin Meis
Benjamin Spottke

Bian Yiyang
Boris Otto
Brian Setz
Carl Corea
Chiara Ghidini
Christoph Drodt
Daning Hu
David Sanchez-Charles
Fabio Patrizi
Fabrizio Maria Maggi

Sponsors

Conference Partner

Platinum Sponsor

Gold Sponsor

Gold Sponsor

Bronze Sponsor

Keynotes

Business Process Management in the Digital Era: Scenarios, Challenges, Technologies

Manfred Reichert

Institute of Databases and Information Systems, Ulm University, Germany
manfred.reichert@uni-ulm.de

Abstract. The Internet of Things (IoT) has become increasingly pervasive in daily life as digitization plays a major role, both in the workplace and beyond. Along with the IoT, additional technologies have emerged, such as augmented reality, mobile and cognitive computing, blockchains or cloud computing, offering new opportunities for digitizing business processes. For example, the Industrial IoT is considered as essential for realizing the Industry 4.0 vision, which targets at the digital transformation of manufacturing processes by integrating smart machines, data analytics, and people at work. Though digitization is a business priority in many application areas, the role of digital processes and their relation with physical (i.e. real-world) ones have not been well understood so far, often resulting in an alignment gap between digital and physical process. In this keynote characteristic scenarios for digitizing processes in a cyber-physical world are illustrated and the challenges to be tackled are discussed. Moreover, a link between the scenarios and contemporary BPM technologies is established, indicating the mutual benefits of combining BPM with IoT and other digital technologies.

Beyond Mining: Theorizing About Processual Phenomena

Brain T. Pentland

Department of Accounting and Information Systems,
Michigan State University, USA
pentland@broad.msu.edu

Abstract. For decades, process miners have been toiling deep in the event logs of digitized organizations. Through this collective experience, the process mining community has developed a powerful set of tools and a compelling set of use cases for those tools (discovering, monitoring, improving, etc.)

In this talk, I want to suggest that some of these same tools may be useful for other entirely different kinds of problems. In particular, recognizing and comparing patterns of action should be useful for theorizing about a wide range of processual phenomena in organizational and social science.

Bringing Middleware to Everyday Developers with Ballerina

Sanjiva Weerawarana

Founder, Chairman and Chief Architect of WSO2, Sri Lanka
sanjiva@wso2.com

Abstract. Middleware plays an important role in making applications secure, reliable, transactional and scalable. Workflow management systems, transaction mangers, enterprise service buses, identity gateways, API gateways, application servers are some of the middleware tools that keep the world running. Yet everyday programmers don't have the luxury (or pain?) of such infrastructure and end up creating fragile systems that we all suffer from.

Ballerina is a general purpose, concurrent, transactional and statically & strongly typed programming language with both textual and graphical syntaxes. Its specialization is integration - it brings fundamental concepts, ideas and tools of distributed system integration into the language and offers a type safe, concurrent environment to implement such applications. These include distributed transactions, reliable messaging, stream processing, workflows and container management platforms. Ballerina's concurrency model is built on the sequence diagram metaphor and offers simple constructs for writing concurrent programs. Its type system is a modern type system designed with sufficient power to describe data that occurs in distributed applications. It also includes a distributed security architecture to make it easier to write applications that are secure by design. This talk will look at how Ballerina makes workflow and other middleware features into inherent aspects of a programming language and how it can help bring middleware to everyday programmers to make all programs better.

Contents

Reflections on BPM

BPM: Foundations, Engineering, Management   3
   Mathias Weske, Marco Montali, Ingo Weber, and Jan vom Brocke

Bringing Middleware to Everyday Programmers with Ballerina.   12
   Sanjiva Weerawarana, Chathura Ekanayake, Srinath Perera,
   and Frank Leymann

Track I: Concepts and Methods in Business Process Modeling and Analysis

Open to Change: A Theory for Iterative Test-Driven Modelling   31
   Tijs Slaats, Søren Debois, and Thomas Hildebrandt

Construction Process Modeling: Representing Activities,
Items and Their Interplay   48
   Elisa Marengo, Werner Nutt, and Matthias Perktold

Feature-Oriented Composition of Declarative Artifact-Centric
Process Models.   66
   Rik Eshuis

Animating Multiple Instances in BPMN Collaborations: From Formal
Semantics to Tool Support   83
   Flavio Corradini, Chiara Muzi, Barbara Re, Lorenzo Rossi,
   and Francesco Tiezzi

Managing Decision Tasks and Events in Time-Aware Business
Process Models. 102
   Roberto Posenato, Francesca Zerbato, and Carlo Combi

Track I: Foundations of Process Discovery

Interestingness of Traces in Declarative Process Mining: The Janus
LTLp$_f$ Approach. 121
   Alessio Cecconi, Claudio Di Ciccio, Giuseppe De Giacomo,
   and Jan Mendling

Unbiased, Fine-Grained Description of Processes Performance
from Event Data. 139
   Vadim Denisov, Dirk Fahland, and Wil M. P. van der Aalst

Abstract-and-Compare: A Family of Scalable Precision Measures
for Automated Process Discovery   158
   Adriano Augusto, Abel Armas-Cervantes, Raffaele Conforti,
   Marlon Dumas, Marcello La Rosa, and Daniel Reissner

Correlating Activation and Target Conditions in Data-Aware Declarative
Process Discovery   176
   Volodymyr Leno, Marlon Dumas, and Fabrizio Maria Maggi

Track II: Alignments and Conformance Checking

Efficiently Computing Alignments: Using the Extended Marking Equation. . .   197
   Boudewijn F. van Dongen

An Evolutionary Technique to Approximate Multiple Optimal Alignments. . .   215
   Farbod Taymouri and Josep Carmona

Maximizing Synchronization for Aligning Observed
and Modelled Behaviour   233
   Vincent Bloemen, Sebastiaan J. van Zelst, Wil M. P. van der Aalst,
   Boudewijn F. van Dongen, and Jaco van de Pol

Online Conformance Checking Using Behavioural Patterns   250
   Andrea Burattin, Sebastiaan J. van Zelst, Abel Armas-Cervantes,
   Boudewijn F. van Dongen, and Josep Carmona

Track II: Process Model Analysis and Machine Learning

BINet: Multivariate Business Process Anomaly Detection Using
Deep Learning   271
   Timo Nolle, Alexander Seeliger, and Max Mühlhäuser

Finding Structure in the Unstructured: Hybrid Feature Set Clustering
for Process Discovery   288
   Alexander Seeliger, Timo Nolle, and Max Mühlhäuser

act2vec, trace2vec, log2vec, and model2vec: Representation Learning
for Business Processes   305
   Pieter De Koninck, Seppe vanden Broucke, and Jochen De Weerdt

Who Is Behind the Model? Classifying Modelers Based on Pragmatic
Model Features   322
   Andrea Burattin, Pnina Soffer, Dirk Fahland, Jan Mendling,
   Hajo A. Reijers, Irene Vanderfeesten, Matthias Weidlich,
   and Barbara Weber

Finding the "Liberos": Discover Organizational Models with Overlaps   339
Jing Yang, Chun Ouyang, Maolin Pan, Yang Yu,
and Arthur H. M. ter Hofstede

Track III: Digital Process Innovation

On the Synergies Between Business Process Management
and Digital Innovation.   359
Amy Van Looy

Effective Leadership in BPM Implementations: A Case Study
of BPM in a Developing Country, Public Sector Context.   376
Rehan Syed, Wasana Bandara, and Erica French

Conceptualizing a Framework to Manage the Short Head and Long Tail
of Business Processes   392
Florian Imgrund, Marcus Fischer, Christian Janiesch,
and Axel Winkelmann

Using Business Process Compliance Approaches for Compliance
Management with Regard to Digitization: Evidence from a Systematic
Literature Review   409
Stefan Sackmann, Stephan Kuehnel, and Tobias Seyffarth

Big Data Analytics as an Enabler of Process Innovation Capabilities:
A Configurational Approach.   426
Patrick Mikalef and John Krogstie

Track III: Method Analysis and Selection

Assessing the Quality of Search Process Models.   445
Marian Lux, Stefanie Rinderle-Ma, and Andrei Preda

Predictive Process Monitoring Methods: Which One Suits Me Best?.   462
Chiara Di Francescomarino, Chiara Ghidini, Fabrizio Maria Maggi,
and Fredrik Milani

How Context-Aware Are Extant BPM Methods? - Development
of an Assessment Scheme   480
Marie-Sophie Denner, Maximilian Röglinger, Theresa Schmiedel,
Katharina Stelzl, and Charlotte Wehking

Process Forecasting: Towards Proactive Business Process Management   496
Rouven Poll, Artem Polyvyanyy, Michael Rosemann,
Maximilian Röglinger, and Lea Rupprecht

Correction to: Using Business Process Compliance Approaches
for Compliance Management with Regard to Digitization:
Evidence from a Systematic Literature Review C1
    Stefan Sackmann, Stephan Kuehnel, and Tobias Seyffarth

Author Index   513

Reflections on BPM

BPM: Foundations, Engineering, Management

Mathias Weske[1]([✉]), Marco Montali[2], Ingo Weber[3], and Jan vom Brocke[4]

[1] Hasso Plattner Institute, University of Potsdam, Potsdam, Germany
mathias.weske@hpi.de
[2] Free University of Bozen-Bolzano, Bolzano, Italy
montali@inf.unibz.it
[3] Data61, CSIRO, Sydney, Australia
ingo.weber@data61.csiro.au
[4] University of Liechtenstein, Vaduz, Liechtenstein
jan.vom.brocke@uni.li

Abstract. This paper reports on the introduction of a track system at the BPM conference series and the experiences made during the organization of BPM 2018, the first issue implementing the track system. By introducing dedicated tracks for foundations, engineering, and management, with dedicated evaluation criteria and program committees, the BPM steering committee aims at providing a fair chance for acceptance to all submissions to the conference. By introducing a management track, the conference reaches out to the management community, which investigates phenomena in business process management from a non-technical perspective that complements the technical orientation of traditional BPM papers. We elaborate on the background of and motivation for the track system, and we discuss the lessons learned in the first iteration of the track structure at BPM 2018.

1 Introduction

This paper reports on the background of and motivation for introducing a track system at the International Conference on Business Process Management (BPM) conference series, and it discusses the experiences gathered during the organization of BPM 2018, the first iteration in the conference series implementing that structure.

The evolution of the BPM conference series towards the track structure is based on two observations. The first observation relates to the increasingly tough reviewing process that we could observe at BPM in recent years. There has been the trend that reviewers were asking not only for a strong technical contribution, but also for a convincing empirical evaluation. While these criteria are applicable to research approaches based on the design science paradigm, these are not well suited for papers looking at foundational aspects.

The second observation is concerned with the breadth of topics discussed at BPM conferences. BPM has its roots in computer science and information

© Springer Nature Switzerland AG 2018
M. Weske et al. (Eds.): BPM 2018, LNCS 11080, pp. 3–11, 2018.
https://doi.org/10.1007/978-3-319-98648-7_1

systems engineering, so that, traditionally, papers presented at BPM conferences have a significant technical contribution. Looking at real-world scenarios in business process management, however, an additional discipline plays an important role: management. Since traditionally BPM papers are expected to have a strong technical contribution, only few management-oriented submissions were presented at the conference.

To provide a fair chance for all papers submitted to BPM conferences, different evaluation criteria have to be employed to review foundational papers, to review papers with an engineering focus, and to review papers that investigate management aspects of business process management. These considerations have triggered the establishment of specific tracks covering foundational (Track I), engineering (Track II), and management aspects (Track III).

The reminder of this paper is organized as follows. The next section elaborates on the background of and motivation for the track structure. We then discuss the lessons learned in the first iteration while organizing BPM 2018, specific to each track. We close with an outlook on future iterations of the conference and concluding remarks.

2 Background and Motivation

Since its inauguration in 2003, the International Conference on Business Process Management has developed to a well-established conference that has shaped the business process management community. It has been a conscious decision by the BPM Steering Committee to position BPM as a conference with a technical focus. This decision proved important to establish the conference as a respected venue for research in computer science aspects of business process management.

As can be expected from an active research community and its flagship conference, the topics being discussed at BPM conferences have changed in an evolutionary manner over the years. The first issues of the conference mainly reported on formal aspects of business processes. With the rise of service oriented computing in the mid 2000s, topics like service composition and quality of service emerged. With the establishment of process mining and the general interest in data analytics, data-driven empirical research has become a major focus of BPM over the last decade.

Two observations can be made, each of which can be regarded as a challenge for the future development of the BPM conference series.

- With the shift in topics, a change in the evaluation criteria employed during the reviewing phase came along; evaluation criteria that are important in data-driven empirical research became the standard. This has led foundational and innovative papers having lower chances of being accepted at the conference, because these can typically not be empirically evaluated in a conclusive way.
- There is a strong stream of research that has not been represented at the conference in adequate strength, and this relates to management. Contributions

in the management field of BPM typically do not have a strong technical contribution. Since the typical reviewer at BPM conferences expect this, only few papers by the management community have been accepted at the conference.

Last year, the steering committee has decided to address these challenges by introducing a track system. It leads to a separation of foundational research, engineering research, and management research. Since papers following different research methods can only be evaluated fairly, if specific evaluation criteria are used, each track comes with a set of dedicated evaluation criteria. To implement these criteria, each track is led by a track chair, and each track has a dedicated program committee. A consolidation chair is responsible for coordinating the processes across the tracks. The reviewing process is enhanced with a dedicated consolidation phase, in which paper acceptance is discussed between tracks.

3   Track Structure

BPM 2018 features three tracks, foundations, engineering, and management. The tracks are characterized with respect to the phenomena studied, the research methods used, and the evaluation criteria employed during the reviewing phase. Notice that the following characterization is taken from the call for papers; we repeat it here for reference, because it tries to characterize the tracks as concisely as possible.

Track I [Foundations] invites papers that follow computer science research methods. This includes papers that investigate the underlying principles of BPM systems, computational theories, algorithms, and methods for modeling and analyzing business processes. This track also covers papers on novel languages, architectures, and other concepts underlying process aware information systems, as well as papers that use conceptual modeling techniques to investigate problems in the design and analysis of BPM systems. Papers in Track I are evaluated according to computer science standards, including sound formalization, convincing argumentation, and, where applicable, proof of concept implementation, which shows that the concepts can be implemented as described. Since papers typically do not have an immediate application in concrete business environments, empirical evaluation does not play a major role in Track I.

Track II [Engineering] invites papers that follow information systems engineering methods. The focus is on the investigation of artifacts and systems in business environments, following the design science approach. Papers in this track are expected to have a strong empirical evaluation that critically tests criteria like usefulness or added value of the proposed artifact. This track covers business process intelligence, including process mining techniques, and the use of process models for enactment, model-driven engineering, as well as interaction with services and deployment architectures like the Cloud. It also covers BPM systems in particular domains, such as digital health, smart mobility, or Internet of Things. Empirical evaluations are important to show the merits of the artifact introduced. A self-critical discussion of threats to validity is expected.

Formalization of problems and solutions should be used where they add clarity or are beneficial in other ways.

Track III [Management] invites papers that aim at advancing our understanding of how BPM can deliver business value, for instance how it builds organizational capabilities to improve, innovate or transform the respective business. Papers that study the application and impact of BPM methods and tools in use contexts based on empirical observation are highly welcome. Areas of interest include a wide range of capability areas that are relevant for BPM, such as strategic alignment, governance, methods, information technology, and human aspects including people and culture. We seek contributions that advance our understanding on how organizations can develop such capabilities in order to achieve specific objectives in given organizational contexts. Papers may use various strategies of inquiry, including case study research, action research, focus group research, big data analytics research, neuroscience research, econometric research, literature review research, survey research or design science research. Papers will be evaluated according to management and information systems standards.

4   Track I: Foundations

The foundations track focuses on papers that follow computer science research methods, with a strong emphasis on the core computing principles and methods underlying the BPM field. This ranges from the investigation of BPM systems and their extension through novel languages and functionalities, to the development of theories, algorithms and methods for (conceptual) modeling of processes and their (formal) analysis. In this respect, typical foundational papers show a strong and sound formalization, with convincing argumentation and rigorous exposition, but do not focus on an immediate, direct application of the presented results in concrete business environments. This is why proof-of-concept implementations are welcome, but it is not expected that they come with an empirical evaluation, as requested in the engineering track.

The accepted papers provide a quite fair coverage of recent developments of the foundations of BPM, with particular emphasis on multi-perspective process models, where decisions, data, temporal aspects, and multiple instances are considered alongside the traditional process dimensions. Some papers concentrate on conventional process modeling notations such as BPMN, while others delve into alternative modelling paradigms, with prominence of declarative, constraint-based approaches. Interestingly, papers employ a quite wide range of techniques, from computational logic to formal and statistical methods. Papers polarized themselves in two phases of the BPM lifecycle: modeling/analysis, and mining. In Track I, 9 papers were finally accepted at the main conference and 5 papers were accepted at the BPM Forum.

By comparing the call for papers with the submissions, reviews, and consequent discussion, the following critical points emerged:

– The evaluation obsession: Even though the call for papers explicitly indicated that evaluation is not the central focus of Track I, the lack of an extensive or on-the-field evaluation has been one of the most frequent reasons to lean towards rejection. During the discussion phase, this has been then clarified, still with resistance in some cases. The track system certainly helped, but more time is needed for the community to get acquainted with the fact that good foundational papers do not necessarily come with an evaluation that is based on data.
– The relevance question: We have observed a very diversified opinion of reviewers on the relationship between relevance of the presented results, and the targeted modeling notation. Some reviewers considered the choice of an unconventional modeling notation, or even alternative notations to the well-established ones, as a reason to consider the contribution weak, irrespective of the actual results therein. This is another point of reflection for the community, given that especially for foundational papers it is very hard to predict today what will be relevant in the future.

All in all, the initial reviews tended to be hypercritical about the submissions. Intense discussions were often needed to single out explicitly also the positive aspects of the contributions, and to assess them in their full generality. This was partly expected, also considering that Track I received quite many more papers than expected, consequently creating a quite heavy load for PC members.

In terms of covered topics, the main unifying theme among many of the accepted papers is the simultaneous consideration of multiple process perspectives. This witnesses the increasing maturity of the field, and the fact that it is finally time to "bring the pieces together". We expect this trend to continue in the coming years. Some accepted papers reflect on previously engineered techniques (in particular for process mining), witnessing that solid foundation papers do not necessarily focus on the formalization of process notations, but can also systematically study Track II contributions, creating a synergy between foundations and engineering.

An open challenge for Track I is how to create a similar kind of synergy towards Track II and Track III, so as to guarantee that strong foundation papers are consequently subject to extensive evaluation both from the engineering and management perspective.

5   Track II: Engineering

The engineering track of the BPM conference focuses on papers that follow the design science approach. In short: a new artifact (algorithm, method, system, or similar) is suggested and rigorously tested. Therefore the track has an emphasis on a strong, self-critical evaluation. The evaluation should expose the artifact to real or realistic conditions, and assess its merits relative to the objective of the design, e.g., under which conditions a new process discovery algorithm is better (or not) than the state of the art. In the call for papers, we also asked for a critical discussion of threats to validity. The implementation of the artifact should typically have the maturity of a prototype, i.e., it can be evaluated in an application context. This is in contrast to the foundations track, where proof-of-concept implementations are sufficient, though not necessary.

In summary, the main topic of the set of accepted papers is conformance checking, with various proposals on improving the efficiency or accuracy, or applying it to online settings. As a new trend, it can be observed that machine learning and deep learning in particular play a big role as underlying technologies. People-specific aspects are considered in two, and complexities of realistic settings in all of the accepted papers.

Comparing the call for papers with the submissions received, we note that a discussion of threats to validity is the exception, not the norm. The criteria in the call for papers were formulated as hard targets, and only two submissions (or less than 5%) were judged as meeting these by the PC with a recommendation of direct acceptance – 7 papers were at first only accepted conditionally. Reviews were in part hypercritical of the submissions, and emphasized flaws more than positive aspects. In the reviews and discussions, the PC members made many constructive remarks, which resulted in the final acceptance of 9 papers, and 5 BPM Forum papers.

A critical question for the community and Track II PCs of coming years is where to place the bar on evaluation strength and quality: if the bar is as high as for top journals, authors are often inclined to submit to these instead; if the bar is too low, validity of the results may not be given, possibly invalidating technical contributions. One step forward would be a broader uptake of validity discussions in submissions. While it is clear that the BPM community has matured, professionalized, and also emancipated itself from other communities since its inception, there is room for further development.

Among the five BPM Forum papers from this track, two discuss the ways to integrate BPM and Internet of Things, which may be an indication that this topic might play a bigger role in the future. Process mining, and conformance checking in particular, are the topics attracting most submissions and accepted papers. Conformance checking only moved in the last few years from foundational works towards a phase of improvement and optimization of approaches, i.e., towards the kinds of contributions that fit Track II well. Business process execution and engineering of process-aware information systems, outside of process mining, did not yield any accepted papers in the main track. Predictive process monitoring, a topic that was present in previous BPM conferences, moved to the management track this year. In summary there is some indication that certain themes actually move from the foundations track to the engineering track and possibly further to the management track, which nicely shows the relationships between the three tracks.

6   Track III: Management

The management track focuses on applications of BPM in organizational settings. Papers investigate the development and impact of BPM capabilities in order to deliver strategic goals, such as, innovation and improvement. Contributions account for the socio-technical nature of organizations, specifically aiming at an alignment of aspects related to technology, business and people in BPM initiatives in various organizational contexts. The disciplinary background, therefore, is that of information systems research, organizational science and management science, and a wide range of research methods established in these disciplines are applied.

In summary, all papers provide valuable insight to highly relevant issues organizations around the globe are concerned with today. These include digital innovation, business transformation, compliance management, and forecasting. The papers also advance prior contributions in that they embrace new technologies, such as data analytics, but also link to human aspects, such as leadership and values, and account for the multi-faceted requirements of different business contexts.

What is more, the papers show how different strategies of inquiry can advance BPM research, including case study research, interviews, literature reviews, experiments, survey research, design-oriented work, and conceptual studies. We find this inspiring, and it shall be particularly promising to compare findings from different strategies of inquiry in the sense of triangulation. The more we learn from different sources the richer our understanding of designing and managing business processes will become, and the more appropriate we will be able to serve the BPM practice. In Track III, 9 papers were finally accepted at the main conference and 4 papers were accepted at the BPM Forum.

We are very pleased that so many colleagues have followed the call to contribute research on management-related aspects of BPM to the conference. Given the plethora of contemporary organizational challenges, it is natural that many areas are yet to be explored, and these areas mark promising fields for future research, such as the use of machine learning, social media, blockchain, robotics, and other technological advancements in practice. It will be intriguing to explore the role of BPM in leveraging the potential of these technologies in real-world organizational settings.

The BPM conference with its new track structure will be an ideal place to present and discuss innovative contributions as they mature from conceptual and technical foundations to engineered artifacts to applications in practice. From a management track perspective, it will be particularly interesting for future research to take contributions from the engineering track and to investigate their application in real-world organizational settings.

7   Lessons Learned

Overall, we can say that the track system works. We received 140 full paper submissions, which is a very good number that exceeds submission numbers in recent issues of the conference. Submissions were very well distributed over the tracks. While keeping the high quality standards at BPM conference, we could accept 27 papers for the main conference and 14 papers at the BPM forum.

Many of these papers would not have made it in the traditional structure, so that our goal of opening the conference to the management community can also be considered fulfilled. Even if the first issue indicates a success, we share some experiences and lessons learned from organizing the conference, mainly related to the reviewing process. Introducing a new structure to a conference and extending a scientific community requires several years and several iterations of the conference to settle. We intend to support this transition by discussing the lessons learned while organizing BPM 2018.

The communication of the track system to the community actually started with the keynote by the first author at BPM 2017, where he motivated and introduced the new conference structure. The track system is the result of discussions within the BPM Steering Committee and with many colleagues of the community. During the preparation of the call for papers for BPM 2018, the PC chairs described the tracks as concisely as possible. This includes the phenomena studied in the tracks and the evaluation criteria used during the reviewing phase, as shown in Sect. 3.

The BPM 2018 publicity chairs did an excellent job in communicating the call for papers on various channels. In addition to the traditional channels BPM has used over the years, we also covered channels used by colleagues from the management discipline. In addition, peers were informed about the new track system through direct emails and many personal conversations, in order to solicit excellent submissions to the conference from the broader BPM field.

The main lessons learned relate to the reviewing phase. We were aware of the challenges when introducing a new structure and incorporating a new discipline to an established conference. Hence, we motivated the track system and we communicated the evaluation criteria and the reviewing standards used at BPM to the program committees of all tracks. Despite the effort to align the level of detail and criticality of reviews over the tracks, at the reviewing deadline, the situation in the tracks looked actually quite different.

On average, reviewers were quite critical in Track I, highly critical in Track II, and not very critical in Track III. Also, differences in the detail and quality of the reviews could be observed. In some cases reviewers based their recommendation on evaluation criteria of a different track, e.g., by asking for empirical evaluations to support foundational results in Track I. We tried to counter these issues by guiding the discussions to value positive aspects of papers, particularly in Track II, and by asking PC members to stick to the evaluation criteria of their specific track. These measures succeeded to a meaningful degree.

To balance between the reviews and the paper or, maybe more accurately, between the reviewers and the authors, we proposed conditional accepts in many cases. On the one hand, this has led to a high percentage of conditional accepts, on the other hand it allowed us to accept quite a large number of papers at the conference. All conditionally accepted papers were finally accepted; in several cases the final version exposed significant improvements over the submitted version. Overall, we accepted 27 papers, 9 in each track. This even distribution was not planned, it emerged as outcome of the discussions we had during the consolidation phase. Based on the reviews and according to the discussions during that phase, we are confident that all accepted papers are excellent and definitely deserve to be presented at a high-quality scientific conference like BPM.

8   Conclusions

In this short paper, we have reported on the experiences gathered during the first iteration of the track system at the BPM conference. While there is agreement that the track system works in general, there are challenges to address in future editions of the conference. While we acknowledge that each year, there are slight differences in the orientation of the topics and – of courses – each reviewing process is different, we still hope that the experiences reported will help in shaping the BPM community, especially related to the reviewing process.

BPM will not abandon its high-quality and selective reviewing standards, but we encourage reviewers to take a slightly more positive attitude towards the submissions of the community. This applies in particular to Track II, which has seen the most critical reviews. The high level of criticality of reviews in that track might also relate to the fact that the evaluation criteria in Track II actually did not change at all, compared to recent issues of the conferences. In contrast, the evaluation criteria in Track I were adapted, and Track III does not ask for a technical contribution. Still, it is pleasing to see excellent papers in the core BPM topics while also witnessing a broadening of topics and research methods addressed at our conference.

Bringing Middleware to Everyday Programmers with Ballerina

Sanjiva Weerawarana[1], Chathura Ekanayake[1], Srinath Perera[1(✉)],
and Frank Leymann[2]

[1] WSO2, Colombo, Sri Lanka
{sanjiva,chathura,srinath}@wso2.com
[2] IAAS, University of Stuttgart, Stuttgart, Germany
frank.leymann@iaas.uni-stuttgart.de

Abstract. Ballerina is a new language for solving integration problems. It is based on insights and best practices derived from languages like BPEL, BPMN, Go, and Java, but also cloud infrastructure systems like Kubernetes. Integration problems were traditionally addressed by dedicated middleware systems such as enterprise service buses, workflow systems and message brokers. However, such systems lack agility required by current integration scenarios, especially for cloud based deployments. This paper discusses how Ballerina solves this problem by bringing integration features into a general purpose programming language.

Keywords: Flow languages · Middleware · Integration technology

1 Introduction

Integration technologies connect various services, APIs and other resources, resulting in meaningful business activities. Since those business activities mirror the real world and interact with the real world, they need to behave accordingly providing security, interruptibility, recoverability, or transactions, respectively.

Often, integration developments are done based on general purpose languages. Those general purpose languages themselves lack the behavior required for integrations mentioned before. Such behavior is provided through middleware. Among examples of those are workflow management systems [1], transaction managers, enterprise service buses, identity gateways, API gateways, and application servers.

However, this behavior is added as an afterthought as middleware. Hence, the corresponding features lack tighter integration with general purpose languages. Therefore, building integration solutions has been a counter-intuitive, complex, and error prone endeavor.

Ballerina[1] is a general purpose, transactional and strongly typed programming language with both, textual and graphical syntaxes. It has the aforementioned behavior natively built in and specialized for the integration domain.

---

[1] https://ballerina.io.

© Springer Nature Switzerland AG 2018
M. Weske et al. (Eds.): BPM 2018, LNCS 11080, pp. 12–27, 2018.
https://doi.org/10.1007/978-3-319-98648-7_2

It brings fundamental concepts, ideas and tools of distributed systems directly into the language and offers a type-safe, parallel environment to implement such applications. The Ballerina environment has first class support for distributed transactions, reliable messaging, stream processing, workflows and container management.

Furthermore, Ballerina offers simple constructs to control parallel executions. It has a type system designed to simplify the development of distributed applications, for example by providing common message formats such as JSON and XML as native data types. It also includes a distributed security architecture to make it easier to write applications that are secure by design.

Ballerina's support of parallelism, interruptibility, recoverability, and transactions reveals its suitability as basis for a workflow management system. Key workflow patterns [2] especially from the control flow, data flow, and exception handling category can be mapped to Ballerina language elements. For example, parallel execution may be controlled by fork and join statements (see Sect. 5.1), or grouping of steps into compensation-based transactions (Sect. 4.3) as well as ACID-based transactions (Sect. 4.2) are immediate features of the language. It is obvious how many BPMN language elements can be mapped to Ballerina in a straightforward manner. In addition, Ballerina goes beyond that by providing immediate support of resiliency (Sect. 4.5) or security features (Sect. 4.1), for example, to workflows.

This paper will look at how Ballerina makes workflow and other middleware behavior an inherent aspect of a programming language, and how Ballerina helps bringing non-functional properties to everyday programmers.

1.1 History of Integration Technology

Most organizations depend on complex IT infrastructure as they are critical to organizations' successful operations. These infrastructures manage resources, employees, customers, business partners, data related to them, and interactions between them. Such IT infrastructures are called "Enterprise Systems".

Enterprise systems are comprised of many components that are often produced by many different vendors. Single business activities need to use several of these components, i.e. these components need to be integrated. Similarly, since these organizations do business with others, these enterprise systems themselves need to be integrated, i.e. they need to work with other organizations' enterprise systems. System integration connects those different, (even geographically) distributed components into a single system supporting an organization's business. We will call "system integration" just "integration" henceforth.

There are several schools of thoughts for building "enterprise systems" architectures. The leading two are Service Oriented Architecture (SOA) [3] and Resource Oriented Architecture (ROA) [4]. Both these approaches depend on interfaces exposed over the network, which we will call services. Services are defined as "a discrete unit of functionality that can be accessed remotely and acted upon and updated independently, such as an act of retrieving a credit card statement online".

Services that can create a ROA design and built using REST principles [4] are called RESTful services. Such service is called a resource. Each part of an enterprise architecture exposes its capabilities as services (in SOA) or resources (in ROA). Enterprise integration realize the functions of the organization by connecting and orchestrating those services, resources, and different data sources.

Middleware that does system integration is often called "Enterprise Service Bus (ESB)" [5]. ESB is used for designing and implementing communication between mutually interacting software applications (e.g. WSO2 Enterprise Service Bus [6]. When authoring integrations, users need to specify integration logic. Earlier ESBs have used XML (WSO2 ESB [6] and Java based Domain Specific Languages (e.g. CAMEL [7])) for this purpose.

Most programming languages, such as Java, C, C#, Java Script, are based on textual syntax structure. Visual programming [8], where users construct the program visually by composing constructs selected from a palette, is an alternative to textual syntax. ESBs often support visual programming using a data flow model variant [9]. Among examples are WSO2 ESB's Studio [6], Mule's Anypoint Editor [10], and Boomi Editor [11].

Earlier architectures supporting integrations were centralized in the sense that ESB(s) host all the integration logic. Later, with the emergence of microservices architecture, focus was given to develop integration logic as independent entities, where each integration flow can be deployed in its own process. Container technologies such as Docker have made this idea practical. Integration logic used within microservices architecture is called micro-integrations. We use the word "integration" to encompass both centralized integration as well as "micro-integrations".

Some of these services are published and consumed within a single trust domain (e.g. a single organization). However, services may be published, shared and consumed beyond a single trust domain. We call them remote APIs, in this context often abbreviated to just APIs. Middleware that manage APIs is called "API Manager" [12]. API management lets the organization have a greater control over creating and publishing APIs, enforcing their usage policies, controlling access, nurturing the subscriber community, and collecting and analyzing usage statistics. Integration logic that is available as an API is called an API composition.

Among related work, software development using visual interfaces is discussed in [8]. UML [13] is one of those visual languages. Sequence Diagrams (SDs) that capture the flow of executions are also a part of UML. Sequence diagrams have been used to model service based applications [14]. Koch et al. [15] describes the use of UML for web engineering. Ermagan et al. [16] describe a new UML profile for modeling services. Furthermore, sequence diagrams have been used for modeling tests cases [17] and High level Petri Nets [18].

1.2   The Importance of Non-functional Behaviors

Primary focus of a programming language is the grammar to specify how various elements of the language can be combined into a valid program. Almost all programming languages provide constructs for data flow (e.g. variable definitions and variable assignments), arithmetic operations, and control flow (e.g. conditional branching and loop). A language also typically supports exception handling, controlling parallel executions and input/output processing.

However, non-functional properties such as availability, resiliency, data integrity, security, observability, recoverability and reliability become important factors when programming in distributed systems environments. In a distributed environment, a single use case may involve many services hosted on different servers. It is possible that one or more of those services become unavailable in certain time periods. In such situations, it should be possible to identify failed services and reroute requests to alternative services if possible or handle the error situation gracefully without causing major disruption to the overall operation. Furthermore, if a critical operation failed due to some reason, the system should guarantee the integrity of itself as well as all interacting systems. For example, assume that a system is handling travel arrangements for a conference involving multiple external services for airline tickets, conference registration and hotel reservations. If an air ticket could not be reserved on required dates after registering for the conference, it may be required to cancel the conference registration, which in turn should adjust expected participants for relevant tracks and seating requirements. If all these functions are handled by services, a failure at invoking the airline service should trigger cancellation actions in conference registration service and all relevant services invoked by conference registration service.

Another important aspect is the reliable delivery of messages in an environment where any component can fail, and the number of messages produced per unit time by a sending component may not match the number of messages that can be consumed per unit time by its receiving component. In such situations, a system that sends a message should have a guarantee that the target system will receive the message even if that system is not available at that moment.

A more critical non-functional requirement from the workflow perspective is the interruptibility, which ensures to resume work from the last checkpoint after a system failure. Depending on the check pointing behavior, a system can avoid redoing expensive operations and prevent unnecessary compensating actions. As an example, assume that an order handling process invokes an inventory service to allocate goods to be shipped. Further, assume that the inventory service involves some manual tasks and hence takes few days to complete. Now if the order handling process crashes (e.g. due to an unexpected shutdown of the server) after receiving the response from the inventory service, whole state of the process would be lost. If interruptibility is not supported, order process has to be started from the beginning after the server restart, which in turn invokes the inventory service again causing few days of delay. Additionally, some compensation action is required to deallocate previously allocated goods. None of these problems would arise if recoverability is supported by the order processing system, which can create a checkpoint immediately after receiving the response from the inventory service, thus allowing the system to resume from that checkpoint after a failure.

Such non-functional properties are not available in most of the current programming languages as they are not developed for systems integration. In order to overcome limitations of current programming languages, integration developers have to use middleware systems such as workflow engines, enterprise service buses, messages brokers or transaction managers. A typical integration use case may require functionality offered by many of these systems. Furthermore, functionality provided by middleware systems alone is not sufficient to address most integration use cases forcing developers to build solutions by combining code written in programming languages with middleware systems. This often results in complex solutions consisting of disparate set of systems that require diverse skill sets and infrastructure.

2 Language Philosophy

Ballerina aims at simplifying integration development by bringing in key features required for systems integration such as efficient communications over multiple protocols, resiliency, recoverability, and transactions, into a programming language. Therefore, developers can develop integration and workflow style solutions in the same way as any other application. For example, if it is necessary to invoke an HTTP endpoint within a program, a developer may create an endpoint with necessary resiliency parameters such as retry interval, retry count and failure threshold, and invoke the endpoint. Ballerina runtime will handle all necessary details such as invoking the endpoint without blocking OS threads, taking specified actions on endpoint failures and facilitating recovery from runtime failures. Integration features offered by Ballerina are discussed in later sections.

Ballerina can be used to develop many integration solutions. Developers can utilize its service publishing capabilities over multiple protocols to develop microservices. They can combine its service publishing capabilities with message processing and endpoint features to develop micro-integrations. Furthermore, microworkflows can be developed using features such as recoverability, compensating transactions and ability receive multiple inputs. All these solutions result in executable Ballerina programs which can be executed in stand-alone mode without any middleware support.

For many years we observed repeatedly that, when faced with a complex integration problem, various stakeholders often draw a sequence diagram to develop a solution to their problem. Hence, a sequence diagram is a suitable visual representation that has a shared meaning among integration programmers and architects.

Ballerina uses UML sequence diagrams (SDs) to build integration solutions. The Ballerina editor includes a visual and a textual view side by side. Programmers can switch between either view in the middle, apply changes in any of the views, and lossless conversion happens immediately. This enables the user to build the integration logic either using a visual syntax or using textual syntax.

Another key aspect is to use automatic analysis and intelligent predictions to reduce the burden on the developer. Unlike middleware implemented separately from general purpose languages, Ballerina has access to the whole program and can fully control its execution. Therefore, it can perform automatic analysis effectively and optimize the execution of a program. Among examples are locks that will be automatically acquired and released allowing maximal concurrency, auto tuning thread pools, and taint checking built into the language.

3   Language Elements

A detailed description of Ballerina can be found in the language specification [14]. In this section, we will explore some of its key features.

A typical program includes a service or a main program. Both main program and resource is built with many statements. Each statement may define data types, call functions, define and create workers, and run control statements such as loops and conditions. Furthermore, Ballerina supports annotations to associate configurations or additional behaviors with language constructs.

3.1   Type System

Ballerina type system includes three types of values: (i) simple values such as booleans and integers, (ii) structured values such as maps and arrays and (iii) behavioral values like functions and streams. Ballerina considers JSON and XML as native structured types simplifying the handling of message payloads used in most integration scenarios.

In addition to the above, *union, optional* and *any* types are supported. *Union* type allows a variable to contain values of more than one type. This is useful for example for a function to return a string or an error depending on the processing outcomes. *Optional* indicates that a variable can be null. Variables of *any* type can be assigned with values of any type defined in Ballerina type system, making it useful for scenarios where the type of the variable is not known at development type.

3.2   Connectors and Endpoints

Interactions with external APIs and services is a fundamental requirement of integration. Ballerina uses connectors to program interactions with external entities. Such entities can be generic HTTP endpoints, databases or specific services such as Twitter, GMail, etc. A connector can be initialized by providing a configuration required for the corresponding external service. For example, a simple HTTP connector initialization would look like:

```
endpoint http:Client taxiEP {
    url: "http://www.qtaxies.com"
};
```

Once initialized, any action supported by the connector can be invoked by providing appropriate parameters as below:

```
taxiEP -> get("/bookings");
```

The connector concept was introduced in order to differentiate network calls from other function invocations, so that Ballerina developer is aware of associated concerns such as latency, probability of failures and security implications. In fact, Ballerina addresses some of these concerns using resiliency features discussed in Sect. 4.5. In addition to the connectors shipped with Ballerina, it is possible to develop custom connectors for any programmable endpoint and use them in the same way as any other existing connector.

3.3   Error Handling

Programming languages handle errors in two ways. The code can either stop the execution of the normal flow or return the details back into upper levels to handle or recover from the error. Languages like Java can support both the above methods by throwing exceptions. However, if it is recoverable, it is expensive to unwind and do exception processing, and it is considered an anti-pattern to throw exceptions in the normal execution flow.

In such cases, languages like Java can return error details (without throwing exceptions), which must be checked at upper level. However, if upper levels forget to check it, it will lead to errors.

Ballerina handles both above cases by introducing a first-class error concept which can both be returned and thrown. Thrown errors behave just like exceptions and cause the call stack to be unwound until a matching catcher is found. However, the language forces the upper levels to check the return value for errors. It is done by returning a union type of type T|error. The error part has to be handled using a match expression or explicitly declaring that the error is not important at assignment via T|_. Unless this is done, the compiler will complain when the program tries to access the value of an union type.

3.4   Ability to Inject Values into a Running Executions

Once a program instance is started, it may be necessary to get additional inputs after completing certain steps. Such inputs can come from other parallel flows of the same program instance or from external systems. The former case can be implemented by most programming languages using shared variables with appropriate synchronization mechanisms. For example, in Java a thread can wait on an object to be notified by another thread once required data is available. Ballerina facilitates parallel executions based on a construct named *worker* as explained in Sect. 5.1. Accordingly, it provides an inter worker communication method to pass variables among workers as below:

```
function f1() {
   worker w1 {
      string city = "NY";
      city -> w2;
```

```
    }
    worker w2 {
        string location;
        location <- w1;
    }
}
```

In the above example, worker *w2* waits for an input from worker *w1* and once the input is available, assigns it to a variable named location.

The more complex scenario is receiving intermediate inputs from external systems. In this scenario, a process instance is already started and it expects an external input in order to continue. Furthermore, there can be many instances of the same program running at the same time. Therefore, the main challenge here is to identify the correct process instance to deliver the message. A program instance can be identified either using a unique instance identifier or using a combination of variables whose values will be unique to an instance. Such combinations of variables are called correlation variables. The latter approach is more flexible as it does not force external systems to be aware of program instance identifiers. Instead, such systems can just send messages including values for correlation variables, which are, in most cases, business variables such as customer identifiers, order numbers, etc. Many workflow systems use this correlation variable based approach for receiving intermediate inputs [19]. Inspired by this, Ballerina brings a similar concept to programming language level by allowing programs to receive intermediate inputs, where the Ballerina runtime environment correlates incoming messages with relevant program instances based on variables defined in the program flow. The below code fragment shows how a Ballerina program can receive the location of a customer:

```
string customer = "smith";
json correlationVars = {"custId": customer};
queue:Message result = custEP -> receive(correlationVars);
string location = result.getTextMessageContent();
```

Any Ballerina program can send messages to such intermediate reception points by providing values for correlation variables as below:

```
string city = "NY";
queue:Message loc = custEP.createTextMessage(city);
json correlationVars = {"custId": "smith" };
loc.setCorrelationID(correlationVars.toString());
custEP -> send(loc);
```

Thus, it is possible to write Ballerina programs that get messages from any source such as HTTP services, JMS or file systems and trigger other Ballerina programs waiting on intermediate inputs. An advantage of this approach is the flexibility on how external systems can send correlated messages. For example, if an external system can only send JMS messages, it is possible to write a Ballerina program to receive JMS messages and trigger a waiting Ballerina program.

4   Non-functional Properties

4.1   Security

Ballerina guarantees security for sensitive parameters using a compiler level taint checking mechanism. There can be functions and connector actions that take sensitive parameters such as SQL queries or file paths. Assigning untrusted values into those parameters can cause major security vulnerabilities such as SQL injection attacks. In order to prevent such vulnerabilities, Ballerina programmers can mark relevant parameters with the @sensitive annotation as below:

```
function selectData(@sensitive string query, string params)
        returns string { .. }
```

If a tainted variable (i.e. a variable that may contain an untrusted value) is assigned to a parameter marked with *@sensitive* annotation, Ballerina compiler will produce an error. Furthermore, sensitivity of parameters is automatically inferred when calling functions. This can be illustrated using the below sample code:

```
function getAddress(string username) {
   string query = "select address from user" +
                  "where uid = '" + username + "'";
   string address = selectData(query, null);
}
```

In this case, *username* parameter of the *getAddress* function is also considered sensitive by the compiler as its value is used to derive the sensitive query parameter of the invoked *selectData* function. Similarly, functions can mark return values as tainted to indicate that they are not safe to be used as sensitive parameters. This is done using the *@tainted* annotation as below:

```
function readUsername() returns @tainted string { ... }
```

4.2   Distributed ACID Transactions

Being an integration language, Ballerina programs have to interact with many external systems including databases, message brokers and other services, which are often required to be grouped into a single unit of work. Therefore, Ballerina programs have to support distributed transactions involving all participating entities to ensure integrity of the corresponding integration scenarios. Ballerina language provides constructs to mark transaction boundaries to coordinate the joint outcome of invoked endpoints based on a 2PC protocol. Below is an example of a transaction within a Ballerina program:

```
transaction with retries = 0, oncommit = commitFunc, onabort = abortFunc
{
    _ = bankDB -> update("UPDATE ACCOUNT SET BALANCE = (BALANCE - ?)" +
```

```
                    "WHERE ID = ?", 1000, 'user1');
    match depositMoney("acc1", 1000) {
        error depositError => {
            abort;
        }
        () => isSuccessful = true;
    }
}
```

In this example, the *update* action of a database connector (i.e. bankDB) is invoked to decrease the amount to be transferred. However, if the *depositMoney* function fails, the transaction is aborted so that Ballerina runtime will rollback the database transaction as well. In addition to coordinating transactions of invoked endpoints, Ballerina programs can participate in distributed transactions as well. If a Ballerina program *B1* invokes another Ballerina program *B2* within a transaction, *B2* automatically participates in the transaction initiated by *B1*. If *B2* also defines a transaction, *B2*'s transaction will not be commited until *B1* is ready to commit. Similarly, if *B2* invokes another Ballerina program *B3*, *B1*'s transaction is propagated to *B3* as well (transaction infection).

4.3 Compensation-Based Transactions

Compensation is another technique of maintaining integrity [20], especially for long running interactions or interactions that do not support ACID semantics. Distributed ACID transactions mentioned above are inherently subject to blocking. This is not be suitable for tasks that span longer time periods (more than several seconds). Instead, compensation-based mechanisms allow each participating entity to commit work immediately. In addition, each such entity has to provide a corresponding compensation action, which will be triggered if the overall task fails. Such compensation mechanisms are implemented in many workflow systems and Ballerina language introduces this concept at programming language level. In workflow languages such as BPMN, compensation actions can be associated with BPMN activities using compensation boundary events. However, programming language statements are too fine grained as compensable units. Therefore, a grouping construct named *scope* is introduced (like in BPEL), so that compensation actions can be associated with a scope containing multiple statements. Scopes can be nested. A construct named *compensate* is introduced to trigger compensations of completed scopes. It can be invoked with or without a scope name. If a scope name is given, only the named scope and its child scopes are compensated. If a scope name is not given, all completed scopes within the current scope are compensated. Compensation actions are triggered in the reverse of the completion order. Following is a Ballerina code snippet with compensations:

```
scope s1 {
    scope s2 {
        ...
```

```
} compensation(var2, var3) { ... }
scope s3 {
     result = ...
} compensation(var4) { ... }
if (result == -1) {
   compensate();
}
} compensation (var1) { ... }
```

In the above example, if the compensation is triggered, compensation actions of scopes *s3* and *s2* are invoked in that order. However, the compensation actions of scope *s1* will not be invoked as *s1* is not completed at the time of compensation.

4.4   Interruptibility

Compiled Ballerina programs run on Ballerina Virtual Machine (BVM) as discussed in Sect. 5. BVM supports interruptibility by persisting the state of running programs. A persisted state is a special kind of checkpoint. If a connector developer (not Ballerina programmer) has indicated that a certain connector action should not be repeated after a recovery, BVM makes a checkpoint whenever a program completes that action. Such actions can include invocations of external services or database operations. For example, if a service is invoked to reserve a ticket, it should not be re-invoked when the invoking program is resumed after a failure. Similarly, if a connector developer has indicated that an action can take long time to complete, BVM makes a checkpoint before a program invokes that action. An example of such action is the waiting for a reception of an intermediate input as mentioned in Sect. 3.4. As such operations can take long time periods, there is more probability of failure at those points. Therefore, checkpointing before starting such operations allows BVM to resume programs from those points in case of a failure. Furthermore, Ballerina has a language element named checkpoint in order to allow programmers to define checkpoints anywhere within a program. If the BVM stops due to any reason (e.g. server crash) and is subsequently restarted, it will resume all program executions from the last available checkpoint.

When a Ballerina program is invoked as an HTTP resource, the program can send a reply back to the client using an HTTP response message. However, if the BVM is restarted, the underlying TCP connection with the client will have been terminated and BVM will not get a connection to send the HTTP response. In this scenario, a client can resend the original request, which will be correlated by the BVM with the corresponding program instance. Then once the program reaches the replying point, it will reply using the new correlated connection. If the program has reached the replying point before receiving the correlating request, the program state will be saved until such request is received. Thus, Ballerina supports request-response behavior for long running flows even after system failures.

4.5   Resiliency

Distributed environments that Ballerina programs are expected run may contain unreliable networks, failure prone servers, overloaded systems, etc. In order to facilitate development of robust programs for such environments, Ballerina provides a set of resiliency features, namely circuit breaking, failing over, load balancing and retrying with timeouts. Circuit breaking allows programmers to associate suspension policies with connectors so that connectors stop sending messages to unresponsive endpoints if suspension criteria is met. For example, an HTTP connector can be configured to suspend further requests for 5 min if more than 5% of requests fail within 30 s.

Similarly, failover configurations can be associated with connectors to select alternative endpoints if one endpoint fails. Load balancing configuration specifies a set of endpoints and a load balancing algorithm, so that requests are distributed among specified endpoints according to the given algorithm. This is useful to avoid overloading backend systems, especially where a dedicated load balancer is not available. Finally, a retry configurations can be defined for connectors to force the connector to retry sending the request in case of a failure.

5   Architectural Aspects

Ballerina is a compiled language. Ballerina compiler takes Ballerina programs as input and generates intermediate code. This intermediate code can be executed in Ballerina Virtual Machine (BVM). The Ballerina compiler performs syntax checks and transforms programs into an intermediate format containing instructions understood by the BVM. The BVM performs the fetch-, decode-, execute-cycle acting as a CPU for Ballerina intermediate code. In addition to executing instructions, the BVM performs tasks such as listening for incoming messages, concurrency control, exception handling and transaction management.

5.1   Thread Model

Ballerina is a parallel language and natively support parallel executions. Each execution of a resource or a main program has implicit workers. However, workers may be created explicitly, they can safely talk to each other, synchronize data, and support complex scenarios such as fork and join based on corresponding language syntax.

A worker can be considered as a sequence of Ballerina instructions with a storage to store variables, input arguments and return values used within those instructions. BVM can execute a worker by assigning it to an OS thread. A worker that needs to be executed synchronously is run in the same thread as its invoker (e.g. synchronous function calls). Asynchronously executed workers are assigned to new threads taken from a thread pool. BVM executes a worker in its assigned OS thread until a blocking instruction is reached (e.g. sleep or connector invocation). At this point, BVM saves the context of the worker in an appropriate callback function and releases the assigned OS thread. Therefore, the OS thread previously assigned for the blocked worker will become available to run an unblocked worker. Once the blocking action returns and its callback is invoked, the callback function gets the saved worker context and lets BVM to execute it by assigning an OS thread.

Each Ballerina function can have one or more workers. Ballerina programmers can define workers explicitly within a function as below:

```
function f1() {
    worker w1 { ... }
    worker w2 { ... }
}
```

In this case, workers *w1* and *w2* will be executed in parallel. If workers are not defined within a function, a default worker is associated with it by the BVM. In addition to the workers associated with functions, Ballerina provides fork/join constructs to trigger parallel flows as below:

```
fork {
    worker w1 { ... }
    worker w2 { ... }
} join (all) { ... }
```

Furthermore, it is also possible to start any function asynchronously by using the start keyword as below:

```
future<int> result = start f2();
...
int value = await result;
```

In this case, the future statement immediately returns without waiting for *f2* to complete. Then the Ballerina program can call await at any point later in the program to wait for the function to complete and get the result.

A common problem in parallel programs is to handle shared data safely. Ballerina supports this via a lock statement. Ballerina goes beyond languages like Java by automatically analyzing the locked data structures, figuring out minimal shared scope, and then locking at that level to allow maximal concurrency.

5.2   Non-blocking I/O

Ballerina supports non-blocking I/O without any additional overhead for programmers. From a programmer's perspective, I/O calls work in a blocking manner, so that the statement immediately after the I/O call is executed only after the I/O call returns with a response. An invoking program can access the response immediately as shown below:

```
var result = clientEndpoint -> get("/get?test=123");
io:println(result);
```

However, according to the thread model discussed in Sect. 5.1, BVM releases the underlying OS thread whenever an I/O call is made and stores the program state in a memory structure along with next instruction pointer. When the result of the I/O call is available, BVM allocates a new thread from its thread pool to continue the saved program state. Therefore, Ballerina programmer sees it as a blocking I/O call although OS threads are not blocked.

5.3 Observability

Observability is a measure of how well internal states of a system can be inferred. Monitoring, logging, and distributed tracing can be used to reveal the internal state of a system to provide observability. Ballerina becomes observable by exposing itself via these three methods to various external systems allowing them to monitor metrics such as request count and response time statistics, analyze logs, and perform distributed tracing. It follows open standards when exposing observability information so that any compatible third party tool can be used to collect, analyze and visualize information.

6 Measurements

Performance is a critical factor for integration software, as a typical deployment may have to connect with many external systems and process thousands of requests per second. These requests may be sent by hundreds of different clients. Two commonly used measurements for evaluating performance are latency and throughput. Latency of a request is the round trip time between sending a request to a system and receiving a response. Throughput is the number of requests that can be processed by a system in a unit time.

We compared these two measurements of Ballerina with those of WSO2 ESB for different concurrency levels. A basic integration scenario of receiving a message from a client over HTTP, sending it to a backend system, receiving the response from the backend and sending the response back to the client is considered for these tests. A Message of size 1kB is used and zero backend processing time is simulated (i.e. backend responds immediately). Tests are conducted on Intel Core i7-3520M 2.90GHz 4 machines with 8 GB RAM and JVM heap size of 2 GB is allocated. Ballerina/WSO2 ESB, client (JMeter[2]) and the backend (WSO2 MSF4J[3]) are run on three separate machines. Results for latency and throughput are shown in Fig. 1(a) and (b) respectively:

According to the results, Ballerina outperforms ESB at all concurrency levels. Ballerina can process a request in 5 ms for 200 concurrent clients whereas ESB takes 10 ms for the same scenario. Performance difference is also significant for throughput, where Ballerina shows around 24000 transactions per second (TPS) for 200 concurrent clients while ESB shows only around 14000 TPS.

---

[2] https://jmeter.apache.org.
[3] https://wso2.com/products/microservices-framework-for-java.

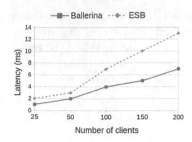

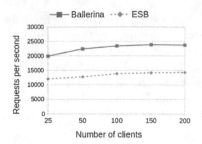

(a) Variation of latency with number of concurrent clients

(b) Variation of throughput with number of concurrent clients

Fig. 1. Comparison of Ballerina with ESB

These results indicate that Ballerina can process requests faster and serve more requests concurrently without considerably degrading performance, which is a desirable property for integration systems.

7   Conclusion and Outlook

The main objective of the Ballerina project is to create a programming language for integration. Therefore, in addition to providing general purpose programming constructs, Ballerina has built-in support for a broad set of integration features. Such features include efficient service invocations, listening for incoming connections, transactions, support for multiple protocols and simplified database access. By recognizing the possible long running nature of certain integrations, some features of workflow systems were absorbed into Ballerina. As a result, persistence based interruptibility is introduced to support recovery from unexpected failures during long-running workflows. Furthermore, compensation is introduced to support long-running transactions. Then, the ability to receive intermediate inputs with correlations were implemented, so that a single Ballerina program can receive messages from any number of channels. Combining all these features, Ballerina can be used as a programming language for programming short running integrations as well as long-running workflows.

As future work, we are planning to improve the current Ballerina workflow implementation to production ready state. In addition, once Ballerina constructs equivalent to other critical BPMN constructs are introduced, a (partial) mapping from BPMN to Ballerina can be performed. By extending this idea, it is possible to develop a BPMN editor, which generates Ballerina code that can run in the BVM.

From the runtime perspective, we are planning to implement the BVM using the LLVM infrastructure in order to increase performance.

References

1. Leymann, F., Roller, D.: Production Workflow: Concepts and Techniques. Prentice Hall PTR, Upper Saddle River (2000)
2. Workflow patterns. http://www.workflowpatterns.com/. Accessed 10 Jun 2018
3. Erl, T.: Service-Oriented Architecture: Concepts, Technology, and Design. Pearson Education India, Noida (2005)
4. Duggan, D.: Enterprise Software Architecture and Design: Entities, Services, and Resources, vol. 10. Wiley, Hoboken (2012)
5. Schmidt, M.-T., Hutchison, B., Lambros, P., Phippen, R.: The enterprise service bus: making service-oriented architecture real. IBM Syst. J. **44**(4), 781–797 (2005)
6. WSO2 Enterprise Service Bus. http://wso2.com/products/enterprise-service-bus/. Accessed 30 May 2018
7. Apache camel. http://camel.apache.org/. Accessed 30 May 2018
8. Sundararajan, P., et al.: Software development using visual interfaces. US Patent 7,793,258, 7 September 2010
9. Hils, D.D.: Visual languages and computing survey: data flow visual programming languages. J. Vis. Lang. Comput. **3**(1), 69–101 (1992)
10. Mule anypoint studio. https://www.mulesoft.com/lp/dl/studio. Accessed 30 May 2018
11. Dell boomi platform. https://boomi.com/platform/integrate/. Accessed 30 May 2018
12. API management. https://en.wikipedia.org/wiki/API_management. Accessed 30 May 2018
13. O.UML, Unified Modeling Language. Object Management Group (2001)
14. Ballerina language specification, v0.970, working draft. https://ballerina.io/res/Ballerina-Language-Specification-WD-2018-05-01.pdf. Accessed 30 May 2018
15. Koch, N., Kraus, A.: The expressive power of UML-based web engineering. In: Second International Workshop on Web-oriented Software Technology (IWWOST 2002), vol. 16. CYTED (2002)
16. Ermagan, V., Krüger, I.H.: A UML2 profile for service modeling. In: Engels, G., Opdyke, B., Schmidt, D.C., Weil, F. (eds.) MODELS 2007. LNCS, vol. 4735, pp. 360–374. Springer, Heidelberg (2007). https://doi.org/10.1007/978-3-540-75209-7_25
17. Sarma, M., Kundu, D., Mall, R.: Automatic test case generation from UML sequence diagram. In: 2007 International Conference on Advanced Computing and Communications, ADCOM 2007, pp. 60–67. IEEE (2007)
18. Alhroob, A., Dahal, K., Hossain, A.: Transforming UML sequence diagram to high level petri net. In: 2010 2nd International Conference on Software Technology and Engineering (ICSTE), vol. 1, pp. V1–260. IEEE (2010)
19. Görlach, K., Leymann, F., Claus, V.: Unified execution of service compositions. In: Proceedings of the 6th IEEE International (2013)
20. Leymann, F.: Supporting business transactions via partial backward recovery in workflow management systems. In: Lausen, G. (ed.) Datenbanksysteme in Büro, Technik und Wissenschaft, pp. 51–70. Springer, Heidelberg (1995). https://doi.org/10.1007/978-3-642-79646-3_4

Track I: Concepts and Methods in Business Process Modeling and Analysis

Open to Change: A Theory for Iterative Test-Driven Modelling

Tijs Slaats[1], Søren Debois[2], and Thomas Hildebrandt[1(✉)]

[1] University of Copenhagen, Copenhagen, Denmark
{slaats,hilde}@di.ku.dk
[2] IT University of Copenhagen, Copenhagen, Denmark
debois@itu.dk

Abstract. We introduce *open tests* to support iterative test-driven process modelling. Open tests generalise the trace-based tests of Zugal et al. to achieve *modularity*: whereas a trace-based test passes if a model exhibits a particular trace, an open test passes if a model exhibits a particular trace *up to* abstraction from additional activities not relevant for the test. This generalisation aligns open tests better with iterative test-driven development: open tests may survive the addition of activities and rules to the model in cases where trace-based tests do not. To reduce overhead in re-running tests, we establishing sufficient conditions for a model update to preserve test outcomes. We introduce open tests in an abstract setting that applies to any process notation with trace semantics, and give our main preservation result in this setting. Finally, we instantiate the general theory for the DCR Graph process notation, obtaining a method for iterative test-driven DCR process modelling.

Keywords: Test-driven modelling · Abstraction · Declarative
DCR graphs

1 Introduction

Test-driven development (TDD) [4,17] is a cornerstone of agile software development [8] approaches such as Extreme programming [3] and Scrum [25]. In TDD, tests drive the software development process. Before writing any code, developers gather and translate requirements to a set of representative tests. The software product is considered complete when it passes all tests.

In [28,29] Zugal et al. proposed applying the TDD approach to process modelling, introducing the concept of *test-driven modelling* (TDM). Like in TDD, the modeller in TDM first defines a set of test cases then uses these test cases to guide the construction of the model. A test case in this setting consists of a trace of activities expected to be accepted by the model.

Work supported in part by the Innovation Fund project EcoKnow (7050-00034A); the first author additionally by the Danish Council for Independent Research project *Hybrid Business Process Management Technologies* (DFF-6111-00337).

© Springer Nature Switzerland AG 2018
M. Weske et al. (Eds.): BPM 2018, LNCS 11080, pp. 31–47, 2018.
https://doi.org/10.1007/978-3-319-98648-7_3

Specifically, Zugal et al. proposed a *test-driven modelling methodology* where the model designer first constructs a set of process executions (traces) that will be used as test cases. The model designer then constructs the process model by repeatedly updating the model to make it satisfy more tests. Once all tests pass, the model is complete.

This methodology *benefits* the end-user by allowing him to focus on specific behaviours of the model that should be allowed in isolation, without having to immediately reason about all possible behaviours. By eventually arriving at a model where all tests pass, he is ensured that all desired behaviour is supported; and should a previously passing test fail after a model update, he knows that this update is wrong.

Process modelling notations can be roughly divided into two classes: declarative notations [10, 14, 16, 20, 22, 23], which model *what* a process should do, e.g. as a formal description the rules governing the process; as opposed to imperative process models, which model *how* a process should proceed, e.g. as a flow between activities. Zugal et al. argued [28, 29] that TDM is particularly useful for the declarative paradigm, where understanding exactly which process executions the model allows and which it does not requires understanding potential non-trivial interplay of rules. In this setting, TDM is helpful to both constructing the model in a principled way (incrementally add declarative rules to satisfy more tests), as well as to recognize when the model is becoming over-constrained (when previously passing tests fail after a model extension). The commercial vendor DCR Solutions has implemented TDM in this sense in their commercial DCR process portal, dcrgraphs.net [18].

Unfortunately, TDM falls short of TDD in one crucial respect: its test cases are insufficiently modular and may cease to adequately model requirements as the model evolves. Consider a requirement "payout can only happen after manager approval", and suppose we have a model passing the test case:

$$\langle \text{Approval, Payout} \rangle \tag{1}$$

Now suppose that, following the iterative modelling approach, we refine the model to satisfy also the only tangentially related requirements that "approval requires a subsequent audit" and "payout cannot happen before an audit". That is, the model would have a trace:

$$\langle \text{Approval, Audit, Payout} \rangle \tag{2}$$

Crucially, while the *requirement* "payout can only happen after manager approval" is still supported by the refined model, the *test case* (1) intended to express that requirement no longer passes.

In the present paper we propose *open tests*: A generalisation of the "test cases" of [28, 29] that is more robust under evolution of the model. Open tests formally generalise [28, 29]: An open test comprises a trace as well as a *context*, a set of activities relevant to that test. This context will always contain the activities of the trace, and will often (but not always) be the set of activities known when the test was defined. For example, if we generalise the test (1) to an open test with the same trace and context {Approval, Payout}, then the extended model which has the trace (2) *passes* this open test, because, *when we ignore the irrelevant activity* Audit, the traces (1) and (2) are identical.

In practice, one will use open tests in a similar way as regular trace-based tests; the core TDM methodology does not change. The difference comes when the specification of a model changes, for example because of changes in real-world circumstances such as laws or business practices, or because one wants to add additional detail to the model. Using simple trace-based tests, one would need to update each individual test to make sure that they still accurately represent the desired behaviour of the system. With open tests one only needs to change those tests whose context contains activities that are directly affected by the changes. Because of the modularity introduced by open tests, any test not directly affected by the changes will continue to work as expected. This also means that open tests are much better suited for *regression testing*: it is possible to make small changes to a model and continue to rely on previously defined tests to ensure that unrelated parts continue to work as intended.

We define both positive and negative open tests. A positive open test passes iff *there exists a model trace whose projection to the context is identical to the test case.* A negative test passes iff *for all model traces, the projection to the context is different from the test case.* Note that positive open tests embody existential properties, and negative open tests universal ones.

We instantiate the approach of open tests for the Dynamic Condition Response (DCR) Graph process notation [10,14] and provide polynomial time methods for approximating which open tests will remain passing after a given model update. This theoretical result makes the approach practical: When the relevant activities are exactly those of the model, an open test is the same as a test of [28,29]. From there, we iteratively update the model verifying at each step in low-order polynomial time that the open tests remain passing.

Altogether, we provide the following contributions:

1. We give a theory of process model testing using open tests (Sect. 2). This theory is general enough that it is applicable to all process notations with trace semantics: it encompasses both declarative approaches such as DECLARE or DCR, and imperative approaches such as BPMN.
2. We give in this general theory sufficient conditions for ensuring preservation of tests (Proposition 16). This proposition is key to supporting iterative test-driven modelling: it explains how a test case can withstand model updates.
3. We apply the theory to DCR graphs, giving a sufficient condition ensuring preservation of tests across model updates (Theorems 30 respectively 33).

Related Work. Test-driven modelling (TDM) was introduced by Zugal et al. in [28,29] as an application of test-driven development to declarative business processes. Their studies [28] indicate in particular the that simple sequential traces are helpful to domain experts in understanding the underlying declarative models. The present approach generalises that of [28,29]: We define and study preservation of tests across model updates, alleviating modularity concerns while preserving the core usability benefit of defining tests via traces.

Connections between refinement, testing-equivalence and model-checking was observed in [6]. But where we consider refinements guaranteeing preservation of the projected language, the connection in [6] uses that a refinement of a state based model (Büchi-automaton) satisfies the formula the state based model was derived from. Our approach (and that of [12]) has strong flavours of refinement. Indeed, the iterative development and abstract testing of system models in the present paper is related to the substantial body of work on abstraction and abstract interpretation, e.g., [1,7,9]. In particular, an open test can be seen as a test on an abstraction of the system under test, where only actions in the context of the test are visible. In this respect, the abstraction is given by string projection on free monoids. We leave for future work to study the ramifications of this relationship and the possibilities of exploiting it in employing more involved manipulations than basic extensions of the alphabet in the process of iterative development, such as, e.g., allowing splitting of actions.

The synergy between static analysis and model checking is also being investigated in the context of programming languages and software engineering [13]. In particular there have been proposals for using static analysis to determine test prioritisation [19,27] when the tests themselves are expensive to run. Our approach takes the novel perspective of analysing the adaptations to a model (or code), instead of analysing the current instance of the model.

2   Open Tests

In this section, we introduce open tests in the abstract setting of trace languages. The definitions and results of this section apply to any process notation that has a trace semantics, e.g., DECLARE [23], DCR graphs [10,12,14], or BPMN [21]. In the following sections, we instantiate the general results of this section specifically for DCR graphs.

Definition 1 (Test case). *A test case (c, Σ) is a finite sequence c over a set of activities Σ. We write $\mathsf{dom}(c) \subseteq \Sigma$ for the set of activities in c, i.e., when $c = \langle e_1 \dots . e_n \rangle$ we have $\mathsf{dom}(c) = \{e_1, \dots, e_n\}$.*

As a running example, we iteratively develop a process for handling reimbursement claims. The process we eventually develop conforms to Sect. 42 of the Danish Consolidation Act on Social Services (Serviceloven) [5]. The process exists to provide a citizen compensation for lost wages in the unfortunate circumstances that he must reduce his working hours to care for a permanently disabled child.

Example 2. Consider the set of activities $\Sigma = \{\mathsf{Apply, Document, Receive}\}$, abbreviating respectively "Apply for compensation", "Document need for compensation", and "Receive compensation". We define a test case t_0:

$$t_0 = (\langle \mathsf{Apply, Document, Receive} \rangle, \Sigma) \tag{3}$$

Intuitively, this test case identifies all traces, over *any* alphabet, whose projection to Σ is exactly $\langle \mathsf{Apply, Document, Receive} \rangle$.

Open test cases come in one of two flavours: a *positive* test requires the presence of (a representation of) a trace, whereas a *negative* test requires the absence of (any representation of) a trace.

Definition 3 (Positive and negative open tests). *An* open test *comprises a test case* $t = (c, \Sigma)$ *and a* polarity *ρ in* $\{+, -\}$*, altogether written* t^+ *respectively* t^-*.*

Example 4. Extending our previous example, define a positive and negative open test as follows:

$$t_0^+ = (\langle \mathsf{Apply}, \mathsf{Document}, \mathsf{Receive} \rangle, \{\mathsf{Apply}, \mathsf{Document}, \mathsf{Receive}\}^+ \tag{4}$$

$$t_1^- = (\langle \mathsf{Receive} \rangle, \{\mathsf{Document}, \mathsf{Receive}\})^- \tag{5}$$

Intuitively, the positive test t_0^+ requires the presence of some trace that projects to exactly $\langle \mathsf{Apply}, \mathsf{Document}, \mathsf{Receive} \rangle$. The negative requires that no trace, when projected to $\{\mathsf{Document}, \mathsf{Receive}\}$, is exactly $\mathsf{Receive}$, that is, this test models the requirement that one may not $\mathsf{Receive}$ compensation without first providing $\mathsf{Documentation}$.

To formalise the semantics of open tests we need a system model representing the possible behaviours of the system under test. In general, we define a system as a set of sequences of activities, that is, a language.

Definition 5. *A system* $S = (L, \Sigma)$ *is a language L of finite sequences over a set of activities Σ.*

We can now define under what circumstances positive and negative open tests pass. First we introduce notation.

Notation. Let ϵ denote the empty sequence of activities. Given a sequence s, write s_i for the ith element of s, and $s|_\Sigma$ defined inductively by $\epsilon|_\Sigma = \epsilon$, $(a.s)|_\Sigma = a.(s|_\Sigma)$ if $a \in \Sigma$ and $(a.s)|_\Sigma = s|_\Sigma$ if $a \notin \Sigma$. E.g, if $s = \langle \mathsf{Apply}, \mathsf{Document}, \mathsf{Receive} \rangle$ is the sequence of test t_0 above, then $s|_{\{\mathsf{Document}, \mathsf{Receive}\}} = \langle \mathsf{Document}, \mathsf{Receive} \rangle$ is the projection of that sequence. We lift projection to sets of sequences point-wise.

Definition 6 (Passing open tests). *Let* $S = (L, \Sigma')$ *be a system and* $t = (c, \Sigma)$ *a test case. We say that:*

1. *S passes the open test t^+ iff there exists $c' \in L$ such that $c'|_\Sigma = c$.*
2. *S passes the open test t^- iff for all $c' \in L$ we have $c'|_\Sigma \neq c$.*

S fails an open test t^ρ iff it does not pass it.

Notice how activities that are not in the context of the open test are ignored when determining if the system passes.

Example 7 (System S, Iteration 1). Consider a system $S = (L, \Sigma)$ with activities $\Sigma = \{\text{Apply, Document, Receive}\}$ and as language L the subset of sequences of Σ^* such that the Receive is always preceded (not necessarily immediately) by Document, and Apply is always succeeded (again not necessarily immediately) by Receive.

Positive tests require existence of a trace that projects to the test case. This system S passes the test t_0^+ for $t_0 = (\langle\text{Apply, Document, Receive}\rangle, \Sigma)$ as defined above, since the sequence $c' = \langle\text{Apply, Document, Receive}\rangle$ in L has $c'|_\Sigma = \langle\text{Apply, Document, Receive}\rangle$.

Negative tests require the absence of any trace that projects to the test case. S also passes the test t_1^- for $t_1 = (\text{Receive}, \{\text{Document, Receive}\})$ since if there were a $c' \in L$ s.t. $c'|_{\{\text{Document,Receive}\}} = \text{Receive}$ that would contradict that Document should always appear before any occurrence of Receive.

Finally, consider the following positive test.

$$t_2^+ = (\text{Apply}, \{\text{Apply, Receive}\})^+$$

The System S fails this test t_2^+, because every sequence in L that contains Apply will by definition also have a subsequent Receive, which would then appear in the projection.

We note that a test either passes or fails for a particular system, never both; and that positive and negative tests are dual: t^+ passes iff t^- fails and vice versa.

Lemma 8. *Let $S = (L, \Sigma)$ be a system and t a test case. Then either (a) S passes t^+ and fails t^-; or (b) S fails t^+ and S passes t^-.*

Example 9 (Iteration 2, Test preservation). We extend our model of Example 7 with the additional requirement that some documentation of the salary reduction is required before compensation may be received. To this end, we refine our system (L, Σ) to a system $S' = (L', \Sigma' = \Sigma \cup \{\text{Reduction}\})$ where Reduction abbreviates "Provide documentation of salary reduction", and L' is the language over Σ'^* that satisfies the original rules of Example 7 and in addition that Receive is always preceded by Reduction.

The explicit context ensures that the tests t_0^+, t_1^-, t_2^+ defined in the previous iteration remain meaningful. The system S' no longer has a trace

$$\langle\text{Apply, Document, Receive}\rangle$$

because Reduction is missing. Nonetheless, S' *still* passes the test t_0^+, because S' *does* have the trace:

$$c' = \langle\text{Apply, Document, Reduction, Receive}\rangle \in L'$$

The projection $c'|_\Sigma = \langle\text{Apply, Document, Receive}\rangle$ then shows that t_0^+ passes S'.

Similarly, S' still passes the test t_1^- since for any $c' \in L'$, if $c'|_\Sigma = \langle\text{Receive}\rangle$ then $c' = \langle c_0, \text{Receive}, c_1\rangle$ for some $c_0, c_1 \in \Sigma'\backslash\Sigma$, but that contradicts the requirement that Document must appear before any occurrence of Receive.

We now demonstrate how open tests may be preserved by model extensions where the trace-based tests of Zugal et al. [28, 29] would not be.

Example 10 (Non-preservation of non-open tests). We emphasize that if we interpret the trace $s = \langle \mathsf{Apply}, \mathsf{Document}, \mathsf{Receive} \rangle$ underlying the test t_0^+ as a test in the sense of [28, 29], that test is *not* preserved when we extend the system from (L, Σ) to (L', Σ'): The original system L has the behaviour s, but the extension L' does not.

Example 11 (Iteration 2, Additional tests). We add the following additional tests for the new requirements of Iteration 2.

$$t_3^- = (\langle \mathsf{Apply}, \mathsf{Document}, \mathsf{Receive} \rangle, \Sigma')^-$$

Note that the trace of t_3 is *the same* as the original test t_0; the two tests differ only in their context. This new test says that in a context where we know about the Reduction activity, omitting it is not allowed.

In these particular examples, the tests that passed/failed in the first iteration also passed/failed in the second. This is not generally the case; we give an example.

Example 12 (Iteration 3). The reduction in salary may be rejected, e.g. if the submitted documentation is somehow unsatisfactory. In this case, compensation must be withheld until new documentation is provided. We model this by adding an activity Rejection to the set of activities Σ' and constrain the language L' accordingly. Now the system will pass the test $t_2^+ = (\mathsf{Apply}, \{\mathsf{Apply}, \mathsf{Receive}\})^+$ defined above, because it has a trace that contains Apply but no Receive: the sequence in which the documentation of reduced salary is rejected.

We now turn to the question how to "run" an open test. Unlike the tests of Zugal et al., running an open test entails more than simply checking language membership. For positive tests we must find a trace of the system with a suitable projection, and for negative tests we must check that no trace has the test trace as projection.

First, we note that if the context of the open test contains all the activities of the model under test, it is in fact enough to simply check language membership.

Lemma 13. *Let $S = (L, \Sigma)$ be a system, let $t = (c, \Sigma')$ be a test case, and suppose $\Sigma \subseteq \Sigma'$. Then: 1. t^+ passes iff $c \in L$. 2. t^- passes iff $c \notin L$.*

Second, we show how checking whether an open test passes or fails in the general case reduces to the language inclusion problem for regular languages.

Proposition 14. *Let $S = (L, \Sigma)$ be a system, let $t = (\langle c_1, \ldots, c_n \rangle, \Sigma')$ be a test case. Define the set of irrelevant activities $I = \Sigma \setminus \Sigma'$ as those activities in the system but not in the test case. Assume wlog $I = \{i_1, \ldots, i_m\}$, and let $\mathsf{ri} = (i_1 | \cdots | i_m)^*$ be the regular expression that matches zero or more irrelevant activities. Finally, define the regular expression $\mathsf{rc} = \mathsf{ri}\, c_1\, \mathsf{ri} \cdots \mathsf{ri}\, c_n\, \mathsf{ri}$. Then:*

1. *S passes the positive test t^+ iff* lang(rc) $\cap L \neq \emptyset$, *and*
2. *S passes the negative test t^- iff* lang(rc) $\cap L = \emptyset$.

Example 15. Consider again S' of Example 9, and the test case t_0^+ of Example 2:

$$S' = (L', \Sigma' = \{\mathsf{Apply}, \mathsf{Document}, \mathsf{Receive}, \mathsf{Reduction}\})$$
$$t_0^+ = (\langle \mathsf{Apply}, \mathsf{Document}, \mathsf{Receive}\rangle, \{\mathsf{Apply}, \mathsf{Document}, \mathsf{Receive}\}^+)$$

In the notation of Proposition 14, we have $I = \{\mathsf{Reduction}\}$, ri = Reduction* and by that Theorem, t_0^+ passes the system S' because S' has non-empty intersection with the language defined by the regular expression:

Reduction* Apply Reduction* Document Reduction* Receive Reduction*

Lemma 13 and Proposition 14 explain how to "run" open tests, they apply directly for any process notation with trace semantics. Language inclusion for regular languages is extremely well-studied: practical methods exist for computing such intersections from both the model-checking and automata-theory communities. For certain models these methods may be sufficient; for example in the case of BPMN where models tend to be fairly strict and allow little behaviour, it could be feasible to always rely on model checking. However, for models that represent large state spaces, which is particularly common for declarative notations, this will not suffice and we will need to reduce the amount of model checking required.

The key insight of open tests is that oftentimes, changes to a model will preserve open test outcomes, obviating the need to re-check tests after the change. The following Proposition gives general conditions for when outcomes of positive (resp. negative) tests for a system S are preserved when the system is changed to a new system S'.

Proposition 16. *Let $S = (L, \Sigma)$ and $S' = (L', \Sigma')$ be systems, and let $t = (c_t, \Sigma_t)$ be a test case. Assume that $\Sigma' \cap \Sigma_t \subseteq \Sigma' \cap \Sigma$. Then:*

1. *If $L'|_\Sigma \supseteq L$ and S passes t^+, then so does S'.*
2. *If $L'|_\Sigma \subseteq L$ and S passes t^-, then so does S'.*

In words, the assumption $\Sigma' \cap \Sigma_t \subseteq \Sigma' \cap \Sigma$ states that the changed system S' does not introduce activities appearing in the context of the test that did not already appear in the original system S. Condition 1 (resp. 2) expresses that positive (resp. negative) tests are preserved if the language of the original system S is included in (resp. including) the language of the changed system S' projected to the activities in the original system. Now, if one can find static properties of changes to process models for a particular notation that implies the conditions of Proposition 16 then these properties can be checked instead of relying on model-checking to infer preservation of tests. We identify such static properties for the Dynamic Condition Response (DCR) graphs [10,12,14] process notation in Sect. 4. First however, we recall the syntax and semantics of DCR graphs in the next section.

3 Dynamic Condition Response Graphs

DCR graphs is a declarative notation for modelling processes superficially similar to DECLARE [22,23] or temporal logics such as LTL [24] in that it allows for the declaration of a set of temporal constraints between activities.

One notable difference is that DCR graphs model also the run-time state of the process using a so-called marking of activities. The marking consists of three finite sets (Ex,In,Re) recording respectively which activities have been executed (Ex), which are currently included (In) in the graph, and which are required (Re) to be executed again in order for the graph to be accepting, also referred to as the pending events. The marking allows for providing semantics of DCR graphs by defining when an activity is enabled in a marking and how the execution of an enabled activity updates the marking. Formally, DCR graphs are defined as follows.

Definition 17 (DCR Graph [14][1]). *A DCR graph is a tuple* $(\mathsf{E}, \mathsf{R}, \mathsf{M})$ *where*

- E *is a finite set of* activities, *the nodes of the graph.*
- R *is the edges of the graph. Edges are partitioned into five kinds, named and drawn as follows: The* conditions $(\to\bullet)$, *responses* $(\bullet\to)$, *inclusions* $(\to+)$, *exclusions* $(\to\%)$ *and* milestones $\to\diamond$.
- M *is the* marking *of the graph. This is a triple* $(\mathsf{Ex}, \mathsf{Re}, \mathsf{In})$ *of sets of activities, respectively the previously executed* (Ex), *the currently pending* (Re), *and the currently included* (In) *activities.*

Next we recall from [14] the definition of when an activity is enabled.

Notation. When G is a DCR graph, we write, e.g., $\mathsf{E}(G)$ for the set of activities of G, $\mathsf{Ex}(G)$ for the executed activities in the marking of G, etc. In particular, we write $\mathsf{M}(e)$ for the triple of boolean values $(e \in \mathsf{Ex}, e \in \mathsf{Re}, e \in \mathsf{In})$. We write $(\to\bullet e)$ for the set $\{e' \in \mathsf{E} \mid e' \to\bullet e\}$, write $(e\bullet\to)$ for the set $\{e' \in \mathsf{E} \mid e \bullet\to e'\}$ and similarly for $(e\to+)$, $(e\to\%)$ and $(\to\diamond e)$.

Definition 18 (Enabled activities [14]). *Let* $G = (\mathsf{E}, \mathsf{R}, \mathsf{M})$ *be a DCR graph, with marking* $\mathsf{M} = (\mathsf{Ex}, \mathsf{Re}, \mathsf{In})$. *An activity* $e \in \mathsf{E}$ *is* enabled, *written* $e \in$ enabled(G), *iff (a)* $e \in \mathsf{In}$ *and (b)* $\mathsf{In} \cap (\to\bullet e) \subseteq \mathsf{Ex}$ *and (c)* $(\mathsf{Re} \cap \mathsf{In}) \cap (\to\diamond e) = \emptyset$.

That is, enabled activities (a) are included, (b) their included conditions have already been executed, and (c) have no pending included milestones.

Executing an enabled activity e of a DCR graph with marking $(\mathsf{Ex}, \mathsf{Re}, \mathsf{In})$ results in a new marking where (a) the activity e is added to the set of executed activities, (b) e is removed from the set of pending response activities, (c) the responses of e are added to the pending responses, (d) the activities excluded by

---

[1] In [14] DCR graphs model constraints between so-called events labelled by activities. To simplify the presentation, we assume in the present paper that each event is labelled by a unique activity and therefore speak only of activities.

e are removed from included activities, and (e) the activities included by e are added to the included activities.

From this we can define the language of a DCR graph as all finite sequences of activities ending in a marking with no activity both included and pending.

Definition 19 (Language of a DCR Graph [14][2]). *Let $G_0 = (\mathsf{E}, \mathsf{R}, \mathsf{M})$ be a DCR graph with marking $\mathsf{M}_0 = (\mathsf{Ex}_0, \mathsf{Re}_0, \mathsf{In}_0)$. A trace of G_0 of length n is a finite sequence of activities $e_0, \dots, e_{n-1}$ such that for all i such that $0 \le i < n$, (i) e_i is enabled in the marking $M_i = (\mathsf{Ex}_i, \mathsf{Re}_i, \mathsf{In}_i)$ of G_i, and (ii) G_{i+1} is a DCR graph with the same activities and relations as G_i but with marking $(\mathsf{Ex}_{i+1}, \mathsf{Re}_{i+1}, \mathsf{In}_{i+1}) = (\mathsf{Ex}_i \cup \{e_i\}, (\mathsf{Re}_i \setminus \{e_i\}) \cup (e_i \bullet \rightarrow), (\mathsf{In}_i \setminus (e_i \rightarrow \%)) \cup (e_i \rightarrow +))$.*

We call a trace of length n accepting if $\mathsf{Re}_n \cap \mathsf{In}_n = \emptyset$. The language $\mathsf{lang}(G_0)$ of G_0 is then the set of all such accepting traces. Write $\hat{G}$ for the corresponding system $\hat{G} = (\mathsf{lang}(G), \mathsf{E})$ (viz. Definition 5). When no confusion is possible, we denote by simply G both a DCR graph and its corresponding system $\hat{G}$.

Example 20 (DCR Iteration 1). We model the Sect. 42 process of Example 2 in Fig. 1a as a DCR graph. This model is simple enough that it uses only the response and condition relations which (in this case) behave the same as the response and precedence constraints in DECLARE. The condition from Document to Receive models the requirement that documentation must be provided before compensation may be received. The response from Apply to Receive models that compensation must eventually be received after it has been applied for. The marking of the graph is $(\emptyset, \emptyset, \{\mathsf{Document}, \mathsf{Receive}\})$, i.e. no activities have been executed, no activities are yet pending responses and all activities are included. (The activity Receive is grey to indicate that it is not enabled).

Example 21 (DCR Iteration 2). Following Example 9, we extend the iteration 1 model of Fig. 1a to the iteration 2 model in Fig. 1b. We model the new requirement that documentation must be provided before compensation may be received by adding a new activity Reduction and a condition relation from Reduction to Receive.

Example 22 (DCR Iteration 3). Following Example 12, we extend the iteration 2 model of Fig. 1b to the iteration 3 model in Fig. 1c. To model the rejection of documentation, we add the activity Rejection and exclude-relations between Rejection and Receive. This models the choice between those two activities: Once one is executed, the other is excluded and no longer present in the model. To model subsequent re-submission, we add an include relation from Reduction to Rejection and Receive: if new documentation of salary is received, the activities Rejection and Receive become included once again, re-enabling the decision whether to reject or pay.

Example 23 (DCR Iteration 3, variant). To illustrate the milestone relation, we show an alternative to the model of Fig. 1c in Fig. 1d. Using a response relation

---

[2] In [14] the language of a DCR graph consists of both finite and infinite sequences. To simplify the presentation, we consider only finite sequences in the present paper.

from Receive to Reduction we model that after rejection, documentation must be resubmitted; and by adding a milestone from Reduction to Receive we model that compensation may not received again while we are waiting for this new documentation.

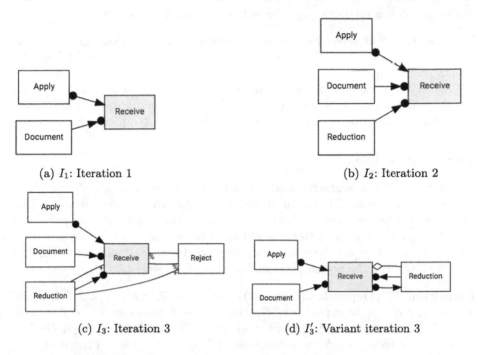

(a) I_1: Iteration 1   (b) I_2: Iteration 2

(c) I_3: Iteration 3   (d) I_3': Variant iteration 3

Fig. 1. DCR Graph models of the Sect. 42 of Examples 2–12.

4   Iterative Test-Driven Modelling for DCR Graphs

In this Section, we show how Proposition 16 applies to DCR graphs, and exemplify how the resulting theory supports iterative test-driven DCR model development by telling us which tests are preserved by model updates. We consider the situation that a graph G' extends a graph G by adding activities and relations. Recall the notation $\mathsf{M}(e) = (e \in \mathsf{Ex}, e \in \mathsf{Re}, e \in \mathsf{In})$.

Definition 24 (Extensions). *Let $G = (\mathsf{E}, \mathsf{R}, \mathsf{M})$ and $G' = (\mathsf{E}', \mathsf{R}', \mathsf{M}')$ be DCR graphs. We say that G' statically extends G and write $G \sqsubseteq G'$ iff $\mathsf{E} \subseteq \mathsf{E}'$ and $\mathsf{R} \subseteq \mathsf{R}'$. If also $e \in \mathsf{E}$ implies $\mathsf{M}(e) = \mathsf{M}'(e)$, we say that G' dynamically extends G and write $G \preceq G'$.*

Our main analysis technique will be the application of Proposition 16. To this end, we need ways to establish the preconditions of that Theorem, that is:

$$\mathsf{lang}(G')|_{\mathsf{E}} \supseteq \mathsf{lang}(G) \qquad (\dagger)$$
$$\mathsf{lang}(G')|_{\mathsf{E}} \subseteq \mathsf{lang}(G) \qquad (\ddagger)$$

Example 25. Consider the graphs I_1 of Fig. 1a and I_2 of Fig. 1b. Clearly $I_1 \sqsubseteq I_2$ since I_2 contains all the activities and relations of I_1. Moreover, since the markings of I_1 and I_2 agree, also $I_1 \preceq I_2$. Similarly, $I_2 \sqsubseteq I_3$ and $I_2 \sqsubseteq I_3'$, where I_3 and I_3' are the graphs of Figs. 1c and 1d. On the other hand, neither $I_3 \sqsubseteq I_3'$, since the former graph has activities not in the latter, nor $I_3' \sqsubseteq I_3$, since the former graph has relations (e.g., the milestone) not in the latter.

We note that DCR activity execution preserves static extensions, i.e. if $G \sqsubseteq G'$ and an activity e is enabled in both G and G' then $G_1 \sqsubseteq G_1'$, if G_1 and G_1' are the results of executing e in G and G' respectively. Dynamic extension is generally *not* preserved by execution, because an execution might make markings between the original and extended graph differ *on the original activities*, e.g., if G' adds an exclusion, inclusion or response constraint between activities of E.

4.1   Positive Tests

We first establish a syntactic condition for a modification of a DCR graph to preserve positive tests. The condition will be, roughly, that the only new relations are either (a) between new activities, or (b) conditions or milestones from new to old activities. For the latter, we will need to be sure we can find a way to execute enough new activities to satisfy such conditions and milestones. To this end, we introduce the notion of dependency graph, inspired by [2].

Definition 26 (Dependency graph). *Let $G = (\mathsf{E}, \mathsf{R}, \mathsf{M})$ be a DCR graph, and let $e, f, g \in \mathsf{E}$ be activities of G. Write $e \to f$ whenever $e \to\bullet f \in \mathsf{R}$ or $e \to\diamond f \in \mathsf{R}$ and $\to^*$ for the transitive closure. The dependency graph $D(G, e)$ for e is the directed graph which has nodes $\{f \mid g \to^* e \wedge g \bullet\!\to^* f\}$ and an edge from node f to node g iff $f \to g$ or $f \bullet\!\to g$ in G.*

With the notion of dependency graph, we can define the notion of "safe" activities, intuitively those that can be relied upon to be executed without having undue side effects on a given (other) set of nodes X. The principle underlying this definition is inspired by the notion of dependable activity from [2].

Definition 27 (Safety). *Let $G = (\mathsf{E}, \mathsf{R}, \mathsf{M})$ be a DCR graph, let $e \in \mathsf{E}$ be an activity of G, and let $X \subseteq \mathsf{E}$ be a subset of the activities of G. We say that e is safe for X iff*

1. *$D(G, e)$ is acyclic,*
2. *no $f \in D(G, e)$ has an include, exclude, or response relation to any $x \in X$.*
3. *for any $f \in D(G, e)$, if f has a condition or milestone to some $f' \in \mathsf{E}$, then f' is reachable from f in $D(G, e)$.*

The notion of safe activity really captures activities that can reliably be discharged if they are conditions or milestones for other activities. We use this to define a notion of transparent process extensions: a process extension which we shall see preserves positive tests.

Definition 28 (Transparent). *Let* $G = (\mathsf{E}, \mathsf{R}, \mathsf{M})$ *and* $G' = (\mathsf{E}', \mathsf{R}', \mathsf{M}')$ *be DCR graphs with* $G \sqsubseteq G'$. *We say that* G' *is* transparent *for* G *iff for all* $e, f \in \mathsf{E}$ *and* $e', f' \in \mathsf{E}'$ *we have:*

1. *if* $e'\mathcal{R}f' \in \mathsf{R}'$ *for* $\mathcal{R} \in \{\rightarrow\bullet, \rightarrow\diamond\}$ *then either* $e'\mathcal{R}f' \in \mathsf{R}$ *or* (a) $e' \notin \mathsf{E}$, (b) e' *is safe for* E, *and* (c) $\mathsf{E}(D(G', e')) \subseteq \mathsf{E}' \setminus \mathsf{E}$,
2. *for* $\mathcal{R} \in \{\rightarrow+, \rightarrow\%, \bullet\rightarrow\}$ *we have* $e\mathcal{R}f \in \mathsf{R}'$ *iff* $e\mathcal{R}f \in \mathsf{R}$.
3. *for* $\mathcal{R} \in \{\rightarrow+, \bullet\rightarrow\}$ *we have if* $e\mathcal{R}e' \in \mathsf{R}'$ *or* $e' \in \mathrm{Re}(G')$ *then* $e' \in \mathsf{E}$

We rephrase these conditions more intuitively. Call an activity $e \in \mathsf{E}$ an *old* activity, and an activity $e' \in \mathsf{E}' \setminus \mathsf{E}$ a *new* activity. The first item then says that we can never add conditions or milestones from old activities and only add a condition or milestone to an old activity when the new activity is safe, that is, we can rely on being able to discharge that milestone or condition. The second item says that we cannot add exclusions, inclusions or responses between old activities. The third says that we also cannot add inclusions or responses from old to new activities, or add a new activity which is initially pending in the marking, which could cause the new graph to be less accepting than the old. Inclusions, exclusions and responses may be added from a new to an old activity; the interplay of condition 1 of Definition 28 and condition 2 of Definition 27 then implies that this can only happen if the new activity is not in the dependency graph of any old activity. The reason is, that such constraints can be *vacuously* satisfied since the new activity at the source of the constraint is *irrelevant* with respect to passing any of the positive tests.

Example 29. It is instructive to see how violations of transparency may lead to non-preservation of positive tests. Consider a DCR graph with activities A, B and no relations. Note that this graph passes the open test $t^+ = (\langle A, B \rangle, \{A, B\})^+$. Consider two possible updates. (i) Adding a relation $B \rightarrow\bullet A$ (a condition between old activities) would stop the test from passing. So would adding an activity C and relations $B \rightarrow\bullet C \rightarrow\bullet A$. (ii) Adding a relation $B \bullet\rightarrow A$ causes t to end in a non-accepting state, stopping the test t from passing.

Theorem 30. *Let* $G \preceq G'$ *with* G' *transparent for* G, *and let* $t^+ = (c_t, \Sigma_t)^+$ *be a positive test with* $\Sigma_t \subseteq \mathsf{E}$. *If* G *passes* t *then so does* G'.

Example 31 (Preservation). Consider the change from the graph I_1 of Fig. 1a to the graph I_2 of Fig. 1b: We have added the activity Reduction and the condition Reduction $\rightarrow\bullet$ Receive. In this case, I_2 is transparent for I_1: The new activity Reduction satisfies Definition 28 part (1c): even though a new condition dependency is added for Receive, the dependency graph for the new Receive remains acyclic. By Theorem 30, it follows that *any* positive test whose context is contained in {Apply, Document, Receive} will pass I_2 if it passes I_1. In particular, we saw in Example 7 that I_1 passes the test t_0^+, so necessarily also I_2 passes t_0^+.

4.2   Negative Tests

For negative tests we must establish the inclusion (‡) stated after Definition 24. This inclusion was investigated previously in [11,12], with the aim of establishing more general refinement of DCR graphs. Definition 24 is a special case of refinement by merging, investigated in the above papers. Hence, we use the sufficient condition for such a merge to be a refinement from [12] to establish a sufficient condition, *exclusion-safety* for an extension to preserves negative tests.

Definition 32 (Exclusion-safe). *Suppose* $G = (\mathsf{E}, \mathsf{R}, \mathsf{M})$ *and* $G' = (\mathsf{E}', \mathsf{R}', \mathsf{M}')$ *are DCR graphs and that* G' *dynamically extends* G. *We say that* G' *is exclusion-safe for* G *iff for all* $e \in \mathsf{E}$ *and* $e' \in \mathsf{E}'$ *we have that:*

1. *if* $e' \rightarrow\% e \in \mathsf{R}'$ *then* $e' \rightarrow\% e \in \mathsf{R}$.
2. *if* $e' \rightarrow+ e \in \mathsf{R}'$ *then* $e' \rightarrow+ e \in \mathsf{R}$.

Theorem 33. *Suppose* $G \preceq G'$ *are DCR graphs with* G' *exclusion-safe for* G, *and suppose* $t^- = (c_t, \Sigma_t)^-$ *is a negative test with with* $\Sigma_t \subseteq \mathsf{E}$. *If* G *passes* t *then so does* G'.

Example 34 (Application). Consider again the change from I_1 to I_2 in Fig. 1a and b. Since neither contains inclusions or exclusions, clearly I_2 is exclusion-safe for I_1. By Theorem 33 it follows that any negative test whose context is contained in {Apply, Document, Receive} which passes I_1 will also pass I_2. In particular, the negative test $t_1^- = (\langle \mathsf{Receive} \rangle, \{\mathsf{Document}, \mathsf{Receive}\})^-$ of Example 7 passes I_1, so by Theorem 33 it passes also I_2. Similarly but less obviously, any negative test with context included in {Apply, Document, Reduction, Receive} which passes I_2 must also pass I_3' (Fig. 1d).

Example 35 (Non-application). The changes two from I_1 to I_2 and from I_2 to I_3 (Figs. 1a, 1b and 1c), where amongst other changes we have added an activity Rejection and a relation Rejection $\rightarrow\%$ Receive, both violate exclusion-safety.

In this case, we can find a negative test that passes I_1 and I_2 but not I_3: $t_2^- = \langle \mathsf{Apply} \rangle, \{\mathsf{Apply}, \mathsf{Receive}\}^-$. It passes both I_1 and I_2, because in both of these, Apply leaves Receive pending, whence one needs to execute also Receive to get a trace of the process. But in I_3, we can use Rejection to exclude the pending Receive. So I_3 has a trace $\langle \mathsf{Apply}, \mathsf{Rejection} \rangle$, and the projection of this trace to the context {Apply, Receive} of our test is the string $\langle \mathsf{Apply} \rangle$: The test fails in I_3.

4.3   Practical Use

To perform iterative test-driven development for DCR graphs using open tests, we proceed as follows. When tests are defined, we normally include all activities of the model under test in the context, and will be able to run them as standard tests, cf. Lemma 13. As we update the model, we verify at each step that the update preserves existing tests using Theorem 30 or 33. Should a model update fail to satisfy the prerequisites for the relevant Theorem, we "re-run" tests using model-checking techniques such as Proposition 14. We refer the reader to [15, 26] for details on model-checking for DCR graphs.

The prerequisites of both Theorem 30 and 33 are effectively computable.

Theorem 36. *Let* $G \preceq G'$ *be DCR graphs. It is decidable in time polynomial in the maximum size of* G, G' *whether (1)* G' *is exclusion-safe for* G *and (2)* G' *is transparent for* G.

5   Conclusion and Discussion

We introduced a general theory for testing abstractions of process models based on a notion of open tests, which extend the test-driven modelling methodology of [28,29]. In particular, we gave sufficient conditions for ensuring preservation of open tests across model updates. We applied the theory to the concrete declarative notation of DCR graphs and gave sufficient static transparency conditions on the updates of a DCR graph, ensuring preservation of open tests.

While the general theory applies to any process notation with trace semantics, the static conditions for transparency will need to be defined for the particular process notation at hand. Consider for example DECLARE [23]. The monotonicity of the semantics implies that adding more constraints will only remove traces from the language, and removing constraints will only add traces to the language. It is thus straightforward to prove that, if the set of activities is not changed, then adding respectively removing a constraint will satisfy part (1) respectively (2) of Proposition 16 and consequently positive respectively negative tests will not need to be re-checked. Further static conditions can be obtained by considering the constraints individually. Here, one source of complexity is the fact that DECLARE allows constraints that implicitly quantify over all possible activities. For instance, if a DECLARE model has the chain succession relation between two activities A and B, then A and B always happen together in the exact sequence $A.B$ with no other activities in-between. Now, if the model is extended by adding condition constraints from A to a new activity C and from C to B, then the test $A.B$ will fail even when considered in the open context $\{A, B\}$. Another source of complexity is simply that DECLARE allows many more constraints than DCR. We leave for future work to further investigate sufficient conditions for transparency for DECLARE.

Acknowledgements. We are grateful to the reviewers for their help not only to improve the presentation but also to identify interesting areas of future work.

References

1. Baeten, J.C.M., van Glabbeek, R.J.: Another look at abstraction in process algebra. In: Ottmann, T. (ed.) ICALP 1987. LNCS, vol. 267, pp. 84–94. Springer, Heidelberg (1987). https://doi.org/10.1007/3-540-18088-5_8
2. Basin, D.A., Debois, S., Hildebrandt, T.T.: In the nick of time: proactive prevention of obligation violations. In: Computer Security Foundations, pp. 120–134 (2016)
3. Beck, K.: Extreme Programming Explained: Embrace Change. Addison-Wesley Professional, Boston (2000)
4. Beck, K.: Test-driven development: by example (2003)
5. Bekendtgørelse af lov om social service. Børne- og Socialministeriet, August 2017
6. Bushnell, D.M.: Research Conducted at the Institute for Computer Applications in Science and Engineering for the Period October 1, 1999 through March 31, 2000. Technical report NASA/CR-2000-210105, NAS 1.26:210105, NASA (2000)
7. Clarke, E.M., Grumberg, O., Long, D.E.: Model checking and abstraction. ACM Trans. Program. Lang. Syst. **16**(5), 1512–1542 (1994)

8. Cockburn, A.: Agile Software Development, vol. 177. Addison-Wesley, Boston (2002)

9. Cousot, P., Cousot, R.: Systematic design of program analysis frameworks. In: Proceedings of the 6th ACM SIGACT-SIGPLAN Symposium on Principles of Programming Languages, pp. 269–282. ACM (1979)

10. Debois, S., Hildebrandt, T.: The DCR workbench: declarative choreographies for collaborative processes. In: Behavioural Types: from Theory to Tools, pp. 99–124. River Publishers, Gistrup (2017)

11. Debois, S., Hildebrandt, T., Slaats, T.: Hierarchical declarative modelling with refinement and sub-processes. In: Sadiq, S., Soffer, P., Völzer, H. (eds.) BPM 2014. LNCS, vol. 8659, pp. 18–33. Springer, Cham (2014). https://doi.org/10.1007/978-3-319-10172-9_2

12. Debois, S., Hildebrandt, T.T., Slaats, T.: Replication, refinement & reachability: complexity in dynamic condition-response graphs. Acta Informatica, 1–32 (2017). https://link.springer.com/article/10.1007%2Fs00236-017-0303-8#citeas

13. Ernst, M.D.: Static and dynamic analysis: synergy and duality. In: ICSE Workshop on Dynamic Analysis, pp. 24–27 (2003)

14. Hildebrandt, T., Mukkamala, R.R.: Declarative event-based workflow as distributed dynamic condition response graphs. In: Post-proceedings of PLACES 2010. EPTCS, vol. 69, pp. 59–73 (2010)

15. Hildebrandt, T., Mukkamala, R.R., Slaats, T., Zanitti, F.: Contracts for cross-organizational workflows as timed dynamic condition response graphs. J. Log. Algebr. Program. 82(5–7), 164–185 (2013)

16. Hull, R., et al.: Introducing the guard-stage-milestone approach for specifying business entity lifecycles. In: Bravetti, M., Bultan, T. (eds.) WS-FM 2010. LNCS, vol. 6551, pp. 1–24. Springer, Heidelberg (2011)

17. Janzen, D., Saiedian, H.: Test-driven development concepts, taxonomy, and future direction. Computer 38(9), 43–50 (2005)

18. Marquard, M., Shahzad, M., Slaats, T.: Web-based modelling and collaborative simulation of declarative processes. In: Motahari-Nezhad, H.R., Recker, J., Weidlich, M. (eds.) BPM 2015. LNCS, vol. 9253, pp. 209–225. Springer, Cham (2015). https://doi.org/10.1007/978-3-319-23063-4_15

19. Mei, H., Hao, D., Zhang, L., Zhang, L., Zhou, J., Rothermel, G.: A static approach to prioritizing junit test cases. IEEE Trans. Softw. Eng. 38(6), 1258–1275 (2012)

20. Object Management Group: Case Management Model and Notation. Technical report formal/2014-05-05, Object Management Group, version 1.0, May 2014

21. Object Management Group BPMN Technical Committee: Business Process Model and Notation, Version 2.0 (2013)

22. Pesic, M., van der Aalst, W.M.P.: A declarative approach for flexible business processes management. In: Eder, J., Dustdar, S. (eds.) BPM 2006. LNCS, vol. 4103, pp. 169–180. Springer, Heidelberg (2006). https://doi.org/10.1007/11837862_18

23. Pesic, M., Schonenberg, H., Van der Aalst, W.M.P.: DECLARE: full support for loosely-structured processes. In: Proceedings of the 11th IEEE International Enterprise Distributed Object Computing Conference, pp. 287–300. IEEE (2007)

24. Pnueli, A.: The temporal logic of programs. In: 18th Annual Symposium on Foundations of Computer Science, pp. 46–57 (1977)

25. Schwaber, K., Beedle, M.: Agile Software Development with Scrum, vol. 1. Prentice Hall, Upper Saddle River (2002)

26. Slaats, T.: Flexible Process Notations for Cross-organizational Case Management Systems. Ph.D. thesis, IT University of Copenhagen, January 2015

27. Zhang, L., Zhou, J., Hao, D., Zhang, L., Mei, H.: Prioritizing JUnit test cases in absence of coverage information. In: Software Maintenance, pp. 19–28. IEEE (2009)
28. Zugal, S., Pinggera, J., Weber, B.: The impact of testcases on the maintainability of declarative process models. In: Halpin, T., et al. (eds.) BPMDS/EMMSAD - 2011. LNBIP, vol. 81, pp. 163–177. Springer, Heidelberg (2011). https://doi.org/10.1007/978-3-642-21759-3_12
29. Zugal, S., Pinggera, J., Weber, B.: Creating declarative process models using test driven modeling suite. In: Nurcan, S. (ed.) CAiSE Forum 2011. LNBIP, vol. 107, pp. 16–32. Springer, Heidelberg (2012). https://doi.org/10.1007/978-3-642-29749-6_2

Construction Process Modeling: Representing Activities, Items and Their Interplay

Elisa Marengo[✉], Werner Nutt, and Matthias Perktold

Faculty of Computer Science, Free University of Bozen-Bolzano, Bolzano, Italy
{elisa.marengo,werner.nutt}@unibz.it, matthias.perktold@hotmail.com

Abstract. General purpose process modeling approaches are meant to be applicable to a wide range of domains. To achieve this result, their constructs need to be general, thus failing in capturing the peculiarities of a particular application domain. One aspect usually neglected is the representation of the items on which activities are to be executed. As a consequence, the model is an approximation of the real process, limiting its reliability and usefulness in particular domains.

We extend and formalize an existing declarative specification for process modeling mainly conceived for the construction domain. In our approach we model the activities and the items on which the activities are performed, and consider both of them in the specification of the flow of execution. We provide a formal semantics in terms of LTL over finite traces which paves the way for the development of automatic reasoning. In this respect, we investigate process model satisfiability and develop an effective algorithm to check it.

Keywords: Multi-instance process modeling
Satisfiability checking of a process model · Construction processes

1 Introduction

Process modeling has been widely investigated in the literature, resulting in approaches such as BPMN, Petri Nets, activity diagrams and data centric approaches. Among the known shortcomings of these approaches: *(i)* they need to be general in order to accommodate a variety of domains, inevitably failing in capturing all the peculiarities of a specific application domain [3,9,12]; *(ii)* they predominantly focus on one aspect between control flow and data, neglecting the interplay between the two [5]; *(iii)* process instances are considered in isolation, disregarding possible interactions among them [1,16].

As a result, a process model is just an abstraction of a real process, limiting its applicability and usefulness in some application domains. This is particularly the case in application domains characterized by *multiple-instance* and *item-dependent* processes. We identify as *multiple-instance* those processes where several process instances may run in parallel on different items, but their execution

© Springer Nature Switzerland AG 2018
M. Weske et al. (Eds.): BPM 2018, LNCS 11080, pp. 48–65, 2018.
https://doi.org/10.1007/978-3-319-98648-7_4

cannot be considered in isolation, for instance because there are synchronization points among the instances or because there are limited resources for the process execution. With *item-dependent* we identify those processes where activities are executed several times but on different items (that potentially differ from activity to activity) and items are different one from the other.

The need of properly addressing multiple-instance and item-dependent processes emerged clearly in the context of some research projects [4,6] in the construction domain. In this context, a process model serves as a synchronization mean and coordination agreement between the different companies simultaneously present on-site. Besides defining the activities to be executed and the dependencies among them, aspects that can be expressed by most of the existing modeling languages, there is the need of specifying for each activity the items on which it has to be executed, where items in the construction domain correspond to locations. In this sense, processes are *item-dependent*. Processes are also *multi-instance* since high parallelism in executing the activities is possible but there is the need to synchronize their execution on the items (e.g, to regulate the execution of two activities in the same location). The aim is not only to model these aspects, but also to provide (automatic) tools to support the modeling and the execution. This requires a model to rely on a formal semantics.

In this paper we address the problem of *multi-instance* and *item-dependent* process modeling, specifically how to specify the items on which activities are to be executed and how the control flow can be refined to account for them. Rather than defining "yet another language", we start from an existing informal language that has been defined in collaboration with construction companies and tested on-site [4,19] in a construction project (Fig. 1). We refined it by taking inspiration from Declare [2], and propose a formal semantics grounded on Linear Temporal Logic (LTL) where formulas are evaluated over finite traces [8].

Concerning the development of automatic tools, we propose an algorithm for *satisfiability* checking, defined as the problem of checking whether, given a process model, there is at least one execution satisfying it. Satisfiability is a prerequisite for the development of further automatic reasoning, such as the generation of (optimized) executions compliant to the model. We also developed a web-based prototype acting as a proof-of-concept for graphical process modeling and satisfiability checking [21]. The developed language and technique can be generalized to other application domains such as manufacturing-as-a-service [17], infrastructure, ship building and to multi-instance domains [16] such as transport/logistic, health care, security, and energy.

The paper presents the related work in Sect. 2, a formalization for process models and for the modeling language in Sect. 3, an excerpt of a real construction process model [7] in Sect. 4, the satisfiability problem in Sect. 5, and our algorithm and an implementation to check it in Sect. 5.2.

2   Related Work

The role of processes in construction is to coordinate the work of a number of companies simultaneously present on-site, that have to perform different kinds of work (activities) in shared locations (items). Coordination should be such that workers do not obstruct each other and such that the prerequisites for a crew to perform its work are all satisfied when it has to start. For example, in the construction of a hotel it should be possible to express that wooden and aluminum windows must be installed respectively in the rooms and in the bathrooms, and that in the rooms the floor must be installed before the windows (not to damage them).

The adoption of IT-tools in construction is lower compared to other industries, such as manufacturing [15]. The traditional and most adopted techniques are the Critical Path Method (CPM) and the Program Evaluation and Review Technique (PERT). They consider the activities as work to be performed, focusing on their duration and the overall duration of a schedule. However, they do not account for the locations where to execute the activities and the location-based relationships between them [14]. As a result, a process representation abstracts from important details causing [22]: *(i)* the communication among the companies to be sloppy, possibly resulting in different interpretations of a model; *(ii)* difficulties in managing the variance in the schedule and resources; *(iii)* imprecise activity duration estimates (based on lags and float in CPM and on probability in PERT); *(iv)* inaccurate duration estimates not depending on the quantity of work to be performed in a location and not accounting for the expected productivity there. As a result, project schedules are defined to satisfy customer or contractual requirements, but are rarely used during the execution for process control [14].

Gantt charts are a graphical tool for scheduling in project management. Being graphical they are intuitive and naturally support the visualization of task duration and precedences among them. However, a Gantt chart already represents a commitment to one particular schedule, without necessarily relying on a process model. A process model explicitly captures the requirements for the allowed executions, thus supporting a more flexible approach: in case of delay or unforeseen events any re-schedule which satisfies the model is a possible one. Besides this limitation, Gantt charts are general-purpose and when applied to construction fail in naturally representing locations (which consequently are also not supported by IT-tools such as Microsoft Project) and have a limited representation of the precedences (only constraining two tasks, rather than, for instance, specifying constraints for tasks by floor or by room and so on). Flow-line diagrams are also a visual approach for process schedules, which explicitly represent locations and production rates. However, they also do not rely on an explicit process model. More recently, BIM-based tools have been developed. They are powerful but also require a big effort and dedicated resources to use the tool and align a BIM model with the construction site [10]. These aspects limit their use by small/medium sized companies.

From the business process literature one finds the standard BPMN. Its notation supports the representation of multi-instance activities and of data objects. However, the connection between the control flow and the data object is under-specified [5]: items and their role in ruling the control flow are not expressed

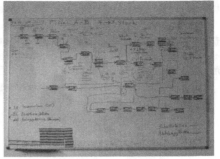

Fig. 1. Process modeling workshop with construction companies and the resulting model.

explicitly. Other approaches [1,3] consider both the flow and the data and the process instances can synchronize via the *data objects*. The Instance Spanning Constraints [16] approach considers multiple instances and the constraints among them. To the best of our knowledge, none of these approaches has been applied to *execution* processes in construction (note that [3] has been applied to construction but not to the execution process, which requires to account for higher level of details). Their adoption would require adaptations and extensions, such as representation of the items and considering them in the execution flow. Similar considerations hold for Declare [2], although it supports a variety of constraint types.

In the context of a research [4] and a construction project [19], a new approach for a detailed modeling and management of construction processes was developed in collaboration with a company responsible for facade construction. An ad-hoc and informal modeling language was defined, with the aim of specifying the synchronization of the companies on-site and used as a starting point for the daily scheduling of the activities. To this aim, both activities and locations had to be represented explicitly. The process model of the construction project, depicted in Fig. 1, was defined collaboratively by the companies participating in the project and was sketched on whiteboards. The resulting process allowed the companies to discuss in advance potential problems and to more efficiently schedule the work. The benefit was estimated in a 8% saving of the man hours originally planned and some synchronization problems (e.g., in the use of the shared crane) were discovered in advance and corresponding delays were avoided.

By applying the approach in the construction project some requirements emerged, mainly related to the ambiguity of the language (which required additional knowledge and disambiguations provided as annotations in natural language). The main requirements were: *(i)* besides the activities to be performed, also the locations where to execute them need to be represented in a *structured* and *consistent* way; *(ii)* to capture the desired synchronization, the specification of the flow of execution must be refined, so that not only activities are considered, but also the locations. For instance, when defining a precedence constraint between two activities it must be clear whether it means that the first must be finished everywhere before the second can start, or whether the precedence applies at a particular location i.e. floor, room and so on; *(iii)* the representation of more details complicates the process management, which would benefit from the development of (automatic) IT supporting tools. In the next section we provide a formalization for process modeling in construction, paving the way for automatic tool development.

3   Multi-instance and Item-Dependent Process Modeling

Our formalism foresees two components for a process model: a *configuration* part, defining the possible activities and items, and a *flow* part, specifying which activity is executed on which item, and the constraints on the execution. We illustrate these parts for the construction domain, although the formalization is domain-independent. As an example, we use an excerpt of a real process for a hotel construction [7].

3.1   Process Model

In this section we describe the two components of a process model.

Configuration Part. The configuration part defines a set of *activities* (e.g., excavation, lay floor) and the *items* on which activities can be performed (e.g., locations).

For the item representation, we foresee a hierarchical structure where the elements of the hierarchy are the *attributes* and for each of them we define a range of possible values. The attributes to consider depend on the domain. In construction a possible hierarchy to represent the locations is depicted in Fig. 2, and the attributes are: *(i) sector* (sr), which represents an area of the construction site such as a separate building (as possible values we consider B1 and B2); *(ii) level* (l), with values underground (u1), zero (f0) and one (f1); *(iii) section* (sn), which specifies the technological content of an area (as possible values we consider room r, bathroom b, corridor c and entrance e); and *(iv) unit* (u), which enumerates locations of similar type (we use it to enumerate the hotel rooms from one to four). Other attributes could be considered, such as *wall*, with values north, south, east and west, to identify the walls within a section (which is important when modeling activities such as cabling and piping). The domain values of an attribute can be ordered, for instance to represent an ascending order on the levels (u1 < f0 < f1). We call *item structure* a hierarchy of attributes representing an item.

In process models, e.g. in manufacturing and construction, it is common to conceptually divide a process into *phases*, where the activities to be performed and the item structure may be different. Common phases in construction are the *skeleton* and *interior*, requiring respectively a coarser representation of the locations and a more fine-grained. Accordingly, in Fig. 2, an item for the skeleton phase is described in terms of sector and level only, with sector B1 having three

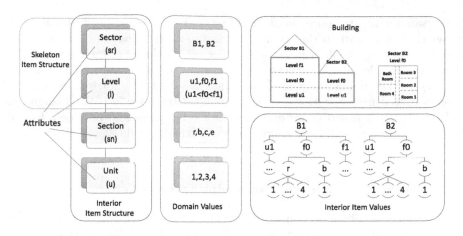

Fig. 2. Representation of the items in the hotel case study. (Color figure online)

levels and sector B2 only two. To express this, we define for each phase the *item structure*, as a tuple of attributes, and a set of *item values* defining the allowed values for the items. Accordingly, the item structure for the skeleton phase is $\langle sector, level \rangle$ and the possible values for sector $B1$ are $\langle B1, u1 \rangle$, $\langle B1, f0 \rangle$, $\langle B1, f1 \rangle$, while for $B2$ they are $\langle B2, u1 \rangle$, $\langle B2, f0 \rangle$. Thus, an item is a sequence of values indexed by the attributes. For the interior item values see Fig. 2.

More formally, a configuration $\mathcal{C}$ is a tuple $\langle At, P \rangle$, where: *(i)* At is the set of attributes each of the form $\langle \alpha, \Sigma_\alpha, \alpha\uparrow \rangle$, where α is the attribute name, Σ_α is the domain of possible values, and $\alpha\uparrow$ is a linear total order over Σ_α (for simplicity, we assume all attributes to be ordered); *(ii)* P is a set of phases, each of the form $\langle Ac, \mathcal{I}_s, \mathcal{I}_v \rangle$, where Ac is a set of activities; $\mathcal{I}_s = \langle \alpha_1, ..., \alpha_n \rangle$ is the item structure for the phase and $\mathcal{I}_v \subseteq \Sigma_{\alpha_1} \times ... \times \Sigma_{\alpha_n}$ is the set of *item values*.

Flow Part. Based on the activities and on the items specified in the configuration part, the flow part specifies on which items the activities must be executed, for instance to express that in all levels of each sector, 'wooden window installation' must be performed in all rooms, while 'aluminum window installation' in the bathrooms. We call *task* an activity a on a set of items I, represented as $\langle a, I \rangle$, where I is a subset of the possible item values for the activity's phase. We use $\langle a, i \rangle$ for an activity on one item i.

Additionally, a process model must define ordering constraints on the execution of the activities. For instance, one may want to specify that the construction of the walls proceeds from bottom to top or that the floor must be installed before the windows. In the original language, the ordering constraints were declarative, i.e., not expressing strict sequences but constraints to be satisfied. By representing the items on which activities are performed in a structured way, it is possible to specify the *scope* at which a precedence between two tasks applies, that is to say whether: *(i)* one task must be finished on all items before progressing with the other; or *(ii)* once a task is finished on an item, the other can be performed on the same item (e.g. once the floor is installed in a room, the installation of the windows can start in that room); or *(iii)* the precedence applies to groups of items (e.g. once the floor is laid everywhere at one level, the windows can be installed at that level). This level of detail was one of the identified requirements (Sect. 2). In the original language from which this approach started, indeed, the scope of precedences was provided as disambiguation notes in natural language.

In Sect. 3.2 we describe our extension of the language and provide a formal semantics in terms of LTL over finite traces. Formally, a flow part $\mathcal{F}$ is a tuple $\langle T, D \rangle$, where T and D are sets of tasks and dependencies.

3.2 A Formal Language for Constraint Specification

In this section we define a language to support the definition of dependencies on task execution, i.e. the process flow. We consider a task execution to have a duration. So, for a task $\langle a, i \rangle$ we represent its execution in terms of a start $(\text{start}(a, i))$ and an end $(\text{end}(a, i))$ event. An execution is then a sequence of states, each of which is defined as the set of start and end events for different tasks, that occurred simultaneously.

Our language defines a number of constructs for the specification of execution and ordering constraints on task execution. For each construct we provide an LTL formula[1] which captures the desired behavior by constraining the occurrence of the start and end events for the activities on the items. For instance, to express a precedence constraint between two tasks, the formula requires the end of the first task to occur before the start of the second. For each of them we also propose a graphical representation meant to support an overall view of the flow part. Figure 3 shows an excerpt of the flow part for our motivating scenario. Intuitively, each *box* is a task where the *color* represents the phase, the *label* the activity and the bottom *matrix* the items. The columns of the matrix are the item values and the rows are the attributes of the item structure. Arrows between the boxes represent binary dependencies among the tasks and can be of different kinds. The language and its graphical representation are inspired by Declare [2].

Execute. As described previously, the flow part specifies a set of tasks which need to be executed. This is captured by specifying an execute dependency $executes(\mathsf{t})$, for each task of the flow part. It is graphically represented by drawing a box and formally an $executes(\mathsf{t})$ constraint for $\mathsf{t} = \langle \mathsf{a}, I \rangle$ is defined as:

$$\forall i \in I \quad \Diamond \ \text{start}(\mathsf{a}, i)$$

Ordered Execution. In some cases it is necessary to specify an order on the execution of an activity on a set of items, for instance, to express that the concrete must be poured from the bottom level to the top one. To express

---

[1] In LTL, $\Box \mathsf{a}$, $\Diamond \mathsf{a}$ and $\bigcirc \mathsf{a}$ mean that condition a must be satisfied *(i) always*, *(ii) eventually* in the future, and *(iii)* in the *next* state. Formula $\mathsf{a} \ \mathsf{U} \ \mathsf{b}$ requires condition a to be true until b.

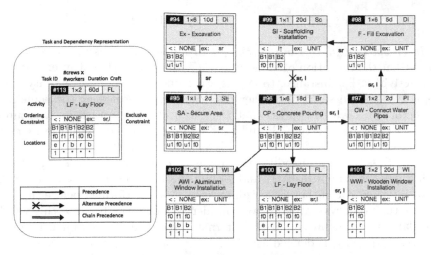

Fig. 3. Excerpt of the process model for the hotel case study. (The * denotes all possible values for the attribute). (Color figure online)

this requirement we define the *ordered execution* construct having the form *ordered_execution*($\langle a, I \rangle, \mathcal{O}$). This constraint specifies that the activity 'a' must be executed on all items in I following the order specified in $\mathcal{O}$. We express $\mathcal{O}$ as a tuple of the form $\langle \alpha_1 o_1, ..., \alpha_m o_m \rangle$ where α_i is an attribute and o_i is an *ordering operator* among $\uparrow$ or $\downarrow$. The expression $\alpha_i \uparrow$ refers to the linear total order of the domain of α (defined in the configuration), while $\alpha_i \downarrow$ refers to its inverse. Given the set of items I, these are ordered according to $\mathcal{O}$ in the following way. The items are partitioned so that the items with the same value for α_1 are in the same partition set. The resulting partition sets are ordered according to $\alpha_1 o_1$. Then, each partition set is further partitioned according to α_2 and each resulting partition set is ordered according to $\alpha_2 o_2$, and so on for the remaining operators. This iterative way of partitioning and ordering defines the ordering relation $<_{\mathcal{O},I}$, based on which precedence constraints are defined to order the execution of the activity a on the items.

As an example, consider the task Concrete Pouring (CP) in Fig. 3. To specify that it must be performed from bottom to top, we graphically use the label $<: l\uparrow$, which corresponds to the constraint *ordered_execution*($\langle CP, I \rangle, l\uparrow$), meaning that the items I are partitioned according to their values for the level (regardless of the sector), and then ordered. As a result, the activity must be performed at level u1 before progressing to f0 (and then f1). Formally, a constraint *ordered_execution*($\langle a, I \rangle, \mathcal{O}$) is:

$$executes(\langle a, I \rangle) \quad \text{and} \quad \forall \langle i_1, i_2 \rangle \in <_{\mathcal{O},I} \; precedes(\langle a, \{i_1\} \rangle, \langle a, \{i_2\} \rangle)$$

Precedes (Auxiliary Construct). The formula above relies on the *precedes*($\langle a, I_a \rangle$, $\langle b, I_b \rangle$), auxiliary construct which requires an activity a to be executed on a set of items I_a before an activity b (potentially the same) is performed on any item in I_b. Formally:

$$\forall i_a \in I_a, i_b \in I_b \quad \neg \mathsf{start}(b, i_b) \ \mathsf{U} \ \mathsf{end}(a, i_a)$$

Not Interrupt (Auxiliary Construct). Another requirement is the possibility to express that the execution of an activity on a set of items is not interrupted by other activities. For instance, to express that once the lay floor activity starts at one level in one sector, no other task can be performed at the same level and sector, we have to express that the execution of the task on a group of items must not be interrupted, and that we group and compare the items by considering their values for sector and level only (abstracting from section and unit). To this aim, we introduce the auxiliary construct *not_interrupt*(T_1, T_2) which applies to sets of tasks T_1 and T_2 and specifies that the two sets of tasks cannot interrupt each other: either all tasks in T_1 are performed before the tasks in T_2 or the other way around (we consider sets of tasks because this will be useful later in the definition of the alternate precedence constraint). Formally, *not_interrupt*(T_1, T_2) is defined as:

$$\forall t_1 \in T_1, t_2 \in T_2 \quad precedes(t_1, t_2) \quad \text{or} \quad \forall t_1 \in T_1, t_2 \in T_2 \quad precedes(t_2, t_1)$$

Projection Operator. To compare two items by considering only some of the attributes of their item structure, we introduce the concept of *scope*. The scope is a sequence of attributes used to compare two items. For instance, given a scope $s = \langle \mathsf{sector}, \mathsf{level} \rangle$, we can say that the items $\langle B1, \mathsf{f1}, \mathsf{room}, 1 \rangle$ and $\langle B1, \mathsf{f1}, \mathsf{bathroom}, 1 \rangle$ are equal under s. In this case, we say that the two items are at the same scope. For the comparison, we define the projection operator to project an item on the attributes in the scope.

Definition 1 (Projection Operator Π_s). *Given an item $i = \langle v_1, .., v_n \rangle$ and a scope $s = \langle \alpha_{j_1}, .., \alpha_{j_m} \rangle$, the projection of i on s is $\Pi_s(i) = \langle v_{j_1}, .., v_{j_m} \rangle$ with $v_{j_h} = \alpha_{j_h}(i)$.*

This means, in particular, that for the empty scope $s = \langle \rangle$, we have $\Pi_{\langle \rangle}(i) = \langle \rangle$, and thus $\forall i, i' \ \Pi_{\langle \rangle}(i) = \Pi_{\langle \rangle}(i')$. When applied to a set of items I, the result of the projection operator with scope s is the set (without duplicates), obtained by applying the projection operator to every item $i \in I$. In other words, it is the set of possible values for the attributes in s, w.r.t. the items in I.

Exclusive Execution. An *exclusive execution* constraint *exclusive_execution* $(\langle a, I_a \rangle, s)$ expresses that once an activity is executed on an item at scope s, no other activity can be performed on items at the same scope. Formally, the task has to be *executed* and for every other task having an item at scope s, the two tasks must not interrupt each other.

For a scope $s = \langle \alpha_{j_1}, .., \alpha_{j_m} \rangle$, let $\pi_s = \langle v_{j_1}, .., v_{j_m} \rangle$ be a tuple of values for the attributes in s. We use the selector operator $\sigma_{\pi_s}(I)$ to select the items in I having the values specified in π_s for the attributes s. Formally, *exclusive_execution*$(\langle a, I_a \rangle, s)$ is:[2]

---

[2] The result of $\Pi_s(I_a)$ is the set of possible values for the attributes in s considering I_a. For each of them we select the items in I_b that are at the same scope, and we apply the *not_interrupt*.

$$executes(\langle \mathsf{a}, I_{\mathsf{a}} \rangle) \quad \text{and} \quad \forall \langle \mathsf{b}, I_{b} \rangle \in T \text{ and } \langle \mathsf{b}, I_{b} \rangle \neq \langle \mathsf{a}, I_{\mathsf{a}} \rangle,$$

$$\forall \pi_{\mathsf{s}} \in \Pi_{\mathsf{s}}(I_{\mathsf{a}}), \forall i_{b} \in \sigma_{\pi_{\mathsf{s}}}(I_{b}) \quad not\_interrupt(\{\langle \mathsf{a}, \sigma_{\pi_{\mathsf{s}}}(I_{\mathsf{a}}) \rangle\}, \{\langle \mathsf{b}, \{i_{b}\} \rangle\})$$

As a special case, when $\mathsf{s} = \langle \rangle$ the execution of the entire task cannot be interrupted. By default, tasks have an exclusive constraint at the finest-granularity level for the items, i.e. two activities cannot be executed at the same time on the same item.

An exclusive execute constraint (except the default at the item scope) is represented with a double border box and the scope is specified in the slot labeled with ex. In Fig. 3, lay floor has an exclusive execution constraint ex:(sr,l) for sector and level.

We now introduce *binary* dependencies that specify ordering constraints between pairs of tasks. By representing also the items, we can specify precedences at different scopes: *(i)* task (a task must be finished on all items before the second task can start); *(ii)* item scope (once the first task is finished on an item, the second task can start on the item); *(iii)* between items at the same scope (e.g. a task must be performed in all locations of a floor before another task can start on the same floor). This is visualized by annotating a binary dependency (an arrow) with the sequence of attributes representing the scope. When no label is provided, the task scope is meant. In Fig. 3, the dependency between concrete pouring and lay floor is at task level, while the one between lay floor and wooden window installation is labeled with sr,l, to represent the scope $\langle$sector, level$\rangle$: given a sector and a level, the activity lay floor must be done in every section and unit before wooden windows installation can start in that sector at that level.

Precedence. A *precedence* dependency $precedence(\langle \mathsf{a}, I_{\mathsf{a}} \rangle, \langle \mathsf{b}, I_{b} \rangle, \mathsf{s})$ expresses that an activity a must be performed on a set of items I_{a} before an activity b starts on items I_{b}. The scope s defines whether this applies at the task, item, or item group.

$$\forall \pi_{\mathsf{s}} \in \Pi_{\mathsf{s}}(I_{\mathsf{a}}) \cap \Pi_{\mathsf{s}}(I_{b}) \quad precedes(\langle \mathsf{a}, \sigma_{\pi_{\mathsf{s}}}(I_{\mathsf{a}}) \rangle, \langle \mathsf{b}, \sigma_{\pi_{\mathsf{s}}}(I_{b}) \rangle)$$

The formula above expresses that for the items at the same scope in I_{a} and I_{b}, activity a must be executed there before activity b. If $\mathsf{s} = \langle \rangle$ activity a must be performed on all its items before activity b can start (task scope).

Alternate Precedence. Let us consider the example of the scaffolding installation and the concrete pouring: once the scaffolding is installed at one level, the concrete must be poured at that level before the scaffolding can be installed to the next level. This alternation is captured by the dependency *alternate*$(\langle \mathsf{a}, I_{\mathsf{a}} \rangle, \langle \mathsf{b}, I_{b} \rangle, \mathsf{s})$, which is in the first place a precedence constraint between a and b. It also requires that once a is started on a group of items at a scope

$(\sigma_{\pi_s}(I_a))$, then b must be performed on its items at the same scope $(\sigma_{\pi_s}(I_b))$, before a can progress on items at a different scope $(\sigma_{\pi_s'}(I_a))$:[3]

$$precedence(\langle a, I_a \rangle, \langle b, I_b \rangle, s) \quad and \quad \forall \pi_s, \pi_s' \in \Pi_s(I_a) \cap \Pi_s(I_b), \pi_s \neq \pi_s'$$
$$not\_interrupt(\{\langle a, \sigma_{\pi_s}(I_a) \rangle, \langle b, \sigma_{\pi_s}(I_b) \rangle\}, \{\langle a, \sigma_{\pi_s'}(I_a) \rangle, \langle b, \sigma_{\pi_s'}(I_b) \rangle\})$$

Graphically, an alternate precedence is represented as an arrow (to capture the precedence), and an X as a source symbol, to capture that the source task cannot progress freely, but it has to wait for the target task to be completed on items at the same scope.

Chain Precedence. Finally, let us consider the case in which the execution of two tasks must not be interrupted by other tasks on items at the same scope. For instance, the tasks excavation and secure area must be performed one after the other and no other tasks can be performed in between for each sector. Note that the dependency types defined before declaratively specify an order on the execution of two tasks but do not prevent other tasks to be performed in-between. To forbid this we define the *chain precedence* dependency $chain(\langle a, I_a \rangle, \langle b, I_b \rangle, s)$, which, as the alternate precedence, builds on top of a precedence dependency. Additionally, it requires that the execution of the two tasks on items at the same scope is not interrupted by other tasks executing on items at the same scope. The formula considers all tasks different from $t_1 = \langle a, I_a \rangle$ and $t_2 = \langle b, I_b \rangle$ sharing items at the same scope. For this it specifies a *not_interrupt* constraint. Formally,

$$precedence(\langle a, I_a \rangle, \langle b, I_b \rangle, s) \quad and \quad \forall \pi_s \in \Pi_s(I_a) \cap \Pi_s(I_b),$$
$$\forall\, t_3 = \langle c, I_c \rangle \in T, \quad s.t.\ t_3 \neq t_1 \quad and \quad t_3 \neq t_2$$
$$\forall i \in \sigma_{\pi_s}(I_c) \quad not\_interrupt(\{\langle c, \{i\} \rangle\}, \{\langle a, \sigma_{\pi_s}(I_a) \rangle, \langle b, \sigma_{\pi_s}(I_b) \rangle\})$$

Graphically, it is represented as a double border arrow.

4   Process Modeling for the Hotel Scenario

The model of the hotel case study consists of roughly fifty tasks. Figure 3 reports an excerpt showing some activities of the skeleton (blue) and interior phases (green). The item structure for the skeleton phase is defined in terms of sector sr and level l, while the interior consists of sector, level, section sn, and unit number u (see Fig. 2).

The task EXCAVATION belongs to the skeleton phase and must be performed on sectors $B1$ and $B2$, in both cases at the underground level u1. The activity SECURE AREA must also be performed in both sectors, but at the underground

---

[3] The projection operator is applied to I_a and I_b and only projections π_s and π_s' that are in common are considered. For every distinct π_s and π_s' either a and b are performed on items at scope π_s without being interrupted by executing a and b on items at scope π_s', or vice versa.

and ground floor f0. For security reasons, once the excavation is finished in one sector, the area must be secured for that sector before any other task can start. This is expressed with a *chain precedence* dependency at scope sr (sector), graphically represented as double bordered arrow. Additionally, while performing the activities, their execution in a sector cannot be interrupted by other activities. This is expressed with an *exclusive execution* at scope sector (*ex:* sr) for both tasks, represented as double border box.

Only after the area has been secured everywhere, the CONCRETE POURING can start. This is expressed with a precedence at the task scope (graphically an arrow without label). The concrete can be poured proceeding from the bottom to the top floor, captured by an *ordered constraint* ($<: \mathsf{l}\uparrow$). Note that this task has an exclusive constraint *ex:* UNIT, i.e. an exclusive constraint at the finest granularity level, which expresses that two activities cannot be performed simultaneously on the same item. For all tasks the same exclusive execution at the item scope is assumed and it is not graphically highlighted. It is represented with a double border box when a coarser scope is specified.

Once the concrete has been poured at the underground level of one sector, the pipes for the water can be connected before the excavation is filled, and after this the task SCAFFOLDING INSTALLATION can start, proceeding from the bottom to the top. Once the scaffolding is installed at one level, the concrete must be poured at that level, in order to be able to install the scaffolding for the next level. This requirement is expressed by the *alternate precedence* at scope sr, l (graphically, with an arrow starting with an X).

When the concrete pouring task is finished everywhere, the lay floor task can start (which belongs to the interior phase). This task must be performed before the installation of the wooden windows, which is foreseen in all rooms, so not to damage them. This is captured by the precedence constraint between LAY FLOOR and WOODEN WINDOW INSTALLATION at scope sr, l. To be more efficient, the lay floor task has an exclusive execution constraint at scope sr, l. Since aluminum windows are less delicate than wooden ones, their installation does not depend on the lay floor task.

To support the graphical definition of a process flow, we implemented a web-based prototype [21] (see the master's thesis [20]), which we used to produce Fig. 3. The prototype and the graphical language support the overall view of a process model.

5   Satisfiability Checking

When developing a tool requiring inputs from the user, one cannot assume that the provided input will be meaningful. Specifically, one cannot assume that a given model is *satisfiable*, that is there exists at least one execution satisfying it. Relying on LTL semantics would allow us to perform the check using model checking techniques. However, as reported in the experiment description (Sect. 5.2) this takes more than 2 min for a satisfiable model with 8 tasks and 9 dependencies. In this section, we describe a more effective algorithm and its performance evaluation by means of experiments.

5.1   How to Check for Satisfiability

An execution is a sequence of states defined in terms of start and end events.

Definition 2 (Execution). *Let E be the set of* start *and* end *events for the tasks of a process model. A process execution ρ of length n is a function $\rho\colon \{1, .., n\} \to 2^E$.*

There are some properties of interest that an execution is expected to satisfy in reality: *(i) start-end order:* a start event is always followed by an end event; *(ii) non-repetition:* an activity cannot be executed more than once on the same item, and start and end events never repeat; *(iii) non-concurrence:* at most one task at a time can be performed on an item. All these properties can be expressed in LTL.[4] We say that an execution ρ is a *well-defined execution* if and only if it satisfies all of these three properties.

Definition 3 (Possible Execution). *A well-defined execution ρ is a* possible execution *for a process model if and only if*

(i) all events occurring in ρ are of activities and items of the configuration part;
(ii) for all tasks $\langle a, I \rangle$ in the flow part, the task $\langle a, i \rangle$ is executed in ρ for all $i \in I$;
(iii) ρ satisfies all the ordering constraints specified in the flow part.

The observation underlying the algorithm that checks statisfiability is that, since all our dependencies relate the end of a task with the start of another one (end-to-start), it holds that if there is a possible execution, then there is one where all activities on an item are atomic, that is, each start event is immediately followed by the corresponding end event. Moreover, if there is such an execution, then there even exists a sequential one where no atomic activities take place at the same time. Therefore, it is sufficient to check for the existence of sequential executions of atomic activities.

The algorithm relies on an auxiliary structure that we call *activity-item* (AI) graph. Intuitively, in an AI-graph we represent each activity to be performed on an item as a node $\langle a, i \rangle$, conceptually representing the execution of a on i. Ordering constraints are then represented as arcs in the graph. This allows us to characterize the satisfiability of a model by the absence of loops in the corresponding AI graph.

Let us first consider *P-models*, which are process models with precedence and ordering constraints only. Given a P-model $\mathcal{M}$, we denote the corresponding AI-graph as $\mathcal{G}_\mathcal{M} = \langle V, A \rangle$, where for each task $t = \langle a, I \rangle$ in the flow part and for each item $i \in I$ there is an AI node $\langle a, i \rangle \in V$, without duplicates; for each precedence and ordering constraint we introduce a number of arcs in A among AI nodes in V. For instance, a precedence constraint between two tasks at the

---

[4] *start-end order:* $\Box(\text{start}(a, i) \to \bigcirc \Diamond \text{end}(a, i))$;
  non-repetition: $\Box(\text{start}(a, i) \to \bigcirc \Box \neg \text{start}(a, i)) \land \Box(\text{end}(a, i) \to \bigcirc \Box \neg \text{end}(a, i))$;
  non-concurrence: $\Box(\text{start}(a, i) \to \neg \text{start}(a', i)\ \mathsf{U}\ \text{end}(a, i))$, where $a \neq a'$.

task scope is translated into a set of arcs, linking each AI node corresponding to the source task to each AI node corresponding to the target task. A precedence constraint at the item scope is translated into arcs between AI nodes of the two activities on the same items.

Theorem 1. *A P-model $\mathcal{M}$ is satisfiable iff the graph $\mathcal{G}_\mathcal{M}$ is cycle-free.*

Proof (Idea). If $\mathcal{G}_\mathcal{M}$ does not contain cycles, the nodes can be topologically ordered and the order is a well-defined execution satisfying all ordering constraints in $\mathcal{M}$. A cycle in $\mathcal{G}_\mathcal{M}$ corresponds to a mutual precedence between two AI nodes, which is unsatisfiable.

We now consider general models, called *G-models*, where all types of dependency are allowed. First, let us consider an exclusive constraint $exclusive\_execution(\langle a, I_a \rangle, s)$. It requires for each scope $\pi_s \in \Pi_s(I_a)$, that the execution of a on the items in the set $\sigma_{\pi_s}(I_a)$ is not interrupted by the execution of other activities at the same scope. Considering an activity b to be executed on an item i_b at the same scope (i.e., $\Pi_s(i_b) = \pi_s$), the exclusive constraint is not violated if the execution of b occurs *before or after* the execution of a on all items in $\sigma_{\pi_s}(I_a)$. We call *exclusive group* a group of AI nodes, whose execution must not be interrupted by another node. We connect this node and the exclusive group with an undirected edge, since the execution of the node is allowed either before or after the exclusive group. Then, we look for an orientation of this edge, such that it does not conflict with other constraints, i.e., it does not introduce cycles. With chain and alternate precedences, we deal in a similar way. Indeed, both require that the execution of two tasks on a set of items is not interrupted by other activities (chain) or by the same activity on other items (alternate).

To represent a *not_interrupt* constraint in a graph we introduce *disjunctive activity-item* graphs (DAI-graphs) inspired by [11]. Given a *G-model* $\mathcal{M}$, the corresponding DAI-graph is $\mathcal{D}_\mathcal{M} = \langle V, A, X, E \rangle$, where $\langle V, A \rangle$ are defined as in the AI-graph of a P-model,[5] $X \subseteq 2^V$ is the set of exclusive groups, and $E \subseteq V \times X$ is a set of undirected edges, called *disjunctive edges*, connecting single nodes to exclusive groups. A disjunctive edge can be *oriented* either by creating arcs from the single node to each node in the exclusive group or vice versa, so that all arcs go in the same direction (i.e., either outgoing from or incoming to the single node). An *orientation* of a DAI-graph is a graph that is obtained by choosing an orientation for each edge. We say that a disjunctive graph is *orientable* if and only if there is an acyclic orientation of the graph.

Theorem 2. *A G-model $\mathcal{M}$ is satisfiable iff the DAI-graph $\mathcal{D}_\mathcal{M}$ is orientable.*

Proof (Idea). If $\mathcal{D}_\mathcal{M}$ is orientable, there exist a corresponding acyclic oriented graph. Then, there exists a well-defined execution satisfying all precedence constraints (see Theorem 1). The non_interrupt constraints are satisfied by construction of $\mathcal{D}_\mathcal{M}$. If $\mathcal{D}_\mathcal{M}$ is not orientable, a topological order satisfying all the constrains does not exist (see [20]).

---

[5] Including also the directed arcs to represent the precedence constraints of the chain and alternate dependencies (see the formalization in Sect. 3.2).

5.2   An Implementation for Satisfiability Checking

Algorithm. To check for the orientability of a DAI-graph $\mathcal{D_M} = \langle V, A, X, E \rangle$ we develop an algorithm that is based on the following observations.

Cycles. If the graph $\mathcal{G_M} = \langle V, A \rangle$ contains a cycle, then $\mathcal{D_M}$ is not orientable.

Simple edges. Disjunctive edges where both sides consist of a single node, called *simple edges*, can be oriented so that they do not introduce cycles (see the thesis [20]).

Resolving. Consider an undirected edge between a node u and an exclusive set of nodes $S \in X$. If there is a directed path from u to a node $v \in S$ (or the other way around), then there is only one way to orient the undirected edge between v and S without introducing cycles. We call this operation *resolving*.

Partitioning. Sometimes, one can partition a DAI-graph $\mathcal{D_M}$ into DAI-subgraphs such that $\mathcal{D_M}$ is orientable if and only if each of these DAI-subgraph is orientable. Then each such subgraph can be checked independently.

Let us discuss the partitioning operation in more detail. Given $\mathcal{D_M} = \langle V, A, X, E \rangle$, we are looking for a partition of the node set V into disjoint subsets, the partition sets, satisfying two conditions: *(i)* nodes of an exclusive group belong all to the same subset; *(ii)* there are no cycles among the subsets, that is, by abstracting each subset to a single node and preserving the arcs connecting different subsets, there is no subset that can be reached from itself. It is possible to prove for such partitions that the original DAI-graph is orientable if and only if each partition set, considered as a DAI-graph, is orientable [20]. Intuitively, if the partition sets do not form a cycle, then they can be topologically ordered, and the AI nodes in each partition set can be executed respecting the order. Since an exclusive group is entirely contained in one partition set, execution according to a topological order also satisfies the exclusive constraint.

To obtain such a partition, we temporarily add for each pair of nodes in an exclusive group two auxiliary arcs, connecting them in both directions. Then we compute the strongly connected components (SCCs) of the extended graph and consider each of them as a partition set. (It may be that there is only one of them.) After that we drop the auxiliary arcs. This construction ensures that *(i)* an exclusive group is entirely contained in a SCC; and *(ii)* there are no cycles among the partition sets because a cycle would cause the nodes to belong to the same partition set.

Below we list our boolean procedure SAT that takes as input a DAI-graph $\mathcal{D}$ and returns "true" iff $\mathcal{D}$ is orientable. It calls the subprocedure NDSAT that chooses a disjunctive edge and tries out its possible orientations.

> **procedure** SAT($\mathcal{D}$)
>   drop all simple edges in $\mathcal{D}$
>   resolve all orientable disjunctive edges in $\mathcal{D}$
>   **if** $\mathcal{D}$ contains a cycle **then**
>   **return** false

else partition $\mathcal{D}$, say into $\mathcal{D}_1, \ldots, \mathcal{D}_n$
   if $\text{NDSAT}(\mathcal{D}_i) = $ true for all $i \in \{1, \ldots, n\}$ **then**
     return true
   else return false

procedure $\text{NDSAT}(\mathcal{D})$
   if $\mathcal{D}$ has a disjunctive edge e **then**
     orient e in the two possible ways, resulting in $\mathcal{D}_+$, $\mathcal{D}_-$
     return $\text{SAT}(\mathcal{D}_+)$ or $\text{SAT}(\mathcal{D}_-)$
   else return true

SAT itself performs only deterministic steps that simplify the input, discover unsatisfiability, or divide the original problem into independent subproblems. After that NDSAT performs the non-deterministic orientation of a disjunctive edge. Since the calls to NDSAT at the end of SAT are all independent, they can be run in parallel.

Experiments. We ran our experiments on a desktop PC with eight cores Intel i7-4770 of 3.40 GHz. We tested the performance of NuSMV, a state-of-the-art model checker, on a process model similar to the one reported in Fig. 3, consisting of 8 tasks and 9 dependencies, resulting in a DAI-graph of 236 nodes. The NuSMV model checker with Bounded Model Checking took 2 min 35 s on a satisfiable model. We also considered different inconsistency scenarios: *(i)* a DAI-graph with a cycle *(ii)* an acyclic DAI graph that is non-orientable; *(iii)* a DAI-graph similar to case *(ii)*, but with an increased number of nodes. On all these unsatisfiable cases, we aborted the check if exceeding 1 h.

Table 1. Experimental results.

Model	Tasks	Dependencies	Nodes	Arcs	Edges	Time (ms)
Satisfiable	8	9	236	9415	524	27
Cyclic	8	9	236	10003	521	5
Non-orientable	12	14	244	9435	574	10
Non-orientable	12	14	424	15131	1740	23
Non-orientable	60	75	2,120	76,103	10,635	598
	120	173	4,240	168,681	42,470	1,189
	180	296	6,360	361,969	95,505	3,682
	300	623	10,600	674,584	265,175	14,199
	360	822	12,720	948,099	381,810	24,223
	480	1,291	16,960	1,436,759	678,680	55,866
	720	2,562	25,440	3,082,925	1,526,820	379,409
	960	4,187	33,920	5,217,426	2,714,160	OOM

We implemented our algorithm in Java and tested it on the same variants of the hotel scenario on which we tested the model checker. The results are reported in Table 1 (top). As can be seen, the implementation outperforms the NuSMV model checker both in the satisfiable and unsatisfiable variants. In order to understand the performance of the implementation w.r.t. the model size, we performed some experiments by increasing the number of tasks and dependencies in the non-orientable variant of the model. We chose this variant because it is the most challenging for the algorithm, which has to find a partition and non-deterministically chose an orientation for the undirected edges. The results are shown in Table 1 (bottom). On a model of 180 tasks, which we believe represents an average real case scenario, the performances are still acceptable (around 4 s). The implementation took around 1 min on a model of 480 tasks, which is acceptable for an offline check. It ran out of memory (OOM) on a model of 960 tasks (too many for most of real cases). More details on the experiments are reported in the thesis [20]. The web-based prototype [21] that we developed implements the satisfiability checking and the export of the NuSMV file of a process model.

6   Conclusions and Future Work

This work presents an approach for process modeling that represents activities, items and accounts for both of them in the control flow specification. We investigate the problem of satisfiability of a model and develop an effective algorithm to check it. The algorithm has been implemented in a proof-of-concept prototype that also supports the graphical definition of a process model [18,21].

The motivation for developing a formal approach for process modeling emerged in the application of non-formal models in real projects [4,19], which resulted in improvements and cost savings in construction process execution. This opens the way for the development of automatic tools to support construction process management. In this paper we presented the statisfiability checking, starting from which we are currently investigating the automatic generation of process schedules, optimal w.r.t. some criteria of interest (e.g., costs, duration). To this aim we are investigating the adoption of constraint satisfaction and (multi-objective) optimization techniques. We are also investigating the use of Petri Nets for planning [13]. We will apply modeling and automatic scheduling to real construction projects in the context of the research project COCkPiT [6].

Acknowledgments. This work was supported by the projects MoMaPC, financed by the Free University of Bozen-Bolzano and by COCkPiT financed by the European Regional Development Fund (ERDF) Investment for Growth and Jobs Programme 2014–2020.

References

1. van der Aalst, W.M.P., Artale, A., Montali, M., Tritini, S.: Object-centric behavioral constraints: integrating data and declarative process modelling. In: Description Logics (2017)
2. van der Aalst, W.M.P., Pesic, M., Schonenberg, H.: Declarative workflows: balancing between flexibility and support. Comput. Sci.-R&D **23**(2), 99–113 (2009)

3. van der Aalst, W.M.P., Stoffele, M., Wamelink, J.: Case handling in construction. Autom. Constr. **12**(3), 303–320 (2003)
4. Matt, D.T., Benedetti, C., Krause, D., Paradisi, I.: Build4future-interdisciplinary design: from the concept through production to the construction site. In: Proceedings of the 1st International Workshop on Design in Civil and Environmental Engineering, KAIST (2011)
5. Calvanese, D., De Giacomo, G., Montali, M.: Foundations of data-aware process analysis: a database theory perspective. In: PODS. ACM (2013)
6. COCkPiT: Collaborative Construction Process Management. www.cockpit-project.com/
7. Dallasega, P., Matt, D., Krause, D.: Design of the building execution process in SME construction networks. In: 2nd International Workshop DCEE (2013)
8. De Giacomo, G., De Masellis, R., Montali, M.: Reasoning on LTL on finite traces: insensitivity to infiniteness. In: AAAI. AAAI Press (2014)
9. Dumas, M.: From models to data and back: the journey of the BPM discipline and the tangled road to BPM 2020. In: BPM. LNCS, vol. 9253, Springer, Berlin (2015)
10. Forsythe, P., Sankaran, S., Biesenthal, C.: How far can BIM reduce information asymmetry in the Australian construction context? Project. Manag. J. **46**(3), 75–87 (2015)
11. Fortemps, P., Hapke, M.: On the disjunctive graph for project scheduling. Found. Comput. Decis. Sci. **22**, 195–209 (1997)
12. Frank, U.: Multilevel modeling - toward a new paradigm of conceptual modeling and information systems design. Bus. Inf. Syst. Eng. **6**(6), 319–337 (2014)
13. Hickmott, S.L., Sardiña, S.: Optimality properties of planning via Petri net unfolding: a formal analysis. In: Proceedings of ICAPS (2009)
14. Kenley, R., Seppänen, O.: Location-Based Management for Construction: Planning Scheduling and Control. Routledge, Abingdon (2006)
15. KPMG International: Building a Technology Advantage. Harnessing the Potential of Technology to Improve the Performance of Major Projects. Global Construction Survey (2016)
16. Leitner, M., Mangler, J., Rinderle-Ma, S.: Definition and enactment of instance-spanning process constraints. In: Wang, X.S., Cruz, I., Delis, A., Huang, G. (eds.) WISE 2012. LNCS, vol. 7651, pp. 652–658. Springer, Heidelberg (2012). https://doi.org/10.1007/978-3-642-35063-4_49
17. Lu, Y., Xu, X., Xu, J.: Development of a hybrid manufacturing cloud. J. Manuf. Syst. **33**(4), 551–566 (2014)
18. Marengo, E., Dallasega, P., Montali, M., Nutt, W.: Towards a graphical language for process modelling in construction. In: CAiSE Forum 2016. CEUR Proceedings, vol. 1612 (2016)
19. Marengo, E., Dallasega, P., Montali, M., Nutt, W., Reifer, M.: Process management in construction: expansion of the Bolzano hospital. In: vom Brocke, J., Mendling, J. (eds.) Business Process Management Cases. MP, pp. 257–274. Springer, Cham (2018). https://doi.org/10.1007/978-3-319-58307-5_14
20. Perktold, M.: Processes in construction: modeling and consistency checking. Master's thesis, Free University of Bozen-Bolzano (2017). http://pro.unibz.it/library/thesis/00012899S_33593.pdf
21. Perktold, M., Marengo, E., Nutt, W.: Construction process modeling prototype. http://bp-construction.inf.unibz.it:8080/ConstructionProcessModelling-beta
22. Shankar, A., Varghese, K.: Evaluation of location based management system in the construction of power transmission and distribution projects. In: 30th International Symposium on Automation and Robotics in Construction and Mining (2013)

Feature-Oriented Composition of Declarative Artifact-Centric Process Models

Rik Eshuis[✉]

Eindhoven University of Technology,
P.O. Box 516, 5600 MB Eindhoven, The Netherlands
h.eshuis@tue.nl

Abstract. Declarative business process models that are centered around artifacts, which represent key business entities, have proven useful to specify knowledge-intensive processes. Currently, declarative artifact-centric process models need to be designed from scratch, even though existing model fragments could be reused to gain efficiency in designing and maintaining such models. To address this problem, this paper proposes an approach for composing model fragments, abstracted into features, into fully specified declarative artifact-centric process models. We use Guard-Stage-Milestone (GSM) schemas as modeling language and let each feature denote a GSM schema fragment. The approach supports feature composition at different levels of granularity. Correctness criteria are defined that guarantee that valid GSM schemas are derived. The approach is evaluated by applying it to an industrial process. Using the approach, declarative artifact-centric process models can be composed from existing model fragments in an efficient and correct way.

1 Introduction

In many business processes, data items are being processed in a non-routine way by knowledge workers. Such data-driven processes are knowledge-intensive [10], since expert knowledge is required to make progress for individual cases. Application domains are case management [1,28,29] and emergency management [10]. *Artifact-centric process models* are a natural fit to model data-driven processes [8,21]. An artifact represents a real-world business entity, such as Product or Order, about which data is processed and for which activities are performed.

Data-driven business processes are much more unpredictable than traditional activity-centric processes. The structure of a data-driven process model depends on the case data that is being processed [10]. Knowledge workers play a key role in determining this structure while performing the process. Given their expertise, they have a lot of freedom in deciding how to process individual cases. This suits a declarative specification of processes [15], in which workers have more freedom in performing processes than with a procedural specification. Several declarative process modeling languages have been proposed in the past years [15], including the Guard-Stage-Milestone (GSM) language for artifacts [9,17].

© Springer Nature Switzerland AG 2018
M. Weske et al. (Eds.): BPM 2018, LNCS 11080, pp. 66–82, 2018.
https://doi.org/10.1007/978-3-319-98648-7_5

Designing declarative artifact-centric process models is currently done in an inefficient way. Such models need to be developed from scratch, even though they could be defined by reusing fragments in existing models. Since similar fragments need to be specified again and again for new artifact-centric process models, the design process is inefficient. Also, maintenance of the different process models is inefficient, since each occurrence of a fragment across different process models needs to be maintained separately, so updates need to be replicated.

To address this problem, this paper proposes a *feature-oriented approach for composing declarative artifact-centric process models*. A feature denotes a fragment of a declarative artifact-centric process model. The approach formally defines how to compose features into declarative artifact-centric process models. From the same set of features, different artifact-centric process models can be composed that share common fragments. The approach supports reuse of those common fragments across different declarative artifact-centric process models.

We use Guard-Stage-Milestone (GSM) [17] as declarative modeling language for business artifacts. GSM is well-defined [9], thus providing a sound basis for the approach. Next, its core modeling constructs have been adopted in the OMG Case Management Model and Notation (CMMN) [7,21]. Therefore, the results can provide a stepping stone towards managing variability for CMMN models.

The approach is inspired by feature-oriented design in software product lines [2,23]. Features are used to distinguish common and variable parts in software artifacts [2] and this way support reuse of software artifacts [5]. Similar to feature-oriented composition for software product lines, the approach supports composition at different levels of granularity [3]. Key difference is that the approach considers the syntactic and semantic level of business artifacts, whereas software product lines only consider the syntactic level of software artifacts [2,23]. Section 7 discusses related work, also from the area of BPM, in more detail.

The paper is organized as follows. Section 2 presents a running example and also introduces GSM schemas [9], which Sect. 3 formally defines. Section 4 defines the feature composition approach while Sect. 5 presents correctness criteria that guarantee that the compositions are valid GSM schemas. Section 6 evaluates the feasibility and potential gain of the approach by applying it to an industrial process that has several variants. Section 7 discusses related work. Section 8 ends the paper with conclusions.

2   Motivating Example

To motivate the use of the feature-oriented composition approach and to informally introduce GSM schemas, we first present an example. It is based on a real-world process from an international high tech company. The example is revisited in Sects. 3 and 4 to illustrate key definitions.

In the process, business criteria for a partner contract are assessed: first data about the partner is gathered and pre-checked, and next a detailed check is performed to decide whether the criteria should be changed or not. If new

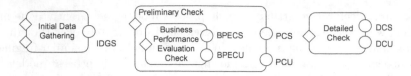

Fig. 1. Base GSM schema Business Criteria Assessment (BCA$_{base}$)

Table 1. Stages and sentries for BCA$_{base}$ in Fig. 1. ';' separates different sentries

Stage	Plus sentries (guards)	Minus sentries (closing)
Initial Data Gathering	E:StartAssessment; E:AdditionalInfo	IDGS
Preliminary Check	IDGS	PCS ; PCU; E:AdditionalInfo
Business Performance Evaluation Check	+Preliminary Check	BPECS ; BPECU; -Preliminary Check
Detailed Check	PCS	DCS ; DCU

information arrives before the business criteria have been assessed, the data is gathered anew and the business criteria check is restarted, if applicable. Figure 1 shows a graphical representation of a GSM schema that models this process. Each rounded rectangle represents a stage, in which work is performed. Each circle represent a milestone, which is a business objective. A milestone is typically achieved by completing the work in a stage, represented by putting the milestone on the right border of the stage. The status of each stage and milestone is represented by a Boolean attribute, which is true (false) if the stage is open (closed) or the milestone is achieved (invalid). Stages and milestones change status if certain conditions, called *sentries*, are met (Tables 1 and 2). Each stage has plus sentries, called guards, that specify when it is opened and minus sentries that specify when it is closed, which could be never (false). Each guard of a stage is visualized as a diamond. Each milestone has plus (minus) sentries that specify when it is achieved (invalidated). Sentries can refer to external or internal events. External events are named events (prefix E:) or completion events of atomic stages (prefix C:). Internal events denote status changes of stages and milestones; a status can become true (prefix +) or false (prefix -). Besides status changes, sentries can refer to the statuses of stages and milestones; for instance, the minus sentry of stage Initial Data Gathering closes the stage if the status of milestone IDGS is true (achieved). Section 3 gives more details on GSM schemas.

The company has offices in different geographic regions, each of which has its own flavor of the process. This results, among others, in three basic variations for the preliminary check in this process (see Table 3). These variations can be combined in a concrete variant, yielding in total eight variants, one of which is the base process in Fig. 1, in which none of the variations has been chosen.

Table 2. Milestones and sentries for BCA$_{base}$ in Fig. 1

Milestone	Full name	Plus sentries (achieving)	Minus sentries (invalidating)
IDGS	Initial Data Gathering Successful	C:Initial Data Gathering	E:AdditionalInfo
BPECS	Business Performance Evaluation Check Successful	C:Business Performance Evaluation Check ∧ BP_good	E:AdditionalInfo
BPECU	Business Performance Evaluation Check Unsuccessful	C:Business Performance Evaluation Check ∧¬BP_good	E:AdditionalInfo
PCS	Pre-checks Successful	BPECS	false
PCU	Pre-checks Unsuccessful	BPECU	false
DCS	Detailed Check Successful	C:Detailed Check ∧ . . .	false
DCU	Detailed Check Unsuccessful	C:Detailed Check ∧ . . .	false

Table 3. Variations of BCA$_{base}$

1	Business performance is checked if the company has more than 300 employees
2	The credit is checked as part of the preliminary check
3	The addressable market is checked as part of the preliminary check

Specifying these eight variants in separate process models leads to redundancy, both regarding the models and the modeling process. In that case modifying one common element like milestone PCS in those eight variants has to be done eight times rather than once. Moreover, drawing each of the eight models has to be done from scratch, since there is no reuse.

To address these problems, we develop an approach based on the technique of feature-oriented composition, well known from Software Product Lines [2]. The same feature can be reused in different compositions. In this paper, a feature denotes a GSM schema, which can be either fully or partially defined (a schema fragment). The GSM schema fragments for variations 1 and 2 in Table 3 are presented and discussed in Sect. 4.

A key challenge is how to compose features that denote GSM schema fragments. In Sect. 4, we define a binary feature composition operator '•' that supports composition at different levels of granularity [2]. A coarse-grained feature composition adds new stages and milestones to a GSM schema. A fine-grained feature composition modifies sentries of existing stages or milestones. Feature composition for GSM schemas requires both granularity levels to be of practical value. For instance, for the variations in Table 3, the coarse-grained feature for variation 2 (defined in Sect. 4) specifies extra work, so it needs to introduce additional stage and milestones to the base schema, while the fine-grained feature for variation 1 (also defined in Sect. 4) needs to modify the plus sentries of Business Performance Evaluation Check and PCS in the base schema.

Designing a new GSM schema variant based on this approach comes down to selecting the relevant features and defining their composition order. The GSM schema variant is then automatically derived by composing the GSM schema fragments corresponding to the selected features in the defined order. This way, the approach shields designers from the complexity of designing GSM schemas and allows them to quickly generate different variants from a set of fragments. Maintaining the variants is efficient, since an update to a shared fragment can be propagated automatically to all the affected variants by recomposing them.

3   GSM Schemas

Syntax. A GSM schema [9] consists of data attributes and status attributes. A status attribute is a Boolean variable that denotes the status of a stage or milestone. For a status attribute of a stage, value *true* denotes that the stage is open, value *false* that the stage is closed. For a status attribute of a milestone, value *true* denotes that the milestone is achieved, value *false* that the milestone is invalid.

Event-Condition-Action rules define for which event under which condition a status attribute changes value (action). The event-condition part of a rule is called a *sentry*. The event of a sentry is optional. We distinguish between external and internal events. An external event signifies a change in the environment. It is either a stage completion event C:S, where S is an atomic stage, as defined below, or a named external event E:n, where n is an event name. An internal event signifies a change in value of a status attribute a: internal event $+a$ denotes that a becomes true, $-a$ that a becomes false. For instance, $+$Preliminary Check in Table 1 is an internal event that signifies that stage Preliminary Check gets opened. The condition of a sentry is a Boolean expression that can refer to data attributes or status attributes. The action of each rule is that a status attribute becomes true or false. Given these two distinct actions, we distinguish between two types of sentries. A *plus sentry* defines when a stage becomes open or a milestone gets achieved. A *minus sentry* defines when a stage is closed or a milestone gets invalid.

Stages and milestones can be nested inside other stages. A milestone cannot contain any other milestone or stage. We require that the nesting relation induces a forest, i.e., the nesting relation is acyclic and if a stage or milestone is nested in two other stages S_1, S_2, then either S_1 is nested in S_2 or S_2 in S_1. Stage completion events only exist for the most nested stages, which are called *atomic*.

We next formally define GSM schemas. We assume a global universe $\mathcal{U}$ of named external events and attributes, partitioned into sets of named external events $\mathcal{U}_E$, data attributes $\mathcal{U}_d$, stage attributes $\mathcal{U}_S$, and milestone attributes $\mathcal{U}_m$.

Definition 1 (GSM schema). *A GSM schema is a tuple* $\Gamma = (\mathcal{A} = \mathcal{A}_d \cup \mathcal{A}_S \cup \mathcal{A}_m, \mathcal{E} = \mathcal{E}_{ext} \cup \mathcal{E}_{cmp}, \preceq, \mathcal{R} = \mathcal{R}_+ \cup \mathcal{R}_-)$*, where*

- $\mathcal{A}_d \subseteq \mathcal{U}_d$ *is a finite set of data attributes;*
- $\mathcal{A}_S \subseteq \mathcal{U}_S$ *is a finite set of stage attributes;*
- $\mathcal{A}_m \subseteq \mathcal{U}_m$ *is a finite set of milestone attributes;*
- $\mathcal{E}_{ext} = \{ \text{E:}n \mid n \in Ev \}$ *is a finite set of named external events, where* $Ev \subseteq \mathcal{U}_E$;
- $\mathcal{E}_{cmp} = \{ \text{C:}S \mid S \in \mathcal{A}_S \wedge S \text{ is atomic} \}$ *is the set of stage completion events;*
- $\preceq \subseteq (\mathcal{A}_S \cup \mathcal{A}_m) \times \mathcal{A}_S$ *is a partial order on stages and milestones that induces a forest, i.e., if* $a_1 \preceq a_2$ *and* $a_1 \preceq a_3$, *then* $a_2 \preceq a_3$ *or* $a_3 \preceq a_2$;
- $\mathcal{R}_+, \mathcal{R}_-$ *are functions assigning to each status attribute* $\mathcal{A}_S \cup \mathcal{A}_m$ *non-empty sets of* sentries *(see Definition 2). For* $a \in \mathcal{A}_S \cup \mathcal{A}_m$, $\mathcal{R}_+(a)$ *is the set of* plus sentries *that define the conditions when to open stage* $a \in \mathcal{A}_S$ *or achieve milestone* $a \in \mathcal{A}_m$, *while* $\mathcal{R}_-(a)$ *is the set of* minus sentries *that define the conditions when to close stage* $a \in \mathcal{A}_S$ *or invalidate milestone* $a \in \mathcal{A}_m$.

Stage S is atomic if there is no other stage $S' \in \mathcal{A}_S$ such that $S' \preceq S$. The definition of $\preceq$ ensures that milestones are atomic by default. Relation $\preceq$ is visually depicted using nesting. For instance, Business Performance Evaluation Check $\preceq$ Preliminary Check and BPECS $\preceq$ Preliminary Check in Fig. 1.

Each sentry φ in set $\mathcal{R}_+(a)$, where $a \in \mathcal{A}_S \cup \mathcal{A}_m$, maps into an Event-Condition-Action rule "$\varphi\,\mathbf{then}+a$", where sentry φ is the Event-Condition part and action $+a$ denotes for $a \in \mathcal{A}_S$ that stage a gets opened and for $a \in \mathcal{A}_m$ that milestone a gets achieved. Each sentry φ in set $\mathcal{R}_-(a)$ maps into a rule "$\varphi\,\mathbf{then}-a$", where action $-a$ denotes for $a \in \mathcal{A}_S$ that stage a gets closed and for $a \in \mathcal{A}_m$ that milestone a gets invalid. For example, the plus sentry for milestone BPECS (cf. Table 2) is C:Business Performance Evaluation Check Successful $\wedge$ BP_good; the corresponding Event-Condition-Action rule is C:Business Performance Evaluation Check Successful $\wedge$ BP_good $\mathbf{then}$ +BPECS. Each sentry in set $\mathcal{R}_+(a)$ or $\mathcal{R}_-(a)$ is sufficient for triggering a status change in the stage or milestone a.

For the definition of sentries, we assume a condition language $\mathcal{C}$ that includes predicates over integers and Boolean connectives. The condition formulas may refer to stage, milestone and data attributes from the universe of attributes $\mathcal{U}$.

Definition 2 (Sentry). *A sentry has the form* $\tau \wedge \gamma$, *where* τ *is the event-part and* γ *the condition-part. The event-part* τ *is either empty (trivially true), a named external event* $E \in \mathcal{U}_E$, *a stage completion event* C:S, *where* $S \in \mathcal{U}_S$ *is an atomic stage, or is an internal event* $+a$ *or* $-a$, *where* $a \in \mathcal{U}_S \cup \mathcal{U}_m$ *is a stage or milestone attribute. The condition* γ *is a Boolean formula in the condition language* $\mathcal{C}$ *that refers to data attributes in* $\mathcal{U}_d$ *and status attributes in* $\mathcal{U}_S \cup \mathcal{U}_m$. *The condition-part can be omitted if it is equivalent to true.*

Note that a sentry in a GSM schema Γ may or may not refer to attributes that are defined in $\mathcal{A}$. This distinction leads to two disjoint classes of GSM schemas.

Definition 3 (Base schema; schema fragment). *Let* $\Gamma = (\mathcal{A}, \mathcal{E}, \preceq, \mathcal{R})$ *be a GSM schema. Then* Γ *is a* base schema *if each sentry in* $\mathcal{R}$ *only refers to data and status attributes in* $\mathcal{A}$ *and events in* $\mathcal{E}$. *Otherwise,* Γ *is a* schema fragment.

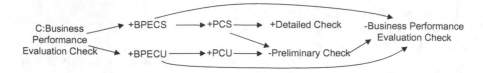

Fig. 2. Event-relativized dependency graph for C:Business Performance Evaluation Check (Fig. 1)

Each sentry in a base schema is "grounded", i.e., it can be evaluated in the context of the schema. A sentry φ in a schema fragment can be non-grounded, i.e., referring to a data or status attribute not defined in the schema fragment. By composing the schema fragment with a base schema or other schema fragments, defined in the next section, sentry φ becomes grounded. The GSM schema presented in Sect. 2 is a base schema; Sect. 4 shows example schema fragments.

Semantics and Well-formedness. In a state of the GSM schema Γ, each attribute in $\mathcal{A}$ has been assigned a value. Initially, all status attributes are false. An incoming external event $E \in \mathcal{E}$, i.e., a stage completion event or a named external event, is processed as follows [9]. Event E can carry in its payload new values for some data attributes; first, these new values are assigned to the relevant data attributes in $\mathcal{A}_d$. Next, one or more Event-Condition-Action rules are fired. Each fired rule changes the value of a status attribute a. If a becomes false, internal event $-a$ is generated; if a becomes true, internal event $+a$ is generated. Generated internal events may trigger additional rules to be fired. Eventually a new state is reached, in which no rules can be applied. In this new state, the next incoming external event can be processed. The full effect of processing an incoming external event is called a Business step, or B-step for short [9,17]. For instance, suppose stage Business Performance Evaluation Check is open and milestones BPECS and BPECU are invalid. If completion event C:Business Performance Evaluation Check with payload (BP_good, *true*) is processed, in the B-step data attribute BP_good is assigned *true*, milestone BPECS is achieved, so stage Business Performance Evaluation Check is closed and milestone PCS is achieved, which means stage Detailed Check gets opened; no further rules can be applied.

The rules need to be processed for each B-step in an order that ensures that each rule for a stage or milestone attribute a is only evaluated if the event- and condition-parts of the rule are stable, so the stage and milestone attributes referenced in those parts do not change value in the B-step after the rule for a has been processed. Therefore, the rules of these referenced stages and milestones are processed before the rule for a. The processing order of rules must be acyclic. To check this, for each external event $E \in \mathcal{E}$ an *event-relativized dependency graph*, denoted $erDG(E)$, can be constructed [9]. The graph contains E plus all internal events (in)directly caused by E. An edge from E or $\odot a_1$, for $\odot \in \{+, -\}$, to $+a_2$ (to $-a_2$) is inserted if a plus (minus) rule for a_2 either (i) contains a_1 in the condition-part, or (ii) contains either E, or $+a_1$, if $\odot = +$, or $-a_1$, if $\odot = -$, in the event-part. Figure 2 shows the event-relativized dependency graph for event C:Business Performance Evaluation Check of BCA_{base}, defined in Sect. 2.

Definition 4. *Let Γ be a GSM schema. Γ is well-formed if for each event $E \in \mathcal{E}$, the event-relativized dependency graph $erDG(E)$ is acyclic [9].*

4   Feature Composition

In this section, we introduce and define feature composition. A feature is a specific functionality of a software artifact that is discernible for an end user [2]. In this paper, each feature denotes a GSM schema. If a feature corresponds to a base GSM schema, we call it *complete*, since it is executable by itself. Otherwise, the feature corresponds to a GSM schema fragment and we call it *partial*, since composition with other features is required to derive a base GSM schema, which can be executed.

To define feature composition, we use a function composition operator '•' [5], also known as superimposition [3]. Let Γ^{comp} be a (partial) feature composition and Γ^{add} a new feature. Both Γ^{comp} and Γ^{add} denote GSM schemas. Then GSM schema $\Gamma^{add} \bullet \Gamma^{comp}$ is the result of adding, also called applying, Γ^{add} to Γ^{comp}. In $\Gamma^{add} \bullet \Gamma^{comp}$, the entire schema of Γ^{add} is embedded into Γ^{comp}. Sentries of stages and milestones that are defined in Γ^{comp} are redefined (overridden by Γ^{add}) in $\Gamma^{add} \bullet \Gamma^{comp}$, if these stages and milestones also are in Γ^{add}.

In some cases, the sentries in the schemas of Γ^{comp} and Γ^{add} for a common stage or milestone should be merged rather than overridden. To specify merging, we use an additional keyword **orig** in conditions of sentries of GSM schemas. Given a feature composition $\Gamma^{add} \bullet \Gamma^{comp}$, if a sentry of Γ^{add} contains keyword **orig**, this refers to the original definition of the sentry according to Γ^{comp}. For instance, if a milestone m has a plus sentry **orig** $\wedge\, x = 10$ in Γ^{add} and m has a plus sentry E:$n \wedge y < 5$ in Γ^{comp}, then **orig** refers to E:$n \wedge y < 5$. Consequently, **orig** can only be used in sentries for stages and milestones that belong to both Γ^{comp} and Γ^{add}.

Example. The example in Sect. 2 has one complete feature, base GSM schema BCA_{base}, and three partial features, one for each variant in Table 3. The sentries for the GSM schema fragments F_1 and F_2 of variants 1 and 2, respectively, are shown in Tables 4 and 5. Feature F_1 only refers to stage, milestone and data attributes that have been defined already in BCA_{base}. In F_1 there are two plus sentries for PCS. The semantically equivalent sentry **orig** $\vee$ (IDGS $\wedge$ employee_count<300) is not allowed, since it is not sentry composable (defined in Sect. 5). The minus sentries in F_1 are not redefined, indicated with **orig**.

The GSM schema for feature F_2 does introduce new attributes (see Fig. 3): stage attribute Check Credit, milestone attributes CCS and CCU, and data attribute rating. Since the base schema does not contain these new stage and milestones, their plus and minus sentries cannot use **orig**. However, these sentries may refer to stages or milestones not present in the fragment. For instance, the plus sentry of new stage Check Credit refers to milestone IDGS, which is not part of F_2.

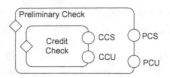

Fig. 3. GSM schema for partial feature Credit Check (F_2)

Definition. In the formal definition of '•', sentries in Γ^{add} may contain the keyword **orig** in the condition-part. Given sentries φ and ψ, sentry $\varphi[\mathbf{orig}/\psi]$ is the result of replacing each occurrence of **orig** in φ by (ψ).

Definition 5 (Feature composition). *Let Γ^{comp} be a base GSM schema and Γ^{add} a GSM schema fragment that is added to Γ^{comp}. Sentries of Γ^{add} can use the keyword **orig** in their condition-parts. Then $\Gamma^{add} \bullet \Gamma^{comp}$ is the GSM schema $\Gamma = (\mathcal{A} = \mathcal{A}_d \cup \mathcal{A}_S \cup \mathcal{A}_m, \mathcal{E} = \mathcal{E}_{ext} \cup \mathcal{E}_{cmp}, \preceq, \mathcal{R} = \mathcal{R}_+ \cup \mathcal{R}_-)$ where*

Table 4. Sentries for partial feature F_1. S = Stage; M = Milestone

Type	Name	Plus sentries	Minus sentries
S	Business Performance Evaluation Check	orig ∧ employee_count ≥ 300	orig
S	Preliminary Check	orig	orig
M	PCS	IDGS ∧ employee_count < 300 ; orig	orig
M	PCU	orig	orig

Table 5. Sentries for partial feature F_2. S = Stage; M = Milestone; CCS = Credit Check Successful; CCU = Credit Check Unsuccessful

Type	Name	Plus sentries	Minus sentries
S	Check Credit	IDGS	CCS ; CCU;-Preliminary Check
S	Preliminary Check	orig	orig
M	CCS	C:Credit Check ∧ rating ≥ 8	+Check Credit
M	CCU	C:Credit Check ∧ rating < 8	+Check Credit
M	PCS	orig ∧ CCS	orig
M	PCU	orig ; CCU	orig

- $\mathcal{A}_d = \mathcal{A}_d^{comp} \cup \mathcal{A}_d^{add}$;
- $\mathcal{A}_S = \mathcal{A}_S^{comp} \cup \mathcal{A}_S^{add}$;
- $\mathcal{A}_m = \mathcal{A}_m^{comp} \cup \mathcal{A}_m^{add}$;
- $\mathcal{E}_{ext} = \mathcal{E}_{ext}^{comp} \cup \mathcal{E}_{ext}^{add}$;
- $\mathcal{E}_{cmp} = \mathcal{E}_{cmp}^{comp} \cup \mathcal{E}_{cmp}^{add}$;
- $\preceq = \preceq^{comp} \cup \preceq^{add}$;
- for each $a \in \mathcal{A}$,

$$\mathcal{R}_+(a) = \begin{cases} \{\varphi[\mathbf{orig}/\psi] \mid \varphi \in \mathcal{R}_+^{add}(a), \psi \in \mathcal{R}_+^{comp}(a)\} & \text{, if } a \in \mathcal{A}^{comp} \cap \mathcal{A}^{add} \\ \mathcal{R}_+^{comp}(a) & \text{, if } a \in \mathcal{A}^{comp} \setminus \mathcal{A}^{add} \\ \mathcal{R}_+^{add}(a) & \text{, if } a \in \mathcal{A}^{add} \setminus \mathcal{A}^{comp} \end{cases}$$

$$\mathcal{R}_-(a) = \begin{cases} \{\varphi[\mathbf{orig}/\psi] \mid \varphi \in \mathcal{R}_-^{add}(a), \psi \in \mathcal{R}_-^{comp}(a)\} & \text{, if } a \in \mathcal{A}^{comp} \cap \mathcal{A}^{add} \\ \mathcal{R}_-^{comp}(a) & \text{, if } a \in \mathcal{A}^{comp} \setminus \mathcal{A}^{add} \\ \mathcal{R}_-^{add}(a) & \text{, if } a \in \mathcal{A}^{add} \setminus \mathcal{A}^{comp} \end{cases}$$

Most lines in the definition above use simple set union. The definition of $\mathcal{R}$ is most involved. The basic principle is that if a stage or milestone a is defined in only one of the two input GSM schemas, the sentries for a in the composition Γ are those of the input schema. If a occurs in both input schemas, the sentries for a in Γ^{add} override the sentries for a in Γ^{comp}. Using $\mathbf{orig}$, the original sentries in Γ^{comp} can be reused in the definition of the overridden sentries in Γ.

The GSM schema resulting from the composition $F_1 \bullet BCA_{base}$ (Table 6) illustrates that using the '$\bullet$' operator, sentries can be merged with the original sentries (the plus sentry of Business Performance Evaluation Check in F_1 has been merged with the one in BCA_{base}) and added to the original sentries (the plus sentry for PCS in F_1 has been added to the plus sentry for PCS in BCA_{base}).

Table 6. Sentries of $F_1 \bullet BCA_{base}$ for the stages and milestones that are both in F_1 and BCA_{base}. S = Stage; M = Milestone

Type	Name	Plus sentries	Minus sentries
S	Business Performance Evaluation Check	+Preliminary Check ∧ employee_count ≥ 300	BPECS; BPECU; -Preliminary Check
S	Preliminary Check	IDGS	PCS ; PCU ; E:AdditionalInfo
M	PCS	IDGS ∧ employee_count < 300 ; BPECS	false
M	PCU	BPECU	false

Discussion. If more than two features are composed, they are ordered in a composition chain. For instance, a possible chain is $F_1 \bullet F_2 \bullet BCA_{base}$. Operator '$\bullet$' (Definition 5) requires that the righthand feature is complete. Therefore '$\bullet$' is right associative.

The ordering of features in a composition chain influences the outcome. In other words, the feature composition operator '$\bullet$' is not commutative. A simple example to illustrate this: consider $\Gamma_{12} = \mathsf{F}_1 \bullet \mathsf{F}_2 \bullet \mathsf{BCA}_{base}$ versus $\Gamma_{21} = \mathsf{F}_2 \bullet \mathsf{F}_1 \bullet$ BCA_{base}. In Γ_{12}, one of the plus sentries for PCS is $\mathsf{IDGS} \wedge \mathsf{employee\_count} < 300$. while in Γ_{21} the corresponding plus sentry is $\mathsf{IDGS} \wedge \mathsf{employee\_count} < 300 \wedge \mathsf{CCS}$. The first sentry allows PCS to become true while CCS is false, which is obviously not desirable. So in this example, feature F_1 should be applied before F_2, so the valid composition order is $\mathsf{F}_2 \bullet \mathsf{F}_1 \bullet \mathsf{BCA}_{base}$.

In general, there are two options to handle this issue: either define correctness conditions that guarantee that '$\bullet$' is commutative or help designers to live with this lack of commutativity. Defining additional correctness conditions that guarantee that '$\bullet$' is commutative would obviously have to rule out features F_1 and F_2, which are perfectly valid. We therefore favor the other option. In particular, it is useful to define additional dependency constraints between features that help designers to manage the correct sequencing of features, if multiple need to be applied to derive an artifact-centric process model. The feature composition chain then has to respect these dependencies.

5   Correctness

The outcome of feature-oriented composition of two GSM schemas may be a structure that is not a GSM schema or not a well-formed GSM schema. We define constraints in this section that ensure that feature-oriented composition produces (well-formed) GSM schemas.

From a syntax point of view, two constraints need to be satisfied. First, the hierarchies of Γ^{add} and Γ^{comp} should be composable to ensure that the resulting relation $\preceq$, defined in Definition 5, is a hierarchy.

Definition 6 (Hierarchy composable). *Let Γ^{comp} be a base GSM schema and Γ^{add} be a GSM schema fragment that is added to Γ^{comp}. The hierarchies of Γ^{comp} and Γ^{add} are composable if the following conditions are met:*

- *For each pair $a_1, a_2 \in (\mathcal{A}_S^{comp} \cup \mathcal{A}_m^{comp}) \cap (\mathcal{A}_S^{add} \cup \mathcal{A}_m^{add})$, where $a_1 \neq a_2$, $a_1 \preceq^{comp} a_2$ iff $a_1 \preceq^{add} a_2$.*
- *For each pair $a_1, a_2 \in \mathcal{A}_S^{add} \cup \mathcal{A}_m^{add}$, if $a_1 \preceq^{add} a_2$ and $a_2 \notin \mathcal{A}_S^{comp} \cup \mathcal{A}_m^{comp}$ then $a_1 \notin \mathcal{A}_S^{comp} \cup \mathcal{A}_m^{comp}$.*
- *For each pair $a_1, a_2 \in \mathcal{A}_S^{add} \cup \mathcal{A}_m^{add}$, if $a_1 \preceq^{add} a_2$ and $a_2 \in \mathcal{A}_S^{comp} \cup \mathcal{A}_m^{comp}$ then a_2 is not atomic in Γ^{comp}.*

The first condition states that for stages and milestones that are shared between Γ^{comp} and Γ^{add} the hierarchy relations $\preceq^{comp}$ and $\preceq^{add}$ are consistent. The second condition rules out that a new stage is inserted by Γ^{add} in the middle of the hierarchy of Γ^{comp}. For instance, the condition is violated for Fig. 4; if the hierarchies are composed, stage S gets two parents. The third

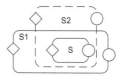

Fig. 4. Base schema $(\mathsf{S}_1, \mathsf{S})$ and fragment schema $(\mathsf{S}_2, \mathsf{S})$ that are not hierarchy composable

condition is that inserting a new stage inside an existing stage S is allowed, provided S is not atomic in Γ^{comp}. If S is atomic in Γ^{comp}, then S generates a completion event C:S, which is not generated if S is not atomic. Hence, the condition ensures that rules triggered by C:S in Γ^{comp} are also triggered in $\Gamma^{add} \bullet \Gamma^{comp}$. Under these conditions, the hierarchy relation in the composition $\Gamma^{add} \bullet \Gamma^{comp}$ is consistent with the hierarchy relations in both Γ^{add} and Γ^{comp}.

The second constraint is that the sentries must be composable if **orig** is used, i.e., merged sentries for $\mathcal{R}$ as defined in Definition 5 satisfy the sentry syntax of Definition 2.

Definition 7 (Sentry composable). *Let Γ^{comp} be a base GSM schema and Γ^{add} be a GSM schema fragment that is added to Γ^{comp}. A sentry $\varphi \in \mathcal{R}^{add}(a)$, where $a \in \mathcal{A}_S^{add} \cup \mathcal{A}_m^{add}$, is sentry composable if the following conditions are met:*

- *If φ contains **orig**, then $a \in \mathcal{A}_S^{comp} \cup \mathcal{A}_m^{comp}$.*
- *If φ contains **orig** and has a non-empty event-part, then*
  - *if $\varphi \in \mathcal{R}_+^{add}(a)$, then each rule in $\mathcal{R}_+^{comp}(a)$ has an empty event-part;*
  - *if $\varphi \in \mathcal{R}_-^{add}(a)$, then each rule in $\mathcal{R}_-^{comp}(a)$ has an empty event-part.*
- *If φ contains **orig** and has an empty event-part, then φ is in conjunctive normal form and **orig** only occurs as conjunct in φ.*
- *If φ references a stage or milestone a that is not in $\mathcal{A}^{add}$, then $a \in \mathcal{A}^{comp}$.*

The first condition in the definition ensures that if a sentry φ for a uses **orig**, then the attribute a exists in the base GSM schema. This way, **orig** can always be substituted by another sentry. The second and third condition ensure that a sentry for a in the base GSM schema and a sentry for a in the schema fragment can be composed properly into a new sentry of the form $\tau \wedge \gamma$. For instance, if a sentry φ in Γ^{add} uses **orig** and is triggered by a completion event C:S, then the sentry of Γ^{comp} referred to by **orig** should have no trigger event. Note that a sentry of the form **orig**$\vee \varphi$, which is ruled out by the third condition, is equivalent to pair of sentries **orig** and φ. Thus, this condition does not diminish expressive power. The fourth condition makes sure that each sentry in the schema fragment becomes grounded.

Under these two constraints, the result of feature composition is guaranteed to be a GSM schema.

Lemma 1. *Let Γ^{comp} be a base GSM schema and Γ^{add} be a GSM schema fragment such that $\Gamma = \Gamma^{add} \bullet \Gamma^{comp}$. If Γ^{comp} and Γ^{add} have composable hierarchies and each rule in Γ^{add} is sentry composable, then Γ is a GSM schema.*

Proof. (Sketch.) We focus on showing that each sentry in Γ satisfies the syntax defined in Definition 2. Suppose a sentry $\varphi = \tau_1 \wedge \gamma_1$ from Γ^{add} contains **orig**. By Definition 5, **orig** is contained in γ_1. By the first condition of Definition 7, $a \in \mathcal{A}_S^{comp} \cup \mathcal{A}_m^{comp}$. We prove that the new sentry is of the form $\tau \wedge \gamma$ (cf. Definition 2). There are two cases.

(a) Sentry φ has a non-empty event-part τ_1. By the second condition of Definition 7, any sentry $\psi = \tau_2 \wedge \gamma_2$ from Γ^{comp} that **orig** is substituted with, has an empty event-part, so the new sentry is $\tau_1 \wedge \gamma_1'$, where $\gamma_1' = \gamma_1[\textbf{orig}/\gamma_2]$.

(b) Sentry φ has an empty event-part τ_1. Let $\psi = \tau_2 \wedge \gamma_2$ be the sentry from Γ^{comp} that **orig** is substituted with. Either the sentry ψ has an empty event-part, in which case the new sentry is $\gamma_1[\mathbf{orig}/\gamma_2]$. Or the sentry ψ has a non-empty event-part. By the third condition of Definition 7, γ_1 is in conjunctive normal form with **orig** as a conjunct. Let γ_{min} be γ_1 minus the conjunct **orig**. The new sentry is $\tau_2 \wedge \gamma_2 \wedge \gamma_{min}$. □

Though Lemma 1 shows under which conditions Γ is a GSM schema, it might be that Γ is not well-formed due to a cycle in an event-relativized dependency graph for some event E. Such a cycle is caused by a different processing order of the rules (in)directly triggered by E in Γ^{comp} and Γ^{add}. To rule out such cycles, we introduce a third constraint: the ordering of status attributes in both Γ^{add} and Γ^{comp} is consistent for each event E, so it is not the case that a_1 before a_2 in Γ^{comp} yet a_2 before a_1 in Γ^{add} while processing E. This can be checked by inspecting all the event-relativized dependency graphs of both Γ^{add} and Γ^{comp}.

Definition 8. (Consistent rule orderings). *Let Γ^{comp} be a base GSM schema and Γ^{add} a GSM schema fragment that is added to Γ^{comp}. Then Γ^{comp} and Γ^{add} have* consistent rule orderings *if for each event E the orderings of status attributes in $erDG_\Gamma^{comp}(E)$ and $erDG_\Gamma^{add}(E)$ are compatible, so for $a_1, a_2 \in \mathcal{A}^{comp} \cap \mathcal{A}^{add}$, a_1 before a_2 in $erDG_\Gamma^{comp}(E)$ implies a_2 not before a_1 in $erDG_\Gamma^{add}(E)$.*

If the third constraint is satisfied too, then composition Γ is guaranteed to be a well-formed GSM schema.

Lemma 2. *Let Γ^{comp} be a well-formed base GSM schema and Γ^{add} be a well-formed GSM schema fragment such that $\Gamma = \Gamma^{add} \bullet \Gamma^{comp}$. If Γ^{comp} and Γ^{add} have composable hierarchies, each sentry in Γ^{add} is sentry composable, and Γ^{comp} and Γ^{add} have consistent rule orderings, then Γ is well-formed.*

Proof. By Lemma 1, Γ is a GSM schema. Suppose Γ is not well-formed. By Definition 4 there is an event E such that $erDG(E)$ contains a cycle between nodes a_1 and a_2. Since Γ^{comp} and Γ^{add} are well-formed, the event-relativized dependency graphs for E in Γ^{comp} and Γ^{add} are acyclic. Hence, in one graph the ordering is a_1 before a_2, in the other a_2 before a_1. Therefore, Γ^{comp} and Γ^{add} have inconsistent rule orderings. □

6   Evaluation

To evaluate the feasibility and potential gain of the approach, we applied the approach to model variants of a real-world process of an international high tech company with offices in different regions of the worlds. In the process the expired due diligence qualification of a business partner of the company is renewed. The company has defined a standard due diligence process, but offices in certain regions can use their own variant of the process. The standard process and the variants had been modeled before in separate GSM schemas [30].

Based on the existing GSM schemas for this process, we defined a base schema DDP_{base} for the standard process and four features that refine the base schema: F_{1a}^{DDP}, F_{1b}^{DDP}, F_2^{DDP}, and F_3^{DDP}; all are available in an online appendix [11]. The first two features are alternatives. Similar to F_1 (Table 4) and F_2 (Table 5), each fragment schema of a feature uses **orig** as sentry or as conjunct of a sentry to specify the connection between the base schema and the fragment schema.

Table 7. Descriptive statistics of due diligence process and its variants

GSM schema	Feature composition	# Stages	# Milestones	# Sentries	Overlap with DDP_{base} in:			
					% Stages	% Milestones	# Sentries	% Sentries
Base schema	DDP_{base}	9	15	60				
Variant 1	$F_{1a}^{DDP} \bullet DDP_{base}$	10	16	66	90	93	59	89
Variant 2	$F_{1b}^{DDP} \bullet DDP_{base}$	11	17	71	82	88	59	83
Variant 3	$F_2^{DDP} \bullet DDP_{base}$	10	16	65	90	93	60	92
Variant 4	$F_3^{DDP} \bullet DDP_{base}$	10	16	66	90	93	60	91

Table 7 shows descriptive statistics of the base schema and four different variants that are derived by applying each of the four features to the base schema. Each variant is equivalent to an existing variant [30]. Note that the base schema is embedded in each of the variants. Hence, there is an overlap between each variant and the base schema. For instance, the first variant shares 9 out of 10 stages and 15 out of 16 milestones with the base schema. Variant 1 and 2 each have a sentry that is derived from a sentry of the base schema; the corresponding sentry in the feature uses conjunct **orig**. Hence, for variants 1 and 2, 59 sentries of the base schema appear in the variant rather than 60.

Table 7 shows that the overlap between the variants is huge. If the variants are modeled separately, this causes maintenance problems. For instance, changing the name of a milestone m shared by all four variants needs to be done four times. By using features, this overlap can be managed efficiently. The milestone m needs to be updated only once, for the base schema. The change is then propagated to the variants derived from the base schema by recomposing them. Another benefit of the approach is that many more variants can be derived than just the ones in Table 7. In total $3 * 2 * 2 = 12$ variants (incl. the base schema) can be derived, without modifying any of the fragment schemas or the base schema.

To conclude, the preliminary evaluation on a real-world process shows that by using the approach, variants in a family of GSM schemas can be expressed as feature compositions. Using features avoids duplicates and thus eases the design and maintenance of declarative artifact-centric process variants.

7   Related Work

For artifact-centric process models, no directly related work on composition of model fragments exists. The general problem of designing artifact-centric process models, either by defining a methodology for specifying declarative business artifacts [6] or by defining an automated synthesis of artifact-centric process models [13,14,20,24] has been addressed. The feature-oriented composition approach facilitates reuse of model fragments and the generation of different but related variants, rather than designing a single artifact-centric process model.

Alternatives to artifact-centric process models are object-aware process models [18,25] and case management models [1,22,28] (though artifact-centric process models can be used for case management too [12,21]). For one of these alternatives, an approach has been defined for composing production case management models out of procedural process models [22]. Composition at a fine-grained level (overriding) is not supported and features are not used.

Variability has been well studied for activity-centric business processes. Recent surveys [4,19] describe the state of the art in variability modeling from the angles of procedural modeling languages [19] and of different phases of the business process life cycle [4], which includes procedural modeling languages. The surveyed mechanisms for modeling variability [4,19] are specific to procedural, flowchart-like languages (e.g. specializing activities [19]). Variability support for declarative, activity-centric business processes has been developed [27], but all variants are encoded in a single declarative process model from which a process variant is generated by hiding activities and omitting constraints, rather than composing a process variant from fragments in a modular fashion using features. Another survey [26] lists papers that have applied variability techniques from software product lines, such as feature models, to activity-centric, procedural business processes. To the best of our knowledge, there is no related work that applies feature composition to declarative or procedural process models.

In earlier work [12], we defined the change operations insertion and deletion for monotonic GSM schemas (each attribute can be written only once during an execution) without hierarchy. An alternative approach for deriving GSM schema variants can be defined by using the GSM change operations in combination with Provop [16], an existing approach for managing variability in procedural process models in terms of change operations. Though that alternative approach allows reducing a base schema, not possible with feature-oriented composition, it is restricted to monotonic GSM schemas without hierarchy, while the feature-oriented composition approach supports non-monotonic, hierarchical GSM schemas.

In software engineering, features have been applied both at the level of modeling languages [23] and programming languages [2]. The feature-oriented design approach defined in Sect. 4 resembles most closely the feature composition approach for software artifacts developed by Batory et al. [5] and extended by Apel et al. [3]. That approach also uses a composition operator (called superimposition [3]) and supports merging of method bodies using an **orig**-like construct. Since merging of GSM schemas can result in incorrect schemas, we consider correctness issues, which are ignored for software artifacts [3,5].

8   Conclusion

The main contribution of this paper is a formally defined, feature-oriented approach for composing declarative artifact-centric process models. The approach defines how to compose features that denote GSM schemas, some of which are partially specified, into completely specified GSM schemas. Correctness criteria are defined that guarantee that valid GSM schemas are derived. The approach supports reuse of model fragments. The approach has been evaluated by applying it to a GSM schema of an industrial process. Using the approach, declarative artifact-centric process models can be designed in an efficient and correct way.

One direction for further work is evaluating the approach in more case studies. Another direction is developing tool support for the approach based on an existing feature composition tool like FeatureHouse [3].

References

1. van der Aalst, W.M.P., Weske, M., Grünbauer, D.: Case handling: a new paradigm for business process support. Data Knowl. Eng. **53**(2), 129–162 (2005)
2. Apel, S., Batory, D.S., Kästner, C., Saake, G.: Feature-Oriented Software Product Lines: Concepts and Implementation. Springer, Heidelberg (2013). https://doi.org/10.1007/978-3-642-37521-7
3. Apel, S., Kästner, C., Lengauer, C.: Language-independent and automated software composition: the featurehouse experience. IEEE Trans. Softw. Eng. **39**(1), 63–79 (2013)
4. Ayora, C., Torres, V., Weber, B., Reichert, M., Pelechano, V.: VIVACE: a framework for the systematic evaluation of variability support in process-aware information systems. Inf. Softw. Technol. **57**, 248–276 (2015)
5. Batory, D.S., Sarvela, J.N., Rauschmayer, A.: Scaling step-wise refinement. IEEE Trans. Softw. Eng. **30**(6), 355–371 (2004)
6. Bhattacharya, K., Hull, R., Su, J.: A data-centric design methodology for business processes. In: Handbook of Research on Business Process Modeling, pp. 503–531 (2009). Chapter 23
7. BizAgi and others: Case Management Model and Notation (CMMN), v1.1. OMG Document Number formal/16-12-01, Object Management Group, December 2016
8. Cohn, D., Hull, R.: Business artifacts: a data-centric approach to modeling business operations and processes. IEEE Data Eng. Bull. **32**(3), 3–9 (2009)
9. Damaggio, E., Hull, R., Vaculín, R.: On the equivalence of incremental and fixpoint semantics for business artifacts with guard-stage-milestone lifecycles. Inf. Syst. **38**, 561–584 (2013)
10. Di Ciccio, C., Marrella, A., Russo, A.: Knowledge-intensive processes: characteristics, requirements and analysis of contemporary approaches. J. Data Semant. **4**(1), 29–57 (2015)
11. Eshuis, R.: Appendix to: Feature-Oriented Composition of Declarative Artifact-Centric Process Models (2018). http://is.ieis.tue.nl/staff/heshuis/foc-appendix.pdf
12. Eshuis, R., Hull, R., Yi, M.: Property preservation in adaptive case management. In: Barros, A., Grigori, D., Narendra, N.C., Dam, H.K. (eds.) ICSOC 2015. LNCS, vol. 9435, pp. 285–302. Springer, Heidelberg (2015). https://doi.org/10.1007/978-3-662-48616-0_18

13. Eshuis, R., Van Gorp, P.: Synthesizing data-centric models from business process models. Computing **98**(4), 345–373 (2016)
14. Fritz, C., Hull, R., Su, J.: Automatic construction of simple artifact-based business processes. In: Proceedings of (ICDT), pp. 225–238 (2009)
15. Goedertier, S., Vanthienen, J., Caron, F.: Declarative business process modelling: principles and modelling languages. Enterp. IS **9**(2), 161–185 (2015)
16. Hallerbach, A., Bauer, T., Reichert, M.: Capturing variability in business process models: the Provop approach. J. Softw. Maint. **22**(6–7), 519–546 (2010)
17. Hull, R., et al.: Introducing the Guard-Stage-Milestone approach for specifying business entity lifecycles. In: Bravetti, M., Bultan, T. (eds.) WS-FM 2010. LNCS, vol. 6551, pp. 1–24. Springer, Heidelberg (2011). https://doi.org/10.1007/978-3-642-19589-1_1
18. Künzle, V., Reichert, M.: PHILharmonicFlows: towards a framework for object-aware process management. J. Softw. Maint. **23**(4), 205–244 (2011)
19. La Rosa, M., van der Aalst, W.M.P., Dumas, M., Milani, F.: Business process variability modeling: a survey. ACM Comput. Surv. **50**(1), 2:1–2:45 (2017)
20. Lohmann, N.: Compliance by design for artifact-centric business processes. Inf. Syst. **38**(4), 606–618 (2013)
21. Marin, M., Hull, R., Vaculín, R.: Data centric BPM and the emerging case management standard: a short survey. In: La Rosa, M., Soffer, P. (eds.) BPM 2012. LNBIP, vol. 132, pp. 24–30. Springer, Heidelberg (2013). https://doi.org/10.1007/978-3-642-36285-9_4
22. Meyer, A., Herzberg, N., Puhlmann, F., Weske, M.: Implementation framework for production case management: modeling and execution. In: Proceedings of EDOC 2014, pp. 190–199. IEEE Computer Society (2014)
23. Pohl, K., Böckle, G., van der Linden, F.: Software Product Line Engineering - Foundations, Principles, and Techniques. Springer, Heidelberg (2005). https://doi.org/10.1007/3-540-28901-1
24. Popova, V., Fahland, D., Dumas, M.: Artifact lifecycle discovery. Int. J. Coop. Inf. Syst. **24**(1), 1–44 (2015)
25. Redding, G., Dumas, M., ter Hofstede, A.H.M., Iordachescu, A.: A flexible, object-centric approach for business process modelling. Serv. Oriented Comput. Appl. **4**(3), 191–201 (2010)
26. dos Santos Rocha, R., Fantinato, M.: The use of software product lines for business process management: a systematic literature review. Inf. Softw. Technol. **55**(8), 1355–1373 (2013)
27. Schunselaar, D.M.M., Maggi, F.M., Sidorova, N., van der Aalst, W.M.P.: Configurable declare: designing customisable flexible process models. In: Meersman, R., et al. (eds.) OTM 2012. LNCS, vol. 7565, pp. 20–37. Springer, Heidelberg (2012). https://doi.org/10.1007/978-3-642-33606-5_3
28. Slaats, T., Mukkamala, R.R., Hildebrandt, T., Marquard, M.: Exformatics declarative case management workflows as DCR graphs. In: Daniel, F., Wang, J., Weber, B. (eds.) BPM 2013. LNCS, vol. 8094, pp. 339–354. Springer, Heidelberg (2013). https://doi.org/10.1007/978-3-642-40176-3_28
29. Swenson, K.D.: Mastering the Unpredictable: How Adaptive Case Management will Revolutionize the Way that Knowledge Workers Get Things Done. Meghan-Kiffer, Tampa (2010)
30. Yi, M.: Managing business process variability in artifact-centric BPM. Master's thesis, Eindhoven University of Technology (2015)

Animating Multiple Instances in BPMN Collaborations: From Formal Semantics to Tool Support

Flavio Corradini, Chiara Muzi, Barbara Re, Lorenzo Rossi,
and Francesco Tiezzi[(✉)]

School of Science and Technology, University of Camerino, Camerino, Italy
{flavio.corradini,chiara.muzi,barbara.re,lorenzo.rossi,
francesco.tiezzi}@unicam.it

Abstract. The increasing adoption of modelling methods contributes
to a better understanding of the flow of processes, from the internal
behaviour of a single organisation to a wider perspective where several
organisations exchange messages. In this regard, BPMN collaboration is
a suitable modelling abstraction. Even if this is a widely accepted notation, only a limited effort has been expended in formalising its semantics, especially for what it concerns the interplay among control features,
data handling and exchange of messages in scenarios requiring multiple
instances of interacting participants. In this paper, we face the problem
of providing a formal semantics for BPMN collaborations including multiple instances, while taking into account the data perspective. Beyond
defining a novel formalisation, we also provide a BPMN collaboration
animator tool faithfully implementing the formal semantics. Its visualisation facilities support designers in debugging multi-instance collaboration models.

1 Introduction

Nowadays, modelling is recognised as an important practice also in supporting the continuous improvement of IT systems. In particular, IT support for
collaborative systems, where participants can cooperate and share information,
demands for a clear understanding of interactions and data exchanges. To ensure
proper carrying out of such interactions, the participants should be provided with
enough information about the messages they must or may send in a given context. This is particularly important when multiple instances of interacting participants are involved. In this regard, BPMN [1] collaboration diagrams result
to be an effective way to reflect how multiple participants cooperate to reach a
shared goal.

Even if widely accepted, a major drawback of BPMN is related to the complexity of the semi-formal definition of its meta-model and the possible misunder-
standing of its execution semantics defined by means of natural text description,
sometimes containing misleading information [2]. This becomes a more prominent issue as we consider BPMN supporting tools, such as animators, simulators

© Springer Nature Switzerland AG 2018
M. Weske et al. (Eds.): BPM 2018, LNCS 11080, pp. 83–101, 2018.
https://doi.org/10.1007/978-3-319-98648-7_6

and enactment tools, whose implementation of the execution semantics may not be compliant with the standard and be different from each other, thus undermining models portability and tools effectiveness.

To overcome these issues, several formalisations have been proposed, mainly focussing on the control flow perspective (e.g., [3–7]). Less attention has been paid to provide a formal semantics capturing the interplay between control features, message exchanges, and data. These perspectives are strongly related, especially when a participant interacts with multi-instance participants. In fact, to achieve successful collaboration interactions, it is required to deliver the messages arriving at the receiver side to the appropriate instances. As messages are used to exchange data between participants, the BPMN standard fosters the use of the content of the messages themselves to correlate them with the corresponding instances. Thus, the data perspective plays a crucial role when considering multi-instance collaborations. Despite this, no formal semantics that considers all together these key aspects of BPMN collaboration models has been yet proposed in the literature.

In this work, we aim at filling this gap by providing an operational semantics of BPMN collaboration models including multi-instance participants, while taking into account the data perspective, considering both data objects and data-based decision gateways. Moreover, we go beyond the mere formalisation, by developing an animator tool that faithfully implements the proposed formal semantics and visualises the execution of multi-instance collaborations. It is indeed well recognised that process animators play an important role in enhancing the understanding of business processes behaviour [8] and that, to this aim, the faithful correspondence with the semantics is essential [9], although it is not always supported [10]. Visualisation of model execution via an animator allows to understand the collaboration history, its current state (also in terms of data-object values) and possible future executions [11]. This is particularly useful in case of models that are not implemented yet [12]. Our tool, called MIDA, supports model designers in achieving a priori knowledge of collaborations behaviour. This can allow them to spot erroneous interactions, which can easily arise when dealing with multiple instances, and hence to prevent undesired executions.

To sum up, the major contributions of this paper are:

- The definition of a formal semantics for BPMN collaborations considering control flow elements, multi-instance pools, data objects and data-based decision gateways. Besides being useful per se, as it provides a precise understanding of the ambiguous and loose points of the standard, a main benefit of this formalisation is that it paves the way for the development of tools supporting model analysis.
- The development of the MIDA tool for animating BPMN collaboration models. MIDA animation features result helpful both in educational contexts, for explaining the behaviour of BPMN elements, and in practical modelling activities, for debugging errors common in multi-instance collaborations.

The rest of the paper is organised as follows. Section 2 provides the motivations underlying the work, and presents our running example. Section 3 introduces the formal framework at the basis of our approach. Section 4 shows how the formal concepts have been practically realised in the MIDA tool. Section 5 compares our work with the related ones. Finally, Sect. 6 closes the paper with lessons learned and opportunities for future work.

2 The Interplay Between Multiple Instances, Messages and Data Objects in BPMN Collaborations

To precisely deal with multiple instances in BPMN collaboration models, it is necessary to take into account the data flow. Indeed, the creation of process *instances* can be triggered by the arrival of *messages*, which contain data. Within a process instance, data is stored in *data objects*, used to drive the instance execution. Values of data objects can be used to fill the content of outgoing messages, and vice versa, the content of incoming messages can be stored in data objects. We clarify below the interplay between such concepts. To this aim, we introduce a BPMN collaboration model, used as a running example throughout the paper, concerning the management of the paper reviewing process of a scientific conference (this is a revised version of the model in [13, Sect. 4.7.2] and [14]). The example concerns the management of a single paper, which is revised by three reviewers; of course, the management of all papers submitted to the conference requires to enact the collaboration for each paper.

The collaboration model in Fig. 1 combines the activities of three participants. The *Program Committee (PC) Chair* organises the reviewing activities. For the sake of simplicity, we assume that the considered conference has only one chair. A *Reviewer* performs the reviewing activity and, since more than one reviewer takes part in this, he/she is modelled as a process instance of a multi-instance pool. Finally, the *Contact Author* is the person who submitted the paper to the conference. The reviewing process is started by the PC chair, who assigns the paper to each reviewer (via a multi-instance sequential activity with loop cardinality set to 3 according to the number of involved reviewers for each paper). The paper is passed to the PC chair process by means of a data input. After all reviews are received, and combined in the *Reviews* data object, the chair starts their evaluation. According to the value of the *Evaluation* data object, the chair prepares the acceptance/rejection letter (stored in the *Letter* data object) or, if the paper requires further discussion, the decision is postponed. Discussion interactions are here abstracted and always result in an accept or reject decision. The chair then sends back a feedback to each reviewer, attaches the reviews to the notification letter, and sends the result to the contact author.

In this scenario, data support is crucial to precisely render the message exchanges between participants, especially because multiple instances of the *Reviewer* process are created. In fact, messages coming into this pool might

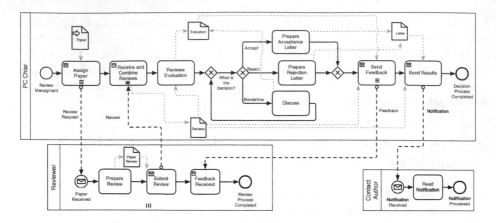

Fig. 1. Paper reviewing collaboration model.

start a new process instance, or be routed to existing instances already underway. Messages and process instances must contain enough information to determine, when a message arrives at a pool, if a new process instance is needed or, if not, which existing instance will handle it. To this aim, BPMN makes use of the concept of *correlation*: it is up to each single message to provide the information that permits to associate the message with the appropriate (possibly new) instance. This is achieved by embedding values, called *correlation data*, in the content of the message itself. Pattern-matching is used to associate a message to a distinct receiving task or event. In our example, every time the chair sends back a feedback to a reviewer, the message must contain information (in our case reviewer name and paper title) to be correlated to the correct process instance of *Reviewer*.

According to the BPMN standard, data objects do not have any direct effect on the sequence flow or message flow of processes, since tokens do not flow along data associations [1, p. 221]. However, this statement is questionable. Indeed, on the one hand, the information stored in data objects can be used to drive the execution of process instances, as they can be referred in the conditional expressions of XOR gateways to take decisions about which branch should be taken. On the other hand, data objects can be connected in input to tasks. In particular, the standard states that *"the Data Objects as inputs into the Tasks act as an additional constraint for the performance of those Tasks. The performers [...] cannot start the Task without the appropriate input"* [1, p. 183]. In both cases, a data object has an implicit indirect effect on the execution, since it can drive the decision taken by a XOR gateway or act as a guard condition on a task. For instance, in our running example, according to the value of the *Evaluation* data object, the conditional expression *What is the decision?* is evaluated and a branch of the XOR split gateway is chosen. As another example, the task *Send Results* can be executed only if an acceptance or rejection letter is stored in the *Letter* data object.

(a)	Paper {title, contact, authors, body} Reviews {title, reviewers, scores, bodies}
	Evaluation {title, decision} Letter {title, evaluation} PaperReview {title, score, body}
(b)	ReviewRequest {title, body} Notification {title, contact, authors, evaluation, scores, bodies}
	Review {reviewerName, title, score, body} Feedback {reviewerName, title, evaluation}

Fig. 2. Structures of data objects (a) and messages (b) of the paper reviewing example.

Concerning the content of data objects, the standard left underspecified its structure, in order to keep the notation independent from the kind of data structure required from time to time. We consider here a generic record structure, assuming that a data object is just a list of fields, characterised by a name and the corresponding value. Of course, a field can be used to represent the state of a data object. More complex XML-like structures, which are out of the scope of this work, can be anyway rendered resorting to nesting. The structure in terms of fields of the data objects used in our running example is specified in Fig. 2(a). Messages are structured as well; the structure of the messages specified in our example is shown in Fig. 2(b). Values can be manipulated and inserted into data object fields via assignments performed by tasks.

Guards, assignments, and structure of data objects and messages are not explicitly reported in the graphical representation of the BPMN model, but are defined as attributes of the involved BPMN elements. We provide information on their definition and functioning in Sect. 3, and show how MIDA users can specify them in Sect. 4.

3   A Formal Account of Multi-instance Collaborations

In this section we formalise the semantics of BPMN collaborations supporting multiple instances. We focus on those BPMN elements, informally presented in the previous section, that are strictly needed to deal with multiple instantiation of collaborations, namely multi-instance pools, message exchange events and tasks, and data objects; additionally, in order to define meaningful collaborations, we also consider some core BPMN elements, whose preliminary formalisation has been given in [15,16].

To simplify the formal treatment of the semantics, we resort to a textual representation of BPMN models, which is more manageable for writing operational rules than the graphical notation. Notice that we do not propose an alternative modelling notation, but we just define a Backus-Naur Form (BNF) syntax of BPMN model structures.

Textual Notation of BPMN Collaborations
We report in Fig. 3 the BNF syntax defining the textual notation of BPMN collaboration models. This syntax only describes the structure of models, without taking into account all those aspects that come into play to describe the model semantics, such as token distribution and messages. In the proposed grammar, the non-terminal symbols C, P and A represent *Collaboration Structures, Process*

$$C ::= \mathsf{pool}(\mathsf{p}, P) \quad | \quad \mathsf{miPool}(\mathsf{p}, P) \quad | \quad C_1 \parallel C_2$$

$$P ::= \mathsf{start}(\mathsf{e}_{enb}, \mathsf{e}_o) \mid \mathsf{startRcv}(\mathsf{m}\!:\!\tilde{\mathsf{t}}, \mathsf{e}_o) \mid \mathsf{end}(\mathsf{e}_i) \mid \mathsf{endSnd}(\mathsf{e}_i, \mathsf{m}\!:\!\tilde{\mathsf{exp}}) \mid \mathsf{terminate}(\mathsf{e}_i)$$

$$| \quad \mathsf{andSplit}(\mathsf{e}_i, E_o) \mid \mathsf{xorSplit}(\mathsf{e}_i, G) \mid \mathsf{andJoin}(E_i, \mathsf{e}_o) \mid \mathsf{xorJoin}(E_i, \mathsf{e}_o)$$

$$| \quad \mathsf{eventBased}(\mathsf{e}_i, (\mathsf{m}_1\!:\!\tilde{\mathsf{t}}_1, \mathsf{e}_{o1}), \ldots, (\mathsf{m}_h\!:\!\tilde{\mathsf{t}}_h, \mathsf{e}_{oh}))$$

$$| \quad \mathsf{task}(\mathsf{e}_i, \mathsf{exp}, A, \mathsf{e}_o) \mid \mathsf{taskRcv}(\mathsf{e}_i, \mathsf{exp}, A, \mathsf{m}\!:\!\tilde{\mathsf{t}}, \mathsf{e}_o) \mid \mathsf{taskSnd}(\mathsf{e}_i, \mathsf{exp}, A, \mathsf{m}\!:\!\tilde{\mathsf{exp}}, \mathsf{e}_o)$$

$$| \quad \mathsf{interRcv}(\mathsf{e}_i, \mathsf{m}\!:\!\tilde{\mathsf{t}}, \mathsf{e}_o) \mid \mathsf{interSnd}(\mathsf{e}_i, \mathsf{m}\!:\!\tilde{\mathsf{exp}}, \mathsf{e}_o) \mid P_1 \parallel P_2$$

$$A ::= \epsilon \quad | \quad \mathsf{d.f} ::= \mathsf{exp}, A$$

Fig. 3. BNF syntax of BPMN collaboration structures.

Structures and *Data Assignments*, respectively. The first two syntactic categories directly refer to the corresponding notions in BPMN, while the latter refers to list of assignments used to specify updating of data objects. The terminal symbols, denoted by the sans serif font, are the typical elements of a BPMN model, i.e. pools, events, tasks and gateways.

We do not provide a direct syntactic representation of *Data Objects*. The evolution of their state during the model execution is a semantic concern (described later in this section). Thus, syntactically, only the connections between data objects and the other elements are relevant. They are rendered by references to data objects within *expressions*, used to check when a task is ready to start (graphically, the task has a connection incoming from the data object), to update the values stored in a data object (graphically, the task has a connection outgoing to the data object), and to drive the decision of a XOR split gateway. A data object is structured as a list of fields; the field f of the data object named d is accessed via d.f.

Intuitively, a BPMN collaboration model is rendered in our syntax as a collection of pools, each one specifying a process. Formally, a collaboration C is a composition, by means of the $\parallel$ operator, of pools either of the form $\mathsf{pool}(\mathsf{p}, P)$ (for single-instance pools) or $\mathsf{miPool}(\mathsf{p}, P)$ (for multi-instance pools), where p is the name that uniquely identifies the pool, and P is the enclosed process. At process level, we use $\mathsf{e} \in \mathbb{E}$ to uniquely denote a *sequence edge*, while $E \in 2^{\mathbb{E}}$ a set of edges. For the convenience of the reader, we refer with e_i to the edge incoming in an element, with e_o to the outgoing edge, and with e_{enb} to the (spurious) edge denoting the enabled status of a start event.

In the data-based setting we consider, messages may carry values. Therefore, a sending action specifies a list of expressions whose evaluation will return a tuple of values to be sent, while a receiving action specifies a template to select matching messages and possibly assign values to data object fields. Formally, a *message* is a pair $\mathsf{m} : \tilde{\mathsf{v}}$, where $\mathsf{m} \in \mathbb{M}$ is the (unique) message name (i.e., the label of the message edge), and $\tilde{\mathsf{v}}$ is a tuple of values, with $\mathsf{v} \in \mathbb{V}$ and $\tilde{\ }$ denoting tuples (i.e., $\tilde{\mathsf{v}}$ stands for $\langle \mathsf{v}_1, \ldots, \mathsf{v}_n \rangle$). Sending actions have as argument a pair of the form $\mathsf{m} : \tilde{\mathsf{exp}}$. The precise syntax of *expressions* is deliberately not specified, it is just assumed that they contain, at least, values v and data object fields d.f. Receiving actions have as argument a pair of the form $\mathsf{m} : \tilde{\mathsf{t}}$, where $\tilde{\mathsf{t}}$ denotes a

---

Overall paper reviewing collaboration scenario:
$\text{pool}(p_{pc}, P_{pc}) \parallel \text{miPool}(p_r, P_r) \parallel \text{pool}(p_{ca}, P_{ca})$

Reviewer process :
$P_r = \text{startRcv}(\text{ReviewRequest}:\tilde{t}_2, e_{15}) \parallel \text{task}(e_{15}, \text{true}, A_6, e_{16}) \parallel$
$\quad\quad \text{taskSnd}(e_{16}, \exp_7, \epsilon, \text{Review}:\tilde{\exp}_8, e_{17}) \parallel$
$\quad\quad \text{taskRcv}(e_{17}, \text{true}, \epsilon, \text{Feedback}:\tilde{t}_3, e_{18}) \parallel \text{end}(e_{18})$

Templates, expressions, assignments :
$\quad \tilde{t}_2 = \langle ?\text{ReviewRequest.title}, ?\text{ReviewRequest.body} \rangle$
$\quad A_6 = \text{PaperReview.title} := \text{ReviewRequest.title},$
$\quad\quad\quad \text{PaperReview.score} := \text{assignscore}(\text{ReviewRequest.body}),$
$\quad\quad\quad \text{PaperReview.body} := \text{writeReview}(\text{ReviewRequest.body})$
$\quad \exp_7 = \text{PaperReview.score} \neq \text{null and PaperReview.body} \neq \text{null}$
$\quad \tilde{\exp}_8 = \langle \text{myName}(), \text{PaperReview.title}, \text{PaperReview.score}, \text{PaperReview.body} \rangle$
$\quad \tilde{t}_3 = \langle \text{myName}(), \text{ReviewRequest.title}, ?\text{Feedback.evaluation} \rangle$

Fig. 4. Textual representation of the running example (an excerpt).

template, that is a sequence of expressions and formal fields used as pattern to select messages received by the pool. Formal fields are data object fields identified by the ?-tag (e.g., ?d.f) and are used to bind fields to values. In order to store the received values and allow their reuse, we associate to each message in the receiving process a data object, whose name coincides with the message name. Data objects are associated to a task by means of a conditional expression, which is a guard enabling the task execution, and a list of *assignments* A, each of which assigns the value of an expression to a data field. When there is no data object as input to a task, the guard is simply true, while if there is no data object in output to a task the list of assignments is empty (ϵ).

The XOR split gateway specifies *guard conditions* in its outgoing edges, used to decide which edge to activate according to the values of data objects. This is formally rendered as a function $G : \mathbb{E} \rightarrow \mathbb{EXP}$ mapping edges to conditional expressions, where $\mathbb{EXP}$ is the set of all expressions that includes the distinguished expression default referring to the *default sequence edge* outgoing from the gateway (it is assigned to at most one edge). When convenient, we will deal with function G as a set of pairs $(e, \exp)$.

The correspondence between the syntax used here to represent multi-instance collaborations and the graphical notation of BPMN is exemplified by means of (an excerpt of) our running example in Fig. 4, while the detailed one-to-one correspondence is reported in the companion technical report [17]. In the textual notation, to support a compositional approach, each sequence (resp. message) edge in the graphical notation is split in two parts: the part outgoing from the source element and the part incoming into the target element; the two parts are correlated by the unique edge name. Notably, even if our syntax would allow to write collaborations that cannot be expressed in BPMN, we only consider those terms that are derived from BPMN models.

Semantics of BPMN Collaborations

The syntax presented so far represents the mere structure of processes and collaborations. To describe their semantics, we mark sequence edges by means of tokens [1, p. 27]. In particular, we enrich the structural information with a notion of execution state, defined by the state of each process instance (given by the marking of sequence edges and the values of data object fields) and the store of the exchanged messages. We call process configurations and collaboration configurations these stateful descriptions, which produce local and global effects, respectively, on the collaboration execution.

Formally, a *process configuration* has the form $\langle P, \sigma, \alpha \rangle$, where: P is a process structure; $\sigma : \mathbb{E} \to \mathbb{N}$ is a *sequence edge state function* specifying, for each sequence edge, the current number of tokens marking it ($\mathbb{N}$ is the set of natural numbers); and $\alpha : \mathbb{F} \to \mathbb{V}$ is the *data state function* assigning values (possibly null) to data object fields ($\mathbb{F}$ is the set of data fields and $\mathbb{V}$ the set of values). We denote by σ_0 (resp. α_0) the edge (resp. data) state where all edges are unmarked (resp. all fields are set to null). The state obtained by updating in σ the number of tokens of the edge e to n, written as $\sigma \cdot [e \mapsto n]$, is defined as follows: $(\sigma \cdot [e \mapsto n])(e')$ returns n if $e' = e$, otherwise it returns $\sigma(e')$. The update of data state α is similarly defined. To simplify the definition of the operational rules, we introduce some auxiliary functions to update states. Function $inc : \mathbb{S}_\sigma \times \mathbb{E} \to \mathbb{S}_\sigma$ (resp. $dec : \mathbb{S}_\sigma \times \mathbb{E} \to \mathbb{S}_\sigma$), where $\mathbb{S}_\sigma$ is the set of edge states, updates a state by incrementing (resp. decrementing) by one the number of tokens marking an edge in the state. They are defined as $inc(\sigma, e) = \sigma \cdot [e \mapsto \sigma(e) + 1]$ and $dec(\sigma, e) = \sigma \cdot [e \mapsto \sigma(e) - 1]$. These functions extend in a natural ways to sets E of edges. Function $reset : \mathbb{S}_\sigma \times \mathbb{E} \to \mathbb{S}_\sigma$, instead, updates an edge state by setting to zero the number of tokens marking an edge in the state: $reset(\sigma, e) = \sigma \cdot [e \mapsto 0]$. We use the *evaluation* relation $eval \subseteq \mathbb{EXP} \times \mathbb{S}_\alpha \times \mathbb{V}$ to evaluate an expression over a data state. This is a relation, not a function, because an expression may contain non-deterministic operators and, in such a case, its evaluation results in one of the possible values for that expression with respect to the given data state. Notation $eval(\mathsf{exp}, \alpha, \mathsf{v})$ states that v is one of the possible values resulting from the evaluation of the expression exp on the data state α. This relation is not explicitly defined, since the syntax of expressions is deliberately not specified; we only assume that $eval(\mathsf{default}, \alpha, \mathsf{v})$ implies $\mathsf{v} = false$ for any α. The relation extends to tuples component-wise. Finally, relation $upd \subseteq \mathbb{S}_\alpha \times \mathbb{A}^n \times \mathbb{S}_\alpha$, where $\mathbb{S}_\alpha$ is the set of data states and $\mathbb{A}$ is the set of assignments, is used to update data object values. Notation $upd(\alpha, A, \alpha')$ states that α' is one of the possible states resulting from the update of α with assignment A. The relation is inductively defined as follows: for any α, $upd(\alpha, \epsilon, \alpha)$; $upd(\alpha, \mathsf{d.f} := \mathsf{exp}, \alpha \cdot [\mathsf{d.f} \mapsto \mathsf{v}])$ with v such that $eval(\mathsf{exp}, \alpha, \mathsf{v})$; and $upd(\alpha, (A_1, A_2), \alpha'')$ with α'' such that $upd(\alpha', A_2, \alpha'')$ and α' such that $upd(\alpha, A_1, \alpha')$.

A *collaboration configuration* has the form $\langle C, \iota, \delta \rangle$, where: C is a collaboration structure; $\iota : \mathbb{P} \to 2^{\mathbb{S}_\sigma \times \mathbb{S}_\alpha}$ is the *instance state function* mapping each pool name ($\mathbb{P}$ is the set of pool names) to a multiset of instance states (ranged over by I and containing pairs of the form $\langle \sigma, \alpha \rangle$); and $\delta : \mathbb{M} \to 2^{\mathbb{V}^n}$ is a *message state function* specifying, for each message name m, a multiset of value tuples representing the messages received along the message edge labelled by m. Function δ can be updated in a way similar to σ, enabling the definition of the following auxiliary functions. Function $add : \mathbb{S}_\delta \times \mathbb{M} \times \mathbb{V}^n \to \mathbb{S}_\delta$ (resp. $rm : \mathbb{S}_\delta \times \mathbb{M} \times \mathbb{V}^n \to \mathbb{S}_\delta$), where $\mathbb{S}_\delta$ is the set of message states, allows updating a message state by adding (resp. removing) a value tuple for a given message name in the state: $add(\delta, \mathsf{m}, \tilde{v}) = \delta \cdot [\mathsf{m} \mapsto \delta(\mathsf{m}) + \{\tilde{v}\}]$ and $rm(\delta, \mathsf{m}, \tilde{v}) = \delta \cdot [\mathsf{m} \mapsto \delta(\mathsf{m}) - \{\tilde{v}\}]$, where $+$ and $-$ are the union and substraction operations on multisets. Finally, the instance state function ι can be updated in two ways: by adding a newly created instance or by modifying an existing one: $newI(\iota, \mathsf{p}, \sigma, \alpha) = \iota \cdot [\mathsf{p} \mapsto \iota(\mathsf{p}) + \{\langle \sigma, \alpha \rangle\}]$ and $updI(\iota, \mathsf{p}, I) = \iota \cdot [\mathsf{p} \mapsto I]$.

Let us go back to our running example. The scenario in its initial state is rendered as the collaboration configuration $\langle (\mathsf{pool}(\mathsf{p}_{pc}, P_{pc}) \parallel \mathsf{miPool}(\mathsf{p}_r, P_r) \parallel \mathsf{pool}(\mathsf{p}_{ca}, P_{ca})), \iota, \delta \rangle$ where: $\iota(\mathsf{p}_{pc}) = \{\langle \sigma, \alpha \rangle\}$ with $\sigma = \sigma_0 \cdot [e_{enb} \mapsto 1]$ and $\alpha = \alpha_0 \cdot [\mathsf{Paper.title}, \ldots, \mathsf{Paper.body} \mapsto title, \ldots, text]$; and $\iota(\mathsf{p}_r) = \iota(\mathsf{p}_{ca}) = \varnothing$. The α function of the p_{pc} instance is initialised with the content of the *Paper* data input.

The operational semantics is defined by means of a *labelled transition system* (LTS), whose definition relies on an auxiliary LTS on the behaviour of processes. The latter is a triple $\langle \mathcal{P}, \mathcal{L}, \to \rangle$ where: $\mathcal{P}$, ranged over by $\langle P, \sigma, \alpha \rangle$, is a set of process configurations; $\mathcal{L}$, ranged over by ℓ, is a set of *labels*; and $\to \subseteq \mathcal{P} \times \mathcal{L} \times \mathcal{P}$ is a *transition relation*. We will write $\langle P, \sigma, \alpha \rangle \xrightarrow{\ell} \langle P, \sigma', \alpha' \rangle$ to indicate that $(\langle P, \sigma, \alpha \rangle, \ell, \langle P, \sigma', \alpha' \rangle) \in \to$. Since process execution only affects the current states, and not the process structure, for the sake of readability we omit the structure from the target configuration of the transition. Similarly, to further improve readability, we also omit α when it is not affected by the transition. Thus, for example, a transition $\langle P, \sigma, \alpha \rangle \xrightarrow{\ell} \langle P, \sigma', \alpha \rangle$ can be written as $\langle P, \sigma, \alpha \rangle \xrightarrow{\ell} \sigma'$. The labels ℓ used by the process transition relation have the following meaning. Label τ denotes an action internal to the process, while $!\mathsf{m}:\tilde{v}$ and $?\mathsf{m}:\tilde{e}t, A$ denote sending and receiving actions, respectively. Notation $\tilde{e}t$ denotes an evaluated template, that is a sequence of values and formal fields. Notably, the receiving label carries information about the data assignments A to be executed, at collaboration level, after the message m is actually received. Label $new\,\mathsf{m}:\tilde{e}t$ denotes taking place of a receiving action that instantiates a new process instance (i.e., it corresponds to the occurrence of a start message event in a multi-instance pool). The meaning of internal actions is as follows: ϵ denotes an internal computation concerning the movement of tokens, while $kill$ denotes taking place of the termination event.

An excerpt of the operational rules defining the transition relation of the processes semantics is given in Fig. 5 (we present here the rules for the BPMN elements used in our running example; we refer to [17] for a complete account). Rule *P-Start* starts the execution of a (single-instance) process when it has been activated (i.e., the enabling edge e_{enb} is marked). The effect of the rule is to increment the number of tokens in the edge outgoing from the start event

$$\langle\mathsf{start}(e_{enb}, e_o), \sigma, \alpha\rangle \xrightarrow{\epsilon} inc(reset(\sigma, e_{enb}), e_o) \quad \sigma(e_{enb}) > 0 \qquad (P\text{-}Start)$$

$$\langle\mathsf{end}(e_i), \sigma, \alpha\rangle \xrightarrow{\epsilon} dec(\sigma, e_i) \quad \sigma(e_i) > 0 \qquad (P\text{-}End)$$

$$\langle\mathsf{startRcv}(m:\tilde{t}, e_o), \sigma, \alpha\rangle \xrightarrow{new\ m:\tilde{e}t} inc(\sigma, e_o) \quad eval(\tilde{t}, \alpha, \tilde{e}t) \qquad (P\text{-}StartRcv)$$

$$\langle\mathsf{xorSplit}(e_i, \{(e, exp)\} \cup G), \sigma, \alpha\rangle \xrightarrow{\epsilon} inc(dec(\sigma, e_i), e) \quad \begin{array}{l} \sigma(e_i) > 0, \\ eval(exp, \alpha, true) \end{array} \quad (P\text{-}XorSplit_1)$$

$$\langle\mathsf{xorSplit}(e_i, \{(e, default)\} \cup G), \sigma, \alpha\rangle \xrightarrow{\epsilon} inc(dec(\sigma, e_i), e) \quad \begin{array}{l} \sigma(e_i) > 0, \\ \forall (e_j, exp_j) \in G\ . \\ eval(exp_j, \alpha, false) \end{array} \quad (P\text{-}XorSplit_2)$$

$$\langle\mathsf{xorJoin}(\{e\} \cup E_i, e_o), \sigma, \alpha\rangle \xrightarrow{\epsilon} inc(dec(\sigma, e), e_o) \quad \sigma(e) > 0 \qquad (P\text{-}XorJoin)$$

$$\langle\mathsf{task}(e_i, exp, A, e_o), \sigma, \alpha\rangle \xrightarrow{\epsilon} \langle inc(dec(\sigma, e_i), e_o), \alpha'\rangle \quad \begin{array}{l} \sigma(e_i) > 0, \\ eval(exp, \alpha, true), \\ upd(\alpha, A, \alpha') \end{array} \quad (P\text{-}Task)$$

$$\langle\mathsf{taskRcv}(e_i, exp, A, m:\tilde{t}, e_o), \sigma, \alpha\rangle \xrightarrow{?m:\tilde{e}t, A} inc(dec(\sigma, e_i), e_o) \quad \begin{array}{l} \sigma(e_i) > 0, \\ eval(exp, \alpha, true), \\ eval(\tilde{t}, \alpha, \tilde{e}t) \end{array} \quad (P\text{-}TaskRcv)$$

$$\langle\mathsf{taskSnd}(e_i, exp', A, m:\tilde{exp}, e_o), \sigma, \alpha\rangle \xrightarrow{!m:\tilde{v}} \langle inc(dec(\sigma, e_i), e_o), \alpha'\rangle \quad \begin{array}{l} \sigma(e_i) > 0, \\ eval(exp', \alpha, true), \\ upd(\alpha, A, \alpha'), \\ eval(\tilde{exp}, \alpha, \tilde{v}) \end{array} \quad (P\text{-}TaskSnd)$$

$$\frac{\langle P_1, \sigma, \alpha\rangle \xrightarrow{\ell} \langle\sigma', \alpha'\rangle}{\langle P_1 \parallel P_2, \sigma, \alpha\rangle \xrightarrow{\ell} \langle\sigma', \alpha'\rangle} \quad \ell \neq kill \qquad (P\text{-}Int_1)$$

Fig. 5. BPMN process semantics.

and to reset the marking of the enabling edge. Rule *P-End* instead is enabled when there is at least one token in the incoming edge of the end event, which is then simply consumed. Rule *P-StartRcv* starts the execution of a process by producing a label denoting the creation of a new instance and containing the information for consuming a received message at the collaboration layer (see rule *C-CreateMi* in Fig. 6). Rule *P-XorSplit_1* is applied when a token is available in the incoming edge of a XOR split gateway and a conditional expression of one of its outgoing edges is evaluated to *true*; the rule decrements the token in the incoming edge and increments the token in the selected outgoing edge. Notably, if more edges have their guards satisfied, one of them is non-deterministically chosen. Rule *P-XorSplit_2* is applied when all guard expressions are evaluated to *false*; in this case the default edge is marked. Rule *P-XorJoin* is activated every time there is a token in one of the incoming edges, which is then moved to the outgoing edge. Rule *P-Task* deals with tasks, possibly equipped with data objects. It is activated only when the guard expression is satisfied and there is a token in the incoming edge, which is then moved to the outgoing edge. The rule also updates the values of the data objects connected in output to the task. Rule *P-TaskRcv* is similar, but it produces a label corresponding to the consumption of a message. In this case, however, the data updates are not executed, because they must be done only after the message is actually received; therefore, the assignment are passed by means of the label to the collaboration layer. Rule *P-TaskSnd* sends a message, updates the data object and moves the incoming token to the outgoing edge. The produced send label is used to deliver the message at the collaboration layer. We consider tasks with atomic execution; relaxation of this requirement is shown in [17]. Finally, rule *P-Int$_1$* deals with interleaving in a standard way for process elements.

Now, the labelled transition relation on collaboration configurations formalises the message exchange and the data update according to the process evolution. The LTS is a triple $\langle \mathcal{C}, \mathcal{L}_c, \rightarrow_c \rangle$ where: $\mathcal{C}$, ranged over by $\langle C, \iota, \delta \rangle$, is a set of collaboration configurations; $\mathcal{L}_c$, ranged over by l, is a set of *labels*; and $\rightarrow_c \subseteq \mathcal{C} \times \mathcal{L}_c \times \mathcal{C}$ is a *transition relation*. We apply the same readability simplifications we use for process configuration transitions. Labels l are as follows: τ is an internal action, $!m:\tilde{v}$ is a sending action, and $?m:\tilde{v}$ and *new* $m:\tilde{v}$ are receiving actions. Notably, at collaboration level the receiving labels just keep track of the received message. To define the collaboration semantics, an auxiliary function is needed: $match(\tilde{et}, \tilde{v})$ is a partial function performing *pattern-matching* on structured data (like in [18]), thus determining if an evaluated template $\tilde{et}$ matches a tuple of values $\tilde{v}$. A successful matching returns a list of assignments A, updating the formal fields in the template; otherwise, the function is undefined.

The relevant operational rules defining the transition relation of the collaboration semantics are given in Fig. 6 (the full account is in [17]). Rule *C-CreateMi* deals with instance creation in the multi-instance case. An instance is created if there is a matching message; as result, the assignments for the received data are performed, and the message is consumed. The created instance is added to the multiset of existing instances of the pool. The (omitted) single-instance case is similar, except that the instance is created only if no instance exists for the considered pool ($\iota(\mathsf{p}) = \varnothing$). The next three rules allow a single pool to evolve according to the evolution of one of its process instances $\langle P, \sigma, \alpha \rangle$. In particular, if the process instance performs an internal action (rule *C-InternalMi*) or a receiving/delivery action (rules *C-ReceiveMi* or *C-DeliverMi*), the pool performs the corresponding action at collaboration layer. As for instance creation, rule *C-ReceiveMi* can be applied only if there is at least one matching message. Recall indeed that at process level the receiving labels just indicate the willingness of a process instance to consume a received message, regardless the actual presence of messages. The delivering of messages is based on the *correlation* mechanism: the correlation data are identified by the template fields that are not formal (i.e., those fields requiring specific matching values). Moreover, when a process performs a sending action, the message state function is updated in order to deliver the sent message to the receiving participant. Finally, rule *C-Int$_1$* permits interleaving the processes execution.

It is worth noticing that the semantics has been defined according to a global perspective. Indeed, the overall state of a collaboration is collected by functions ι

and δ of its configuration. On the other hand, the global semantics of a collaboration configuration is determined, in a compositional way, by the local semantics of the involved processes, which evolve independently from each other. The use of a global perspective simplifies *(i)* the technicalities required by the formal definition of the semantics, and *(ii)* the implementation of the animation of the overall collaboration execution. The compositional definition of the semantics, anyway, would allow to easily pass to a purely local perspective, where state functions are kept separate for each process.

4   The MIDA Animation Tool

In this section, we present our BPMN animator tool MIDA (*Multiple Instances and Data Animator*) and provide details about its implementation and use. MIDA is based on the Camunda *bpmn.io* web modeller. More precisely, we have integrated our formal framework into the *bpmn.io* token simulation plug-in. The MIDA tool, as well as its source code, user guide and examples, are freely available from http://pros.unicam.it/mida/.

MIDA is a web application written in JavaScript. Its graphical interface, shown in Fig. 7, is conceived as a modelling environment. It allows users to create BPMN models using all the facilities of the Camunda modeller. In particular, data/message structures, guards and assignments can be specified by using the *Property Panel*, which permits accessing element attributes. This information is stored in appropriate elements of the standard XML representation of the BPMN model. When the animation mode is activated, by clicking the corresponding button, one or more instances of the desired processes can be fired. To do this,

$$\frac{\langle P,\sigma_0,\alpha_0\rangle \xrightarrow{new\,m:\tilde{e}\tilde{t}} \langle\sigma',\alpha'\rangle \quad \tilde{v}\in\delta(m) \quad match(\tilde{e}\tilde{t},\tilde{v})=A \quad upd(\alpha',A,\alpha'')}{\langle\mathsf{miPool}(\mathsf{p},P),\iota,\delta\rangle \xrightarrow{new\,m:\tilde{v}} \langle newI(\iota,\mathsf{p},\sigma',\alpha''),rm(\delta,m,\tilde{v})\rangle} \ (C\text{-}CreateMi)$$

$$\frac{\iota(\mathsf{p})=\{\langle\sigma,\alpha\rangle\}+I \quad \langle P,\sigma,\alpha\rangle \xrightarrow{\tau} \langle\sigma',\alpha'\rangle}{\langle\mathsf{miPool}(\mathsf{p},P),\iota,\delta\rangle \xrightarrow{\tau} \langle updI(\iota,\mathsf{p},\{\langle\sigma',\alpha'\rangle\}+I),\delta\rangle} \ (C\text{-}InternalMi)$$

$$\frac{\iota(\mathsf{p})=\{\langle\sigma,\alpha\rangle\}+I \quad \langle P,\sigma,\alpha\rangle \xrightarrow{?m:\tilde{e}\tilde{t},A} \langle\sigma',\alpha'\rangle}{\tilde{v}\in\delta(m) \quad match(\tilde{e}\tilde{t},\tilde{v})=A' \quad upd(\alpha',(A',A),\alpha'')} \ (C\text{-}ReceiveMi)$$
$$\overline{\langle\mathsf{miPool}(\mathsf{p},P),\iota,\delta\rangle \xrightarrow{?m:\tilde{v}} \langle updI(\iota,\mathsf{p},\{\langle\sigma',\alpha''\rangle\}+I),rm(\delta,m,\tilde{v})\rangle}$$

$$\frac{\iota(\mathsf{p})=\{\langle\sigma,\alpha\rangle\}+I \quad \langle P,\sigma,\alpha\rangle \xrightarrow{!m:\tilde{v}} \langle\sigma',\alpha'\rangle}{\langle\mathsf{miPool}(\mathsf{p},P),\iota,\delta\rangle \xrightarrow{!m:\tilde{v}} \langle updI(\iota,\mathsf{p},\{\langle\sigma',\alpha'\rangle\}+I),add(\delta,m,\tilde{v})\rangle} \ (C\text{-}DeliverMi)$$

$$\frac{\langle C_1,\iota,\delta\rangle \xrightarrow{l} \langle\iota',\delta'\rangle}{\langle C_1\parallel C_2,\iota,\delta\rangle \xrightarrow{l} \langle\iota',\delta'\rangle} \ (C\text{-}Int_1)$$

Fig. 6. BPMN collaboration semantics.

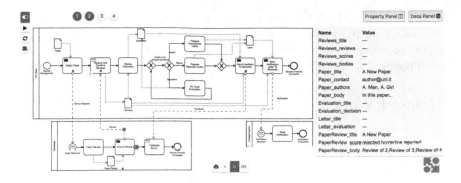

Fig. 7. MIDA web interface.

users have to press the *play* button depicted over each fireable start event. This creates a new token labelled with a number uniquely representing a process instance. Tokens will cross the model following the rules induced by our formal semantics. The execution of a process instance terminates once all its tokens cannot move forward. We refer to the MIDA's user guide for more details on the practical use of the tool.

MIDA animation features may be an effective support to business process designers in their modelling activities, especially when multi-instance collaborations are involved. Indeed, in this context, the choice of correlation data is an error-prone task that is a burden on the shoulders of the designers. For example, let us consider the *Reviewer* participant in our running scenario; if the template within the task for receiving the feedback would not properly specify the correlation data (e.g., $\tilde{t}_3 = \langle$?Feedback.reviewerName, ?Feedback.title, ?Feedback.evaluation$\rangle$), the feedback messages could not be properly delivered. Indeed, each *Reviewer* instance would be able to match any feedback message, regardless the reviewer name and the paper title specified in the message. Thus, the feedback messages could be mixed up. Fortunately, MIDA allows to detect, and hence solve, this correlation issue. Similarly, MIDA helps designers to detect issues concerning the exchange of messages. In fact, malformed or unexpected messages may introduce deadlocks in the execution flow, which can be easily identified by looking for blocked tokens in the animation. For instance, in the running example a feedback message without the evaluation field would be never consumed by a receiving task of the *Reviewer* instances. Finally, since our animation is based on data object values, also issues due to bad data handling can be detected using MIDA. For instance, let us suppose that the *Discuss* task in *PC Chair* would not be in a loop, but it would have its outgoing edge directly connected to the XOR join in its right hand side. After the execution of the *Discuss* task, the task *Send Feedback* would be performed, and the task *Send Results* would be activated. However, the guard of the latter task would not be satisfied, because the *Letter* data object would not be properly instantiated. This would cause a deadlock, which can be found out by using MIDA.

To sum up, the MIDA tool can support designers in debugging their multi-instance collaboration models, as it permits to check the evolution of data, messages and processes marking while executing the models step-by-step. Like in code debugging, the identification of the bug is still in charge of the human user.

5   Related Work

In this section we discuss the most relevant attempts in formalising multiple instances and data for BPMN models. We then compare MIDA with other animation tools.

On Formalising Multiple Instances and Data. Many works in the literature attempted to formalise the core features of BPMN. However, most of them (see, e.g., [3–7]) do not consider multiple instances and data, which are the focus of our work. Considering these features in BPMN collaborations, relevant works are [19–22]. Meyer et al. in [19] focus on process models where data objects are shared entities and the correlation mechanism is used to distinguish and refer data object instances. Use of data objects local to (multiple) instances, exchange of messages between participants, and correlation of messages are instead our focus. In [20], the authors describe a model-driven approach for BPMN to include the data perspective. Differently from us, they do not provide a formal semantics for BPMN multiple instances. Moreover, they do not use data in decision gateways. Moreover, Kheldoun et al. propose in [21] a formal semantics of BPMN covering features such as message-exchange, cancellation, multiple instantiation of sub-processes and exception handling, while taking into account data flow aspects. However, they do not consider multi-instance pools and do not address the correlation issue. Semantics of data objects and their use in decision gateways is instead proposed by El-Saber and Boronat in [22]. Differently from us, this formal treatment does not include collaborations and, hence, exchange of messages and multiple instances. Considering other modelling languages, YAWL [23] and high-level Petri nets [24] provide direct support for the multiple instance patterns. However, they lack support for handling data. In both cases, process instances are characterised by their identities, rather than by the values of their data, which are however necessary to correlate messages to running instances.

Regarding choreographies, relevant works are [25–27]. Gómez-López et al. [25] study the choreography problem derived from the synchronisation of multiple instances necessary for the management of data dependencies. Knuplesch et al. [26] introduces a data-aware collaboration approach including formal correctness criteria. However, they define the data perspective using data-aware interaction nets, a proprietary notation, instead of the wider accepted BPMN. Improving data-awareness and data-related capabilities for choreographies is the goal of Hahn et al. [27]. They propose a way to unify the data flow across participants with the data flow inside a participant. The scope of data objects is global to the overall choreography, while we consider data objects with scope local to participant instances, as prescribed by the BPMN standard. Apart from the specific differences mentioned above, our work differs from the others for the focus on collaboration diagrams, rather than on choreographies. This allows us to specifically deal with multiple process instantiation and messages correlation.

Finally, concerning the correlation mechanism, the BPMN standard and, hence, our work have been mainly inspired by works in the area of service-oriented computing (see the relationship between BPMN and WS-BPEL [28] in [1, Sect. 14.1.2]). In fact, when a service engages in multiple interactions, it is generally required to create an instance to concurrently serve each request, and correlate subsequent incoming messages to the created instances. Among the others, the COWS [18] formalism captures the basic aspects of SOC systems, and in particular service instantiation and message correlation à la WS-BPEL. From the formal point of view, correlation is realised by means of a pattern-matching function similar to that used in our formal semantics.

Business Process Animation. Relevant contributions about animation of business processes are proposed by Allweyer and Schweitzer [12], and by Signavio and Visual Paradigm. Differently from us, in their implementations they do not fully support the interplay between multiple instances, messages and data. Allweyer and Schweitzer propose a tool for animating BPMN models that, however, only considers processes, as it discards message exchanges, both semantically and graphically. In addition, gateway decisions are performed manually by users during the animation, instead of depending on data. The animator of the Signavio modeller allows users to step through the process element-by-element and to focus completely on the process flow. However, it discards important elements, such as message flows and data objects. Hence, Signavio animates only non-collaborative processes, without data-driven decisions, which instead are key features of our approach. Finally, Visual Paradigm provides an animator that supports also collaboration diagrams. This tool allows users to visualise the flow of messages and implements the semantics of receiving tasks and events, but it does not animate data evolution and multiple instances.

6   Concluding Remarks

This paper aims at answering the following research questions:

RQ1: What is the precise semantics of multi-instance BPMN collaborations?
RQ2: Can supporting tools assist designers to spot erroneous behaviours related to multiple instantiation and data handling in BPMN collaborations?

The answer to RQ1 is mainly given in Sect. 3, where we provide a novel operational semantics clarifying the interplay between control features, data, message exchanges and multiple instances. The answer to RQ2 is instead given in Sect. 4, where we propose MIDA, an animator tool, based on our formal semantics, that provides the visualisation of the behaviour of a collaboration by taking into account the data-based correlation of messages to process instances. We have shown, on our running example, that MIDA supports the identification of erroneous interactions, due e.g. to incorrect data handling or wrong message correlation.

We conclude the paper by discussing lessons learned, and the assumptions and limitations of our approach, also touching upon directions for future work.

Lessons Learned. The BPMN standard has the flavour of a framework rather than of a concrete language, because some aspects are not covered by it, but left to the designer [13]. For example, the standard left underspecified the internal structure of data objects: *"Data Object elements can optionally reference a DataState element [...] The definition of these states, e.g., possible values and any specific semantics are out of scope of this specification"* [1, p. 206]. This gap left by the BPMN standard must be filled in order to concretely deal with data in our formalisation, and hence in the animation of BPMN collaboration models. To this aim, we consider a generic record structure for data objects. Similarly, the expression language operating on data is left unspecified by the standard. This is not an issue for the formalisation, but the expression language has to be instantiated in the concrete implementation of the animator. In MIDA, for the sake of simplicity, we resort to the expression language of JavaScript, as this is the programming language used for implementing the tool. It conveniently allows, for example, to define expression operators that randomly select a value from a given set, which are used to define non-deterministic behaviours in our running example (see, e.g., operator assignscore() used by the *Prepare Review* task).

In addition, the lack of a formal semantics in the standard may lead to different interpretations of the tricky features of BPMN. In this work we aim at clarifying the interplay between multiple instances, messages and data objects. In particular, the standard provides an informal description of the mechanism used to correlate messages and process instances [1, p. 74], which we have formalised and implemented by following the solution adopted by the standard for executable business processes [28].

Assumptions and Limitations. Our formal semantics focusses on the communication mechanisms of collaborative systems, where multiple participants cooperate and share information. Thus, we have left out those features of BPMN whose formal treatment is orthogonal to the addressed problem, such as timed events and error handling. To keep our formalisation more manageable, multi-instance parallel tasks, sub-processes and data stores are left out too, despite they can be relevant for multi-instance collaborations. We discuss below what would be the impact of their addition to our work.

Let us first consider multi-instance tasks. The sequential instances case, as shown in the formalisation of our running example, can be simply dealt with as a macro; indeed, it corresponds to a task enclosed within a 'for' loop. The parallel case, instead, is more tricky. It is a common practice to consider it as a macro as well, which can be replaced by tasks between AND split and join gateways [3,23], assuming to know at design time the number of instances to be generated. However, this replacement is no longer admissible when this kind of element is used within multi-instance pools [17], thus requiring a direct definition of the formal semantics of multi-instance parallel tasks.

Similar reasoning can be done for sub-processes, which again are not mere macros. In fact, in general, simply flattening a process by replacing its sub-process elements by their expanded processes results in a model with different behaviour. This because a sub-process, for example, delimits the scope of the enclosed data objects and confines the effect of termination events. Therefore, it would be necessary to explicitly deal with the resulting multi-layer perspective, which adds complexity to the formal treatment. The formalisation would become even more complex if we consider multi-instance sub-processes, which would require an extension of the correlation mechanism.

Moreover, we do not consider BPMN data stores, used to memorise shared information that will persist beyond process instance completion. Providing a formalisation for data stores would require to extend collaboration configurations with a further state function, dedicated to data stores. Moreover, the treatment of data assignments would become more intricate, as it would be necessary to distinguish updates of data objects from those of data stores, which affect different data state functions in the configuration.

Finally, values of data objects can be somehow "constrained" by assignments. Indeed, as mentioned above in the *Lessons learned* paragraph, assignment expressions can restrict the set of possible values that can be assigned to a data object field. Moreover, guard expressions of tasks or XOR split gateways can check if data object values respect given conditions. However, such constraints imposed on data object values are currently "hidden" in the expressions and, hence, in their evaluation. Assignments could be extended with an explicit definition of constraints in order to ease their specification and make more evident the effects of assignments on data values.

Future Work. We plan to continue our programme to effectively support modelling and animation of BPMN multi-instance collaborations, by overcoming the above limitations. More practically, we intend to enlarge the range of functionalities provided by MIDA, especially for what concerns the data perspective, and improve its usability. Moreover, we plan to exploit the formal semantics, and its implementation, to enable the verification of properties using, e.g., model checking techniques.

References

1. OMG: Business Process Model and Notation (BPMN V 2.0) (2011)
2. Suchenia, A., Potempa, T., Ligęza, A., Jobczyk, K., Kluza, K.: Selected approaches towards taxonomy of business process anomalies. In: Pełech-Pilichowski, T., Mach-Król, M., Olszak, C.M. (eds.) Advances in Business ICT: New Ideas from Ongoing Research. SCI, vol. 658, pp. 65–85. Springer, Cham (2017). https://doi.org/10. 1007/978-3-319-47208-9_5
3. Dijkman, R.M., Dumas, M., Ouyang, C.: Semantics and analysis of business process models in BPMN. Inf. Softw. Technol. **50**(12), 1281–1294 (2008)

4. Decker, G., Dijkman, R., Dumas, M., García-Bañuelos, L.: Transforming BPMN diagrams into YAWL nets. In: Dumas, M., Reichert, M., Shan, M.-C. (eds.) BPM 2008. LNCS, vol. 5240, pp. 386–389. Springer, Heidelberg (2008). https://doi.org/10.1007/978-3-540-85758-7_30

5. Wong, P.Y.H., Gibbons, J.: A process semantics for BPMN. In: Liu, S., Maibaum, T., Araki, K. (eds.) ICFEM 2008. LNCS, vol. 5256, pp. 355–374. Springer, Heidelberg (2008). https://doi.org/10.1007/978-3-540-88194-0_22

6. Börger, E., Thalheim, B.: A method for verifiable and validatable business process modeling. In: Börger, E., Cisternino, A. (eds.) Advances in Software Engineering. LNCS, vol. 5316, pp. 59–115. Springer, Heidelberg (2008). https://doi.org/10.1007/978-3-540-89762-0_3

7. Van Gorp, P., Dijkman, R.: A visual token-based formalization of BPMN 2.0 based on in-place transformations. Inf. Softw. Technol. 55(2), 365–394 (2013)

8. Hermann, A., et al.: Collaborative business process management - a literature-based analysis of methods for supporting model understandability. In: WI, pp. 286–300 (2017)

9. Becker, J., Kugeler, M., Rosemann, M.: Process Management: A Guide for the Design of Business Processes. Springer, Heidelberg (2013). https://doi.org/10.1007/978-3-540-24798-2

10. Emens, R., Vanderfeesten, I., Reijers, H.A.: The dynamic visualization of business process models: a prototype and evaluation. In: Reichert, M., Reijers, H.A. (eds.) BPM 2015. LNBIP, vol. 256, pp. 559–570. Springer, Cham (2016). https://doi.org/10.1007/978-3-319-42887-1_45

11. Momotko, M., Nowicki, B.: Visualisation of (distributed) process execution based on extended BPMN. In: DEXA, pp. 280–284. IEEE (2003)

12. Allweyer, T., Schweitzer, S.: A tool for animating BPMN token flow. In: Mendling, J., Weidlich, M. (eds.) BPMN 2012. LNBIP, vol. 125, pp. 98–106. Springer, Heidelberg (2012). https://doi.org/10.1007/978-3-642-33155-8_8

13. Weske, M.: Business Process Management. Springer, Heidelberg (2007). https://doi.org/10.1007/978-3-540-73522-9

14. Corradini, F., Fornari, F., Muzi, C., Polini, A., Re, B., Tiezzi, F.: On avoiding erroneous synchronization in BPMN processes. In: Abramowicz, W. (ed.) BIS 2017. LNBIP, vol. 288, pp. 106–119. Springer, Cham (2017). https://doi.org/10.1007/978-3-319-59336-4_8

15. Corradini, F., Polini, A., Re, B., Tiezzi, F.: An operational semantics of BPMN collaboration. In: Braga, C., Ölveczky, P.C. (eds.) FACS 2015. LNCS, vol. 9539, pp. 161–180. Springer, Cham (2016). https://doi.org/10.1007/978-3-319-28934-2_9

16. Corradini, F., Muzi, C., Re, B., Rossi, L., Tiezzi, F.: Global vs. local semantics of BPMN 2.0 Or-Join. In: Tjoa, A.M., Bellatreche, L., Biffl, S., van Leeuwen, J., Wiedermann, J. (eds.) SOFSEM 2018. LNCS, vol. 10706, pp. 321–336. Springer, Cham (2018). https://doi.org/10.1007/978-3-319-73117-9_23

17. Corradini, F., Muzi, C., Re, B., Rossi, L., Tiezzi, F.: Animating multiple instances in BPMN collaborations. Technical report, University of Camerino (2018). http://pros.unicam.it/mida/

18. Pugliese, R., Tiezzi, F.: A calculus for orchestration of web services. J. Appl. Log. 10(1), 2–31 (2012)

19. Meyer, A., Pufahl, L., Fahland, D., Weske, M.: Modeling and enacting complex data dependencies in business processes. In: Daniel, F., Wang, J., Weber, B. (eds.) BPM 2013. LNCS, vol. 8094, pp. 171–186. Springer, Heidelberg (2013). https://doi.org/10.1007/978-3-642-40176-3_14

20. Meyer, A., et al.: Data perspective in process choreographies: modeling and execution. In: Techn. Ber. BPM Center Report BPM-13-29 (2013). BPMcenter.org
21. Kheldoun, A., Barkaoui, K., Ioualalen, M.: Formal verification of complex business processes based on high-level Petri nets. Inf. Sci. **385–386**, 39–54 (2017)
22. El-Saber, N.A.: CMMI-CM compliance checking of formal BPMN models using Maude. Ph.D. thesis, Department of Computer Science (2015)
23. Wohed, P., et al.: Pattern-based analysis of UML activity diagrams. Beta, Research School for Operations Management and Logistics, Eindhoven (2004)
24. Van Der Aalst, W.M., Ter Hofstede, A.H.: YAWL: yet another workflow language. Inf. Syst. **30**(4), 245–275 (2005)
25. Gómez-López, M.T., Pérez-Álvarez, J.M., Varela-Vaca, A.J., Gasca, R.M.: Guiding the creation of choreographed processes with multiple instances based on data models. In: Dumas, M., Fantinato, M. (eds.) BPM 2016. LNBIP, vol. 281, pp. 239–251. Springer, Cham (2017). https://doi.org/10.1007/978-3-319-58457-7_18
26. Knuplesch, D., et al.: Data-aware interaction in distributed and collaborative workflows: modeling, semantics, correctness. In: CollaborateCom, pp. 223–232. IEEE (2012)
27. Hahn, M., Breitenbücher, U., Kopp, O., Leymann, F.: Modeling and execution of data-aware choreographies: an overview. Comput. Sci.-Res. Dev. **33**, 1–12 (2017)
28. OASIS WS BPEL TC: Web Services Business Process Execution Language Version 2.0. Technical report, OASIS, April 2007

Managing Decision Tasks and Events in Time-Aware Business Process Models

Roberto Posenato[(✉)], Francesca Zerbato, and Carlo Combi

Dipartimento di Informatica, Università degli Studi di Verona, Verona, Italy
roberto.posenato@univr.it

Abstract. Time-aware business process models capture processes where temporal properties and constraints have to be suitably managed to achieve proper completion. Temporal aspects also constrain how decisions are made in processes: while some constraints hold only along certain paths, decision outcomes may be restricted to satisfy temporal constraints. In this paper, we present *time-aware* BPMN processes and discuss how to: (i) add temporal features to process elements, by considering also the impact of events on temporal constraint management; (ii) characterize decisions based on when they are made and used within a process; (iii) specify and use two novel kinds of decisions based on how their outcomes are managed; (iv) deal with intertwined temporal and decision aspects of time-aware BPMN processes to ensure proper execution.

1 Introduction and Motivation

Time-awareness is undeniably a crucial property of business processes [13,24]. In the last years, temporal features of process models have been widely considered and studied with a focus on different intertwined aspects. Among them, we mention the modeling and checking of temporal constraints at design time [6, 13,17,23], the management of uncertainty for task duration [9], the modular design of time-aware processes [24], the specification of time patterns [20], and the modeling of temporal constraints in Business Process Model and Notation (BPMN) [5,8,12].

In general, temporal features of process models have to be dealt with by considering how they relate to the semantics of process elements. Particularly, decision tasks and events [22] are important concepts to consider jointly with temporal constraints, as they represent points in the process flow where information is acquired and used to determine the following flow of process execution. Indeed, information about decisions is used by exclusive gateways to choose one among alternative execution flows, based on the evaluation of conditions previously set by decision tasks or related to event occurrence.

Thus, during execution time-aware processes have to face two different kinds of uncertainty, one related to activity duration, which is known only after activity completion, the other one stemming from the outcomes of decision tasks and

© Springer Nature Switzerland AG 2018
M. Weske et al. (Eds.): BPM 2018, LNCS 11080, pp. 102–118, 2018.
https://doi.org/10.1007/978-3-319-98648-7_7

from events that determine which process path to follow. Such uncertainties are solved only when tasks have been executed and events have occurred.

At design time, given a time-aware business process model, it is desirable to know whether it is possible to execute it in a correct way, by considering all the possible combinations of activity durations and decision outcomes. However, such durations and outcomes are not under the control of a process engine. In this scenario, if an engine can plan the execution of future steps considering only the history of already executed elements and made decisions, and guaranteeing that all the specified temporal constraints are satisfied, we say that the process is *dynamically controllable*.

In this paper, we propose a new *time-aware* well-structured process model based on BPMN [22] to handle the subtle relations between temporal constraints and decisions and show how to check if process cases can be executed successfully. The main novelties of our approach can be summarized as follows: (i) we add temporal features to process elements, considering also the impact of event occurrence on temporal constraint management, (ii) we discuss the relation between the making and the use of decisions, (iii) we conceptually distinguish two novel types of decisions, (iv) we describe how to deal with both the uncertainty related to the effective duration of executed activities and the uncertainty related to decisions made during the process execution, (v) we define the notion of *dynamic controllability (DC)* for such processes, and (vi) we show a mapping of time-aware well-structured BPMN processes onto suitable temporal constraint networks to check the dynamic controllability of such processes.

1.1   A Motivating Example Taken from the Clinical Domain

As a motivating scenario, let us consider the management of patients diagnosed with knee osteoarthritis (OA). The process diagram of Fig. 1 shows some important treatment steps, excerpt from widely adopted clinical practice guidelines [2]. The core of the diagram is designed by using BPMN [22], which is enriched with different kinds of temporal constraints, such as activity/gateway durations, event waiting times, sequence flow delays, and relative temporal constraints.

Knee OA is a common degenerative joint disease involving cartilage and nearby tissues [2].

In Fig. 1 we focus on pharmacologic treatment, thus leaving nonpharmacologic treatment represented as collapsed subprocess NonPhTr. Moreover, we specify a type only for decision tasks: those representing human decision-making are given type *user*, while tasks enclosing a detailed decision logic are of type *business rule*.

Prior to prescribing treatments, a physician in charge must *Check Contraindications* (business rule task T_0) to commonly administered drugs, such as paracetamol, Non-Steroidal Anti-Inflammatory Drugs (NSAIDs), and opioids. Being potentially life-threatening, absolute contraindications are precisely defined in clinical guidelines to avoid misinterpretation (hence the use of a business rule task).

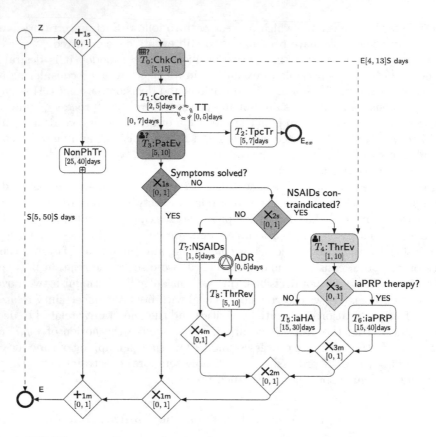

Fig. 1. BPMN process diagram showing the main steps for treating knee osteoarthritis. If not specified, the granularity of temporal ranges in the diagrams is minute.

Depending on the assessed drug tolerance, different treatments are prescribed, in a stepwise manner. The *Core Treatment* (task T_1) is essentially based on paracetamol. To improve pain control, topical preparations may be added during core treatment. In Fig. 1 the need for topical treatment is represented by non-interrupting event TT, which leads to the *Topical Treatment* itself (task T_2). However, paracetamol is often insufficient, even if combined with topical treatment. Thus, a *Patient Evaluation* (task T_3) is scheduled afterwards. If symptoms persist (gateway $\times_{1s}$), an advanced treatment must be prescribed. The choice (gateway $\times_{2s}$) of the best treatment for the patient depends on the contraindications evaluated in T_0. In case of contraindications, intra-articular therapy is preferred. Among existing alternatives, intra-articular hyaluronic acid (iaHA) and platelets-rich plasma (iaPRP) have shown similar efficacy [16]. Therefore, during *Therapy Evaluation* (user task T_4) the physician can choose the therapy among the available ones. This decision differs from those made in T_0 and T_3 since not all of the possible outcomes are required to be available at runtime to guarantee that the process can be executed successfully. Indeed, it is sufficient that at least one of the available therapies iaHA and iaPRP can be chosen. Then, gateway $\times_{3s}$ uses the decision made in T_4 to route the process flow towards either T_5 or T_6. If there are no contraindications, NSAIDs drugs (task T_7) may be administered. However, severe adverse drug reactions (ADRs) may occur while taking NSAIDs: if an ADR is reported, the treatment is immediately interrupted. In Fig. 1, this scenario is captured by signal boundary event ADR, whose occurrence interrupts task T_7 and leads to *Therapy Re-evaluation* (task T_8).

The example shows how process execution relies also on temporal and decision aspects. On the one hand temporal constraints must be satisfied to guarantee that the process is completed successfully. On the other hand, decision tasks determine which process path is preferred over another.

2   Characterization of Time-Aware BPMN Processes

In this section, we introduce temporal aspects, distinguish the two types of decisions that characterize the novel time-aware BPMN and discuss their relations. Then, we propose the notion of dynamic controllability for time-aware BPMN processes. Hereinafter, we consider only well-structured processes as they offer several advantages in terms of comprehension, modularity, and robustness [6, 11, for a detailed discussion].

2.1   Specification of Temporal Properties and Constraints

Here, we enrich BPMN processes by adding a temporal dimension to a relevant subset of BPMN elements and suitable temporal constraints based on concepts presented in [6, 9, 23]. The obtained *time-aware* BPMN fosters the temporal characterization of tasks, gateways, events, sequence flow edges, and time lags between process elements. We describe the introduced temporal aspects by referring to the example of Fig. 1 and borrowing the notions of "activity/event activation" and "event triggering/handling" from the BPMN standard [22].

- **Activities** have a duration attribute represented as a range $[x, y]G$ with $0 < x < y < \infty$, where x/y is the minimum/maximum allowed time span for an activity to go from state "started" to "completed" [20] and G (Granularity) stands for the time unit used (e.g., seconds,...). At run-time, the real duration of an activity cannot be fixed by the process engine, but only observed after who is in charge of executing it completes the activity (*contingent duration*). Process engine takes into account the real duration to properly enact the following elements. Who is in charge of executing the activity must observe the two bounds x and y.
  For example, T_3 has a duration $[5, 10]$ *min*: physicians in charge of T_3 may take between 5 and 10 min to execute it.
- **Intermediate catching events** have a temporal property, $[x, y]G$ with $0 \leq x \leq y < \infty$, representing the minimum/maximum amount of time during

which they may be triggered (*event waiting time*). When x is 0, it means that the event may be triggered as soon as it is activated. y is the upper bound on the amount of time allowed for event triggering that prevents the process to wait infinitely for the occurrence of an event. Since the triggering of an event is not controlled by the process engine, the actual event waiting time is only known at run-time. When a catching event is attached to the boundary of an activity, its waiting time is implied by the duration of the activity. Specifically, if activity T has a duration $[x, y]mG$ and a boundary event e, then the event waiting time for e must be $[0, y']mG$ where $y' < y$ and mG is the minimum time granularity considered in the process. This ensures that e cannot occur at the T completion instant as required by [22]. For practicality, in case of coarser granularity the model admits $y' \leq y$ assuming that y' is always before y after the conversion to the minimum granularity. As an example, task T_7, having duration $[1, 5]$ days, and boundary event ADR, having waiting time $[0, 5]$ days: in this case, since days are coarser than minutes (the minimum granularity), the upper bounds coincide. If e is *non-interrupting* (e.g. TT in Fig. 1), BPMN requires that the duration of the associated activity includes the duration of all non-interrupting event handlers [22]. As discussed later, such specification can be satisfied by combining the described temporal properties, thus allowing designers to think about the elementary temporal characterization of activities.

– **Gateways and sequence flow** also have a duration range of the form $[l, u]$ G, with $0 \leq l \leq u \leq \infty$. However, in this case, the process engine plans the real execution time for such elements by choosing a suitable value of the range. A range associated to a sequence flow edge connecting an element A to an element B is called *sequence flow delay* because it represents the possibility for the process engine to delay the enacting of B after A is completed. For example, between T_1 and T_3 there is a delay of $[0, 7]$ days: considering the decision in T_0 and the completion time of T_1, the engine could reduce this delay even to 0 to guarantee that following tasks be executed without constraint violations. If a designer does not set a duration, it is assumed to be $[0, \infty]mG$.

– **Relative constraints** are depicted in Fig. 1 as dashed edges that connect any two process nodes [9]. Relative constraints limit the time distance between the starting/ending instants of two elements and have the form $I_S[u, v]I_F G$, where I_S is the starting (S)/ending (E) instant of the first element, while I_F is the starting/ending instant of the second one [9]. For example, the time distance between the end of task T_0 and the beginning of task T_4 is given by E$[4, 13]$S days meaning that, if iaHA or iaPRP are needed, the decision of which one to prescribe must be made after at least 4 days and before 13 days from the completion of T_0. To deal with event instantaneity, we choose to always adopt the notation I_S to denote the triggering instant of one event. For example, constraint S$[5, 50]$S days represents the overall minimum and maximum process durations and holds between events Z and E.

2.2   Specification of Decisions: A Novel View on Decision Outcomes

The modeling decisions associated to processes is becoming increasingly important. Here, we offer a novel view on decision tasks, based on where and how their outcomes are used in a process.

In our proposal, decisions are made in decision tasks and any following exclusive gateway may use the outcome of such decisions to route the process flow [22]. In this way, there is a greater flexibility compared to the assumption that decisions are made by a decision task immediately preceding the exclusive gateway [1]. Moreover, allowing a decision to be made at any place prior to an exclusive gateway, may increase temporal flexibility during process execution. For example, T_0 determines the therapies contraindicated for a certain patient, thus affecting which process path is taken at gateway $\times_{2s}$. If the outcome of T_0 is that NSAIDs are contraindicated then, to guarantee that at least one of T_5 and T_6 can be chosen, the delay $[0, 7]$days between T_1 and T_3 must be set to 3 days at the most, for not precluding the possibility to satisfy also the overall duration constraint $S[5, 50]S$ days. Otherwise, a delay of 7 days can be allowed. Decision tasks and corresponding gateways are given the same coloring scheme to highlight the connection between where decisions are made and where they are used. W.l.o.g, we assume that exclusive gateways are binary, i.e., they only have two alternative outgoing sequence flows. Indeed, if a decision has n alternative outcomes, these can be evaluated by setting a proper sequence of $\lceil \log n \rceil$ exclusive gateways.

Beside decision tasks and exclusive gateways, interrupting boundary events, such as ADR of Fig. 1, may also represent decisions as their occurrence determines the enactment of an alternative, exception flow. However, being their triggering instantaneous, interrupting boundary events always represent points in the process where a decision is made and used at the same time.

Beside their position, decisions may be distinguished at a conceptual level based on the availability of their outcomes during process run-time. In general, a process should be guaranteed to be executable for any possible combination of decision outcomes also with respect to the temporal constraints. Since such requirement can be very strict, sometimes it reasonable to relax it by admitting that some paths are not executable at run-time if temporal constraints cannot be satisfied. In other words, it is reasonable to reduce the possible outcomes of some decisions in order to guarantee that the process can be completed successfully.

In this regard, we propose a novel perspective aimed to conceptually distinguish decisions based on how their outcomes may be chosen at run-time.

Some decisions represent the response of the process to conditions that are dictated by the context in which the process is executed, such as data-based conditions or event occurrence. At run-time the process must always be guaranteed to run **any** of such alternative outcomes. We refer to decisions of this kind as *observations*: *A decision is called* observation *when the number of its possible outcomes cannot be reduced at run-time*. Task T_0 makes an observation: Physicians must determine which drugs are contraindicated based on well-documented evidence. For every possible outcome (alternative flows in X_{2s}), the process must be executable for any possible duration of other process tasks.

Conversely, for some decisions it is possible to limit their outcomes at run-time if this can help to execute the process successfully. In this case, the choice of the outcomes is still arbitrary, but the set of possible outcomes can be reduced considering the past execution of previous elements. It must be ensured that **at least one** outcome is always allowed. To denote that the decision is guided by the limitation of its outcomes we refer to decisions of this kind as *guided decisions*: *A decision is a* guided decision *when its possible outcomes can be reduced at run-time to comply with temporal constraints.* In Fig. 1, T_4 makes a guided decision. Since iaHA and iaPRP have similar efficacy and safety, physicians may suggest one or the other without the need for both to always be available when the decision is made. At run-time, if T_4 has been enacted 13 days after T_0 (in case that TT is triggered at day 5 and T_2 lasts 7 days), then T_6 cannot be allowed as its execution could violate the upper bound of the process duration constraint S[5, 50]S; therefore, T_4 can only select iaHA task.

In Fig. 1, symbol **?** next to the task type icon denotes decision tasks that make observations, and symbol **!** denotes tasks that make guided decisions. Decisions based on the occurrence of boundary events are always *observations*.

2.3 Controllability of a Time-Aware BPMN Process

From a temporal perspective, *executing* a time-aware BPMN process P means: (i) to schedule the starting time of all elements, (ii) to set the duration of gateways and sequence flow delays, and (iii) to determine which are the allowed outcomes of a guided decision before enacting the corresponding decision task. The values of observations in P are not known in advance as they are incrementally revealed over time as decision tasks are executed. Similarly, the durations of activities are only known as the activities complete. Therefore, a *dynamic* execution of P must *react* to observations and contingent durations in real time. A *viable* execution is one that guarantees that all *relevant* constraints–those holding in the paths being executed–will be satisfied no matter which observation outcomes and durations are revealed over time. A time-aware BPMN process with a dynamic and viable strategy is called *dynamically controllable* (DC).

3   Dynamic Controllability Checking

In this section, we show how to determine if a time-aware BPMN schema is DC. First, we introduce a temporal-constraint model, called Conditional Simple Temporal Network with Uncertainty and Decision (CSTNUD) [26], that results to be a well-founded model for representing and reasoning about temporal constraints; then, we show how to verify the dynamic controllability of a process model P using a corresponding CSTNUD S_P.

3.1   A Short Introduction to CSTNUD

In general, a temporal-constraint network can be viewed as a graph in which nodes represent real-valued variables and edges represent binary constraints on variables. The kind of binary constraint that can be attached to edges characterizes the network and its expressive power. For example, in a *Simple Temporal Network* (STN) [10] $(\mathcal{T}, \mathcal{C})$, where $\mathcal{T}$ is a set of real-valued variables, called *time-points,* and $\mathcal{C}$ is a set of binary constraints, each constraint has the form $(Y - X \leq \delta)$, where $X, Y \in \mathcal{T}$ and $\delta \in \mathbb{R}$. When it is possible to assign a value to each time-point of a STN such that all constraints are satisfied, then the STN is said to be *consistent.*

By *executing* a temporal-constraint network we mean that the assignment of values to time-points is made by an (executing) engine incrementally following an execution strategy. An execution strategy determines the schedule to apply. For example, if an STN S has a solution, the *earliest execution strategy* schedules S assigning to each time-point its earliest possible execution time.

In [7], Hunsberger et al. proposed Conditional Temporal Network with Uncertainty (CSTNU), that extends STNs including *scenarios* [14] and *contingent links* [21]. A scenario specifies which time-points and constraints to consider during an execution and it is represented by a conjunction of propositional literals. The value of each proposition is unveiled during the execution (environment decides its value). A contingent link is a special kind of temporal constraint having the form, (A, x, y, C), where A and C are time-points, and $0 < x < y < \infty$. Typically, it assumed that a contingent link is activated when A is executed. Then, the value for setting C is decided by the environment not by the engine. However, C is guaranteed to execute such that the temporal difference, $C - A$, is between x and y, i.e., the contingent link is satisfied. Contingent links are used to represent actions with uncertain durations.

In [3] the authors propose CSTN with Decisions (CSTND), a generalization of STN with scenarios (CSTN) that allows some of the propositional variables to be assigned not by the environment, but by the engine executing the network. In [26] CSTND are generalized incorporating contingent links. The resulting network is called a CSTNU with Decisions (CSTNUD).

In the following we combine and extend concepts from earlier work [3, 7, 14, 26].

Definition 1 (Label representing scenario). *Let $\mathcal{P}$ be a set of propositional letters. A label ℓ over $\mathcal{P}$ is a conjunction, $\ell = l_1 \wedge \cdots \wedge l_k$, of literals $l_i \in \{p_i, \neg p_i\}$ on distinct variables $p_i \in \mathcal{P}$. The empty label is denoted by $\square$. For labels, ℓ_1 and ℓ_2, if $\ell_1 \models \ell_2$, we say ℓ_1 entails ℓ_2. $\mathcal{P}^*$ denotes the set of all labels over $\mathcal{P}$.*

Definition 2 (CSTNUD). *A Conditional STN with Uncertainty and Decision is a tuple, $\langle \mathcal{T}, \mathcal{P}, \mathcal{CP}, \mathcal{DP}, \mathcal{C}, \mathcal{OT}, \mathcal{O}, \mathcal{L} \rangle$, where:*

- *$\mathcal{T}$ is a finite set of temporal variables or time-points;*
- *$\mathcal{P}$ is a finite set of propositional variables, i.e., boolean variables;*
- *$(\mathcal{CP}, \mathcal{DP})$ is a partition of $\mathcal{P}$ into contingent propositional variables (**observations**) $\mathcal{CP}$ and controllable propositional variables (**decisions**) $\mathcal{DP}$;*

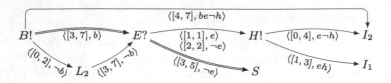

Fig. 2. An example of CSTNUD. b, e and h are relative to $B!, E?$ and $H!$, respectively.

- $\mathcal{C}$ is a finite set of labeled constraints, *each of the form,* $(l \leq Y - X \leq u, \ell)$, *where:* $X, Y \in \mathcal{T}$; $l \leq u$; $l, u \in \mathbb{R}$; *and* $\ell \in \mathcal{P}^*$;
- $\mathcal{OT} \subseteq \mathcal{T}$ *is the set of* disclosing time-points; *and*
- $\mathcal{O}: \mathcal{P} \rightarrow \mathcal{OT}$ *is a bijection that associates each* $p \in \mathcal{P}$ *to a disclosing time-point* $\mathcal{O}(p) \in \mathcal{OT}$, *i.e., to a time-point that, when executed, determines the disclosure of value of the associated proposition variable. If* $p \in \mathcal{CP}$, *then its* $\mathcal{O}(p)$ *is called an* observation time-point; *otherwise a* decision time-point.
- $\mathcal{L}$ *is a set of* contingent links *each of the form* (A, x, y, C, ℓ), *where:* $0 < x < y < \infty$; $A, C \in \mathcal{T}$ *are the* activation *and* contingent *time-points;* $\ell \in \mathcal{P}^*$; *and distinct contingent links have distinct contingent time-points.*

When an observation time-point is executed, the environment assigns a truth value to the corresponding observation; however, when a decision time-point is executed, the decision is assigned a truth value by the engine. $P!$ represents the decision time-point associated to decision p, while $Q?$ the observation time-point associated to observation q. As shown in [4], w.l.o.g. in Definition 2 only constraints are labeled, not time-points.

Viewed as a graph, a CSTNUD edge represents either a labeled constraint or a contingent link. In particular, each edge having the form $X \xrightarrow{\langle [l, u], \ell \rangle} Y$ represents a labeled constraint, $(l \leq Y - X \leq u, \ell)$, and it is called also *standard edge*; each edge having the form $A \xrightarrow{\langle [x, y], \ell \rangle} C$ represents the labeled contingent link (A, x, y, C, ℓ), and it is called *contingent*. The pair $\langle [l, u], \ell \rangle$ is called *labeled range/value*. If between two time-points there exist more labeled constraints, the standard edge connecting them has more labeled ranges, one for each labeled constraint.

Figure 2 shows a CSTNUD having 7 time-points of which 2 are contingents and 3 are disclosing time-points. Contingent link $(B!, 3, 7, E!, b)$ is activated only if decision b is true, while contingent link $(E?, 3, 5, S, \neg e)$ is activated only if observation e is false. For example, if $(B!, 3, 7, E!, b)$ is executed (b true), and it lasts 7, and the observation e results true, for executing the network without violating the constraint between $B!$ and I_2 is necessary to set decision h to false.

In [26], the execution semantics of a CSTNUD is given as a two-player game in which Pl_1 models the executing agent and Pl_2 models the environment, assumed as the most powerful possible player. A game runs in turns: at any time instant t, there exist two turns: the Pl_1 turn, $T_1(t)$, and the Pl_2 one, $T_2(t)$, occurring after $T_1(t)$. At each turn, a player may decide to make k move(s), with $0 \leq k < \infty$. A Pl_1 move is either the execution of a non-contingent time-point X or the assignment of a truth-value to a decision d. A Pl_2 move is either the execution of a contingent time-point C or the assignment of a truth-value to a observation p. Pl_2 is guaranteed to always have full information on what Pl_1 has done before. During the game, the conjunction of truth-values of propositional variables is represented by the label ℓ_{cps}. Pl_1 wins the game when there are no more time-points to execute and for each constraint $(l \leq Y - X \leq u, \ell) \in \mathcal{C}$ such that $\ell_{cps} \models \ell$, then the execution times $S(X)$ of X and $S(Y)$ of Y satisfy the constraint $l \leq S(Y) - S(X) \leq u$. Pl_2 wins otherwise. We denote by σ_i a winning strategy if Pl_i wins the game by following σ_i. Informally, a CSTNUD has the dynamic controllability property if Pl_1 has a winning strategy that is based only on the history of past moves made in the game. The history is defined in terms of *execution sequence*, the ordered sequence of executed time-points and assigned propositions [26]. Usually, $Z_1(Z_2)$ represents an execution sequence of Pl_1 (Pl_2) and $\sigma_1(Z_1, t)(\sigma_2(Z_2, t))$ represents a moved-based strategy that tells a player to make a move at time instant t only if the move is applicable at t [26].

Definition 3 (Dynamic Controllability [26]). *A CSTNUD is* dynamically controllable *(DC) if Pl_1 has a winning strategy such that for any $t > 0$ and any pair of execution sequences Z_1, Z_2, if $\sigma_2(Z_1, t') = \sigma_2(Z_2, t')$ for $0 \leq t' < t$, then $\sigma_1(Z_1, t) = \sigma_1(Z_2, t)$.*

In [26] a DC checking algorithm for CSTNUDs based on Timed Game Automata (TGA) is proposed, while a DC checking algorithm based on constraint propagation for a sub class of CSTNUDs is presented in [3].

3.2   Mapping Time-Aware BPMN onto CSTNUD

To verify the dynamic controllability of a process model P, it is convenient to transform it into an CSTNUD S_P using the transformation rules depicted in Tables 1 and 2. Such rules are described in the proof of Theorem 1. The obtained S_P may be checked for DC by applying one of the available algorithms for DC checking [3, 26]. The following theorem shows that the process model P results to be DC if and only if S_P is DC.

Theorem 1. *Given a time-aware BPMN process P, there exists a CSTNUD S_P such that P is dynamically controllable if and only if S_P is DC.*

Proof. W.l.o.g., we assume that all temporal ranges have the same base granularity mG. In case that P contains ranges with different time granularities, it is possible to convert them to mG. Tables 1 and 2 give the mapping of the elements that can be used to transform a time-aware BPMN fragment into the corresponding CSTNUD. By applying the proposed mappings to P, one can simply verify that the obtained S_P represents all precedence relations and temporal constraints of P. Let us consider each mapping in detail.

- **Task/Subprocess.** Each task A is transformed into two CSTNUD time-points, A_S and A_E, representing its start and end instants. The duration range, $[x, y]$, is converted to the contingent link (A_S, x, y, A_E, ℓ). The label ℓ is determined

Table 1. Mapping of time-aware BPMN fragments to CSTNUDs.

Time-aware BPMN fragment	Corresponding CSTNUD	Time-aware BPMN fragment	Corresponding CSTNUD
Start/End event	$z \rightarrow$ $\rightarrow E$	**Task/Subprocess**	$\rightarrow A_S \xrightarrow{\langle [x,y], \ell \rangle} A_E \rightarrow$
Decision Task (observation)	$\rightarrow A_S \xrightarrow{\langle [x,y], \ell \rangle} A_E$ $\langle [0,0], \ell \rangle$ $\leftarrow A_{\lceil \log n \rceil}? \leftarrow \cdots \leftarrow A_1?$	**Decision Task (guided decision)**	$\rightarrow A_S \xrightarrow{\langle [x,y], \ell \rangle} A_E$ $\langle [0,0], \ell \rangle$ $\leftarrow A_{\lceil \log n \rceil}! \leftarrow \cdots \leftarrow A_1!$
AND Split	$\rightarrow +_S \xrightarrow{\langle [l,u], \ell \rangle} +_E$	**AND Join**	$\rightarrow +_S \xrightarrow{\langle [l,u], \ell \rangle} +_E \rightarrow$
XOR Split (true/false, associated to p proposition)	$\rightarrow +_S \xrightarrow{\langle [l,u], \ell \rangle} +_E \xrightarrow{\langle [0,\infty], \ell p \rangle} \xrightarrow{\langle [0,\infty], \ell \neg p \rangle}$	**XOR Join** (true/false, associated to p proposition)	$\xrightarrow{\langle [0,\infty], \ell p \rangle} \xrightarrow{\langle [0,\infty], \ell \neg p \rangle} +_S \xrightarrow{\langle [l,u], \ell \rangle} +_E \rightarrow$
Sequence flow delay $[l,u]$	$\xrightarrow{\langle [l,u], \ell \rangle}$	**Intermediate Event** $I[x,y]$	$\rightarrow I_S \xrightarrow{\langle [x,y], \ell \rangle} I_E? \rightarrow$
Boundary Interrupting Event	$B_S \xleftarrow{\langle [0,0], \ell \rangle} A_S$ $\langle [0,y'], \ell b \rangle \downarrow \quad \downarrow \langle [x,y], \ell \rangle$ $B_E? \qquad A_E$ $\langle [0, \epsilon_2], \ell b \rangle \downarrow$ $BP_S \xrightarrow{\langle [0,\infty], \ell \neg b \rangle}$ $\langle [w,k], \ell b \rangle \downarrow$ $BP_E \xrightarrow{\langle [0, \epsilon_3], \ell b \rangle} \times \rightarrow$	**Boundary Non-Interrupting Event**	$B_S \xrightarrow{\langle [0,0], \ell \rangle} A_S$ $\langle [0,y'], \ell b \rangle \downarrow \quad \downarrow \langle [x,y], \ell \rangle$ $B_E? \qquad A_E$ $\langle [0, \epsilon_2], \ell b \rangle \downarrow$ BP_S $\langle [w,k], \ell b \rangle \downarrow \langle [0,\infty], \ell b \rangle$ $BP_E \xrightarrow{\langle [0,\infty], \ell b \rangle}$ $\langle [0, \epsilon_1], \ell \rangle \downarrow$ E_{BP}

If not specified, the scenario is assumed to be ℓ and the temporal range of an edge is $\langle [0, \infty], \ell \rangle$.

considering the (possible) XORSplit/XORJoin gateways that are present in the path, Π, from the start event to A in P: (i) Initialize $\ell = \square$; (ii) For each (possible) XORSplit $\times_i$ in Π associated to proposition p, add p or $\neg p$ to ℓ according to the branch present in Π; (iii) For each (possible) XORJoin $\times_i$ in Π associated to proposition p, remove any p literal from ℓ. The obtained label represents the scenario in S_P where A_S and A_E have to be executed and their contingent link observed. The mapping of a subprocess onto a CSTNUD is equivalent.

- **Decision Task (observation).** The conversion is analogous to the one of a task as for its duration attribute. As regards the observation made, it is necessary to represent all the possible outcomes by adding to S_P a suitable number of observation time-points. In particular, if an observation of a decision task A in P can assume n distinct values, then, in S_P there must be $\lceil \log n \rceil$ propositions, associated to new $A?_1, \ldots, A?_{\lceil \log n \rceil}$ observation time-points. In this way, each A outcome is represented by a proper combination of truth-values of such $\lceil \log n \rceil$ propositions. $A?_1, \ldots, A?_{\lceil \log n \rceil}$ are added in sequence after the CSTNUD time-point A_E. The temporal distance between $A?_1, \ldots, A?_{\lceil \log n \rceil}$ and A_E is set to 0 to constrain that the observation values are available at the same instant in which A_E is executed.

- **Decision Task (guided decision).** The conversion is analogous to the one of a decision task making an observation. In this case, however, the possible outcomes of a decision task are represented using decision time-points instead of observation ones to capture the semantics associated to guided decisions.
- **ANDSplit/ANDJoin** gateways. The conversion is analogous to the one of a task. In this case, however, duration attribute $[x, y]$ is converted to a standard edge as gateways are executed by the process engine.
- **XORSplit/XORJoin** gateways. The conversion is analogous to the one of an ANDSplit/ANDJoin as regards its duration attribute. As for scenario, if ℓ is the scenario in which a XORSplit is present, then all converted elements located in its *true* outgoing flow will have a label entailing ℓp, while all converted elements located in its *false* outgoing flow will have a label entailing $\ell \neg p$, where p is the proposition associated to the considered XORSplit. In case of XORJoin, the process is reverse: the scenario label is updated removing p literal.
- **Sequence Flow Delay.** A sequence flow edge having temporal delay $[l, u]$ is converted to a standard edge having $\langle [l, u], \ell \rangle$ as labeled range.
- **Intermediate Event.** Since the temporal range of an intermediate event represents the waiting time allowed for event triggering, its mapping is analogous to the one of a task. Moreover, the end time-point $I_E?$ is also an observation time-point associated to a proper proposition i for representing if the event occurred (true value). In this way, the semantics of the CSTNUD fragment is the following: the execution of $I_E?$ reveals if the event occurred or not and, in case of occurrence, the execution time of $I_E?$ is the exact instant in which the event occurred.
- **Boundary Interrupting Event.** This conversion can be split in two parts that work in parallel, one for task A and one for interrupting event B. $y' < y$ as required by BPMN [22]. Task A and event B are mapped using the previous mappings for tasks and intermediate events, respectively. The contingent links associated to A and B must start at the same time: in Table 1, B_S is at distance 0 from A_S. In S_P, after $B_E?$, there are time-points and constraints related to the temporal characterization of subprocess BP (representing exception handling) that are labeled by ℓb to represent the fact that they must be considered only in case B occurs. Both A_E and BP_E are connected to time-point ✘ that represents the original exclusive gateway ✘ assumed instantaneous for simplicity.

Since the value of an observation cannot be constrained in any way, the obtained CSTNUD fragment allows also the representation (and reasoning) of cases that are not possible in a real process run. For example, in the CSTNUD it is possible that (i) the contingent link associated to B lasts more than the contingent link associated to A, (ii) the proposition b is set true and, therefore, (iii) all temporal constraints associated to the interruption branch must be observed. But this case cannot occur in a real run of P because all interrupting events must occur prior to task completion [22]. Therefore, the dynamic controllability of this CSTNUD fragment guarantees the dynamic controllability of the original fragment containing Boundary Interrupting Event even for execution cases that can never occur. On the other hand, since it is necessary to guarantee the controllability for any possible combination of task duration and event occurrence, all real cases are simpler cases of the real worst case in which A completes at its maximum and B occurs at the last possible instant and BP completes at its maximum, that it is captured by the this conversion. In case that there exist some relative constraints (explained below) involving the start/end of task BP or the end of task A, in their corresponding CSTNUD edges, labels have to be adjusted considering literals $b/\neg b$ to guarantee that the edges are considered in the right scenarios. For example, in the edge associated to a relative constraint involving the end instant of A, the label must contain also $\neg b$ because the constraint has to be considered only when A is not interrupted.

- **Boundary Non-Interrupting Event.** The conversion is analogous to the one of a boundary interrupting event. In this case, however, there is a small complication given by the fact that the BPMN semantics dictates that, in case of event occurrence, task A (cf. Table 1) must wait the completion of event handler represented by BP for reaching state "complete" [22]. Thus, to properly represent such temporal constraint in the CSTNUD, it is necessary to add a join time-point $+$ after A_E. $+$ has to be executed after A_E and BP_E in case the event occurred (proposition b is true) and immediately after A_E in case the event did not occur. Such behavior is ensured by associating two temporal ranges to the edge $A_E \rightarrow +$: $\langle [0, \infty], \ell b \rangle$ represents the delay to observe in case the event occurred (b is true), while $\langle [0,0], \ell \rangle$ represents the 0 delay in case the event did not occur. After $+$, there is the delay $\langle [0, \epsilon_1], \ell \rangle$ corresponding to the sequence flow delay associated to the flow outgoing of A in the BPMN fragment.
- **Relative Constraints.** Let us consider a relative constraint $\langle I_F \rangle [l, u] \langle I_S \rangle$ between two elements A and B, where I_F and I_S represent the kind of instants to be considered, i.e., S or E. In Table 2, A and B are tasks for space reasons, but they can be any combination of tasks, subprocesses, gateways and events. A relative constraint is converted to a CSTNUD edge between the time-points associated to instants A_{I_F} and B_{I_S} with labeled range $\langle [l, u], \ell \rangle$. If ℓ_A / ℓ_B is the scenario where A/B are mapped, then ℓ must satisfy $\ell \models \ell_A \ell_B$, i.e., relative constraints can be defined only in consistent scenarios.

As introduced above, a time-aware BPMN process is dynamically controllable if it is possible to execute it by satisfying all relevant constraints while reacting in real time to (i) the observation values that occur, (ii) tasks/subprocess durations, and (iii) event occurrences.

According to the two-player semantics of CSTNUDs, a CSTNUD is dynamically controllable if it is possible to execute it in a way such that, no matter

Table 2. Mapping of relative constraints between BPMN elements to CSTNUDs. For brevity, only tasks are exemplified but it is possible to consider any pair of elements.

Time-aware BPMN fragment	Equivalent CSTNUD	Time-aware BPMN fragment	Equivalent CSTNUD
Start(A)–Start(B) A · B · S$[l,u]$S	$A_S \xrightarrow{\langle [l,u], \ell \rangle} B_S$ A_E · B_E	**End(A)–End(B)** A · B · E$[l,u]$E	A_S · B_S $A_E \xrightarrow{\langle [l,u], \ell \rangle} B_E$
Start(A)–End(B) A · B · S$[l,u]$E	$A_S \searrow^{\langle [l,u], \ell \rangle} B_S$ A_E · B_E	**End(A)–Start(B)** A · B · E$[l,u]$S	A_S · B_S $A_E \nearrow^{\langle [l,u], \ell \rangle} B_E$

Tasks A and B can be in different paths but not in not-compatible branches.

how the execution of any contingent link turns out and any observations turns out (Pl$_2$ execution strategy), it is possible to set a sequence of decisions and to schedule all non-contingent time-points in real time (Pl$_1$ strategy) satisfying all relevant constraints.

Considering the provided mapping and, in particular, the mapping of process fragments containing boundary events, it is a matter of definitions to verify that, given a process diagram P and its corresponding CSTNUD S_P, the dynamic controllability in S_P implies the dynamic controllability in P and vice-versa. □

4 Discussion and Related Work

Our approach is general and can be extended to other BPMN elements. For example, delays can be easily applied also to message flows while the concepts of duration and relative constraints can be applied to loop activities with some conditions. Moreover, we could easily include absolute temporal constraints. Our proposal of decision modeling can also be used to represent inclusive gateways. It would simply require to extend the mapping shown in Table 1.

The adoption of the CSTNUD model seems to be the more promising for checking a time-aware BPMN process considering techniques different from TGA as done in other proposals [5, 25]. Indeed, while in [26] the author proposed a first CSTNUD DC checking algorithm based on TGA (The decision problem of reachability-time games in TGA with at least two clock is in EXP [15]), in [3] the problem of checking the DC of a CSTNUD was proven to be PSPACE-complete. Moreover, the authors proposed more efficient algorithms for checking the DC of some subclasses of CSTNUDs and this efficiency seems to be preserved for more general CSTNUD instances [3].

A significant analysis of temporal constraint modeling for process-aware information systems is presented in [20], where language-independent time patterns are defined, formalized, and their relevance is empirically shown.

As for modeling temporal constraints by means of BPMN or related extensions, relevant proposals are presented in [5, 8, 12, 23, 25]. In [25], the authors proposed an extension of BPMN for representing activity/process durations and resource constraints and show how to check if a process satisfies or not business requirements by using TGA. In [5] authors adopted the same verification approach for another extension of BPMN including more kinds of temporal constraints, while in [12] the authors proposed an encoding of timed business processes into the Maude language for automatically verifying properties of a simpler extension of BPMN where only task timeouts and sequence flow delays can be expressed. All these proposals lack of a formal characterization of decisions and do not address how such decisions and temporal constraints can be modeled in the corresponding TGA/Maude language, which is indeed an important challenge. Finally, they do not consider events and their temporal characterization. In [8] the authors presented a modular approach to express various nuances of activity duration by combing BPMN elements in suitable blocks aimed to enhance re-use. Despite duration violations are managed, the authors did not consider the formal verification of the proposed cases. In [23], authors devised a process model enriched with temporal conditions in the formulation of conditional constructs, in particular XOR-splits and loops, and provided the related notions of schedule and controllability. The idea that a XOR-split/loop can check a temporal condition is quite interesting but they do not consider the issue of verifying process cases containing such rich elements.

In [18] authors introduced controlled violations (based on relaxation variables) of activity durations and proposed an approach based on constraint satisfaction to determine the best schedule for process while minimizing the cost of violations. Their approach does not consider events and different kinds of decisions.

Finally, in [19] the authors discuss *user decisions* made in knowledge-intensive processes by comparing them to decisions based on history data which cannot be freely made. Despite sharing the same concept of "freedom to decide", the proposed guided decisions limit their outcomes at run-time only when it is necessary to guarantee a successful execution.

5   Conclusions

In this paper, we presented a novel extension of time-aware BPMN where events can be temporally characterized and decisions are distinguished into two types.

As regards temporal characterization, we proposed to distinguish between durations that can be limited by a process engine and durations that become known only at run-time. As regards decisions, we proposed that they can be made and used in different points in the process to allow a greater flexibility with respect to temporal constraints. Moreover, they can be of two kinds: observations, for which the process has to be guaranteed to run for all possible outcomes, and guided decisions, for which the number of possible outcomes can be reduced at run-time to ensure a successful execution.

At application level, it is important to have processes that can be executed reacting to already executed activities, event occurrences and past decisions; we formalized this property as dynamic controllability for time-aware BPMN processes. For verifying whether a BPMN process is dynamically controllable, we propose to map it onto a corresponding CSTNUD instance, whose dynamic controllability is likely to be checked by algorithms beyond the common adopted TGA model checking approach.

References

1. Batoulis, K., Weske, M.: Soundness of decision-aware business processes. In: Carmona, J., Engels, G., Kumar, A. (eds.) BPM 2017. LNBIP, vol. 297, pp. 106–124. Springer, Cham (2017). https://doi.org/10.1007/978-3-319-65015-9_7
2. Bruyère, O., et al.: An algorithm recommendation for the management of knee osteoarthritis in Europe and internationally. Semin. Arthritis Rheum. **44**(3), 253–263 (2014). https://doi.org/10.1016/j.semarthrit.2014.05.014
3. Cairo, M., et al.: Incorporating decision nodes into conditional simple temporal networks. In: 24th International Symposium on Temporal Representation and Reasoning (TIME 2017), pp. 9:1–9:18 (2017). https://doi.org/10.4230/LIPIcs.TIME.2017.9
4. Cairo, M., Hunsberger, L., Posenato, R., Rizzi, R.: A streamlined model of conditional simple temporal networks. In: 24th International Symposium on Temporal Representation and Reasoning (TIME 2017), vol. 90, pp. 10:1–10:19 (2017). https://doi.org/10.4230/LIPIcs.TIME.2017.10
5. Cheikhrouhou, S., Kallel, S., Guermouche, N., Jmaiel, M.: Enhancing formal specification and verification of temporal constraints in business processes. In: IEEE International Conference on Services Computing (SCC 2014), pp. 701–708 (2014). https://doi.org/10.1109/SCC.2014.97
6. Combi, C., Gambini, M., Migliorini, S., Posenato, R.: Representing business processes through a temporal data-centric workflow modeling language: an application to the management of clinical pathways. IEEE Trans. Syst. Man Cybern.: Syst. **44**(9), 1182–1203 (2014). https://doi.org/10.1109/TSMC.2014.2300055
7. Combi, C., Hunsberger, L., Posenato, R.: An algorithm for checking the dynamic controllability of a conditional simple temporal network with uncertainty - revisited. In: Filipe, J., Fred, A. (eds.) ICAART 2013. CCIS, vol. 449, pp. 314–331. Springer, Heidelberg (2014). https://doi.org/10.1007/978-3-662-44440-5_19
8. Combi, C., Oliboni, B., Zerbato, F.: Modeling and handling duration constraints in BPMN 2.0. In: 32nd ACM Symposium on Applied Computing (SAC 2017), pp. 727–734 (2017). https://doi.org/10.1145/3019612.3019618
9. Combi, C., Posenato, R.: Controllability in temporal conceptual workflow schemata. In: Dayal, U., Eder, J., Koehler, J., Reijers, H.A. (eds.) BPM 2009. LNCS, vol. 5701, pp. 64–79. Springer, Heidelberg (2009). https://doi.org/10.1007/978-3-642-03848-8_6
10. Dechter, R., Meiri, I., Pearl, J.: Temporal constraint networks. Artif. Intell. **49**(1–3), 61–95 (1991). https://doi.org/10.1016/0004-3702(91)90006-6
11. Dumas, M., García-Bañuelos, L., Polyvyanyy, A.: Unraveling unstructured process models. In: Mendling, J., Weidlich, M., Weske, M. (eds.) BPMN 2010. LNBIP, vol. 67, pp. 1–7. Springer, Heidelberg (2010). https://doi.org/10.1007/978-3-642-16298-5_1
12. Durán, F., Salaün, G.: Verifying timed BPMN processes using maude. In: Jacquet, J.-M., Massink, M. (eds.) COORDINATION 2017. LNCS, vol. 10319, pp. 219–236. Springer, Cham (2017). https://doi.org/10.1007/978-3-319-59746-1_12

13. Eder, J., Panagos, E., Rabinovich, M.: Time constraints in workflow systems. In: Bubenko, J., Krogstie, J., Pastor, O., Pernici, B., Rolland, C., Sølvberg, A. (eds.) Seminal Contributions to Information Systems Engineering: 25 Years CAiSE, pp. 191–205. Springer, Heidelberg (2013). https://doi.org/10.1007/978-3-642-36926-1_15

14. Hunsberger, L., Posenato, R., Combi, C.: A sound-and-complete propagation-based algorithm for checking the dynamic consistency of conditional simple temporal networks. In: 22nd International Symposium on Temporal Representation and Reasoning (TIME 2015), pp. 4–18 (2015). https://doi.org/10.1109/TIME.2015.26

15. Jurdziński, M., Trivedi, A.: Reachability-time games on timed automata. In: Arge, L., Cachin, C., Jurdziński, T., Tarlecki, A. (eds.) ICALP 2007. LNCS, vol. 4596, pp. 838–849. Springer, Heidelberg (2007). https://doi.org/10.1007/978-3-540-73420-8_72

16. Kon, E., et al.: Platelet-rich plasma intra-articular injection versus hyaluronic acid viscosupplementation as treatments for cartilage pathology. Arthrosc. Relat. Surg. 27(11), 1490–1501 (2011). https://doi.org/10.1016/j.arthro.2011.05.011

17. Kumar, A., Barton, R.: Controlled violation of temporal process constraints-models, algorithms and results. Inf. Syst. 64, 410–424 (2017). https://doi.org/10.1016/j.is.2016.06.003

18. Kumar, A., Sabbella, S.R., Barton, R.R.: Managing controlled violation of temporal process constraints. In: Motahari-Nezhad, H.R., Recker, J., Weidlich, M. (eds.) BPM 2015. LNCS, vol. 9253, pp. 280–296. Springer, Cham (2015). https://doi.org/10.1007/978-3-319-23063-4_20

19. Künzle, V., Reichert, M.: PHILharmonicFlows: towards a framework for object-aware process management. J. Softw.: Evol. Process 23(4), 205–244 (2011)

20. Lanz, A., Reichert, M., Weber, B.: Process time patterns: a formal foundation. Inf. Syst. 57, 38–68 (2016). https://doi.org/10.1016/j.is.2015.10.002

21. Morris, P.H., Muscettola, N., Vidal, T.: Dynamic control of plans with temporal uncertainty. In: International Joint Conference on Artificial Intelligence (IJCAI 2001), pp. 494–502 (2001)

22. Object Management Group: Business Process Model and Notation (BPMN) Version 2.0 (2011). http://www.omg.org/spec/BPMN/2.0. Accessed 07 June 2018

23. Pichler, H., Eder, J., Ciglic, M.: Modelling processes with time-dependent control structures. In: Mayr, H.C., Guizzardi, G., Ma, H., Pastor, O. (eds.) ER 2017. LNCS, vol. 10650, pp. 50–58. Springer, Cham (2017). https://doi.org/10.1007/978-3-319-69904-2_4

24. Posenato, R., Lanz, A., Combi, C., Reichert, M.: Managing time-awareness in modularized processes. Softw. Syst. Model. (2018). https://doi.org/10.1007/s10270-017-0643-4

25. Watahiki, K., Ishikawa, F., Hiraishi, K.: Formal verification of business processes with temporal and resource constraints. In: IEEE International Conference on Systems, Man, and Cybernetics, pp. 1173–1180 (2011). https://doi.org/10.1109/ICSMC.2011.6083857

26. Zavatteri, M.: Conditional simple temporal networks with uncertainty and decisions. In: 24th International Symposium on Temporal Representation and Reasoning (TIME 2017), pp. 23:1–23:17 (2017). https://doi.org/10.4230/LIPIcs.TIME.2017.23

Track I: Foundations of Process Discovery

Interestingness of Traces in Declarative Process Mining: The Janus LTLp$_f$ Approach

Alessio Cecconi[1]([✉])(iD), Claudio Di Ciccio[1](iD), Giuseppe De Giacomo[2](iD), and Jan Mendling[1](iD)

[1] Vienna University of Economics and Business, Vienna, Austria
{alessio.cecconi,claudio.di.ciccio,jan.mendling}@wu.ac.at
[2] Sapienza University of Rome, Rome, Italy
degiacomo@dis.uniroma1.it

Abstract. Declarative process mining is the set of techniques aimed at extracting behavioural constraints from event logs. These constraints are inherently of a reactive nature, in that their activation restricts the occurrence of other activities. In this way, they are prone to the principle of ex falso quod libet: they can be satisfied even when not activated. As a consequence, constraints can be mined that are hardly interesting to users or even potentially misleading. In this paper, we build on the observation that users typically read and write temporal constraints as if-statements with an explicit indication of the activation condition. Our approach is called Janus, because it permits the specification and verification of reactive constraints that, upon activation, look forward into the future and backwards into the past of a trace. Reactive constraints are expressed using Linear-time Temporal Logic with Past on Finite Traces (LTLp$_f$). To mine them out of event logs, we devise a time bi-directional valuation technique based on triplets of automata operating in an on-line fashion. Our solution proves efficient, being at most quadratic w.r.t. trace length, and effective in recognising interestingness of discovered constraints.

Keywords: Process mining · Declarative processes · Temporal logics Separation theorem · Automata theory

1 Introduction

Declarative process mining is the set of techniques aimed at extracting and validating temporal constraints out of event logs, i.e., multi-sets of finite traces. The semantics of these constraints are typically expressed using Linear Temporal Logic on Finite Traces (LTL$_f$) over activities occurring in traces. Examples of declarative process modelling languages are DECLARE [23] and DCR Graphs [11]. In DECLARE, for instance, RESPONSE(a, b) is a constraint applying the parametric template RESPONSE to activities a and b. It imposes that if a occurs in a trace, then b must occur eventually afterwards. PRECEDENCE(c, d) states that if d occurs in a trace, then c must have occurred before.

Constraints are inherently of a reactive nature in that the activation of certain conditions, e.g., the occurrence of a or d, exert restrictions on the occurrence of other

© Springer Nature Switzerland AG 2018
M. Weske et al. (Eds.): BPM 2018, LNCS 11080, pp. 121–138, 2018.
https://doi.org/10.1007/978-3-319-98648-7_8

activities, that is b afterwards and c beforehand in the examples. For this reason, they are prone to the principle of *ex falso quod libet*: not activated constraints are satisfied. This is a serious problem for the ambition of process mining to provide a precise understanding into the behaviour of the process. Returning not activated constraints is hardly interesting to the user or even misleading. A trace, e.g., in which neither a nor d occur, satisfies both RESPONSE(a, b) and PRECEDENCE(c, d). To avoid such spurious results, approaches have been introduced to detect *satisfaction vacuity*. These are based on the parsing of the underlying LTL_f formulae [14], tailored to language-specific templates [18], and checking the state-change of accepting automata [20]. Those solutions hardly provide results that meet the intuition of users who read and write constraints [7], as they respectively provide different activation criteria depending on how formulae are written [14], are restricted to the sole standard DECLARE templates [18], and are bound to the underlying automata semantics, thus, e.g., considering the occurrence of c, and not d, as an activation for PRECEDENCE(c, d) [20].

Against this background, we introduce the Janus approach. First, with this approach, the user can explicitly define activation conditions directly within the constraint formula, similarly to what devised for Compliance Rule Graphs in [17]. The rationale is that users themselves specify what the activation is, thereby making explicit what is of interest and what is not. Because constraints are meant to be activated upon specified conditions, we refer to them as *reactive constraints*. Second, we define a *degree of interestingness* that is computed on traces for such constraints. These constraints are expressed using Linear-time Temporal Logic with Past on Finite Traces($LTLp_f$) [2, 16, 22] in a specific form that singles out the activating event, conditions on its past and conditions on its future. Third, in order to mine these constraints out of event logs, we automatically derive triples of automata based on the well-known separation theorem [8], for the analysis of past prefix, current event, and future suffix of the analysed trace, every time a new activation occurs. Fourth, we integrate these concepts into a time bi-directional constraint evaluation algorithm operating in an on-line fashion, which inspired us to call this approach Janus. Our solution proves efficient with at most quadratic complexity w.r.t. trace length, and is effective in terms of separating interesting from not interesting constraints.

The paper is structured as follows. Section 2 revisits preliminaries, in particular $LTLp_f$ and how to separate its formulae through the *separation theorem*. Section 3 presents the Janus approach, defining the *reactive constraints*, their *interestingness degree*, and how to verify them using automata theory. Section 4 presents the algorithm to retrieve the interestingness degree in an on-line fashion. Section 5 evaluates our proposal in comparison with other state of the art process miners, both using synthetic and real-life logs. Section 6 concludes the paper and identifies directions for future research.

2   Preliminaries

Event Logs. In this paper, we consider a *trace* as a sequence events $t = \langle e_1, \ldots, e_n \rangle$, where $n \in \mathbb{N}$ is the length of the trace, and events e_i belong to an alphabet of symbols Σ. Function $t(i)$ returns event e_i occurring at *instant* i in the trace. With $t_{[i:j]}$ we identify a segment (henceforth, sub-trace) extracted from trace t and ranging from instant i to

Table 1. Some DECLARE constraints expressed as Reactive Constraints

Constraint	LTL$_f$ expression [3]	RCon	Separation degree
PARTICIPATION(a)	$\Diamond$a	$t_{Start} \square\!\!\rightarrow \Diamond$a	1
INIT(a)	a	$t_{Start} \square\!\!\rightarrow$ a	1
END(a)	$\square\Diamond$a	$t_{End} \square\!\!\rightarrow$ a	1
RESPONDEDEXISTENCE(a,b)	$\Diamond$a $\rightarrow \Diamond$b	a $\square\!\!\rightarrow (\Diamond$b $\vee \lozenge\!\!\!\!-$b$)$	2
RESPONSE(a,b)	$\square$(a $\rightarrow \Diamond$b)	a $\square\!\!\rightarrow \Diamond$b	1
ALTERNATERESPONSE(a,b)	$\square$(a $\rightarrow \Diamond$b) $\wedge \square$(a $\rightarrow \bigcirc(\neg$a **W** b))	a $\square\!\!\rightarrow \bigcirc(\neg$a **U** b)	1
CHAINRESPONSE(a,b)	$\square$(a $\rightarrow \Diamond$b) $\wedge \square$(a $\rightarrow \bigcirc$b)	a $\square\!\!\rightarrow \bigcirc$b	1
PRECEDENCE(a,b)	$\neg$b **W** a	b $\square\!\!\rightarrow \lozenge\!\!\!\!-$a	1
ALTERNATEPRECEDENCE(a,b)	$(\neg$b **W** a$) \wedge \square$(b $\rightarrow \bigcirc(\neg$b **W** a))	b $\square\!\!\rightarrow \ominus(\neg$b **S** a)	1
CHAINPRECEDENCE(a,b)	$(\neg$b **W** a$) \wedge \square(\bigcirc$b $\rightarrow$ a)	b $\square\!\!\rightarrow \ominus$a	1

instant j, where $1 \leqslant i \leqslant j \leqslant n$. t_R is the reverse of a trace t, i.e., such that $t(i) = t_R(n - i + 1)$ for every $1 \leqslant i \leqslant n$. An *event log* $L = \{\!\!\{ t_1, t_2, \ldots, t_m \}\!\!\} \in \mathbb{M}(\Sigma^*)$, or *log* for short, is a multi-set of traces t_j with $1 \leqslant j \leqslant m$, $m \in \mathbb{N}$. We shall use a compact notation denoting the *multiplicity* of traces with a superscript, e.g., t_1^5 means that t_1 occurs 5 times in L.

Example 1. Let $L = \{\!\!\{ t_1^{25}, t_2^{15}, t_3^{10}, t_4^{20}, t_5^5, t_6^{20}, t_7^5 \}\!\!\}$ be an event log of 100 traces, defined on the alphabet of events $\Sigma = \{$a, b, $\ldots$, i$\}$, where $t_1 = \langle$d, f, a, f, c, a, f, b, a, f$\rangle$, $t_2 = \langle$f, e, d, c, b, a, g, h, i$\rangle$, $t_3 = \langle$a, d, a, a, a, a, a, a, a, a, a, a, a, a, a, a, a, a, a, a, a, a, c$\rangle$, $t_4 = \langle$d, b, a, e$\rangle$, $t_5 = \langle$a, d, a, c, a$\rangle$, $t_6 = \langle$b, c, d, e$\rangle$, $t_7 = \langle$b, c, a$\rangle$. We have that $t_1(7) =$ f, $t_{1[3:7]} = \langle$a, f, c, a, f$\rangle$, $t_{2R} = \langle$i, h, g, a, b, c, d, e, f$\rangle$.

DECLARE. DECLARE is a declarative process modelling language and notation [23]. It defines a set of standard templates based on [7], abstracting from the actual semantics used to describe them. In that way the complexity of the underneath logic is hidden to the user. The template parameters are tasks occurring as events in traces. For example, CHAINPRECEDENCE is a binary template stating that the occurrence of the second task imposes the first one to occur immediately before. PRECEDENCE loosens that condition requiring the second task to occur any time before the first one. DECLARE semantics are rooted in temporal logics. Its standard template set, a part of which is listed in Table 1, is meant to be extended with custom ones that suit best the user needs [23].

DECLARE *constraints* are templates whose parameters are assigned with tasks, e.g., PRECEDENCE(b, a) applies the PRECEDENCE template on tasks a, the *activation*, and b. Constraints are verified over events of traces. Those that do not *violate* a constraint, *satisfy* it. In Example 1, e.g., all events of t_7 satisfy PRECEDENCE(b, a) whereas CHAINPRECEDENCE(b, a) is violated by $t_7(3)$. Notice the principle of *ex falso quod libet*: both PRECEDENCE(b, a) and CHAINPRECEDENCE(b, a) are satisfied by t_6, where a, namely the *activation*, never occurs. This is arguably misleading, and calls for an approach that considers the reactive nature of constraints, i.e., such that it *(i)* singles out activations and *(ii)* dictates the conditions to check upon activation in the future and past of the trace. In the literature, constraints are formulated in LTL$_f$ as of [3], as listed in Table 1.

To cater for temporal modalities referring to the past, we resort on an extension of the syntax of LTL$_f$, namely the one of Linear-time Temporal Logic with Past on Finite Traces (LTLp$_f$).

LTLp$_f$. Well-formed LTLp$_f$ formulae are built from an alphabet $\Sigma \supseteq \{a\}$ of propositional symbols and are closed under the boolean connectives, the unary temporal operators $\bigcirc$(next) and $\ominus$(previous), the binary temporal operators **U**(Until) and **S** (Since):

$$\varphi ::= a|(\neg\varphi)|(\varphi_1 \wedge \varphi_2)|(\bigcirc\varphi)|(\varphi_1 \, \mathbf{U} \, \varphi_2)|(\ominus\varphi)|(\varphi_1 \, \mathbf{S} \, \varphi_2).$$

From these basic operators it is possible to derive: classical boolean abbreviations *True*, *False*, $\vee$, $\rightarrow$; constant t_{End}, verified as $\neg\bigcirc True$, denoting the last instant of the trace; constant t_{Start}, verified as $\neg\ominus True$, denoting the first instant of the trace; $\Diamond\varphi$ as *True* $\mathbf{U}\,\varphi$ indicating that φ holds true eventually before t_{End}; $\varphi_1 \, \mathbf{W} \, \varphi_2$ as $(\varphi_1 \, \mathbf{U} \, \varphi_2) \vee \Box\varphi_1$, which relaxes **U** as φ_2 may never hold true; $\diamondsuit\varphi$ as *True* $\mathbf{S}\,\varphi$ indicating that φ holds true eventually in the past after t_{Start}; $\Box\varphi$ as $\neg\Diamond\neg\varphi$ indicating that φ holds true from the current instant till t_{End}; $\boxminus\varphi$ as $\neg\diamondsuit\neg\varphi$ indicating that φ holds true from t_{Start} to the current instant.

We remark that, w.l.o.g., we consider here the non-strict semantics of **U** and **S** [12].

Given a trace t, a LTLp$_f$ formula φ is satisfied in a given instant i $(1 \leqslant i \leqslant n)$ by induction of the following:

$t, i \models True$; $t, i \not\models False$; $t, i \models$ a iff $t(i)$ is assigned with a;

$t, i \models \neg\varphi$ iff $t, i \not\models \varphi$; $t, i \models \varphi_1 \wedge \varphi_2$ iff $t, i \models \varphi_1$ and $t, i \models \varphi_2$;

$t, i \models \bigcirc\varphi$ iff $i < n$ and $t, i + 1 \models \varphi$; $t, i \models \ominus\varphi$ iff $i > 1$ and $t, i - 1 \models \varphi$;

$t, i \models \varphi_1 \, \mathbf{U} \, \varphi_2$ iff $t, j \models \varphi_2$ with $i \leqslant j \leqslant n$, and $t, k \models \varphi_1$ for all k s.t. $i \leqslant k < j$;

$t, i \models \varphi_1 \, \mathbf{S} \, \varphi_2$ iff $t, j \models \varphi_2$ with $1 \leqslant j \leqslant i$, and $t, k \models \varphi_1$ for all k s.t. $j < k \leqslant i$.

In the following, we will classify $\bigcirc$, $\Box$, $\Diamond$, **U** as *future operators*, $\ominus, \boxminus, \diamondsuit$, **S** as *past operators*, and the following pairs of operators as *mirror images*: (i) $\bigcirc$ and $\ominus$, (ii) $\Box$ and $\boxminus$, (iii) $\Diamond$ and $\diamondsuit$, (iv) **U** and **S**. We shall also name as *mirror image* of formula φ the temporal formula obtained by replacing all its operators with their mirror images [25], henceforth denoted as φ_M.

Definition 1 (Pure temporal formula [8]**).** *A LTLp$_f$ formula φ is named:* **pure past** *($\varphi^{\blacktriangleleft}$) if it contains only past operators;* **pure present** *($\varphi^{\blacktriangledown}$) if it contains no temporal operators at all;* **pure future** *($\varphi^{\blacktriangleright}$) if it contains only future operators.*

For example, $\varphi^{\blacktriangleleft} = \boxminus(a\,\mathbf{S}(\diamondsuit b))$, $\varphi^{\blacktriangledown} = a \wedge b \vee c$, and $\varphi^{\blacktriangleright} = \Box(\Diamond a \vee (\bigcirc b)\,\mathbf{U}\,c)$ are pure past, pure present, and pure future formulae, respectively. We argue that separating formulae into ones that refer to the sole past, future, or present, allows for a bi-directional on-line analysis of sub-traces at activation time. The separation theorem, first introduced in [8], proves that such a separation can be obtained.

Theorem 1 (Separation theorem (adapted from [8]**)).** *Any propositional temporal formula written with* **U**, **S**, $\bigcirc$, *and $\ominus$ operators can be rewritten as a boolean combination of pure temporal formulae.*

The constructive proof of Theorem 1 in [8] provides a syntactic procedure to pull out $\ominus$, S from the scope of $\bigcirc$, U in LTLp$_f$ formulae, and vice-versa. It thus provides the base substitution rules such that their recursive application brings to the decomposition of a LTLp$_f$ formula in pure temporal formulae. We capture this notion as follows.

Definition 2 (Temporal separation, separated formula). *Let* φ *be a LTLp$_f$ formula over* Σ. *A temporal separation is a function* $\mathscr{S} : LTLp_f \rightarrow 2^{LTLp_f \times LTLp_f \times LTLp_f}$. *Indicating* $\varphi^{\blacktriangleleft}$, $\varphi^{\blacktriangledown}$, *and* $\varphi^{\blacktriangleright}$ *respectively as pure past, present, and future formulae, defined over* Σ *as per Definition 1,* $\mathscr{S}(\varphi) = \{(\varphi^{\blacktriangleleft}, \varphi^{\blacktriangledown}, \varphi^{\blacktriangleright})_1, \dots, (\varphi^{\blacktriangleleft}, \varphi^{\blacktriangledown}, \varphi^{\blacktriangleright})_m\}$ *is such that*

$$\varphi \equiv \bigvee_{j=1}^{m} (\varphi^{\blacktriangleleft} \wedge \varphi^{\blacktriangledown} \wedge \varphi^{\blacktriangleright})_j.$$

We call separated formula *any element in the co-domain of* $\mathscr{S}(\varphi)$.

For example, $(\ominus b \vee \Diamond c) \equiv ((\ominus b) \wedge (\textit{True}) \wedge (\textit{True})) \bigvee ((\textit{True}) \wedge (\textit{True}) \wedge (\Diamond c))$.

Automata. A (deterministic finite-state) automaton is a rooted finite-state labelled transition system $A = (\Sigma, S, \delta, s_0, S_F) \in \mathcal{A}$, where: Σ is the alphabet S is a finite non-empty set of states; $\delta : S \times \Sigma \rightarrow S$ is a (total) transition function; s_0 is the initial state; $S_F \subseteq S$ is the set of accepting states. An automaton *accepts* a trace $t = \langle e_1, \dots, e_n \rangle$ if a walk of tuples $\langle (s_0, e_1, s_1), \dots, (s_{n-1}, e_n, s_n) \rangle$, namely a *replay*, exists such that $s_i = \delta(s_{i-1}, e_i)$ for $1 \leqslant i \leqslant n$ and $s_n \in S_F$. We shall name s_i in tuple (s_{i-1}, e_i, s_i) as *current state* of the replay at instant i. The set of all traces accepted by A is named *language* of A, denoted as $\mathscr{L}(A)$. Figure 1 depicts an automaton whose language consists of all traces of length $n \geqslant 2$ having b as its sec-

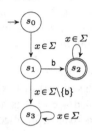

Fig. 1. An automaton verifying $\bigcirc$ b

ond event. Considering the event log of Example 1, it accepts only $t_4 = \langle d, b, a, e \rangle$. Reportedly approaches exist that build automata verifying any formula of Linear Temporal Logic(LTL) [26], Linear-time Temporal Logic with Past(LTLp) [9,24], or LTL$_f$ [2] on traces: such automata accept all and only the traces such that all events satisfy a formula φ. We indicate them as A_φ. The automaton of Fig. 1, e.g., verifies $\bigcirc$ b. Notice that automata verification considers whole traces as either satisfying or violating a formula.

3 The Janus Approach

Here we present the concepts upon which our approach is built, beginning with the core notion of Reactive Constraints (RCons). RCons are meant to bear the interestingness semantics, because the role of activation is singled out from the rest of exerted conditions: only if the activation α "triggers" the constraint *and* a LTLp$_f$ formula φ is verified on the trace, then its fulfilment is interesting.

Definition 3 (Reactive Constraint (RCon)). *Given an alphabet Σ, let $\alpha \in \Sigma \cup$ $\{t_{Start}, t_{End}\}$ be an* activation, *and φ be a $LTLp_f$ formula over Σ. A Reactive Constraint (RCon) Ψ is a pair (α, φ) hereafter denoted as $\Psi \triangleq \alpha \mathrel{\square\!\!\rightarrow} \varphi$. We denote the set of all RCons over Σ as $\mathscr{R}$.*

In the following, we will assume traces, automata, $LTLp_f$ formulas and RCons to be all defined over a shared alphabet Σ. Constraints of declarative process languages such as DECLARE can be expressed as RCons, as shown in Table 1. For example, the RCon corresponding to PRECEDENCE(d, a) is a $\mathrel{\square\!\!\rightarrow} \Diamond d$. Activations in constraints are identified according to the classification of [5]. RCons can include non-standard DECLARE constraints such as a $\mathrel{\square\!\!\rightarrow} (\ominus b \lor \Diamond c)$, which imposes that if an event a occurs in a trace, either b immediately precedes it, or c eventually occurs after. We say that a *activates* the RCon.

We remark that although φ is a $LTLp_f$ formula, semantics of RCons detach from classical $LTLp_f$ in that every occurrence of α triggers a new verification of φ on the trace.

Definition 4 (Activator, triggering trace). *Given a trace $t \in \Sigma^*$ of length n and an instant i s.t. $1 \leq i \leq n$, an RCon $\Psi = \alpha \mathrel{\square\!\!\rightarrow} \varphi$ is* activated *at i if $t, i \models \alpha$. Event $t(i)$ is then named* activator *of Ψ. A trace in which at least an activator of Ψ exists, is* triggering *for Ψ.*

Consider, e.g., the event log from Example 1 and $\Psi = a \mathrel{\square\!\!\rightarrow} \Diamond d$, i.e., PRECEDENCE(d, a) in DECLARE. Ψ is activated in trace $t_4 = \langle d, b, a, e \rangle$ by $t_4(3)$; in trace $t_5 = \langle a, d, a, c, a \rangle$ it is activated by $t_5(1)$, $t_5(3)$, and $t_5(5)$; in trace $t_6 = \langle b, c, d, e \rangle$ it is never activated; in trace $t_7 = \langle b, c, a \rangle$ is activated by $t_7(3)$. Therefore t_4, t_5, and t_7 are triggering for Ψ.

Definition 5 (Interesting fulfilment). *Given a trace $t \in \Sigma^*$ of length n, an instant i s.t. $1 \leq i \leq n$, and an RCon $\Psi = \alpha \mathrel{\square\!\!\rightarrow} \varphi$, Ψ is* interestingly fulfilled *at i if $t, i \models \alpha$ and $t, i \models \varphi$. The RCon is* violated *at instant i if $t, i \models \alpha$ and $t, i \not\models \varphi$. Otherwise, the RCon is* unaffected.

Consider again Example 1 and $\Psi = a \mathrel{\square\!\!\rightarrow} \Diamond d$. In trace t_4 the RCon is interestingly fulfilled by $t_4(3)$; in trace t_5 it is interestingly fulfilled by $t_5(3)$ and $t_5(5)$, and violated by $t_5(1)$; in trace t_6 it is unaffected at all instants, i.e., neither interestingly fulfilled nor violated; in trace t_7 it is violated by $t_7(3)$. We remark that instants at which a DECLARE constraint is satisfied can be such that the corresponding RCon is unaffected, i.e., when $t, i \models \varphi$ but $t, i \not\models \alpha$.

3.1 Measuring the Interesting Fulfilments of a Reactive Constraint on a Log

Because each activator triggers a new check for the RCon, the degree of interestingness of a trace is analysed at the level of events as follows.

Definition 6 (Interestingness degree). *Given a trace $t \in \Sigma^*$ and an RCon $\Psi = \alpha \; \square\!\!\!\rightarrow$ φ, we define the interestingness degree function $\zeta : \mathscr{R} \times \Sigma^* \to [0, 1] \subseteq \mathbb{R}$ as follows:*

$$\zeta(\Psi, t) = \begin{cases} \dfrac{|\{i : t, i \models \alpha \text{ and } t, i \models \varphi \}|}{|\{i : t, i \models \alpha\}|} & \text{if } |\{i : t, i \models \alpha\}| \neq 0; \\ 0 & \text{otherwise.} \end{cases}$$

Intuitively the interestingness degree measures the percentage of activations leading to (interesting) fulfilment within the trace. For instance, the interestingness degree of $\Psi = a \; \square\!\!\!\rightarrow \; \diamond d$ in $t_5 = \langle a, d, a, c, a \rangle$ is $\zeta(\Psi, t_5) = 0.667$, because it is interestingly fulfilled by 2 activators out of 3. All the 20 activators of Ψ in $t_3 = \langle a, d, a, \ldots, a, c \rangle$ lead to interesting fulfilment, except one (the first a event). Therefore $\zeta(\Psi, t_3) = 0.950$.

Next, we introduce the computation of two measures for the whole event log, adapting the classical definition of [1]: *support* measures how often the constraint is (interestingly) satisfied in the whole event log; *confidence* quantifies how the constraint is satisfied in the triggering traces in the event log.

Definition 7 (Support). *Given an event log $L = \{\!|t_1, t_2, \ldots, t_m|\!\} \in M(\Sigma^*)$, and an RCon $\Psi \in \mathscr{R}$, we define as support of Ψ on L the function $\sigma : \mathscr{R} \times M(\Sigma^*) \to [0, 1] \subseteq \mathbb{R}$ calculated as the average of interestingness degree values of Ψ over all traces $t_j \in L$, with $1 \leqslant j \leqslant m$:*

$$\sigma(\Psi, L) = \begin{cases} \dfrac{\sum_{j=1}^{m} \zeta(\Psi, t_j)}{|L|} & \text{if } |L| > 0; \\ 0 & \text{otherwise.} \end{cases}$$

Definition 8 (Confidence). *Given an event log $L = \{\!|t_1, t_2, \ldots, t_m|\!\} \in M(\Sigma^*)$ and an RCon $\Psi \in \mathscr{R}$, let $\widetilde{L} = \{\!|\widetilde{t}_1, \widetilde{t}_2, \ldots, \widetilde{t}_p|\!\}$ with $1 \leqslant p \leqslant m$ be the portion of L that consists of all the traces triggering Ψ. We define the confidence of Ψ on L the function $\kappa : \mathscr{R} \times M(\Sigma^*) \to [0, 1] \subseteq \mathbb{R}$ as the average of interestingness degree values of Ψ over the triggering traces $\widetilde{t}_j \in \widetilde{L}$:*

$$\kappa(\Psi, L) = \begin{cases} \dfrac{\sum_{j=1}^{p} \zeta(\Psi, \widetilde{t}_j)}{|\widetilde{L}|} & \text{if } |\widetilde{L}| > 0; \\ 0 & \text{otherwise.} \end{cases}$$

As seen above, e.g., the interestingness degree of the RCon of PRECEDENCE(d, a), $\Psi = a \; \square\!\!\!\rightarrow \; \diamond d$, is $\zeta(\Psi, t_5) = \frac{2}{3} = 0.667$. Averaging the interestingness degree of PRECEDENCE(d, a) on all traces (including their multiplicities), we obtain the support $\sigma(\Psi, L)$, which amounts to $\frac{(1 \times 25) + (1 \times 15) + (0.95 \times 10) + (1 \times 20) + (0.67 \times 5) + (0 \times 20) + (0 \times 5)}{100} =$ 0.728. The value of confidence $\kappa(\Psi, L)$ is $\frac{(1 \times 25) + (1 \times 15) + (0.95 \times 10) + (1 \times 20) + (0.67 \times 5) + (0 \times 5)}{80} =$ 0.910, thus higher than support, because Ψ is not activated in trace t_6, hence the lower denominator. Table 2 shows in detail the interestingness degree, support, and confidence of constraints PRECEDENCE(d, a) and a $\square\!\!\!\rightarrow (\ominus b \vee \diamond c)$ on the event log seen in Example 1.

Considering each activation independently to compute interestingness degree, support and confidence, allows for higher resilience to noise in event logs. This is particularly evident in t_3, in which 19 occurrences of a, the activation, are preceded by d, thus fulfilling the constraint. Only the first a leads to violation. Considering the whole trace as fulfilling or not, as in [18, 19], would lead to an interestingness degree of 0, instead of

Table 2. Interestingness degree, support, and confidence of RCons on an example log

Trace	Multi.	PRECEDENCE(d,a)			$a \rightarrowtail (\ominus b \vee \Diamond c)$		
		Activ.'s	Int.fulfil.'s	ς	Activ.'s	Int.fulfil.'s	ς
t_1	$\langle d,f,a,f,c,a,f,b,a,f\rangle$	25	3	3	1	3	2	0.667
t_2	$\langle f,e,d,c,b,a,g,h,i\rangle$	15	1	1	1	1	1	1
t_3	$\langle a,d,a,a,\ldots,a,a,c\rangle$	10	20	19	0.950	20	20	1
t_4	$\langle d,b,a,e\rangle$	20	1	1	1	1	1	1
t_5	$\langle a,d,a,c,a\rangle$	5	3	2	0.667	3	2	0.667
t_6	$\langle b,c,d,e\rangle$	20	0	0	0	0	0	0
t_7	$\langle b,c,a\rangle$	5	1	0	0	1	0	0
Support				0.728			0.650
Confidence				0.910			0.813

0.95, thus decreasing support and confidence too. Because the constraints returned by discovery techniques are those that have a support and confidence above user-defined thresholds, that could lead to a loss of information.

3.2 An Automata Approach to Reactive Constraint Verification

Once an RCon $\Psi = \alpha \rightarrowtail \varphi$ is activated, its fulfilment relies on the verification of the LTLp$_f$ formula φ at the instant of activation. As seen in Theorem 1 it is possible to separate a LTLp$_f$ formula into sub-formulae, each containing either only past, only future, or no temporal operators. Therefore its verification can be decoupled by splitting the trace in two independent sub-traces: one from the beginning to the activator, with which the sole past operators are verified, and one from the activator on, concerning only future operators.

Lemma 1 *Given a pure past formula $\varphi^{\blacktriangleleft}$, a pure present formula $\varphi^{\blacktriangledown}$, a pure future formula $\varphi^{\blacktriangleright}$, a trace $t \in \Sigma^*$ of length n and an instant i s.t. $1 \leqslant i \leqslant n$, the following hold true:*
$$t, i \models \varphi^{\blacktriangleleft} \equiv t_{[1,i]}, i \models \varphi^{\blacktriangleleft}; \quad t, i \models \varphi^{\blacktriangledown} \equiv t_{[i,i]}, i \models \varphi^{\blacktriangledown} \quad t, i \models \varphi^{\blacktriangleright} \equiv t_{[i,n]}, i \models \varphi^{\blacktriangleright}.$$

The lemma follows from definition of LTLp$_f$ semantics. For example, evaluating $\varphi^{\blacktriangleleft} = \diamond d$ on $t_2 = \langle f, e, d, c, b, a, g, h, i\rangle$ from Example 1 at instant $i = 6$ is equivalent to evaluating it on $t_{2[1:6]} = \langle f, e, d, c, b, a\rangle$ at i, because $\varphi^{\blacktriangleleft}$ concerns only the prefix of t_2. Instead $\varphi^{\blacktriangleright} = \Diamond c$ at i concerns only the suffix, $t_2[6:9] = \langle a, g, h, i\rangle$.

Theorem 2 (Trace sub-valuation). *Given a LTLp$_f$ formula φ, a trace $t \in \Sigma^*$ of length n and an instant i s.t. $1 \leqslant i \leqslant n$, we have that $t, i \models \varphi$ if and only if $t_{[1:i]}, i \models \varphi^{\blacktriangleleft}$, $t_{[i:i]}, i \models \varphi^{\blacktriangledown}$, and $t_{[i:n]}, i \models \varphi^{\blacktriangleright}$ for at least a $(\varphi^{\blacktriangleleft}, \varphi^{\blacktriangledown}, \varphi^{\blacktriangleright}) \in \mathcal{S}(\varphi)$.*

The proof follows from Theorem 1 and Lemma 1. Consider, e.g., the RCon $\Psi = a \rightarrowtail (\ominus b \vee \Diamond c)$. It follows that $\varphi = \ominus b \vee \Diamond c$ and $\mathcal{S}(\varphi) = \{(\ominus b, True, True), (True, True, \Diamond c)\}$. Applying Theorem 2 on $t_2 = \langle f, e, d, c, b, a, g, h, i\rangle$ from Example 1, we have that $t_2, 6 \models \varphi$ if *(i)* $\langle f, e, d, c, b, a\rangle, 6 \models \ominus b$ or *(ii)* $\langle a, g, h, i\rangle, 6 \models \Diamond c$, aside of *True* formulae which are trivially satisfied. Because the first holds true, we conclude that φ is satisfied by $t_2(6)$.

Table 3. Graphical representation of the separated automata set of a $\square\mapsto$ ($\ominus$b $\vee$ $\diamond$c)

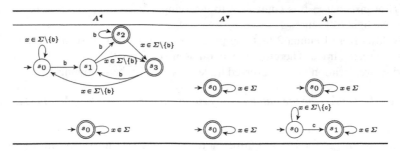

As seen in Sect. 2, a formula φ can be verified on a trace t by checking whether t is accepted by automaton A_φ. Given a LTLp$_f$ formula φ and its temporal separation $\mathscr{S}(\varphi)$, we thus introduce the notion of separated automata set, namely a set of triples of automata, each verifying a triple of pure temporal formulae in $\mathscr{S}(\varphi)$.

Definition 9 (Separated automata set (sep.aut.set)). *Given a LTLp$_f$ formula φ we define as separated automata* sep.aut.set $\mathcal{A}^{\blacktriangleleft\blacktriangledown\blacktriangleright} \in 2^{\mathcal{A}\times\mathcal{A}\times\mathcal{A}}$ *the set of triples* $A^{\blacktriangleleft\blacktriangledown\blacktriangleright} = (A^\blacktriangleleft, A^\blacktriangledown, A^\blacktriangleright) \in \mathcal{A}\times\mathcal{A}\times\mathcal{A}$ *such that* $A^\blacktriangleleft = A_{\varphi^\blacktriangleleft}$, $A^\blacktriangledown = A_{\varphi^\blacktriangledown}$, *and* $A^\blacktriangleright = A_{\varphi^\blacktriangleright}$ *for every* $(\varphi^\blacktriangleleft, \varphi^\blacktriangledown, \varphi^\blacktriangleright) \in \mathscr{S}(\varphi)$. *We denote as* separation degree D *the number of triples of* $\mathcal{A}^{\blacktriangleleft\blacktriangledown\blacktriangleright}$.

Considering the latest example $A^{\blacktriangleleft\blacktriangledown\blacktriangleright} = \{(A_{\ominus b}, A_{True}, A_{True}), (A_{True}, A_{True}, A_{\diamond c})\}$, the separation degree is 2. Table 1 lists the separation degrees of some RCons of DECLARE. In light of the above, we derive from Theorem 2 the following.

Theorem 3 (Trace sub-valuation through automata). *Given a LTLp$_f$ formula φ, its sep.aut.set $A^{\blacktriangleleft\blacktriangledown\blacktriangleright}$, a trace $t \in \Sigma^*$ of length n, and an instant i s.t. $1 \leqslant i \leqslant n$, we have that $t, i \models \varphi$ if and only if $t_{[1:i]} \in \mathscr{L}(A^\blacktriangleleft)$, $t_{[i:i]} \in \mathscr{L}(A^\blacktriangledown)$ and $t_{[i:n]} \in \mathscr{L}(A^\blacktriangleright)$ for at least a $(A^\blacktriangleleft, A^\blacktriangledown, A^\blacktriangleright) \in A^{\blacktriangleleft\blacktriangledown\blacktriangleright}$.*

In the example, the application of Theorem 3 entails that a $\square\mapsto$ ($\ominus$b $\vee$ $\diamond$c) is interestingly fulfilled by $t_2(6)$ if $t_{2[1:6]} = \langle$f, e, d, c, b, a$\rangle \in \mathscr{L}(A_{\ominus b})$ or $t_{2[6:9]} = \langle$a, g, h, i$\rangle \in \mathscr{L}(A_{\diamond c})$.

Past Automata Reversion. To the best of our knowledge, there is no available technique to build automata that verify LTLp$_f$ formulae. We thus exploit Theorem 2 to rely on the readily available techniques for LTL$_f$, i.e., without past operators, as described in [2]. To this extent, we rely on mirror images and reversed automata.

Lemma 2. *Let $t \in \Sigma^*$ be a trace of length n and t_R its reverse. Given a pure past formula $\varphi^\blacktriangleleft$, and its mirror image $\varphi^\blacktriangleleft_M$, we have that $t, n \models \varphi^\blacktriangleleft$ iff $t_R, 1 \models \varphi^\blacktriangleleft_M$.*

The proof follows from the semantics of future and past operators of LTLp$_f$ provided in Sect. 2. For instance, verifying $\varphi^\blacktriangleleft = \diamond d$ on $t_2 = \langle$f, e, d, c, b, a, g, h, i$\rangle$ from Example 1 at instant $i = 9$ is equivalent to verifying $\varphi^\blacktriangleleft_M = \diamond d$ on $t_{2R} =$

$\langle i, h, g, a, b, c, d, e, f \rangle$ at $i = 1$. Notice that this holds for sub-traces too, thus verifying $\varphi^{\blacktriangleleft}$ on t_2 at instant $i = 6$ is equivalent to verifying $\varphi_M^{\blacktriangleleft}$ over $t_{2[1:6]R} = \langle a, b, c, d, e, f \rangle$ at $i = 1$ in the light of Lemma 1.

It follows from Lemma 2 that any pure past formula can be seen as a pure future one on a reversed trace. Therefore the automaton verifying the mirror image of $\varphi^{\blacktriangleleft}$ can be used for verification on the reversed trace, as stated in the following.

Corollary 1. *Let $A_{\varphi_M^{\blacktriangleleft}}$ be the automaton verifying $\varphi_M^{\blacktriangleleft}$. Then $t, n \models \varphi^{\blacktriangleleft}$ iff $t_R \in \mathscr{L}\left(A_{\varphi_M^{\blacktriangleleft}}\right)$.*

Notice that $\varphi_M^{\blacktriangleleft}$ is a pure future formula, therefore $A_{\varphi_M^{\blacktriangleleft}}$ can be built by applying the technique of [2] for LTL$_f$. Furthermore, it is possible to transform the obtained automaton in order to read directly the original trace t thanks to the property of *closure under reversion* of regular languages [13].

Definition 10 (Reversed automaton [13]). *Given a trace $t \in \Sigma^*$, its reverse t_R, and the automaton $A \in \mathcal{A}$, the reversed automaton $\overleftarrow{A} \in \mathcal{A}$ is an automaton such that A accepts t if and only if $\overleftarrow{A}$ accepts t_R, i.e., $t \in \mathscr{L}(A)$ iff $t_R \in \mathscr{L}\left(\overleftarrow{A}\right)$.*

From Lemma 2 and Corollary 1 we derive the following.

Theorem 4 (Valuation through reversed automaton of mirror image). *Let $\varphi^{\blacktriangleleft}$ be a pure past formula and $\varphi_M^{\blacktriangleleft}$ its mirror image. Let $A_{\varphi_M^{\blacktriangleleft}} \in \mathcal{A}$ be the automaton verifying $\varphi_M^{\blacktriangleleft}$. Given a trace $t \in \Sigma^*$ of length n, we have that: $t, n \models \varphi^{\blacktriangleleft}$ iff $t \in \mathscr{L}\left(\overleftarrow{A}_{\varphi_M^{\blacktriangleleft}}\right)$.*

Consider the RCon $\Psi = a \square\!\!\rightarrow (\ominus b \vee \Diamond c)$ and the pure past formula of its sep.aut.set $\varphi^{\blacktriangleleft} = \ominus b$. It is activated by $t_2(6)$. Its mirror image is $\varphi_M^{\blacktriangleleft} = \bigcirc b$. With automaton $A_{\varphi_M^{\blacktriangleleft}}$, depicted in Fig. 1, we can verify $\varphi_M^{\blacktriangleleft}$ over trace $t_{2[1:6]R} = \langle a, b, c, d, e, f \rangle$ at $i = 1$, thereby verifying $\varphi^{\blacktriangleleft}$ over $t_{2[1:6]}$ as per Lemma 2. Thanks to Theorem 4, $\varphi_M^{\blacktriangleleft}$ can be verified on $t_{2[1:6]}$ with the reversed automaton $\overleftarrow{A}_{\varphi_M^{\blacktriangleleft}}$, depicted in the top-left corner of Table 3.

We remark that in this way the pure past formulae of sep.aut.sets can be verified by parsing sub-traces from the beginning of the trace till the activator event.

4   The Janus Algorithm

Algorithm 1 shows the pseudo-code of our on-line technique to compute the interestingness degree $\zeta(\Psi, t)$ of a RCon Ψ with respect to a trace t. Its fundamental data structure is a set of pairs, each associating an automaton to its current state, as the replay of the trace proceeds. We call it *Janus state* and denote it as $\mathcal{J}$.

More specifically, only future automata of sep.aut.sets are considered. The naïve approach would indeed parse trace t, and apply the check of Theorem 3 whenever Ψ is activated. This is an impractical solution, because it requires the replay of prefix $t_{[1:i]}$ and suffix $t_{[i:n]}$ on the respective automata at every instant of activation i. We save computation time by keeping track of the past valuation state, so that at each activation it is already known, thus improving on the sub-valuation of pure past formulae $\varphi^{\blacktriangleleft}$. We rely on the fact that automata preserve the history of replays in the reached state: if at

---

Algorithm 1: Computing the interestingness degree of an RCon $\Psi = \alpha \boxminus\rightarrow \varphi$ on trace t, given the sep.aut.set $\mathcal{A}^{\blacktriangleleft\blacktriangledown\blacktriangleright}$ of φ

---

1   $O \leftarrow$ empty bag ;
2   **foreach** event $t(i) \in t$ **do**
3   **foreach** $A^{\blacktriangleleft} \in \mathcal{A}^{\blacktriangleleft\blacktriangledown\blacktriangleright}$ **do** perform transition $t(i)$ on $A^{\blacktriangleleft}$;
4   **if** $t(i) = \alpha$ **then**   // Activation triggered
5          $\mathcal{J} \leftarrow$ empty set of pairs ;   // $\mathcal{J}$ stores the replay-state of future automata for one activation
6   **foreach** $(A^{\blacktriangleleft}, A^{\blacktriangledown}, A^{\blacktriangleright}) \in \mathcal{A}^{\blacktriangleleft\blacktriangledown\blacktriangleright}$ **do**
7   **if** $A^{\blacktriangleleft}$ is in an accepting state and $A^{\blacktriangledown}$ accepts $\langle t(i) \rangle$ **then**
8   Take $A^{\blacktriangleright} = (\Sigma, S^{\blacktriangleright}, \delta^{\blacktriangleright}, s_0^{\blacktriangleright}, S_F^{\blacktriangleright})$ and add $(s_0^{\blacktriangleright}, A^{\blacktriangleright})$ to $\mathcal{J}$
9   add $\mathcal{J}$ to O;   // O collects replay-states for all activations
10   **foreach** $\mathcal{J} \in O$ **do**
11   **foreach** $(s^{\blacktriangleright}, A^{\blacktriangleright}) \in \mathcal{J}$ **do** $s^{\blacktriangleright} \leftarrow \delta^{\blacktriangleright}(s^{\blacktriangleright}, t(i))$ // Perform $t(i)$ on $A^{\blacktriangleright}$ and save state ;

12   **if** $|O| > 0$ **then return** $\dfrac{\left|\left\{ \mathcal{J} \in O : \text{ at least a } (s^{\blacktriangleright}, A^{\blacktriangleright}) \in \mathcal{J} \text{ is s.t. } s^{\blacktriangleright} \in S_F^{\blacktriangleright} \right\}\right|}{|O|}$ **else return** 0 ;

---

instant i the current state of $A^{\blacktriangleleft}$ is s_i, then at $i + 1$ it is known that $s_{i+1} = \delta(s_i, t(i + 1))$. To this extent, the algorithm requires all past automata $A^{\blacktriangleleft}$ to be already reversed as per Theorem 4.

The runtime of the algorithm will be explained considering as input: *(i)* $t = t_1 = \langle$d, f, a, f, c, a, f, b, a, f$\rangle$ from Example 1, *(ii)* $\Psi = $ a $\boxminus\rightarrow$ ($\ominus$b $\vee$ $\Diamond$c), and *(iii)* the sep.aut.set $\left\{ \left(A_{\ominus b}^{\blacktriangleleft}, A_{True}^{\blacktriangledown}, A_{True}^{\blacktriangleright} \right), \left(A_{True}^{\blacktriangleleft}, A_{True}^{\blacktriangledown}, A_{\Diamond c}^{\blacktriangleright} \right) \right\}$ of φ, with past automaton $A_{\ominus b}^{\blacktriangleleft}$ already reversed as depicted in Table 3. Let i denote the current instant in the following.

At lines 1–2,a bag of Janus states O is initialised, to store the states of current replays for the verification of pure future formulae. Every replay is triggered by the occurrence of an activation, thus every Janus state refers to an activator. Thereupon, trace t is parsed one event at a time starting with the left-first one, e.g., $t(1) = $ d. We remark that no knowledge is assumed on the subsequent events of the trace, as per the on-line setting.

At line 3, past automata replay t performing each transition $t(i)$ as it is read, i.e., not waiting for the activation to occur. By contrast, future automata will begin with independent replays at each occurrence of an activator. Therefore every $A^{\blacktriangleleft}$ automaton starts a replay from $i = 1$. At line 4, the activator is captured, as, e.g., when i is equal to 3, 6, and 9, i.e., when $t(i) = $ a. Consequently at line 5 a new Janus state $\mathcal{J}$ is initialised to store information on the new replay. At line 7 it is checked that, for every triple in the sep.aut.set, *(i)* $A^{\blacktriangleleft}$ is in an accepting state, and *(ii)* $A^{\blacktriangledown}$ accepts the trace made of the current event. If this is the case the replay on the future automaton of the triple, $A^{\blacktriangleright}$, can start. At line 8 the pair consisting of $A^{\blacktriangleright}$ and its initial state $s_0^{\blacktriangleright}$ is added to $\mathcal{J}$. In the example, the triple $\left(A_{\ominus b}^{\blacktriangleleft}, A_{True}^{\blacktriangledown}, A_{True}^{\blacktriangleright} \right)$ at $i = 3$ and $i = 6$ has $A_{\ominus b}^{\blacktriangleleft}$ not accepting the prefix $t_{[1:i]}$ because $t(i - 1) \neq $ b. On the contrary, the replay of $A^{\blacktriangleright}$ can start at $i = 9$. For what the triple $\left(A_{True}^{\blacktriangleleft}, A_{True}^{\blacktriangledown}, A_{\Diamond c}^{\blacktriangleright} \right)$ is concerned, $A_{\Diamond c}^{\blacktriangleright}$ always starts a new replay because $A_{True}^{\blacktriangleleft}$ and $A_{True}^{\blacktriangledown}$ accept any trace.

We remark that for $A^{\blacktriangledown}$ no state is retained because the scope of a pure present formula is limited to a single event. Notice that because every triple $(A^{\blacktriangleleft}, A^{\blacktriangledown}, A^{\blacktriangleright}) \in \mathcal{A}^{\blacktriangleleft\blacktriangledown\blacktriangleright}$ represents a conjunctive formula, if $A^{\blacktriangleleft}$ is not in an accepting state then the activation leads the entire triple to violation, regardless of the replay on $A^{\blacktriangleright}$.

The new Janus state $\mathcal{J}$ is added to O at line 9. In the example, at $i = 9$, $O = \left\{\left\{\left(s_1, A_{\diamond c}^{\blacktriangleright}\right)\right\}, \left\{\left(s_0, A_{\diamond c}^{\blacktriangleright}\right)\right\}, \left\{\left(s_0, A_{\diamond c}^{\blacktriangleright}\right), \left(s_0, A_{True}^{\blacktriangleright}\right)\right\}\right\}$. At lines 10–11 all states in every $\mathcal{J} \in O$ are updated by executing the current transition on their respective future automata. For example, upon the reading of $t(5) = c$, $\left\{\left(s_0, A_{\diamond c}^{\blacktriangleright}\right)\right\}$ is updated to $\left\{\left(s_1, A_{\diamond c}^{\blacktriangleright}\right)\right\}$.

An event in a trace interestingly fulfils an RCon if, upon activation, at least a triple $(A^{\blacktriangleleft}, A^{\blacktriangledown}, A^{\blacktriangleright})$ is such that $A^{\blacktriangleleft}$, $A^{\blacktriangledown}$, and $A^{\blacktriangleright}$ all accept the respective sub-traces, as per Theorem 3. By construction, this holds true if at least a future automaton in $\mathcal{J}$ is in its accepting state at the end of the replay. To measure the interestingness degree of the RCon, we thus compute the ratio between the number of all such Janus states and the cardinality of O at line 12. For instance, at $i = 10$, $O = \{\mathcal{J}_1, \mathcal{J}_2, \mathcal{J}_3\} = \left\{\left\{\left(s_1, A_{\diamond c}^{\blacktriangleright}\right)\right\}, \left\{\left(s_0, A_{\diamond c}^{\blacktriangleright}\right)\right\}, \left\{\left(s_0 A_{\diamond c}^{\blacktriangleright}\right), \left(s_0, A_{True}^{\blacktriangleright}\right)\right\}\right\}$ and $|O| = 3$. Only $\mathcal{J}_1$ and $\mathcal{J}_3$ contain automata in accepting states, therefore $\zeta(\Psi, t) = \frac{2}{3} = 0.667$.

Computational Cost. To compute the asymptotic computational cost, we consider the worst case scenario, occurring when at each instant *(i)* Ψ is activated, and *(ii)* all past and present automata are in an accepting state. In such a case, given an event log L, a trace t of length t, an RCon $\Psi = \alpha \mathbin{\square\!\!\rightarrow} \varphi$ for which the sep.aut.set $\mathcal{A}^{\blacktriangleleft\blacktriangledown\blacktriangleright}$ of φ is generated with separation degree D, the cost of verifying Ψ on t is:

$$\sum_{j=1}^{n} (D + (n - j)D) = nD + D \sum_{j=1}^{n} (n - j) = Dn\left(1 + \frac{(n-1)}{2}\right).$$

The cost is linear in the number of activations, as each one requires a single replay of the trace for every automaton. For each activation only trace suffixes are replayed, owing to our optimisation over past formulae, hence the $^1/_2$ factor. For $D \ll n$, the upper-bound is $O(n^2)$, which is comparable to state-of-the-art techniques [6, 19]. Because every trace of L is parsed singularly, the cost is $O(|L|)$. Finally, denoting as m the maximum amount of parameters of a template, the cost is $O(|\Sigma|^m)$ because constraints are verified for every permutation of symbols in alphabet Σ. For standard DECLARE, e.g., $m = 2$.

5   Evaluation

A proof-of-concept implementation of our technique has been developed for experimentation. It is available for download at https://github.com/Oneiroe/Janus. We compare Janus to other declarative process mining techniques, highlighting specific properties and scenarios through synthetic logs. Thereafter, we evaluate our tool against a real-world event log and compare the output to a reference model.

5.1   State-of-the-art Declarative Process Mining Approaches

The following state-of-the-art declarative process discovery algorithms have been considered for comparison: *(i)* Declare Maps Miner (DMM) [18], the first declarative process miner, based on the replay of traces on automata; *(ii)* Declare Miner2(DM2) [19], adopting DECLARE-specific heuristics to improve on original DMM performance; *(iii)*

Table 4. Characteristics of declarative process mining approaches

	Extendibility	On-line	Interestingness	Granularity
Declare Maps Miner [18]	✓	✓	Ad hoc for DECLARE	Traces
Declare Miner 2 [19]	✗	✗	Ad hoc for DECLARE	Traces
MINERful [6]	✗	✗	✗	Events over log
MINERful Vacuity Checker [20]	✓	✓	Vacuity	Traces
Janus	✓	✓	✓	Events over traces

MINERful(Mf) [6], building a knowledge base of task co-occurrence statistics, so as to mine DECLARE constraints via queries; *(iv)* MINERful Vacuity Checker (Mf-Vchk) [20], extending Mf to include semantical vacuity detection. Criteria for comparison reflect the goals of our research. They are: *(i) extendibility* over custom templates beyond standard DECLARE, *(ii)* capability of performing the analysis *on-line*, *(iii)* characterisation of constraint *interestingness*, and *(iv) granularity* of support and confidence measures with respect to the event log. Table 4 summarizes the outcome of our comparison.

Extendibility. DECLARE has been introduced as a language to be extended by users with the addition of custom templates [23]. DMM allows indeed for the check of any LTL$_f$ formula. On the other hand Mf and DM2 work only with the DECLARE standard template set. Any new constraint outside this scope needs to be hard-coded into the framework. Janus allows for the check of any LTLP$_f$ formula expressing an RCon.

On-Line Analysis. By design, only DMM and Mf-Vchk can be employed in *on-line* settings like run-time monitoring, as well as Janus, whilst DM2 and Mf operate off-line.

Interestingness. The core rationale of *interestingness* is to distinguish whether a constraint is not only satisfied but also activated in the trace. DMM and DM2 provide ad-hoc solutions only for the DECLARE standard template set, because for each template the activation condition is hard-coded. Mf checks only the satisfaction of constraints. Mf-Vchk instead relies on a semantical vacuity detection technique independent from the specific constraint or language. Nevertheless it provides misleading results with constraints involving implications in the past such as PRECEDENCE. In Janus, RCons are such that the activation is singled out in their formulation itself and its occurrence treated as a trigger in their semantics, so as to overcome those issues by design and address interestingness.

Granularity. DMM and DM2 calculate the support as the percentage of fully compliant traces. Similarly, Mf-Vchk calculates it as the percentage of traces that non-vacuously satisfy the constraint. Mf calculates support as the percentage of activations leading to constraint satisfaction over the entire event log, therefore the analysis is at the level of single events. Janus support mediates between them as it is computed as the average of interestingness degree values over the traces in the log.

5.2   Comparative Evaluation over Synthetic Logs

Synthetic logs have been constructed to show the behaviour of mining techniques in specific scenarios highlighting how the differences seen in Sect. 5.1 influence the discovery outcome. Table 5 summarizes the results.

Table 5. Results of experiments over synthetic logs for comparative evaluation

		False positives			Numerous activations in trace	Partial satisfaction
	Supp.	Triggering tr.s	Discarded tr.s	Violating tr.s	Supp.	Supp.
Mf	1.000	1000	0	0	0.848	0.941
Mf-VChk	0.881	881	119	0	0.100	0.875
DMM/DMM2	0.231	231	769	0	0.100	0.875
Janus	0.231	231	769	0	0.100	0.979

False Positives. Columns 2–5 of Table 5 show the results of an experiment conducted on a synthetic log of 1000 traces, simulating a model consisting only of PRECEDENCE(d, a). The event log, built with the MINERful log generator [4], has a high amount of non-triggering traces for that constraint: 231 traces contain a and d, 650 only d, 119 neither a nor d. The goal of this experiment is to show the distinction between interestingness and non-vacuity. Mf has natively no vacuity nor interestingness detection mechanisms. Both DMM and Janus recognise correctly the triggering traces, and discard the remaining ones when computing support. As said, the vacuity detection approach of Mf-Vchk shows instead misleading results with constraints involving a time-backward implication. In this case, it recognises as non-vacuous those traces that contain d, with or without any a. From a vacuity-detection perspective, it is logically correct because if d occurs, the constraint is satisfied regardless of the occurrence of a. Nevertheless, it is misleading because the activation of PRECEDENCE(d, a) is a, not d. Traces without a-events satisfy the constraint without any occurrence of the activation, and should not be considered as interesting – hence the name of the experiment, "False positives". Janus prevents false positives thanks to the semantics of Reactive Constraints.

Numerous Activations in Trace. Column 6 of Table 5 shows the support of PRECEDENCE(d, a) on event log $L = \{\!|t_1^1, t_2^9|\!\}$, with $t_1 = \langle \mathsf{da} \ldots \mathsf{a} \rangle$ and $t_2 = \langle \mathsf{a} \rangle$. Trace t_1 satisfies the constraint and consists of a sequence of 50 a-events following a single d, but its multiplicity in the log is 1. Trace t_2 violates the constraint, and its multiplicity is 9. The goal is to show the misleading influence of the number of activations in a trace on support calculation. DMM, DM2, Mf-Vchk, and Janus are not influenced by

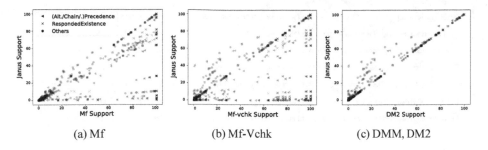

| (a) Mf | (b) Mf-Vchk | (c) DMM, DM2 |

Fig. 2. Comparison of support values between Janus and state-of-the-art algorithms

the number of activations in t_1 because they compute support as an average over whole traces. Mf is instead highly influenced by t_1, because it contains the majority of the constraint activations, despite the higher multiplicity of t_2. We argue that the rate of interesting fulfilments per trace, not the their total amount in the event log, should be considered. Janus follows this rationale.

Partial Satisfaction. Column 7 of Table 5 shows the support of PRECEDENCE(d, a) on event log $L = \{\!|t_1^1, t_2^4, t_3^3|\!\}$, with $t_1 = \langle a, d, a, a, a, a, a\rangle$, $t_2 = \langle d, a, a\rangle$, and $t_3 = \langle d, a\rangle$.

Activations lead to fulfilment in all cases except the first event of t_1. Slight violations are common in real-life logs, thus there is the risk of losing valuable information if entire traces are discarded for that. The goal is to show how the discovery techniques behave in such situations. MF overcomes this issue owing to its support calculation at the granularity level of events. DMM and Mf-Vchk instead discard completely those traces because of the single violation. Janus assigns a lower interestingness degree to t_1, but does not discard it.

With synthetic event logs we have shown that Janus is capable of discerning interesting fulfilments, is resilient to noise, and balances the number of activations with the multiplicity of traces for support calculation. Next we show insights on the application of Janus on a real-world event log.

5.3   Evaluation on a Real-Life Event Log

Healthcare processes are known to be highly flexible [11], thus being suitable for declarative process modelling approaches [23]. The real-life log referred to as *Sepsis* [1] has been thus analysed. It reports the trajectories of patients showing symptoms of sepsis in a Dutch hospital. The model mined by Janus with support threshold equal to 10% and confidence threshold equal to 94% consists of the following constraints:

---

[1] https://doi.org/10.4121/uuid:915d2bfb-7e84-49ad-a286-dc35f063a460.

INIT(ER Registration)	ALT.PRECEDENCE(ER Registration, ER Triage)	ALT.PRECEDENCE(Admission NC, Return ER)
ALT.PRECEDENCE(ER Triage, ER Sepsis Triage)	ALT.PRECEDENCE(ER Triage, Return ER)	PRECEDENCE(ER Triage, Admission NC)
RESPONDEDEXISTENCE(IV Antibiotics, Lactic Acid)	ALT.PRECEDENCE(Admission IC, CRP)	PRECEDENCE(ER Triage, Admission IC)
RESPONDEDEXISTENCE(IV Liquid, IV Antibiotics)	ALT.PRECEDENCE(Leucocytes, Release A)	ALT.RESPONSE(ER Registration, ER Triage)
RESPONDEDEXISTENCE(IV Liquid, Lactic Acid)	ALT.PRECEDENCE(ER Triage, Release A)	ALT.RESPONSE(ER Registration, Leucocytes)
PRECEDENCE(ER Registration, CRP)	ALT.PRECEDENCE(CRP, Return ER)	ALT.PRECEDENCE(ER Sepsis Triage, IV Antibiotics)
PRECEDENCE(ER Registration, Leucocytes)	ALT.PRECEDENCE(Admission IC, Leucocytes)	ALT.PRECEDENCE(Leucocytes, Return ER)
PRECEDENCE(ER Registration, Admission IC)	ALT.PRECEDENCE(CRP, Release A)	ALT.RESPONSE(ER Triage, ER Sepsis Triage)
ALT.RESPONSE(ER Registration, CRP)		

We remark that all constraints comply with the normative process model reported in [21], which was designed iteratively with the help of domain experts. To check conformance of the mined model with the event log we have utilised the Declare Replayer plug-in [15], yielding fitness equal to 98%.

Figure 2 compares the output of Janus with the one of other declarative process mining techniques. Figure 2(a) shows that in absence of a dedicated mechanism, support is a metric that is not sensitive to interestingness. Indeed Mf assigns a support of 1.0 to several constraints to which Janus attributes a value ranging from 0.0 to 1.0, whereby the latter accounts for interestingness as shown throughout this paper. The semantical vacuity detection of Mf-Vchk partially solves this issue, as it can be noticed in Fig. 2(b). Still there are constraints assigned with a support of about 1.0 by Mf-Vchk against a value spanning over the whole range from 0.0 to 1.0 as per Janus. As expected, they are those constraints that have a time-backward component. For example, the support of CHAINPRECEDENCE(Leucocytes, Release C) on the log is 0.948 for Mf-Vchk, and 0.008 for Janus. However, other tasks than Leucocytes can immediately precede Release C as per the reference model of [21]. The constraint is indeed rarely activated, i.e., Release C occurs in 0.024% of the traces in the log, as opposed to Leucocytes, occurring in 0.964% of the traces and thus causing the misleading result of Mf-Vchk. Figure 2(c) shows that the DECLARE-tailored vacuity detection technique of DMM and DM2 prevents uninteresting constraints to be returned. Nevertheless, the support computation based solely on trace satisfaction of DMM and DM2 makes the assigned support be a lower bound for Janus.

The time taken by Janus to mine the event log amounted to 9s on a machine equipped with an Intel Core i5-7300U CPU at 2.60 GHz, quad-core, Ubuntu 17.10 operating system. The timing is in line with existing approaches that tackle vacuity detection: 3 s for DM2, 9 s for Mf-Vchk, and 270 s for DMM on the same machine.

6   Conclusion

In this paper, we have described Janus, an approach to model and discover Reactive Constraints on-line from event logs. With the Janus approach users can single out activations in the constraint expressions, namely the triggering conditions. Thereby interesting constraint fulfilments are discerned from the uninteresting ones by checking whether the activation occurred. Experimental evidence on synthetic and real-life event logs confirms its compatibility with previous DECLARE discovery techniques, yet improving on the capability of identifying interestingly fulfilled constraints.

For our research outlook we aim at enriching RCons by allowing for activations expressed as temporal logic formulae rather than propositional symbols, inspired by [3] and [17]. It is in our plans to integrate the MONA tool [10] for the automatic generation of automata for verification, as suggested in [27]. We will also study the application of Janus on run-time monitoring and off-line discovery tasks. Furthermore, future work will include the analysis of other declarative languages such as DCR graphs [11], and of multi-perspective approaches encompassing time, resources, and data. To that extent, we will employ metrics beyond support and confidence.

Acknowledgements. The work of Alessio Cecconi, Claudio Di Ciccio, and Jan Mendling has been funded by the Austrian FFG grant 861213 (CitySPIN) and from the EU H2020 programme under MSCA-RISE agreement 645751 (RISE_BPM). Giuseppe De Giacomo has been partially supported by the Sapienza project "Immersive Cognitive Environments".

References

1. Adamo, J.: Data Mining for Association Rules and Sequential Patterns - Sequential and Parallel Algorithms. Springer, New York (2001). https://doi.org/10.1007/978-1-4613-0085-4
2. De Giacomo, G., De Masellis, R., Montali, M.: Reasoning on LTL on finite traces: insensitivity to infiniteness. In: AAAI, pp. 1027–1033 (2014)
3. De Giacomo, G., Vardi, M.Y.: Linear temporal logic and linear dynamic logic on finite traces. In: IJCAI, pp. 854–860. Association for Computing Machinery (2013)
4. Di Ciccio, C., Bernardi, M.L., Cimitile, M., Maggi, F.M.: Generating event logs through the simulation of declare models. In: Barjis, J., Pergl, R., Babkin, E. (eds.) EOMAS 2015. LNBIP, vol. 231, pp. 20–36. Springer, Cham (2015). https://doi.org/10.1007/978-3-319-24626-0_2
5. Di Ciccio, C., Maggi, F.M., Montali, M., Mendling, J.: Resolving inconsistencies and redundancies in declarative process models. Inf. Syst. **64**, 425–446 (2017)
6. Di Ciccio, C., Mecella, M.: On the discovery of declarative control flows for artful processes. ACM Trans. Manag. Inf. Syst. (TMIS) **5**(4), 24 (2015)
7. Dwyer, M.B., Avrunin, G.S., Corbett, J.C.: Patterns in property specifications for finite-state verification. In: ICSE, pp. 411–420. IEEE (1999)
8. Gabbay, D.: The declarative past and imperative future. In: Banieqbal, B., Barringer, H., Pnueli, A. (eds.) Temporal Logic in Specification. LNCS, vol. 398, pp. 409–448. Springer, Heidelberg (1989). https://doi.org/10.1007/3-540-51803-7_36
9. Gastin, P., Oddoux, D.: LTL with past and two-way very-weak alternating automata. In: Rovan, B., Vojtáš, P. (eds.) MFCS 2003. LNCS, vol. 2747, pp. 439–448. Springer, Heidelberg (2003). https://doi.org/10.1007/978-3-540-45138-9_38
10. Henriksen, J.G., Jensen, J., Jørgensen, M., Klarlund, N., Paige, R., Rauhe, T., Sandholm, A.: Mona: monadic second-order logic in practice. In: Brinksma, E., Cleaveland, W.R., Larsen, K.G., Margaria, T., Steffen, B. (eds.) TACAS 1995. LNCS, vol. 1019, pp. 89–110. Springer, Heidelberg (1995). https://doi.org/10.1007/3-540-60630-0_5
11. Hildebrandt, T., Mukkamala, R.R., Slaats, T.: Declarative modelling and safe distribution of healthcare workflows. In: Liu, Z., Wassyng, A. (eds.) FHIES 2011. LNCS, vol. 7151, pp. 39–56. Springer, Heidelberg (2012). https://doi.org/10.1007/978-3-642-32355-3_3
12. Hodkinson, I.M., Reynolds, M.: Separation-past, present, and future. In: We Will Show Them!, vol. 2, pp. 117–142 (2005)

13. Hopcroft, J.E., Motwani, R., Ullman, J.D.: Introduction to Automata Theory, Languages, and Computation, 3rd edn. Pearson/Addison Wesley, Boston (2007)
14. Kupferman, O., Vardi, M.Y.: Vacuity detection in temporal model checking. Int. J. Softw. Tools Technol. Transf. **4**(2), 224–233 (2003)
15. de Leoni, M., Maggi, F.M., van der Aalst, W.M.P.: An alignment-based framework to check the conformance of declarative process models and to preprocess event-log data. Inf. Syst. **47**, 258–277 (2015)
16. Lichtenstein, O., Pnueli, A., Zuck, L.: The glory of the past. In: Parikh, R. (ed.) Logic of Programs 1985. LNCS, vol. 193, pp. 196–218. Springer, Heidelberg (1985). https://doi.org/10.1007/3-540-15648-8_16
17. Ly, L.T., Rinderle-Ma, S., Knuplesch, D., Dadam, P.: Monitoring business process compliance using compliance rule graphs. In: Meersman, R., Dillon, T., Herrero, P., Kumar, A., Reichert, M., Qing, L., Ooi, B.-C., Damiani, E., Schmidt, D.C., White, J., Hauswirth, M., Hitzler, P., Mohania, M. (eds.) OTM 2011. LNCS, vol. 7044, pp. 82–99. Springer, Heidelberg (2011). https://doi.org/10.1007/978-3-642-25109-2_7
18. Maggi, F.M., Bose, R.P.J.C., van der Aalst, W.M.P.: Efficient discovery of understandable declarative process models from event logs. In: Ralyté, J., Franch, X., Brinkkemper, S., Wrycza, S. (eds.) CAiSE 2012. LNCS, vol. 7328, pp. 270–285. Springer, Heidelberg (2012). https://doi.org/10.1007/978-3-642-31095-9_18
19. Maggi, F.M., Di Ciccio, C., Di Francescomarino, C., Kala, T.: Parallel algorithms for the automated discovery of declarative process models. Inf. Syst. **74**, 136–152 (2018)
20. Maggi, F.M., Montali, M., Di Ciccio, C., Mendling, J.: Semantic vacuity detection in declarative process mining. In: La Rosa, M., Loos, P., Pastor, O. (eds.) BPM 2016. LNCS, vol. 9850, pp. 158–175. Springer, Cham (2016). https://doi.org/10.1007/978-3-319-45348-4_10
21. Mannhardt, F., Blinde, D.: Analyzing the trajectories of patients with sepsis using process mining. In: RADAR+ EMISA, vol. 1859, pp. 72–80 (2017)
22. Markey, N.: Past is for free: on the complexity of verifying linear temporal properties with past. Electr. Notes Theor. Comput. Sci. **68**(2), 87–104 (2002)
23. Pesic, M.: Constraint-based workflow management systems: shifting control to users (2008)
24. Ramakrishna, Y.S., Moser, L.E., Dillon, L.K., Melliar-Smith, P.M., Kutty, G.: An automata-theoretic decision procedure for propositional temporal logic with since and until. Fundam. Inform. **17**(3), 271–282 (1992)
25. Reynolds, M.: The complexity of temporal logic over the reals. Ann. Pure Appl. Log. **161**(8), 1063–1096 (2010)
26. Vardi, M.Y., Wolper, P.: An automata-theoretic approach to automatic program verification. In: LICS, pp. 322–331. IEEE Computer Society (1986)
27. Zhu, S., Tabajara, L.M., Li, J., Pu, G., Vardi, M.Y.: Symbolic LTLf synthesis. In: IJCAI, pp. 1362–1369 (2017)

Unbiased, Fine-Grained Description of Processes Performance from Event Data

Vadim Denisov[1(✉)], Dirk Fahland[1], and Wil M. P. van der Aalst[1,2]

[1] Eindhoven University of Technology, Eindhoven, The Netherlands
{v.denisov,d.fahland}@tue.nl
[2] Department of Computer Science, RWTH Aachen, Aachen, Germany
wvdaalst@pads.rwth-aachen.de

Abstract. Performance is central to processes management and event data provides the most objective source for analyzing and improving performance. Current process mining techniques give only limited insights into performance by aggregating all event data for each process step. In this paper, we investigate process performance of all process behaviors without prior aggregation. We propose the *performance spectrum* as a simple model that maps all observed flows between two process steps together regarding their performance over time. Visualizing the performance spectrum of event logs reveals a large variety of very distinct *patterns of process performance* and performance variability that have not been described before. We provide a taxonomy for these patterns and a comprehensive overview of elementary and composite performance patterns observed on several real-life event logs from business processes and logistics. We report on a case study where performance patterns were central to identify systemic, but not globally visible process problems.

Keywords: Process mining · Performance analysis · Visual analytics

1 Introduction

Performance analysis is an important element in process management relying on precise knowledge about actual process behavior and performance to enable improvements [11]. Descriptive performance analysis has been intensively studied within process mining, typically by annotating discovered or hand-made models with time-related information from event logs [1–3,23] as illustrated in Fig. 1 (left). These descriptive models provide aggregate measures for performance over the entire data such as average or maximum waiting times between two process steps. Models for predicting waiting times until the next step or remaining case duration learned from event data distinguish different performance classes or distribution functions based on case properties [4,5,12,15].

However, these techniques assume the timed-related observations to be taken from stationary processes that are executed in isolation, i.e., that distribution

© Springer Nature Switzerland AG 2018
M. Weske et al. (Eds.): BPM 2018, LNCS 11080, pp. 139–157, 2018.
https://doi.org/10.1007/978-3-319-98648-7_9

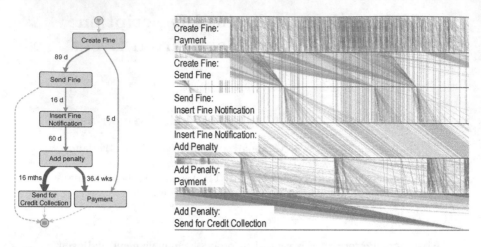

Fig. 1. Performance analysis using a graph-based model (left) and the performance spectrum (right); see online version for colored figures.

functions describing performance of a case do not change over time and do not depend on other cases. These assumptions are often made by a lack of a more precise understanding of the (changes in) process performance across cases and over time.

In this paper, we consider the problem of descriptive analytics of process behavior and performance *over time*. In particular, we aim to provide a *comprehensive* description of raw process behavior without enforcing prior aggregation of data. Note that Fig. 1 (left) only shows aggregates performance and no temporal patterns or changes over time. The objective of this comprehensive description is to identify patterns, trends, and properties of interest without the representational bias of an algorithm or a particular formal model.

We approach the problem through *visual analytics* which employs structuring of data in a particular form that, when visualized, allows offloading the actual data processing to the human visual system [7] to identify patterns of interest for subsequent analysis. We propose a new simple model for event data, called the *performance spectrum* and a corresponding visualization. Figure 1 (right) shows the performance spectrum of the data used to the discover the model in Fig. 1 (left) over a 20 month period. The performance spectrum describes the event data in terms of *segments*, i.e., pairs of related process steps; the performance of each segment is measured and plotted for any occurrences of this segment over time and can be classified, e.g., regarding the overall population.

The visualization in Fig. 1 (right) shows that different cases perform very differently due to systematic and unsystematic *variability of performance in the different steps over time* and synchronization of multiple cases. We implemented this visualization in an *interactive* exploration tool (ProM package "Performance Spectrum") allowing for zooming, filtering, and performance classification and aggregation of the data.

Exploring the performance spectrum of real-life logs typically reveals numerous, novel patterns in process performance and behavior as shown in Fig. 1 (right) that cannot be seen in process models as in Fig. 1 (left). To enable documenting and conceptualizing these patterns for further analysis, we propose a *taxonomy* for describing elementary patterns in the performance spectrum. We evaluated the performance spectrum and the taxonomy on 12 real-life logs of business and logistics processes. Numerous elementary patterns as well as larger patterns composed of elementary ones recur throughout different event logs. We show how these patterns reveal novel insights into the interplay of control-flow, resource, and time perspective of processes. The performance spectrum of real-life logs reveals that performance in a case may be dependent on the performance of other cases, performance generally varies over time (non-stationary), and many processes exhibit temporary or permanent concept drift. We report on a case study performed with Vanderlande Industries to identify control-flow problems in very large logistics processes. Further, we found that each process has a characteristic *signature* of the patterns in its performance spectrum and that similar signatures indicate extremely similar processes not only in control-flow but also in the performance perspective.

The remainder of this paper is structured as follows. We discuss work related to performance analysis in Sect. 2. We formally define the performance spectrum in Sect. 3 and introduce the taxonomy for patterns in the performance spectrum in Sect. 4. We report on our evaluation on real-life event logs in Sect. 5 and discuss our findings and future work in Sect. 6.

2   Related Work

Analysis of process performance from *event data* can be divided into descriptive, predictive, and visual analysis, which we summarize here; see [12] for an extensive discussion.

Commonly, process performance is described by *enhancing* a given or discovered process model with information about durations of activities (nodes in a model) or waiting times between activities (edges in a model) [1]. In the visualization, each node and edge can be annotated with one aggregate performance measure (avg., sum, min, max) for all cases passing through this node or edge, as illustrated in Fig. 1 (left). Visualization of performance on a model is more accurate if the discovery algorithm takes the underlying performance information into account [8,18]. A non-fitting log can be aligned to a model to visualize performance [2]. More detailed visualization of performance requires more dimensions. Wynn et al. [23] plot different process variants (with different performance) into a 3-dimensional space "above" the process model. Transition system discovery allows to split occurrences of an activity based on its context and visualize performance each context separately [3].

Performance prediction for the *remaining time* until completion of a given case can be predicted by regression models [5], by annotating transition system states with remaining times [4], by learning a clustering of transition system states [6], by combining models for prediction of the next activity in a case with regression models [12]. Completion time of the next activity can be predicted by training an LSTM neural network [22], or by learning process models with arbitrary probability density functions for time delays through non-parametric regression from event logs [14] that can also be used for learning simulation models to predict performance [15,17]. These models predict performance of a single case based on case-specific features. Performance of cases synchronizing on shared resources can be analyzed through simulation models [16] or from queuing models [19] learned from event logs. Synchronization in batch activities can be studied through queue models [13], or through aggregating event logs into a matrix [10].

The above techniques assume that probability densities for time delays are *stationary* for the whole process (do not change over time) or only depend on the individual case (*isolation* between cases). Techniques for *describing the performance of all cases* construct simpler models through stronger aggregation [3]. Also, the recent temporal network representation abstracts non-stationary changes in performance over time [18]. Techniques for predicting *performance of a single case* construct more complex models for higher precision, e.g., [22]; precision increases further when more assumption are dropped, e.g., different distribution functions [15], queues [19], non-stationarity [12]. No current model learning technique can describe process performance without making assumptions about the data. However, the results of this paper show that in particular stationarity and isolation of cases do not hold in the performance perspective.

Assumptions and representational bias of models can be avoided through visualization and visual analytics [7]. Dotted Chart [21] plots all events per case (y-axis) over time (x-axis) allowing to observe arrival rates and seasonal patterns over time. Story graphs [20] plot a case as poly-line in a plane of event types (y-axis) and time (x-axis) allowing to observe patterns of similar cases wrt. behavior and performance over time but convolutes quickly with many crossing lines.

In Sect. 3 we propose a model and visualization that avoids the problems of [20] in describing the *performance of each process step* without assumptions about the data (except having a log of discrete events). The visualization shall *reveal where a process violates typical assumption about performance* such as non-stationarity or cases influencing each other; we provide a *taxonomy* to describe these phenomena in Sect. 4.

3   Performance Spectrum

We first establish some basic notations for events and logs, and then introduce our model to describe the performance of any observable dynamic process over time.

Let A be a set of *event classifiers*; A is usually the set of activity names, but it may also be the set of resource names, or a set of locations. Let $(T, \leq, +, \cdot, 0)$ be a totally ordered set (with addition $+$, multiplication $\cdot$, and 0) of timestamps, e.g., the rational numbers $\mathbb{Q}$. An *event* $e = (a, t) \in (A \times T)$ describes that observation a occurred at time t. A *trace* $\sigma \in (A \times T)^*$ is a finite sequence of related events. An event log $L \in \mathbb{B}((A \times T)^*)$ is a multi-set of traces. For $\sigma = \langle e_1, \ldots, e_n \rangle$, we write $|\sigma| = n$ and $\sigma_i = e_i, i = 1, \ldots, n$.

The goal of the performance spectrum is to visualize the performance of process steps over time. We call $(a, b) \in A \times A$ a *process segment* describing a step from activity a to activity b, hand-over of work from resource a to b or the movement of goods from location a to b. We first formalize the *performance spectrum* for a single process segment, and then lift this to *views* on a process.

Each occurrence of a segment (a, b) in a trace $\langle \ldots, (a, t_a), (b, t_b), \ldots \rangle$ allows to measure the time between occurrences of a and b. A histogram $H = H(a, b, L) \in \mathbb{B}(T)$ describes how often all the *time differences* $t_b - t_a$ between a and b have been observed in L. In contrast, the *performance spectrum* $\mathbb{S}(a, b, L)$ collects the actual *time intervals* (t_a, t_b) observed in L. To aid recognition of patterns, we allow users to classify each interval (t_a, t_b) wrt. other observations. The specific classification depends on the analysis at hand, for example, the actual duration $t = t_b - t_a$, or whether $t = t_b - t_a$ is in the 25%-quartile of the histogram H, or other properties such as remaining time until case completion. Generally, a *performance classification* function $\mathbb{C} \in T \times T \times \mathbb{B}((A \times T)^*) \to C$ maps any interval (t_a, t_b) into a class $\mathbb{C}(t_a, t_b, L) = c \in C$. Figure 1 classifies intervals based on the quartile of the duration in the histogram.

Definition 1 (Detailed performance spectrum). *The* performance spectrum *of a segment (a, b) is the bag of all its observation intervals in a trace σ (in a log L): $\mathbb{S}(a, b, \sigma) = [(t_a, t_b) \mid \exists_{1 \leq i < |\sigma|}(a, t_a) = \sigma_i, (b, t_b) = \sigma_{i+1}] \in \mathbb{B}(T \times T)$; we lift $\mathbb{S}$ to L by bag union $\mathbb{S}(a, b, L) = \sum_{\sigma \in L}(L(\sigma) \cdot \mathbb{S}(a, b, \sigma))$. The detailed performance spectrum of a segment (a, b) in log L wrt. performance classification $\mathbb{C}$ is $\mathbb{S}^{\mathbb{C}}(a, b, L) = [(t_a, t_b, c) \mid (t_a, t_b) \in \mathbb{S}(a, b, L), c = \mathbb{C}(t_a, t_b, L)] \in \mathbb{B}(T \times T \times C)$.*

Figure 1 (right) visualizes the detailed performance spectrum $S = \mathbb{S}(a, b, L)$ of six different segments. For each segment (a, b) we fix coordinates y_a and y_b on the y-axis and plot each classified observation $(t_a, t_b, c) \in \mathbb{S}^{\mathbb{C}}(a_i, b_i, L)$ as a line from (t_a, y_a) to (t_b, y_b). In Fig. 1 each line is colored based on the quartile of the duration $t_b - t_a$.

The detailed performance spectrum visualizes variability of durations in a segment across cases and time. To capture and visualize also the amount of cases of particular performance over time, we define an *aggregate performance spectrum*. We *group* segments into bins of a user-chosen *period p* depending on whether they *start*, *stop*, or are *pending* in a bin, and then aggregating the observations (t_a, t_b, c) in each bin wrt. their class c (for *finitely many* classes C), akin to relational algebra or SQL operations.

Definition 2 (Aggregated performance spectrum). *Let $S = \mathbb{S}(a, b, L)$ be a detailed performance spectrum with finite performance classes $C = \{c^1, \ldots, c^k\}$. Let period $p \in T$ and grouping $g \in \{start, stop, pending\}$. The binning of S wrt. p and g is the sequence of multisets $\langle b_0, b_1, \ldots \rangle$ such that for $i = 0, 1, \ldots$ holds*

$- b_i = [(t_a, t_b, c) \in S \mid i \cdot p \leq t_a < (i + 1) \cdot p]$ *if $g = start$,*

- $b_i = [(t_a, t_b, c) \in S \mid i \cdot p \leq t_b < (i+1) \cdot p]$ if $g = stop$, and
- $b_i = [(t_a, t_b, c) \in S \mid i \cdot p \leq t_b$ and $t_a < (i+1) \cdot p]$ if $g = pending$ (the segment started before the end of the bin, and ends after the start of the bin).

The aggregation of S wrt. p and g is the sequence $agg_g(S, p)$ of vectors $\langle v_0, v_1, \ldots \rangle$ where each $v_i = (v_i^1, \ldots, v_i^j, \ldots, v_i^k) \in \mathbb{N}^k$ counts how often performance class c^j occurred in bin b_i: $v_i^j = |\{(t_a, t_b, c^j) \mid (t_a, t_b, c^j) \in b_i\}|$. The aggregated performance spectrum of a segment (a, b) in a log L is then $\mathbb{S}_{g,p}^{\mathbb{C}}(a, b, L) = agg_g(\mathbb{S}^{\mathbb{C}}(a, b, L), p)$.

An aggregated performance spectrum A of one segment (a, b) can be visualized as a series of stacked bar-charts as shown in Fig. 7 where the k-th bar starts at x-coordinate $k \cdot p$ and has width p; the bottom-line of the series of bar-charts is at y-coordinate y_b and the height of all bars is normalized wrt. $y_b - y_a$.

Visualizing the performance spectrum of multiple process segments on a 2D plane requires some compromises. As the x-axis of the plane is used for visualizing time, we can only visualize control-flow by mapping segments along the single dimension of the y-axis. This forces to visualize even alternative segments (a, b) and (b, c) in a sequential manner. To give the user control over this sequentialization we let a user specify the *(sub-)trace variants* Var that shall be mapped (one after the other) onto the y-axis as in Fig. 1 (right). The notion of a *view* provides all parameters for a performance spectrum.

Definition 3 (View). *A view $V = (Var, \mathbb{C}, g, p)$ is a set Var of (sub-)trace variants $Var = \{\sigma^0, \ldots, \sigma^k\} \subseteq A^*$, a performance classification $\mathbb{C}$, a grouping $g \in \{start, stop, pending, none\}$ and period $p \in T$. The segment sequence of variant $\sigma^i = \langle a_1^i, a_2^i, \ldots, a_{n_i}^i \rangle \in Var$ is $seg(\sigma^i) = \langle (a_1^i, a_2^i), \ldots, (a_{n_i-1}^i, a_{n_i}^i) \rangle$. The segment sequence of all variants Var is their concatenation, i.e., $seg(Var) = seg(\sigma^1)seg(\sigma^2) \ldots seg(\sigma^k)$.*

For example, for traces $\langle a, b, c, d, e \rangle, \langle a, b, f, d, e \rangle, \langle a, b, c, b, f, e \rangle$, the variants $Var = \{\langle b, c, d, e \rangle, \langle f, d, e \rangle\}$ yield the segment sequence $seg(Var) = \langle (b, c), (c, d), (d, e), (f, d), (d, e) \rangle$.

Let L be a log, $V = (Var, \mathbb{C}, g, p)$ be a view. The *performance spectrum* of L wrt. V with $g = none$ is the sequence of the detailed performance spectra along the segment sequence $seg(Var)$: $\mathbb{S}(L, V) = \langle \mathbb{S}^{\mathbb{C}}(a, b, L) \rangle_{(a,b) \in seg(Var)}$. The *aggregated performance spectrum* of L wrt. V with $g \neq none$ is the sequence of aggregated spectra $\mathbb{S}_{g,p}(L, V) = \langle \mathbb{S}_{g,p}^{\mathbb{C}}(a, b, L) \rangle_{(a,b) \in seg(Var)}$.

For the visualization in Fig. 1, the segments $seg(Var)$ are mapped to the y-axis in order of $seg(Var)$ in equidistant steps for some length $ydist$: in the i-th segment $(a_i, b_i) = seg(Var)_i$, a_i and b_i get y-coordinates $y_{a,i} = i \cdot ydist$ and $y_{b,i} = (i+1) \cdot ydist$. By default any two consecutive segments touch at $y_{b,i} = y_{a,i+1}$; an extra gap can be added whenever $b \neq a$. Figure 7 visualizes the view of an aggregate performance spectrum.

An optimal definition of Var for a given log is outside the scope of this paper, and we assume user input. Yet we identified some principles. There are two canonical trace variants Var for views on a log L. The *minimal* variant defines the most frequent variant in L, visualizing all its process segments consecutively. The *maximal* variant includes all individual observed process segments $Var_{max}(L) = \{\langle(a,b)\rangle \mid \langle \ldots, (a,t_a), (b,t_b), \ldots \rangle \in L\}$ in no specific order. Mapping segments consecutively along the y-axis allows to follow the flow of multiple cases over time as shown in Fig. 1 (right). Choices in a process can be handled by defining two alternative trace variants in the view; thereby segments (a,b) occurring multiple times in Var are replicated (with the entire performance spectrum), allowing to see the flow of the variant through this segment in the performance context of other variants. Handling loops and concurrency requires event log preprocessing. Loops can be unrolled through label refinement [9] in the log. In case of concurrency, analyzing the performance of segment (a,b) with Definition 1 requires filtering from the log all activities concurrent to a and b. In case studies with Vanderlande, $Var_{max}(L)$ in combination with a hierarchical naming scheme of events allowed to visually group and analyze related segments even in very complex processes (of 10000s of segments).

4   Performance Patterns

Performance spectra of processes may contain an overwhelming amount of information and are – for the untrained eye – more difficult to read and interpret than known visualizations. However processes with similar performance characteristics show similar *patterns* in their performance spectra, and vice versa, similar patterns mean similar performance characteristics. Such patterns introduce a higher abstraction level over 'plain' performance spectra, thereby aiding in description and analysis of performance. Next, we illustrate the idea of patterns in the performance spectrum distinguishing *elementary* and *composite* patterns. We provide a *taxonomy* of elementary patterns in Sect. 4.2. We discuss composite patterns in Sect. 4.3, but posit their systematization in future work.

4.1   Elementary Patterns

Intuitively, a performance pattern is a specific configuration of the lines and bars in a performance spectrum that (1) is visually distinct within a larger part of the spectrum, (2) describes a particular *performance scenario* (of *multiple* cases over time), and (3) repeats when this scenario repeats. An *elementary pattern* relates to a single segment and cannot be broken down further without loss of its meaning.

The elementary pattern shown in Fig. 2 occurred in segment (*Insert Fine Notification, Add penalty*) of the *Road Traffic Fines Management* (RF) log[1] and consists of many parallel inclined lines of the same color, cor-

Fig. 2. The elementary pattern shows a FIFO behavior with constant waiting time

responding to multiple observations distributed over time. Non-crossing lines

[1]   https://doi.org/10.4121/uuid:270fd440-1057-4fb9-89a9-b699b47990f5.

show a strict FIFO order and identical inclinations show a constant waiting time for all cases. Variation in density of the lines (and in the height of the bars of the aggregated performance spectrum) shows continuous, varying workload throughout the entire log. Patterns with such characteristics are typical for highly standardized automated activities with strict time constraints. Note that existing models describe the performance of this segment as "constant" delay of 60d.

We consider the pattern to be "elementary" in the sense that we cannot decompose it further without losing its key qualities: single segment, strict FIFO with constant time, workload is continuous and varying.

4.2 Taxonomy of Elementary Patterns

We observed a great variety of elementary patterns and combinations of patterns in the performance spectra of real-life processes (see Sect. 5). That makes it impossible to provide a comprehensive catalog. Nevertheless, we are able to provide a comprehensive taxonomy of *parameters* of elementary patterns. It allows us to completely and unambiguously describe performance of a process over time in a way that patterns that correspond to similar performance scenarios have identical descriptions and identical descriptions of patterns mean similar performance scenarios, while changing the value of any parameter in a pattern would mean a different performance scenario.

The taxonomy provides parameters to characterize the Shape of lines and bars in a process in a particular Scope over time; line density and bar height describe Workload while their color describes Performance. The parameter values form a hierarchy which is shown together with typical patterns having these characteristics in Fig. 3. We provide a unique short-hand value [in brackets] for each parameter, to allow succinct notation of patterns.

Scope parameters capture the place of pattern in the performance spectrum.

- size: one segment [1 seg], one subsequence [1 sub-seq], several subsequences [>1 sub-seq]
- occurrence: globally [glob], as a local instance [loc]
- repetitions (for patterns occurring in local instance): once [once], regular [reg], periodic [per=T], arbitrary [arb],
- overlap (for repeating patterns): overlapping [overlap], non-overlapping
- duration: absolute value [D=T]

Size describes the pattern length from the control-flow perspective: a single segment, a single subsequence or several subsequences of event classifiers. Although all elementary patterns have size 1 seg, we include this parameter in the taxonomy for compatibility with composite patterns. A pattern occurrence can be either global, when it occurs continuously throughout a segment without clear boundaries, otherwise it distinctly occurs as a local instance. Pattern instances

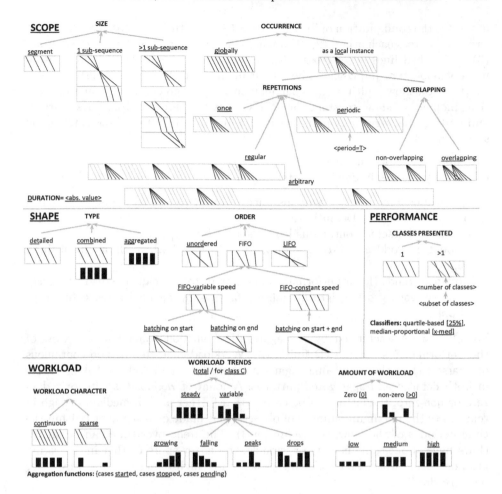

Fig. 3. Taxonomy of elementary patterns

may occur once or repeat (1) periodically in particular intervals T, (2) regularly, i.e., seemingly systematic but not periodic, or (3) arbitrarily. Repeated pattern instances can be overlapping or non-overlapping in time. Parameter duration describes the absolute duration over time (e.g. as an interval in seconds).

Shape parameters describe the appearance of lines and bars in the visualization of the performance spectrum.

– type: detailed [det], aggregated [agg], combined [comb]
– order: unordered [unord], LIFO [LIFO], FIFO with variable time [FIFO-var], FIFO with constant time [FIFO-const], batching on start [batch(s)], batching on end [batch(e)], batching on start and end [batch(s+e)]

A pattern described just in terms of lines (bars) of a detailed (aggregated) performance spectrum is detailed (aggregated); if it requires both it is combined. Order describes the configuration of lines in a detailed pattern: (1) unordered when lines irregularly cross each other, (2) LIFO when lines end in reversed order of starting, (3) FIFO when lines never cross. (3b) Non-crossing lines of variable inclination mean variable time [FIFO-var], where multiple lines starting (or ending) in a very short period show multiple cases batching on start (or on end). (3c) Lines of identical inclination show constant time [FIFO-const], where multiple lines starting and ending in a very short period (with no lines before/after) show batching on start and end.

Workload describes the height of bars in aggregated or combined patterns, and the density of lines in detailed patterns over time.

- aggregation function: segment instances started [start], stopped [stop], cases pending [pend], see Definition 2
- workload character: continuous [cont], sparse [sparse]
- amount of workload: zero [0], non-zero [>0], low [low], medium [med], high [high]
- workload trends (for a performance class or in total): can be steady [steady], variable [var], growing [grows], falling [falls], showing peaks [peak] or drops [drop]

Workload is characterized by the aggregation function defined in the view of the performance spectrum (Definition 3). Workload character can be continuous or sparse (when there are longer gaps between lines or bars), and it is visible in both detailed and aggregated patterns. Amount of workload is categorized as zero or non-zero, the latter can be categorized further as low, medium or high in relation to the maximum number of observations made on a segment (within the time period p of the view, see Definition 3). The trend over time can be steady (bars have about same height) or variable, the latter splits further into steadily growing, falling workload or showing peaks (a few high bars surrounded by lower bars) or drops.

Performance is described in terms of the performance classes present in the pattern with respect to the classifier $\mathbb{C}$ of the view (Definition 3) chosen by the user.

- classes presented: 1, > 1, number of classes, subset of classes
- Classifiers: various, we discuss quartile-based [25%] (e.g., all observations belonging to the 26%–50% quartile), median-proportional [x·med] (e.g., all observations 2–3 times longer than the median duration)

In the visualization of the performance spectrum, classes are coded by colors. A monochrome pattern has 1 class presented while a multi-colored one has > 1 classes presented.

Now we show how the taxonomy describes the elementary patterns E1–E3 found in the RF log and highlighted in Fig. 4 (left). Pattern E1 occurs in a single segment in local instances with a duration of 6 months, instances repeat regularly and overlap; the detailed pattern shows batching on end in a continuous workload for 4 performance classes in a quartile-based classifier. Using the short-hand

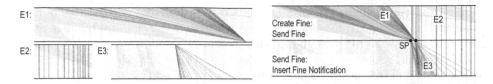

Fig. 4. Three elementary patterns E1, E2, and E3 (left) and two occurrences of a composite pattern consisting of E1–E3 (right).

notation, we write E1 = [Scope(seg,loc,reg,overlap,D=6mo), Shape(det,batch(e)), Work(cont), Perf(25%,4 classes)]. Similarly, we can characterize E2 = [Scope(seg,glob), Shape(det,FIFO-const), Work(sparse), Perf(25%,1 class)] and E3 = [Scope(seg,loc,reg,overlap,D=1mo), Shape(det,batch(s)), Wo(cont), Perf(25%,4 classes)].

In case of creating a catalog of elementary patterns, some additional information can be added to pattern descriptions: a unique identifier and name and a meaning depending on the domain and the chosen event classifier, e.g., resources in a business process, or physical locations of a material handling system.

4.3 Composite Patterns

In the previous sections, we described performance of single process segments through elementary patterns. However, the performance spectrum of real-life processes gives rise to *composite* patterns comprised of several elementary ones. While a full taxonomy is beyond the scope of this paper, we outline some basic principles for describing composite patterns by relating elementary patterns to each other in their context.

The context of a pattern P1 as shown in Fig. 5(a) consists of (1) observations earlier and later than P1 in the same process segment, (2) observations before and after P1 in the control flow perspective, and (3) a distinct pattern P2 occurring simultaneously to P1 in the same segment. Using this context, the taxonomy can be extended with further parameters. For instance, observations before and after can be used to characterize performance of a pattern in context and the performance variants contained in the same timed period as shown in Fig. 5(b).

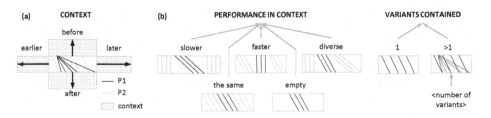

Fig. 5. (a) Pattern context, and (b) Context parameters

Figure 4 (right) shows two instances of a composite pattern consisting of the elementary patterns E1, E2, and E3, described in Sect. 4.2. E1 and E3 align at a synchronization point SP, that shows synchronization of multiple cases in a "sand clock" pattern, while the cases in E2 do not synchronize with the cases in E1 or E3: we can clearly see 2 variants of behavior contained (E1+E3 and E2). The performance context of the composite pattern is diverse.

The taxonomy of Fig. 3 and the new parameters of Fig. 5 only partially describe composite patterns. In particular, a comprehensive taxonomy for precisely describing the alignment of patterns to each other in their context is subject of future work.

5   Evaluation

We implemented the transformation of logs into detailed and aggregated performance spectra and their visualization through an interactive ProM plug-in in package "Performance Spectrum".[2] We applied our implementation on 11 real-life event logs from business processes (BPI12, BPI14, BPI15(1-5), BPI17, Hospital Billing, RF)[3] and on 1 real-life log from logistics (BHS) provided by Vanderlande. We illustrate how the performance spectrum provides detailed insights into performance for RF; for BHS we report on a case study for identifying performance problems; and we summarize performance characteristics of the 11 business process logs.

5.1   Road Traffic Fine Management Process (RF)

Event Log and View. The RF event log consists of 11 activities, more than 150.000 cases and 550.000 events over a period of 12 years. We analyze the 3 trace variants R1-R3 of Fig. 6, which cover $> 80\%$ of the events in the log, by defining a view for the sub-sequences {⟨*Create Fine, Payment*⟩, ⟨*Create Fine, Send Fine, Insert Fine Notif., Add   penalty, Payment*⟩, ⟨*Add penalty, Send for CC*⟩} and quartile-based performance classes.

First, we discuss the detailed patterns P1-P5 that can be observed in the performance spectrum of a 2-years period in Fig. 6 which represents behavior typical for the entire 12-years period. All cases start from activity *Create Fine* and continue either with activity *Payment* (variant R1) or activity *Send Fine* (R2 and R3).

P1: Segment S1 *Create Fine:Payment* globally contains many traces of variable duration, which are continuously distributed over time and can overtake each other, i.e., P1 = [Scope(seg,glob), Shape(det,unord), Work(cont), Perf(25%,4 classes)]. We can clearly observe that traffic offenders pay at various speeds.

P2: The performance spectrum of Fig. 6 shows that the sub-trace ⟨ *Create Fine, Send Fine Insert Fine notification*⟩ shared by R2 and R3 contains the

[2] source code and further documentation available at https://github.com/processmining-in-logistics/psm.

[3] available at https://data.4tu.nl/repository/collection:event_logs_real.

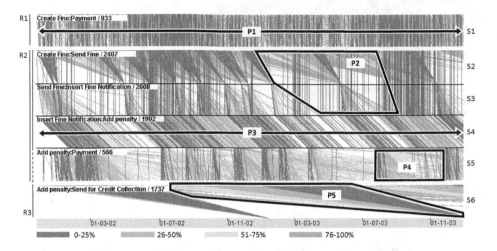

Fig. 6. Detailed performance spectrum of Road Traffic Fines Management log for years 2002 and 2003 for trace variants R1: ⟨*Create Fine, Payment*⟩, R2: ⟨*Create Fine, Send Fine, Insert Fine Notif., Payment*⟩, and R3: ⟨*Create Fine, Send Fine, Insert Fine Notif., Add penalty, Send for CC*⟩.

composite pattern P2 which we already discussed in Sect. 4.3. P2 consists of two *different* performance variants. The "sand clock" pattern of E1+E3 of Sect. 4.1 shows that cases are accumulated over a period of 6 months; the period until *Insert fine notification* varies from zero up to 4 months. Cases in pattern E2 of Sect. 4.1 are not synchronized but processed instantly.

P3: The two variants E1+E3 and E2 vanish in the next segment S3 *Insert Fine Notif.:Add penalty* where all cases show a strong FIFO behavior: P3 = [Scope(seg,glob), Shape(det,FIFO-const), Work(cont), Perf(25%,2 classes)]; the switch from CEST to CET in October shows as a slower performance class in Fig. 6. After *Add penalty*, R2 continues with *Payment* (S5 in Fig. 6) and R3 continues with *send for CC* (S6 in Fig. 6).

P4: On segment S5 *Add penalty:Payment* we surprisingly observe emergent batching on start despite the absence of batching on end in the preceding segment S4. The "sand-clock" batching in P2 results in groups of "fast" cases which are forwarded by the FIFO pattern P3 and together create a new batching on start pattern P4 (similar to E2) in segment S5 that can take months to years to complete the *Payment*.

P5: The alternative segment S6 *Add penalty:Send for CC* shows batching on end every 12 months for cases that entered the batch 20 to 6 months prior: P5 = [Scope(seg,loc,per= 12mo,D=20mo), Shape(det,batch(e)), Work(cont), Perf(25%,4 classes)]. The 6-month delay revealed by P5 is mandated by Italian law. A unique pattern for this process occurs in segment *Add Penalty:Send Appeal to Prefecture* in Fig. 7(b) where a batch on end pattern occurs only once with a duration of 10 years.

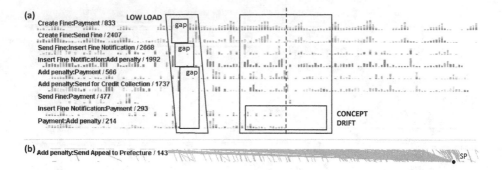

Fig. 7. Aggregated performance spectrum of Road Traffic Fines Management log (2000–2012)

Aggregated Patterns are shown in Fig. 7(a), where every bar shows how many segments start every month. Here we can see patterns related to workload, for example, in the first quarter of 2004 we can see a gap pattern of 3 months, gap=[Scope(seg,loc,once,D=3mo), Shape(agg,batch(e)), Work(0)]. This gap pattern propagates to subsequent segments creating a composite pattern surrounded by context with much higher load. Figure 7(a) also reveals concept drift in the control-flow perspective: the medium non-zero workload in segments *Insert Fine Notification:Payment* and *Payment:Add penalty* drops to 0 in 2007 (up to some outliers).

5.2 Baggage Handling System of a Major European Airport (BHS)

Event Log and View. In this case study, we analyzed flows of bags through a Vanderlande-built baggage handling system (BHS). In the event log, each case corresponds to one bag, events are recorded when bags pass sensors on conveyors, and activity names describe locations of sensors in the system. For 1 day of operations, an event log contains on average 850 activities, 25.000–50.000 cases and 1–2 million events.

To provide examples of the BHS performance spectrum and patterns, we selected conveyor subsequence $\langle a1, a2, a3, a4, a5, s \rangle$ that moves bags from Check-In counter $a1$ to a main sorter entry point s. Cases starting in

Fig. 8. The path from Check-In counter $a1$ to sorter entry point s

$a1$ correspond to the BHS registering that a passenger put a bag onto the belt of the Check-In counter. We chose this particular part because (1) any BHS has such paths and (2) it shows many typical performance patterns of a BHS. The diagram of the corresponding system part in Fig. 8 shows that more bags join from other Check-In counters on the way in points $a2$-5. We first discuss elementary detailed patterns in the performance spectrum and then show how their compositions explain complex system behavior.

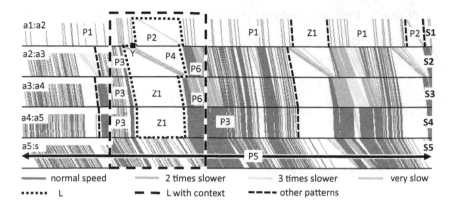

Fig. 9. Performance spectrum of bags movements between Check-In counters and the main sorter

The detailed performance spectrum in Fig. 9 shows events over the period of 1 h in a median-proportional performance classification. In the first segment S1 *a1:a2* we can observe pattern P1 (FIFO, constant waiting time, variable workload, normal performance) and P2 (batching on start and end with very slow performance). Empty zone Z1 shows zero workload. In BHS, FIFO behavior is typical for conveyors, where bags cannot overtake each other, and variable workload is typical for manual operations: a counter's arrival process depends on a passenger flow and their service times, which vary from passenger to passenger. Despite conveyors having constant speed S1 shows not only P1 but also P2 and Z1: some conveyors were temporarily stopped and all bags on them 'share' the same delay, as in P2.

By looking at S1 alone, we cannot explain causes of the delays in those pattern instances. But as segments in a BHS are synchronized through movement of physical objects on conveyors, we can identify the cause by following the control-flow of Fig. 8. After P4 in S2 we observe Z1 in S3 and S4, both having non-zero workload earlier (P3) and later (P3,P6), followed by non-zero workload P5 in S5 (FIFO, constant waiting time, high workload, normal performance). This gives rise to pattern L and its context highlighted in Fig. 9. Reading pattern L from S4 backwards gives the following interpretation: the conveyors in S3 and/or S4 stopped operation, so bags from S2 could not move further to S3. When S2 was stopped, S1 also was stopped (point Y), because bags could not enter S2. The slow cases of P2 and P4 are the bags waiting on a stopped conveyor. This is called a *die-back* scenario, where delays or non-operation (in S3,S4) propagate backwards through the system. When S3 and S4 return to operations, waiting bags of S1 and S2 (and from other parts that are not included in Fig. 9) resume their movement. The two times slower performance in P6 shows that S2 and S3 are at their capacity limits in this restart phase until all workload decreased. Figure 9 shows that Pattern L repeats regularly during the day.

Table 1. Presence of selected pattern classes in real-life event logs.

	BPI12	BPI14	BPI15-1	BPI15-2	BPI15-3	BPI15-4	BPI15-5	BPI17	Hospital	H-billing	Road fine
unord,low		glob	glob	glob	glob	glob	gob		glob	glob	glob
unord,high								glob		glob	glob
FIFO								glob		glob	glob
FIFO+unord									reg	glob	
FIFO (weekly)	glob	glob			arb			glob			
batching		arb						per		per	reg
workload spikes								arb			reg
concept drift		once	once	once	arb	arb	arb			once	reg
sparse work	reg	reg	glob*	glob*	glob*	glob*	glob*		glob	glob	

Using the same reasoning as explained above, we identified the root cause of critical performance problems in the BHS of a large European airport which could not be identified with existing process mining tools. Our analysis took one week and was confirmed as correct by experts of Vanderlande who required several weeks of manual data analysis and years-long experience to identify the root cause.

5.3 Comparison of Event Logs

We compared the 11 real-life business process event logs regarding the types of performance patterns they contain. We visualized the performance spectrum of each log and noted the properties of the immediately visible patterns (in terms of the taxonomy of Sect. 4.2), see https://github.com/vadimmidavvv/psm for details.

Table 1 shows the results. We identified combined patterns of unordered behavior with low and high workload; detailed patterns of FIFO behavior, also overlaid with an unordered variant, FIFO+unord, and occurring only Mon-Sat, FIFO(weekly), and various forms of batching. The aggregate patterns showed workload spikes, concept drift, and sparse work.

The cells in Table 1 indicate for each log the occurrence and repetition values of the patterns according to the taxonomy of Fig. 3. The logs differ strongly in the presence and repetition of patterns, indicating that very different performance scenarios occur in these processes. Interestingly, the BPI15 logs which all relate to the same kind of process that is being executed in different organizations all show very similar patterns: glob* for sparse work means that sparse work co-occurs globally in a synchronized way: a large number of segments show behavior during exactly the same days.

6   Conclusion

In this paper, we proposed the performance spectrum as a novel visualization of process performance data in event logs. We project each process step (from one activity to the next) in a log over time. By making time explicit and avoiding aggregation, the performance spectrum reveals non-stationarity of performance and synchronization of different cases over time. We provided a taxonomy to isolate and describe various performance phenomena in terms of distinct elementary and composite patterns. Applying the technique on 12 real-life event logs validated its usefulness in exploration of data for identifying expected and unusual performance patterns and in confirming that process performance is neither stationary nor are cases isolated from each other. Future research is to automatically identify performance patterns from event logs and annotating process models with identified patterns. We believe the insights obtained through visual analysis to be useful in further research on performance prediction: improve queueing-based predictions based on FIFO-related patterns, aid discovery and identification of batching activities, aid in developing improved prediction models, simulation models, and prescriptive models that incorporate insights on non-stationary, or cross-case conformance checking of performance models. The identified patterns suggest also the need for performance-based filtering and sorting of event data.

Our technique is currently limited by the fact that process logic has to be flattened into sequences along the y-axis of the visualization, lack of support for concurrency and choices, and the very large variety of composite patterns cannot be described well by our taxonomy. Future work comprises the extension of the taxonomy, enhancement of process models with performance patterns, identifying "optimal" views for a particular analysis questions, and improved visualizations to handle concurrency and choices.

Acknowledgements. The research leading to these results has received funding from Vanderlande Industries in the project "Process Mining in Logistics". We thank Elena Belkina for support in the tool development.

References

1. van der Aalst, W.M.P.: Process Mining - Data Science in Action, 2nd edn. Springer, Heidelberg (2016). https://doi.org/10.1007/978-3-662-49851-4
2. van der Aalst, W.M.P., Adriansyah, A., van Dongen, B.F.: Replaying history on process models for conformance checking and performance analysis. Wiley Interdiscip. Rev.: Data Min. Knowl. Discov. **2**(2), 182–192 (2012)
3. van der Aalst, W.M.P., Pesic, M., Song, M.: Beyond process mining: from the past to present and future. In: Pernici, B. (ed.) CAiSE 2010. LNCS, vol. 6051, pp. 38–52. Springer, Heidelberg (2010). https://doi.org/10.1007/978-3-642-13094-6_5
4. van der Aalst, W.M.P., Schonenberg, H., Song, M.: Time prediction based on process mining. Inf. Syst. **36**, 450–475 (2011)
5. van Dongen, B.F., Crooy, R.A., van der Aalst, W.M.P.: Cycle time prediction: when will this case finally be finished? In: Meersman, R., Tari, Z. (eds.) OTM 2008. LNCS, vol. 5331, pp. 319–336. Springer, Heidelberg (2008). https://doi.org/10.1007/978-3-540-88871-0_22

6. Folino, F., Guarascio, M., Pontieri, L.: Discovering context-aware models for predicting business process performances. In: Meersman, R., et al. (eds.) OTM 2012. LNCS, vol. 7565, pp. 287–304. Springer, Heidelberg (2012). https://doi.org/10.1007/978-3-642-33606-5_18

7. Keim, D., Andrienko, G., Fekete, J.-D., Görg, C., Kohlhammer, J., Melançon, G.: Visual analytics: definition, process, and challenges. In: Kerren, A., Stasko, J.T., Fekete, J.-D., North, C. (eds.) Information Visualization. LNCS, vol. 4950, pp. 154–175. Springer, Heidelberg (2008). https://doi.org/10.1007/978-3-540-70956-5_7

8. Leemans, S.J.J., Fahland, D., van der Aalst, W.M.P.: Using life cycle information in process discovery. In: Reichert, M., Reijers, H.A. (eds.) BPM 2015. LNBIP, vol. 256, pp. 204–217. Springer, Cham (2016). https://doi.org/10.1007/978-3-319-42887-1_17

9. Lu, X., Fahland, D., van den Biggelaar, F.J.H.M., van der Aalst, W.M.P.: Handling duplicated tasks in process discovery by refining event labels. In: La Rosa, M., Loos, P., Pastor, O. (eds.) BPM 2016. LNCS, vol. 9850, pp. 90–107. Springer, Cham (2016). https://doi.org/10.1007/978-3-319-45348-4_6

10. Martin, N., Swennen, M., Depaire, B., Jans, M., Caris, A., Vanhoof, K.: Retrieving batch organisation of work insights from event logs. Decis. Support Syst. **100**, 119–128 (2017)

11. Maruster, L., van Beest, N.R.T.P.: Redesigning business processes: a methodology based on simulation and process mining techniques. Knowl. Inf. Syst. **21**(3), 267–297 (2009)

12. Polato, M., Sperduti, A., Burattin, A., de Leoni, M.: Time and activity sequence prediction of business process instances. Computing 1–27 (2018). https://doi.org/10.1007/s00607-018-0593-x

13. Pufahl, L., Bazhenova, E., Weske, M.: Evaluating the performance of a batch activity in process models. In: Fournier, F., Mendling, J. (eds.) BPM 2014. LNBIP, vol. 202, pp. 277–290. Springer, Cham (2015). https://doi.org/10.1007/978-3-319-15895-2_24

14. Rogge-Solti, A., van der Aalst, W.M.P., Weske, M.: Discovering stochastic petri nets with arbitrary delay distributions from event logs. In: Lohmann, N., Song, M., Wohed, P. (eds.) BPM 2013. LNBIP, vol. 171, pp. 15–27. Springer, Cham (2014). https://doi.org/10.1007/978-3-319-06257-0_2

15. Rogge-Solti, A., Weske, M.: Prediction of business process durations using non-markovian stochastic petri nets. Inf. Syst. **54**, 1–14 (2015)

16. Rozinat, A., Mans, R.S., Song, M., van der Aalst, W.M.P.: Discovering simulation models. Inf. Syst. **34**, 305–327 (2009)

17. Senderovich, A., et al.: Data-driven performance analysis of scheduled processes. In: Motahari-Nezhad, H.R., Recker, J., Weidlich, M. (eds.) BPM 2015. LNCS, vol. 9253, pp. 35–52. Springer, Cham (2015). https://doi.org/10.1007/978-3-319-23063-4_3

18. Senderovich, A., Weidlich, M., Gal, A.: Temporal network representation of event logs for improved performance modelling in business processes. In: Carmona, J., Engels, G., Kumar, A. (eds.) BPM 2017. LNCS, vol. 10445, pp. 3–21. Springer, Cham (2017). https://doi.org/10.1007/978-3-319-65000-5_1

19. Senderovich, A., Weidlich, M., Gal, A., Mandelbaum, A.: Queue mining for delay prediction in multi-class service processes. Inf. Syst. **53**, 278–295 (2015)

20. Shrestha, A., Miller, B., Zhu, Y., Zhao, Y.: Storygraph: extracting patterns from spatio-temporal data. In: ACM SIGKDD Workshop IDEA 2013, pp. 95–103. ACM (2013)

21. Song, M., van der Aalst, W.M.: Supporting process mining by showing events at a glance. In: Proceedings of the 17th Annual Workshop on Information Technologies and Systems (WITS), pp. 139–145 (2007)
22. Tax, N., Verenich, I., La Rosa, M., Dumas, M.: Predictive business process monitoring with LSTM neural networks. In: Dubois, E., Pohl, K. (eds.) CAiSE 2017. LNCS, vol. 10253, pp. 477–492. Springer, Cham (2017). https://doi.org/10.1007/978-3-319-59536-8_30
23. Wynn, M.T., et al.: ProcessProfiler3D: a visualisation framework for log-based process performance comparison. Decis. Support Syst. **100**, 93–108 (2017)

Abstract-and-Compare: A Family of Scalable Precision Measures for Automated Process Discovery

Adriano Augusto[1,2]($\boxtimes$), Abel Armas-Cervantes[2], Raffaele Conforti[2], Marlon Dumas[1], Marcello La Rosa[2], and Daniel Reissner[2]

[1] University of Tartu, Tartu, Estonia
{adriano.augusto,marlon.dumas}@ut.ee
[2] University of Melbourne, Melbourne, Australia
{abel.armas,raffaele.conforti,marcello.larosa}@unimelb.edu.au

Abstract. Automated process discovery techniques allow us to extract business process models from event logs. The quality of models discovered by these techniques can be assessed with respect to various criteria related to simplicity and accuracy. One of these criteria, namely *precision*, captures the extent to which the behavior allowed by a process model is observed in the log. While several measures of precision have been proposed, a recent study has shown that none of them fulfills a set of five axioms that capture intuitive properties behind the concept of precision. In addition, existing precision measures suffer from scalability issues when applied to models discovered from real-life event logs. This paper presents a family of precision measures based on the idea of comparing the k-th order Markovian abstraction of a process model against that of an event log. We demonstrate that this family of measures fulfils the aforementioned axioms for a suitably chosen value of k. We also empirically show that representative exemplars of this family of measures outperform a commonly used precision measure in terms of scalability and that they closely approximate two precision measures that have been proposed as possible ground truths.

1 Introduction

Contemporary enterprise information systems store detailed records of the execution of the business processes they support, such as records of the creation of process instances (a.k.a. *cases*), the start and completion of tasks, and other events associated with a case. These records can be extracted as event logs consisting of a set of traces, each trace itself consisting of a sequence of events associated with a case. Automated process discovery techniques [3] allow us to extract process models from such event logs. The quality of process models discovered in this way can be assessed with respect to several quality criteria related to simplicity and accuracy.

Two commonly used criteria for assessing accuracy are fitness and precision. *Fitness* captures the extent to which the behavior observed in an event log is

© Springer Nature Switzerland AG 2018
M. Weske et al. (Eds.): BPM 2018, LNCS 11080, pp. 158–175, 2018.
https://doi.org/10.1007/978-3-319-98648-7_10

allowed by the discovered process model (i.e. Can the process model generate every trace observed in the event log?). Reciprocally, *precision* captures the extent to which the behavior allowed by a discovered process model is observed in the event log. A low precision indicates that the model under-fits the log, i.e. it can generate traces that are unrelated or only partially related to traces observed in the log, while a high precision indicates that it over-fits (i.e. it can only generate traces in the log and nothing more).[1]

While several precision measures have been proposed, a recent study has shown that none of them fulfils a set of five axioms that capture intuitive properties behind the concept of precision [15]. In addition, most of the existing precision measures suffer from scalability issues when applied to models discovered from real-life event logs.

This paper presents a family of precision measures based on the idea of comparing the kth-order Markovian abstraction of a process model against that of an event log using a graph matching operator. We show that the proposed precision measures fulfil four of the aforementioned axioms for any k, and all five axioms for a suitable k dependent on the log. In other words, when measuring precision, we do not need to explore the entire state space of a process model but only its state space up to a certain memory horizon.

The paper empirically evaluates exemplars of the proposed family of measures using: (i) a synthetic collection of models and logs previously used to assess the suitability of precision measures, and (ii) a set of models discovered from 20 real-life event logs using three automated process discovery techniques. The synthetic evaluation shows that the exemplar measures closely approximate two precision measures that have been proposed as ground truths. The evaluation based on real-life logs shows that for values of up to $k = 5$, the kth-order Markovian precision measure is considerably more efficient than a commonly used precision measure, namely alignments-based ETC precision [1].

The rest of the paper is structured as follows. Section 2 introduces existing precision measures and the axioms defined in [15]. The family of Markovian precision measures is presented in Sect. 3 and evaluated in Sect. 4. Finally, Sect. 5 draws conclusions and directions for future work.

2   Background and Related Work

One of the earliest precision measures was proposed by Greco et al. [8], based on the *set difference* (SD) between the model behavior and the log behavior, each represented as a set of traces. This measure is a direct operationalization of the concept of precision, but it is not applicable to models with cycles since the latter have an infinite set of traces.

---

[1] A third accuracy criterion in automated process discovery is *generalization*: the extent to which the process model captures behavior that, while not observed in the log, is implied by it.

Later, Rozinat and van der Aalst [14] proposed the *advanced behavioral appropriateness* (ABA) precision. The ABA precision is based on the comparison between the sets of activity pairs that sometimes but not always follow each other, and the set of activity pairs that sometimes but not always precede each other. The comparison is performed on the sets extracted both from the model and the log behaviors. The ABA precision does not scale to large models and it is undefined for models with no routing behavior (i.e. models without concurrency or conflict relations) [15].

De Weerdt et al. [7] proposed the *negative events* precision measure (NE). This method works by inserting inexistent (so-called negative) events to enhance the traces in the log. A negative event is inserted after a given prefix of a trace if this event is never observed preceded by that prefix anywhere in the log. The traces extended with negative events are then replayed on the model. If the model can parse some of the negative events, it means that the model has additional behavior. This approach is however heuristic: it does not guarantee that all additional behavior is identified.

Muñoz-Gama and Carmona [13] proposed the *escaping edges* (ETC) precision. Using the log behavior as reference, it builds a *prefix automaton* and, while replaying the process model behavior on top of it, counts the number of *escaping edges*, i.e. edges not in the prefix automaton which represent extra behavior of the process. Subsequently, to improve the robustness of the ETC precision for logs containing non-fitting traces, the ETC precision evolved into the *alignments-based ETC* precision (ETC_a) [1] where the replay is guided by alignments.

Despite its robustness, ETC_a does not scale well to real-life datasets. To address this issue, Leemans et al. [12] proposed the *projected conformance checking* (PCC) precision. This precision, starting from the log behavior and the model behavior builds a projected automaton (an automaton where a reduced number of activities are encoded) from each of them, i.e. A_l and A_m. These two automata are then used to generate a third automaton capturing their common behavior, i.e. $A_{l,m}$. The precision value is then computed as the ratio between the number of outgoing edges of each state in $A_{l,m}$ and the number of outgoing edges of the corresponding states occurring in A_m.

Finally, van Dongen et al. [16] proposed the *anti-alignment* precision (AA). This measure analyses the anti-alignments of the process model behavior to assess the model's precision. An anti-alignment of length n is a trace in the process model behavior of length at most equal to n, which maximizes the Levenshtein distance from all traces in the log.

In a recent study, Tax et al. [15] proposed five axioms to capture intuitive properties behind the concept of precision advising that any precision measure should fulfill these axioms. We start by introducing preliminary concepts and notations, and then proceed to present the five axioms.

Definition 1 [Trace]. *Given a set of activity labels Σ, we define a trace on Σ as a sequence $\tau_\Sigma = \langle t_1, t_2, \ldots, t_{n-1}, t_n \rangle$, such that $\forall 1 \leq i \leq n, t_i \in \Sigma$.[2] Furthermore, we denote with τ_i the activity label in position i, and we use the symbol Γ_Σ to refer to the universe of traces on Σ. With abuse of notation, hereinafter we refer to any $t \in \Sigma$ as an activity instead of an activity label.*

Definition 2 [Subtrace]. *Given a trace $\tau = \langle t_1, t_2, \ldots, t_{n-1}, t_n \rangle$, with the notation $\tau^{i \to j}$, we refer to the subtrace $\langle t_i, t_{i+1}, \ldots, t_{j-1}, t_j \rangle$, where $0 < i < j \leq n$. We extend the subset operator to traces, i.e., given two traces τ and $\hat{\tau}$, $\hat{\tau}$ is contained in τ, shorthanded as $\hat{\tau} \subset \tau$, if and only if (iff) $\exists i, j \in \mathbb{N} \mid \tau^{i \to j} = \hat{\tau}$.*

Definition 3 [Process Model Behavior]. *Given a process model P (regardless of its representation) and being Σ the set of its activities. We refer to the model behavior as $\mathcal{B}_P \subseteq \Gamma_\Sigma$, where $\forall \langle t_1, t_2, \ldots, t_{n-1}, t_n \rangle \in \mathcal{B}_P$ there exists an execution of P that allows to execute the sequence of activities $\langle t_1, t_2, \ldots, t_{n-1}, t_n \rangle$, where t_1 is the first activity executed, and t_n the last.[3]*

Definition 4 [Event Log Behavior]. *Given a set of activities Σ, an event log L is a finite multiset of traces defined over Σ. The event log behavior of L is defined as $\mathcal{B}_L = support(L)$.[4]*

Definition 5 [Precision Axioms].

- **Axiom-1.** *A precision measure is a deterministic function $prec : \mathcal{L} \times \mathcal{P} \to \mathbb{R}$, where $\mathcal{L}$ is the universe of event logs, and $\mathcal{P}$ is the universe of processes.*
- **Axiom-2.** *Given two process models P_1, P_2 and a log L, if the behavior of L is contained in the behavior of P_1, and this latter is contained in the behavior of P_2, the precision value of P_1 must be equal to or greater than the precision value of P_2. Formally, if $\mathcal{B}_L \subseteq \mathcal{B}_{P_1} \subseteq \mathcal{B}_{P_2} \implies prec(L, P_1) \geq prec(L, P_2)$.*
- **Axiom-3.** *Given two process models P_1, P_2 and a log L, if the behavior of L is contained in the behavior of P_1, and P_2 is the flower model, the precision value of P_1 must be greater than the precision value of P_2. Formally, if $\mathcal{B}_L \subseteq \mathcal{B}_{P_1} \subset \mathcal{B}_{P_2} = \Gamma_\Sigma \implies prec(L, P_1) > prec(L, P_2)$.*
- **Axiom-4.** *Given two process models P_1, P_2 and a log L, if the behavior of P_1 is equal to the behavior of P_2, the precision values of P_1 and P_2 must be equal. Formally, if $\mathcal{B}_{P_1} = \mathcal{B}_{P_2} \implies prec(L, P_1) = prec(L, P_2)$.*
- **Axiom-5.** *Given a process model P and two event logs L_1, L_2, if the behavior of L_1 is contained in the behavior of L_2, and the behavior of L_2 is contained in the behavior of P, the precision value of the model measured over L_2 must be equal to or greater than the precision value measured over L_1. Formally, if $\mathcal{B}_{L_1} \subseteq \mathcal{B}_{L_2} \subseteq \mathcal{B}_P \implies prec(L_2, P) \geq prec(L_1, P)$.*

Tax et al. [15] showed that none of the existing measures fulfils all the axioms.

---

[2] To enhance the readability, in the rest of this paper we refer to τ_Σ as τ, omitting the set Σ.

[3] In the case $\mathcal{B}_P = \Gamma_\Sigma$, P corresponds to the flower model.

[4] The support of a multiset is the set containing the distinct elements of the multiset.

3   Markovian Abstraction-Based Precision (MAP)

This section presents a family of precision measures based on k^{th}-order Markovian abstractions. Intuitively, precision measures try to estimate how much of the behavior captured in a process model can be found in the behavior recorded in an event log. The computation of our precision measures can be divided into three steps: (i) abstraction of the behavior of a process model, (ii) abstraction of the behavior recorded in an event log, and (iii) comparison of the two behavioral abstractions. We start by defining the kth-order Markovian abstraction, as well as its features, and then introduce the algorithm to compare a pair of Markovian abstractions. Finally, we show that our precision measures satisfy four of the five precision axioms, while the fifth axiom is also satisfied for specific values of k.

3.1   Markovian Abstraction

A kth-order Markovian abstraction (M^k-abstraction) is a graph composed by a set of states (S) and a set of edges ($E \subseteq S \times S$). In an M^k-abstraction, every state $s \in S$ represents a (sub)trace of at most length k, e.g. $s = \langle b, c, d \rangle$, while two states $s_1, s_2 \in S$ are connected via an edge $e = (s_1, s_2) \in E$ iff s_1 and s_2 satisfy the following three properties: (i) the first activity of the (sub)trace represented by s_1 can occur before the (sub)trace represented by s_2, (ii) the last activity of the (sub)trace represented by s_2 can occur after the (sub)trace represented by s_1, and (iii) the two (sub)traces represented by s_1 and s_2 overlap with the exception of their first and last activity, respectively, e.g. $e = (\langle b, c, d \rangle, \rangle c, d, e \rangle)$. Every state of an M^k-abstraction is unique, i.e. there are no two states representing the same (sub)trace. An M^k-abstraction is defined w.r.t. a given order k, which defines the size of the (sub)traces encoded in the states. An M^k-abstraction contains a fresh state (denoted as $-$) representing the sink and source of the M^k-abstraction. Intuitively, every state represents either a trace of length less than or equal to k or a subtrace of length k, whilst every edge represents an existing subtrace of length $k+1$ or a trace of length less than or equal to $k+1$. Thus, M^k-abstraction captures how all the traces of the input behavior evolves in chunks of length k. The definitions below show the construction of a M^k-abstraction from a given $\mathscr{B}_X$, and a fundamental property of the M^k-abstractions to show that our precision measure fulfils the 5 precision axioms.

Definition 6 [k^{th}-order Markovian Abstraction]. *Given a set of traces $\mathscr{B}_X$, the k-order Markovian Abstraction is the graph $M^k_X = (S, E)$ where S is the set of the states and $E \subseteq S \times S$ is the set of edges, such that*

- $S = \{-\} \cup \{\tau : \tau \in \mathscr{B}_X \land |\tau| \le k\} \cup \{\tau^{i \to j} : \tau \in \mathscr{B}_X \land |\tau| > k \land |\tau^{i \to j}| = k\}$
- $E = \{(-, \tau) : \tau \in S \land |\tau| \le k\} \cup \{(\tau, -) : \tau \in S \land |\tau| \le k\} \cup \{(-, \tau) : \exists \hat{\tau} \in \mathscr{B}_X \ s.t. \ \tau = \hat{\tau}^{1 \to k}\} \cup \{(\tau, -) : \exists \hat{\tau} \in \mathscr{B}_X \ s.t. \ \tau = \hat{\tau}^{(|\hat{\tau}|-k+1) \to |\hat{\tau}|}\} \cup \{(\tau', \tau'') : \tau', \tau'' \in S \land \tau' \oplus \tau''_{|\tau''|} = \tau'_1 \oplus \tau'' \land \exists \hat{\tau} \in \mathscr{B}_X \ s.t. \ \tau'_1 \oplus \tau'' \subseteq \hat{\tau}\}^5$

---

5 The operator $\oplus$ is the *concatenation* operator.

**Theorem 1 *[Equality and Containment Inheritance].* ** *Given two sets of traces $\mathscr{B}_X$ and $\mathscr{B}_Y$, and their respective M^k- abstractions $M_X^k = (S_X, E_X)$ and $M_Y^k = (S_Y, E_Y)$, any equality or containment relation between $\mathscr{B}_X$ and $\mathscr{B}_Y$ is inherited by E_X and E_Y. I.e., if $\mathscr{B}_X = \mathscr{B}_Y$ then $E_X = E_Y$, or if $\mathscr{B}_X \subset \mathscr{B}_Y$ then $E_X \subseteq E_Y$.*

Proof. (Sketch) This follows by construction. Specifically, every edge $e \in E_X$ represents either a subtrace $\tau^{x \rightarrow y} : \tau \in \mathscr{B}_X \wedge |\tau^{x \rightarrow y}| = k + 1$, or it represents a trace $\tau : \tau \in \mathscr{B}_X \wedge |\tau| < k + 1$. The last implies that from the same sets of traces the corresponding M^k-abstractions contain the same sets of edges.   □

Note, however, that the theorem above cannot say anything for the traces in $\mathscr{B}_Y \setminus \mathscr{B}_X$, i.e. adding new traces to $\mathscr{B}_X$ does not imply that new edges are added to E_X. As a result the relation $\mathscr{B}_X \subset \mathscr{B}_Y$ guarantees only $E_X \subseteq E_Y$, instead of $E_X \subset E_Y$.

Note that, M^1-abstraction is equivalent to a *directly-follows graph* (a well-known behavior abstraction used as starting point by many process discovery approaches [5,10,17,18]). Instead, if k approaches to infinite then M^∞-abstraction is equivalent to listing all the traces. The M^k-abstraction of a process model can be built from its reachability graph by replaying it. The time complexity of such operation strictly depends on k, and it ranges from polynomial time ($k = 1$) to double exponential time for greater values of k. Instead, the M^k-abstraction of an event log can be built always in polynomial time, since the log behavior is a finite set of traces.

Table 1. Log L^*.

Traces
$\langle a, a, b \rangle$
$\langle a, b, b \rangle$
$\langle a, b, a, b, a, b \rangle$

One can tune the level of behavioral approximation by varying the order k of the M^k-abstraction. For example, let us consider the event log L^* as in Table 1, and the Process-X (P_x) in Fig. 1c. Their respective M^1-abstractions: $M_{L^*}^1$ and $M_{P_x}^1$ are shown in Fig. 2d and c. We can notice that $M_{L^*}^1 = M_{P_x}^1$, though $\mathscr{B}_{P_x}$ is infinite whilst $\mathscr{B}_{L^*}$ is not. This is an example on how the M^1-abstraction can over-approximate the behavior it represents. However, the increase of k can lead to more accurate representations (decreasing the degree of over-approximation) and thus to behavioral differences between behaviorally-similar abstractions, e.g., L^* and P_x, can be detected, see Fig. 3d and c. We remark that for k equal to the length of the longest trace in the log, the behavioral abstraction of this latter is exact. However, a similar reasoning cannot be done for the model behavior, since its longest trace may be infinite.

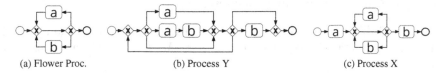

(a) Flower Proc.   (b) Process Y   (c) Process X

Fig. 1. Examples of processes in the BPMN language.

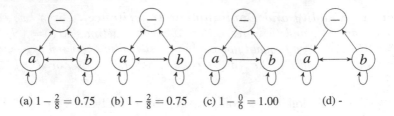

(a) $1 - \frac{2}{8} = 0.75$ (b) $1 - \frac{2}{8} = 0.75$ (c) $1 - \frac{0}{6} = 1.00$ (d) -

Fig. 2. From left to right: the M^1-abstraction of the Flower Process, Process-Y, Process-X and the event log L^*. The respective labels report the value of their MAP^1.

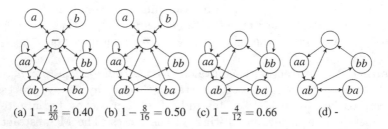

(a) $1 - \frac{12}{20} = 0.40$ (b) $1 - \frac{8}{16} = 0.50$ (c) $1 - \frac{4}{12} = 0.66$ (d) -

Fig. 3. From left to right, the M^2-abstraction of the Flower Process, Process-Y, Process-X and the event log L^*. The respective labels report the value of their MAP^2.

3.2 Comparing Markovian Abstractions

The third and final step of our precision measure is the comparison of the M^K-abstractions of the process model and the event log. In short, given two M^k-abstractions, we compare them using a weighted edge-based graph matching algorithm.

Definition 7 [Weighted Edge-based Graph Matching Algorithm (GMA)]. *A Weighted Edge-based Graph Matching Algorithm (GMA) is an algorithm that receives as input two graphs $G_1 = (N_1, E_1)$ and $G_2 = (N_2, E_2)$, and outputs a mapping function $\mathscr{I}_C : E_1 \to (E_2 \cup \{\varepsilon\})$. The function $\mathscr{I}_C$ maps pairs of edges matched by a graph matching algorithm or, if no mapping was found, the edges in E_1 are mapped to ε, i.e., $\forall e_1, e_2 \in E_1 : \mathscr{I}_C(e_1) = \mathscr{I}_C(e_2) \Rightarrow (e_1 = e_2) \lor (\mathscr{I}_C(e_1) = \varepsilon \land \mathscr{I}_C(e_2) = \varepsilon)$. A GMA is characterised by an underlying cost function $C : E_1 \times (E_2 \cup \{\varepsilon\}) \to [0, 1]$, s.t. $\forall e_1 \in E_1$ and $\forall e_2 \in E_2 \implies C(e_1, e_2) \in [0, 1]$ and $\forall e_1 \in E_1 \implies C(e_1, \varepsilon) = 1$. Hereinafter we refer to any GMA as its mapping function $\mathscr{I}_C$.*

Given a GMA $\mathscr{I}_C$, an event log L and a process P as inputs, the kth-order Markovian abstraction-based precision (hereby MAP^k) is estimated applying Eq. 1.

$$MAP^k(L, P) = 1 - \frac{\Sigma_{e \in E_P} C(e, \mathscr{I}_C(e))}{|E_P|} \qquad (1)$$

The selected GMA for the implementation of our MAP^k is an adaptation of the Hungarian method [9], where: the cost of a match between two edges is defined as the average of the Levenshtein distance between the source states and the target states; and the final matching is the one minimising the total costs of the matches.

Figure 1 shows three models in BPMN notation. Their respective Markovian abstractions are captured in Figs. 2a–c and 3a–c, for $k = 1$ and $k = 2$. We can observe that by increasing k, the quality of the behavior approximation decreases. Consequently, the MAP^k achieves a finer result.

Note that each of the proposed precision measures fulfills the properties of an ordinal scale. Specifically, given an event log $\mathscr{L}$ and for a given k, MAP^k induces an order over the possible process models that fit log $\mathscr{L}$. This property is desirable given that the purpose of a precision measure is to allow us to compare two possible process models in terms of their additional behavior.

3.3   Proofs of the 5-Axioms

We now turn our attention to show that our Markovian abrastraction-based precision measure fulfils the axioms presented in Sect. 2. For the remaining part of the section, let L_x be a log, P_x be a process model, and $M_{L_x}^k = (S_{L_x}, E_{L_x})$ and $M_{P_x}^k = (S_{P_x}, E_{P_x})$ be the M^k-abstractions of the log and the model, respectively.

Axiom-1. $MAP^k(L, P)$ is a deterministic function. Given a log L and a process P, The construction of M_L^k and M_P^k is fully deterministic for $\mathscr{B}_P$ and $\mathscr{B}_L$ (see Definition 6). Furthermore, being the graph matching algorithm $\mathscr{I}_C$ deterministic, and being $MAP^k(L, P)$ function of E_L, E_P and $\mathscr{I}_C$ (see Eq. 1), it follows that $MAP^k(L, P)$ is also deterministic with codomain $\mathbb{R}$.

Axiom-2. Given two processes P_1, P_2 and an event log L, s.t. $\mathscr{B}_L \subseteq \mathscr{B}_{P_1} \subseteq \mathscr{B}_{P_2}$, then $MAP^k(L, P_1) \geq MAP^k(L, P_2)$. First, the following relation holds, $E_L \subseteq E_{P_1} \subseteq E_{P_2}$ (see Theorem 1). Then, we distinguish two possible cases:

1. if $E_{P_1} = E_{P_2}$, then it follows straightforward $MAP^k(L, P_1) = MAP^k(L, P_2)$, because $MAP^k(L, P)$ is a deterministic function of E_L, E_P and $\mathscr{I}_C$ (see Axiom-1 proof and Eq. 1).
2. if $E_{P_1} \subset E_{P_2}$, then $E_L \subset E_{P_2} \wedge (|E_{P_2}| - |E_{P_1}|) > 0$. In this case, we show that $MAP^k(L, P_2) - MAP^k(L, P_1) < 0$ is always true, as follows.

$$1 - \frac{\sum_{e_2 \in E_{P_2}} C(e_2, \mathscr{I}_C(e_2))}{|E_{P_2}|} - \left(1 - \frac{\sum_{e_1 \in E_{P_1}} C(e_1, \mathscr{I}_C(e_1))}{|E_{P_1}|}\right) =$$
$$\frac{\sum_{e_1 \in E_{P_1}} C(e_1, \mathscr{I}_C(e_1))}{|E_{P_1}|} - \frac{\sum_{e_2 \in E_{P_2}} C(e_2, \mathscr{I}_C(e_2))}{|E_{P_2}|} < 0$$

For each edge e_1 that can be found both in E_{P_1} and E_L, the cost $C(e_1, \mathscr{I}_C(e_1))$ is 0, being $\mathscr{I}_C(e_1) = e_1$. Instead, for each edge e_1 that can be found in E_{P_1} but not in E_L, the cost $C(e_1, \mathscr{I}_C(e_1))$ is 1, being $\mathscr{I}_C(e_1) = \varepsilon$. It follows that the total cost of matching E_{P_1} over L is $\sum_{e_1 \in E_{P_1}} C(e_1, \mathscr{I}_C(e_1)) = |E_{P_1}| - |E_L|$. A similar reasoning can be done for the matching of E_{P_2} over $_L$. Indeed, $\forall e_2 \in E_{P_2} \cap E_L \implies C(e_2, \mathscr{I}_C(e_2)) = 0$ and $\forall e_2 \in E_{P_2} \setminus E_L \implies C(e_2, \mathscr{I}_C(e_2)) = C(e_2, \varepsilon) = 1$, therefore $\Sigma_{e_2 \in E_{P_2}} C(e_2, \mathscr{I}_C(e_2)) = |E_{P_2}| - |E_L|$.

Applying these results to the above inequality, it turns into the following:

$$\frac{|E_{P_1}| - |E_L|}{|E_{P_1}|} - \frac{|E_{P_2}| - |E_L|}{|E_{P_2}|} = \frac{|E_L|\,(|E_{P_1}| - |E_{P_2}|)}{|E_{P_1}||E_{P_2}|} < 0$$

This latter is always true, since the starting hypothesis of this second case is $(|E_{P_1}| - |E_{P_2}|) < 0$.

Axiom-3. Given two processes P_1, P_2 and an event log L, s.t. $\mathscr{B}_L \subseteq \mathscr{B}_{P_1} \subset \mathscr{B}_{P_2} = \Gamma_\Sigma$ then $MAP^k(L, P_1) > MAP^k(L, P_2)$. For any $k \in \mathbb{N}$, the relation $MAP^k(L, P_1) \geq MAP^k(L, P_2)$ holds for Axiom-2. The case $MAP^k(L, P_1) = MAP^k(L, P_2)$ occurs when $M_{P_2}^k$ over-approximates the behavior of P_2, i.e. $\mathscr{B}_{P_1} \subset \mathscr{B}_{P_2}$ and $E_{P_1} = E_{P_2}$. Nevertheless, for any $\mathscr{B}_{P_1}$ there always exists a k^* s.t. $E_{P_1} \subset E_{P_2}$. This is true since being $\mathscr{B}_{P_1}$ strictly contained in $\mathscr{B}_{P_2}$, there exists a trace $\hat{\tau} \in \mathscr{B}_{P_2}$ s.t. $\hat{\tau} \notin \mathscr{B}_{P_1}$. Choosing $k^* = |\hat{\tau}|$, the $M_{P_2}^{k^*}$ would produce an edge $\hat{e} = (-, \hat{\tau}) \in E_{P_2}$ s.t. $\hat{e} \notin E_{P_1}$ because $\hat{\tau} \notin \mathscr{B}_{P_1}$ (see also Definition 6).[6] Consequently, for any $k \geq k^*$, we have $E_{P_1} \subset E_{P_2}$ and $MAP^k(L, P_1) > MAP^k(L, P_2)$ holds, being this latter the case 2 of Axiom-2.

Axiom-4. Given two processes P_1, P_2 and an event log L, s.t. $\mathscr{B}_{P_1} = \mathscr{B}_{P_2}$ then $MAP^k(L, P_1) = MAP^k(L, P_2)$. If $\mathscr{B}_{P_1} = \mathscr{B}_{P_2}$, then $E_{P_1} = E_{P_2}$ (see Theorem 1). It follows straightforward that $MAP^k(L, P_1) = MAP^k(L, P_2)$ (see proof Axiom-1 and Eq. 1).

Axiom-5. Given two event logs L_1, L_2 and a process P, s.t. $\mathscr{B}_{L_1} \subseteq \mathscr{B}_{L_2} \subseteq \mathscr{B}_P$, then $MAP^k(L_2, P) \geq MAP^k(L_1, P)$. Consider the two following cases:

1. if $\mathscr{B}_{L_1} = \mathscr{B}_{L_2}$, then $E_{L_1} = E_{L_2}$ (see Theorem 1). It follows $MAP^k(L_2, P) = MAP^k(L_1, P)$, because $MAP^k(L, P)$ is a deterministic function of E_L, E_P and $\mathscr{I}_C$ (see Axiom-1 proof and Eq. 1).

2. if $\mathscr{B}_{L_1} \subset \mathscr{B}_{L_2}$, then $E_{L_1} \subseteq E_{L_2}$ (see Theorem 1). In this case, the graph matching algorithm would find matchings for either the same number or a larger number of edges between M_P^k and $M_{L_2}^k$, than between M_P^k and $M_{L_1}^k$ (this follows from $E_{L_1} \subseteq E_{L_2}$). Thus, a smaller or equal number of edges will be mapped to ε in the case of $MAP^k(L_2, P)$ not decreasing the value for the precision,i.e., $MAP^k(L_2, P) \geq MAP^k(L_1, P)$.

In Axiom-3 we showed that there exists a specific value of k, namely k^*, for which $MAP^{k^*}(L_x, P_x)$ satisfies Axiom-3 and we identified such value being $k^* = |\hat{\tau}|$, where $\hat{\tau}$ can be any trace of the set difference $\Gamma_\Sigma \setminus \mathscr{B}_{P_x}$. In the following, we show how to identify the minimum value of k^* such that all the 5-Axioms are satified. To identify the lowest value of k^*, we have to consider the traces $\hat{\tau} \in \Gamma_\Sigma$ such that does not exists a $\tau \in \mathscr{B}_{P_x}$ where $\hat{\tau} \subseteq \tau$. If a trace $\hat{\tau} \in \Gamma_\Sigma$ that is not a sub-trace of any other trace of the process model behavior $(\mathscr{B}_{P_x})$ is found, by setting $k^* = |\hat{\tau}|$ would mean that in the M^{k^*}-abstraction of Γ_Σ there will be a state $\hat{s} = \hat{\tau}$ and an edge $(-, \hat{\tau})$ that are not captured by the M^{k^*}-abstraction of

---

[6] Formally, $\exists \hat{\tau} \in \mathscr{B}_{P_2} \setminus \mathscr{B}_{P_1}$, s.t. for $k^* = |\hat{\tau}| \implies \exists(-, \hat{\tau}) \in E_{P_2} \setminus E_{P_1}$.

$\mathscr{B}_{P_x}$. This difference will allow us to distinguish the process P_x from the flower model (i.e. the model having a behavior equal to Γ_Σ), satisfying in this way the Axiom-3. At this point, considering the set of the lengths of all the subtraces not contained in any trace of $\mathscr{B}_{P_x}$, $Z = \{|\widehat{\tau}| \ : \ \widehat{\tau} \in \Gamma_\Sigma \ \wedge \ \nexists \, \tau \in \mathscr{B}_{P_x} \mid \widehat{\tau} \subseteq \tau\}$, we can set the lower-bound of $k^* \geq min(Z)$.

Note that the value of k^* is equal to 2 for any process model with at least one activity that cannot be executed twice in a row. If we have an activity $\widehat{t}$ that cannot be executed twice in a row, it means that $|\langle \widehat{t}, \widehat{t} \rangle| \in Z$ and thus we can set $k^* = 2$. In practice, $k^* = 2$ satisfies all the 5-Axioms in real-life cases, since it is very common to find process models that have the above topological characteristic.

4   Evaluation

In this section, we report on a two-pronged evaluation we performed to assess the following two objectives: (i) comparing our family of precision measures to state-of-the-art precision measures; and (ii) analysing the role of the parameter k.

To do so, we implemented the Markovian Abstraction-based Precision (MAP^k) as a standalone open-source tool[7] and used it to carry out a qualitative evaluation on synthetic data and a quantitative evaluation on real-life data.[8] All experiments were executed on an Intel Core i5-6200U @2.30 GHz with 16 GB RAM running Windows 10 Pro (64-bit) and JVM 8 with 12 GB RAM (8 GB Stack and 4 GB Heap).

4.1   Qualitative Evaluation

In a previous study, van Dongen et al. [16] showed that their anti-alignment precision was able to improve on a range of state-of-the-art precision measures. To qualitatively assess our MAP^k, we decided to repeat the experiment carried out in [16] using the same synthetic dataset. Table 2 and Fig. 4 show the synthetic event log and a model, called "original model", that was used to generate eight variants: a *single trace* model capturing the most frequent trace; a model incorporating all *separate traces*; a *flower model* of all activities in the log; a model with activities G and H in parallel (*Opt. G* $\|$ *Opt. H*, see Fig. 5); one with G and H in self-loop ($\circlearrowright G$, $\circlearrowright H$, Fig. 6); a model with D in self-loop ($\circlearrowright D$, Fig. 7); a model with all activities in parallel (*All parallel*); and a model where all activities are in round robin (*Round robin*, Fig. 8). Using each log-model pair, we compared our precision measure MAP^k to the precision measures discussed in Sect. 2 (these include those evaluated by van Dongen et al. [16]), namely: traces set difference precision (SD), alignment-based ETC precision (ETC$_a$), negative events precision (NE), projected conformance checking (PCC), anti-alignment precision (AA). We left out the advanced behavioral appropriateness (ABA) as

---

[7] Available at http://apromore.org/platform/tools.

[8] The public data used in the experiments can be found at https://doi.org/10.6084/m9.figsharc.6376592.v1.

it is not defined for some of the models in this dataset. We limited the order k to 7, because it is the length of the longest trace in the log. Setting an order greater than 7 would only (further) penalise the cyclic behavior of the models, which is not necessary to assess the models' precision.

Table 2. Test log [16].

Traces	#
$\langle A, B, D, E, I \rangle$	1207
$\langle A, C, D, G, H, F, I \rangle$	145
$\langle A, C, G, D, H, F, I \rangle$	56
$\langle A, C, H, D, F, I \rangle$	23
$\langle A, C, D, H, F, I \rangle$	28

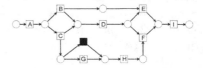

Fig. 4. Original model [16].

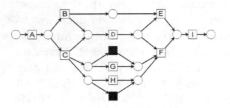

Fig. 5. Opt. G || Opt. H model [16].

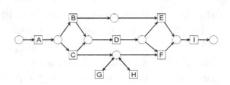

Fig. 6. ↺G, ↺H model [16].

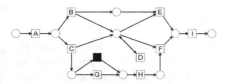

Fig. 7. ↺D model [16].

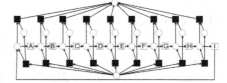

Fig. 8. Round robin model [16].

Table 3 reports the results of our qualitative evaluation.[9] To discuss these results, we use two precision measures as a reference, as these have been advocated as possible ground truths of precision, though none of them satisfies the axioms in [15]. The first one is AA. This measure has been shown [16] to be intuitively more accurate than other precision measures. The second one is SD, as it closely operationalizes the definition of precision by capturing the exact percentage of model behavior that cannot be found in the log. As discussed in Sect. 2 though, this measure can only be computed for acyclic models, and uses a value of zero for cyclic models by design.

From the results in Table 3, we can observe that MAP^1 does not penalise enough the extra behavior of some models, such as the *original* model, which cannot be distinguished from the *single trace* and the *separate traces* models

---

[9] Some values differ from those in [16] as we used each measure's latest implementation.

Table 3. Comparison of different precision measures over synthetic dataset (* indicates a rounded-down value: $0.000^* > 0.000$).

Process variant	Model traces (#)	SD	ETC_a	NE	PCC	AA	MAP^1	MAP^2	MAP^3	MAP^4	MAP^5	MAP^6	MAP^7
Original model	6	0.833	0.900	0.995	1.000	0.871	1.000	0.909	0.880	0.852	0.852	0.852	0.852
Single trace	1	1.000	1.000	0.893	1.000	1.000	1.000	1.000	1.000	1.000	1.000	1.000
Separate traces	5	1.000	1.000	0.985	0.978	1.000	1.000	1.000	1.000	1.000	1.000	1.000	1.000
Flower model	986,410	0.000	0.153	0.117	0.509	0.000	0.189	0.024	0.003	0.000	0.000	0.000	0.000
Opt. G || Opt. H	12	0.417	0.682	0.950	0.974	0.800	0.895	0.645	0.564	0.535	0.535	0.535	0.535
↻G, ↻H	362	0.000	0.719	0.874	0.896	0.588	0.810	0.408	0.185	0.080	0.034	0.015	0.006
↻D	118	0.000	0.738	0.720	0.915	0.523	0.895	0.556	0.349	0.223	0.145	0.098	0.069
All parallel	362,880	0.000	0.289	0.158	0.591	0.033	0.210	0.034	0.006	0.001	0.000^*	0.000^*	0.000^*
Round robin	27	0.000	0.579	0.194	0.594	0.000	0.815	0.611	0.496	0.412	0.350	0.306	0.274

(all have a precision of 1). Also, the values of MAP^1 are far away from those of both AA and SD (with the exception of the simplest models, i.e. *single trace* and *separate traces*). As we increase k, MAP^k tends to get closer to AA and to SD, barring a few exceptions. In particular, the more is the cyclic behavior allowed by a model, the quicker MAP^k tends to zero. In this respect, let us consider the cyclic models in our datasets: (i) the *flower* model, (ii) the ↻G, ↻H model (Fig. 6), (iii) the ↻D model (Fig. 7), and (iv) the *round robin* (Fig. 8). The value of our precision measure tends to zero faster in the *flower* model (k=3) than in the other cyclic models, because the *flower* model allows the greatest amount of cyclic behavior, due to all the possible combinations of activities being permitted. At $k = 7$ this is consistent with both SD and AA. Similarly, our measure tends to zero slower in the *round robin* model because this model is very strict on the order in which activities can be executed, despite having infinite behavior. In fact, it only allows the sequence $\langle A, B, C, D, F, G, H, I \rangle$ to be executed, with the starting activity and the number of repetitions being variable. This is taken into account by our measure, since even with $k = 7$ we do not reach a value of zero for this model, as opposed to SD and AA. This allows us to discriminate the *round robin* model from other models with very large behavior such as the *flower* model. This is not possible with SD and AA, because both models have a precision of zero in these two measures. As for the other two cyclic models in our dataset, MAP^k tends to zero with speeds between those of the *flower* model and the *round robin* model, with the ↻G, ↻H model being faster to drop than the ↻D, due to the former allowing more cyclic behavior than the latter. Similar considerations as above apply to these two models: even at $k = 7$ their precision does not reach zero, which allows us to distinguish these models from other models such as the *all parallel* model, which has a very large behavior (360K+ distinct traces). While in SD the precision of these two models is set to zero by design, for AA these two models have a precision greater than zero, though the ↻G, ↻H model has a higher precision than the ↻D model (0.588 vs. 0.523). This is counter-intuitive, since the former model allows more model behavior not permitted by the log (in terms of number of different traces) than the latter model does. In addition, AA penalizes more the *round robin* model, despite this has less model behavior than the two models with self-loop activities. Altogether, these results show that the higher the k, the more the behavioral differences that our measure can catch and penalise.

In terms of ranking (see Table 4), our measure is the most consistent with the ranking of the models yielded by both SD (for acyclic models) and AA (for all models), than all other measures. As discussed above, the only differences with AA are in the swapping of the order of the two models with self loops, and in the order of the *round robin* model. Note that given that both the round robin and the flower model have a value of zero in AA, the next model in the ranking (*all parallel*) is assigned a rank of 3 instead of 2 in MAP^k. This is just the way the ranking is computed and is not really indicative of a ranking inconsistency between the two measures. Another observation is that the ranking yielded by our family of metrics remains the same for $k > 1$. This indicates that as we increase k, while the extent of behavioral differences we can identify and penalize increases, this is not achieved at the price of changing the ranking of the models.

Table 4. Models ranking yielded by the precision measures over the synthetic dataset.

Process variant	SD	ETC_a	NE	PCC	AA	MAP^1	MAP^2	MAP^3	MAP^4	MAP^5	MAP^6	MAP^7
Original model	7	7	9	8	7	7	7	7	7	7	7	7
Single trace	8	8	6	8	8	7	8	8	8	8	8	8
Separate traces	8	8	8	7	8	7	8	8	8	8	8	8
Flower model	1	1	1	1	1	1	1	1	1	1	1	1
Opt. G ‖ Opt. H	6	3	7	6	6	5	6	6	6	6	6	6
⟲G, ⟲H	1	5	5	4	5	3	3	3	3	3	3	3
⟲D	1	6	4	5	4	5	4	4	4	4	4	4
All parallel	1	2	2	2	3	2	2	2	2	2	2	2
Round robin	1	4	3	3	1	4	5	5	5	5	5	5

On average it took less than a second per model to compute MAP^k, except for the *all parallel* model, for which it took 3.8 s at $k = 7$, due to the large number of distinct traces yielded by this model.

4.2   Quantitative Evaluation

In our second evaluation, we used two datasets for a total of 20 logs. The first dataset is the collection of real-life logs publicly available from the 4TU Centre for Research Data, as of March 2017.[10] Out of this collection, we retained twelve logs related to business processes, as opposed to e.g. software development processes. These include the *BPI Challenge* (BPIC) logs (2012-17), the

---

[10] https://data.4tu.nl/repository/collection:event_logs_real.

Road Traffic Fines Management Process (RTFMP) log, and the *SEPSIS* log. These logs record executions of business processes from a variety of domains, e.g. healthcare, finance, government and IT service management. In seven logs (BPIC14, the BPIC15 collection, and BPIC17), we applied the filtering technique proposed in [6] to remove infrequent behavior. The second dataset is composed of eight proprietary logs sourced from several companies in the education, insurance, IT service management and IP management domains. Table 5 reports the characteristics of both datasets, highlighting the heterogeneous nature of the data.

Table 5. Descriptive statistics of the real-life logs (public and proprietary).

Log		BPIC12	BPIC13$_{cp}$	BPIC13$_{inc}$	BPIC14$_f$	BPIC15$_{1f}$	BPIC15$_{2f}$	BPIC15$_{3f}$	BPIC15$_{4f}$	BPIC15$_{5f}$
Total Traces		13,087	1,487	7,554	41,353	902	681	1,369	860	975
Dist. Traces		33.4	12.3	20	36.1	32.7	61.7	60.3	52.4	45.7
Total Events		262,200	6,660	65,533	369,485	21,656	24,678	43,786	29,403	30,030
Dist. Events		36	7	13	9	70	82	62	65	74
Tr. length	(min)	3	1	1	3	5	4	4	5	4
	(avg)	20	4	9	9	24	36	32	34	31
	(max)	175	35	123	167	50	63	54	54	61

Log		BPIC17$_f$	RTFMP	SEPSIS	PRT1	PRT2	PRT3	PRT4	PRT6	PRT7	PRT9	PRT10
Total Traces		21,861	150,370	1,050	12,720	1,182	1,600	20,000	744	2,000	787,657	43,514
Dist. Traces		40.1	0.2	80.6	8.1	97.5	19.9	29.7	22.4	6.4	0.01	0.01
Total Events		714,198	561,470	15,214	75,353	46,282	13,720	166,282	6,011	16,353	1,808,706	78,864
Dist. Events		41	11	16	9	9	15	11	9	13	8	19
Tr. length	(min)	11	2	3	2	12	6	6	7	8	1	1
	(avg)	33	4	14	5	39	8	8	8	8	2	1
	(max)	113	2	185	64	276	9	36	21	11	58	15

First, we discovered different process models from each log, using three state-of-the-art automated process discovery methods [3]: Split Miner [4] (SM), Inductive Miner [11] (IM), and Structured Heuristics Miner [2] (SHM). Then, we measured the precision for each model with our MAP^k measure, by varying the order k in the range 2–5. Unfortunately, we were not able to use any of the previous reference measures, because SD does not work for cyclic models (all models discovered by IM were cyclic) and AA does not scale to real-life models [16]. Thus, we resorted to ETC_a as a baseline, since this is, to date, the most-scalable and widely-accepted precision measure for automated process discovery in real-life settings [3].

Table 6 shows the results of the quantitative evaluation. In line with the former evaluation, the value of MAP^k decreases when k increases. However, being the behavior of the real-life models more complex than the one of the synthetic models, for some logs (e.g. the BPIC15 logs), it was not possible to compute

MAP^4 and MAP^5 for the models discovered by IM. This was due to scalability issues, as the models discovered by IM exhibit flower-like behavior (with more than 50 distinct activities per flower construct). This is reflected by the very low values of MAP^2 and MAP^3 for IM. However, we recall that by design, for small values of k, MAP^k compares small chunks of the model behavior to small chunks of the log behavior. Thus, low values of MAP^k can already indicate poorly-precise models. ETC_a and MAP^5 agreed on the precision ranking 50% of the times. This result is consistent with our qualitative evaluation. Also in-line with the former evaluation, ETC_a showed to be very tolerant to infinite model behavior, regardless of the type of such behavior. The clearest example supporting this flaw is the SEPSIS log case. The models discovered by IM and SM are shown in Figs. 9 and 10. We can see that more than the 80% of the activities in the IM model are skippable and over 60% of them are inside a long loop, resembling a flower construct with some constraints, e.g. the first activity is always the same. Instead, the model discovered by SM, even if cyclic, does not allow many variants of behavior. Consequently, for the IM model, the value of MAP^k drastically drops when increasing k from 2 to 3, whilst it remains 1 for the SM model. In contrast, ETC_a reports a precision of 0.445 for IM, which is counter-intuitive considering the flower-like model.

As discussed in Sect. 3, $k = 2$ is sufficient to satisfy all the 5-Axioms in practice. However, as we also observe from the results of this second experiment, higher values of k lead to finer results for MAP^k. In fact, the notable drops of value from $k = 2$ to $k = 3$ (e.g. in SEPSIS, BPIC17$_f$ and PRT9), confirm that the 5-Axioms are a necessary but not sufficient condition for a reliable precision measure [15].

Finally, Tables 7 and 8 report statistics on the time performance of MAP^k and ETC_a. We divided the results by public and private logs to allow the reproducibility of the experiments for the set of public logs. We can see that MAP^k scales well to real-life logs, being quite fast for models with a reasonable state-space size (i.e. with non-flower constructs), as those produced by SM and SHM, while ETC_a remains slower even when compared to MAP^5. However, as expected, by increasing k the performance of MAP^k reduces sharply for flower-like models, as those produced by IM.

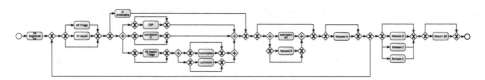

Fig. 9. Model discovered by IM from the SEPSIS log.

Table 6. Comparison of MAP^k results with $k = 2$–5 using three discovery methods on 20 real-life logs.

Log	BPIC12			BPIC13$_{cp}$			BPIC13$_{inc}$			BPIC14$_f$			BPIC15$_{1f}$		
Miner	SM	IM	SHM	SM	IM	SHM	SM	IM	SHM	SM	IM	SHM	SM	IM	SHM
ETC_a	0.762	0.502	-	0.974	1.000	0.992	0.979	0.558	0.978	0.673	0.646	-	0.880	0.566	-
MAP^2	1.000	0.089	0.083	1.000	1.000	1.000	1.000	1.000	1.000	1.000	0.775	0.285	1.000	0.020	0.016
MAP^3	1.000	0.014	0.021	1.000	1.000	1.000	1.000	1.000	1.000	1.000	0.754	0.168	1.000	0.003	0.005
MAP^4	0.546	0.002	0.010	1.000	1.000	1.000	1.000	0.990	1.000	1.000	0.750	0.116	1.000	-	0.002
MAP^5	0.234	-	-	1.000	1.000	1.000	1.000	0.861	1.000	1.000	0.718	-	1.000	-	-

Log	BPIC15$_{2f}$			BPIC15$_{3f}$			BPIC15$_{4f}$			BPIC15$_{5f}$			BPIC17$_f$		
Miner	SM	IM	SHM	SM	IM	SHM	SM	IM	SHM	SM	IM	SHM	SM	IM	SHM
ETC_a	0.901	0.556	0.594	0.939	0.554	0.671	0.910	0.585	0.642	0.943	0.179	0.687	0.846	0.699	0.620
MAP^2	1.000	0.024	0.899	1.000	0.035	0.872	1.000	0.017	0.810	1.000	0.007	0.826	0.764	0.604	0.170
MAP^3	1.000	0.003	0.629	1.000	0.004	0.561	1.000	0.002	0.546	1.000	-	0.584	0.533	0.399	0.080
MAP^4	1.000	-	0.380	1.000	-	0.310	1.000	-	0.333	1.000	-	0.371	0.376	0.268	0.039
MAP^5	1.000	-	0.212	1.000	-	0.154	1.000	-	0.189	1.000	-	0.226	0.255	0.172	0.019

Log	RTFMP			SEPSIS			PRT1			PRT2			PRT3		
Miner	SM	IM	SHM	SM	IM	SHM	SM	IM	SHM	SM	IM	SHM	SM	IM	SHM
ETC_a	1.000	0.700	0.952	0.859	0.445	0.419	0.985	0.673	0.768	0.737	-	-	0.914	0.680	0.828
MAP^2	1.000	0.554	0.323	1.000	0.226	0.227	1.000	1.000	0.796	1.000	0.873	1.000	1.000	0.970	0.978
MAP^3	1.000	0.210	0.093	1.000	0.051	0.072	1.000	1.000	0.578	1.000	0.633	1.000	1.000	0.843	0.652
MAP^4	1.000	0.084	0.027	1.000	0.009	0.021	1.000	1.000	0.386	1.000	0.240	0.438	1.000	0.643	0.328
MAP^5	1.000	0.039	0.008	1.000	-	-	1.000	1.000	0.241	1.000	-	0.151	1.000	0.529	0.157

Log	PRT4			PRT6			PRT7			PRT9			PRT10		
Miner	SM	IM	SHM	SM	IM	SHM	SM	IM	SHM	SM	IM	SHM	SM	IM	SHM
ETC_a	0.995	0.753	0.865	1.000	0.822	0.908	0.999	0.726	0.998	0.999	0.611	0.982	0.972	0.790	-
MAP^2	1.000	1.000	1.000	1.000	0.938	0.984	1.000	0.922	0.973	1.000	0.602	0.680	1.000	0.065	-
MAP^3	1.000	1.000	1.000	1.000	0.916	0.946	1.000	0.709	0.742	1.000	0.277	0.294	1.000	0.007	-
MAP^4	1.000	1.000	0.972	1.000	0.622	0.641	1.000	0.596	0.700	1.000	0.121	0.098	0.666	0.001	-
MAP^5	1.000	1.000	0.854	1.000	0.314	0.318	1.000	0.556	0.673	1.000	0.062	0.029	0.434	0.000*	-

Table 7. Time performance statistics (in seconds) using the twelve public logs (+ indicates a result obtained on a subset of the twelve logs, due to some of the models not being available).

	Split Miner				Inductive Miner				Struct. Heuristics Miner			
Precision	Avg	Max	Min	Total	Avg	Max	Min	Total	Avg	Max	Min	Total
ETC_a	60.0	351.9	0.3	720.3	84.2	642.7	0.1	1009.8	34.0	101.4	0.2	305.9
MAP^2	1.9	7.3	0.1	23.2	5.4	15.2	0.1	65.3	6.2	24.4	0.4	74.3
MAP^3	2.0	7.7	0.1	22.5	109.6	426.7	0.1	1205.7	18.5	59.9	0.2	203.7
MAP^4	3.7	16.9	0.2	44.7	927.9[+]	3970.5[+]	0.1[+]	6495.0[+]	102.8	476.2	0.1	1233.7
MAP^5	7.3	24.7	0.2	87.9	-	-	-	-	29.8[+]	102.2[+]	0.2[+]	238.1[+]

Table 8. Time performance statistics (in seconds) using the eight proprietary logs.

Precision	Split Miner				Inductive Miner				Struct. Heuristics Miner			
	Avg	Max	Min	Total	Avg	Max	Min	Total	Avg	Max	Min	Total
ETC_a	16.1	106.5	0.2	129.1	16.4	99.2	0.2	114.9	74.3	350.2	0.7	520.1
MAP^2	4.8	32.1	0.1	38.3	6.3	35.6	0.1	50.7	10.6	57.6	0.1	85.2
MAP^3	7.3	51.3	0.1	58.5	11.4	42.6	0.1	91.2	11.7	55.1	0.1	93.6
MAP^4	9.3	58.8	0.1	74.5	121.8	604.7	0.4	974.5	60.9	382.4	0.4	486.9
MAP^5	15.3	71.8	0.1	122.4	711.1	4841.7	0.8	4977.6	75.1	267.8	0.7	525.8

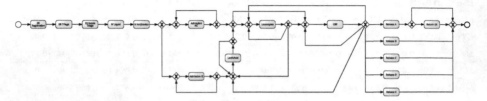

Fig. 10. Model discovered by SM from the SEPSIS log.

5   Conclusion

This paper presented a family of precision measures based on the idea of comparing the kth-order Markovian abstraction of a process model against that of an event log using graph matching algorithms. We showed that this family of precision measures, namely MAP^k, fulfils four of the five axioms of precision of [15] for any value of k and all five axioms for a suitable value of k, dependent on the event log. The empirical evaluation on real-life logs shows that the execution times of the MAP^k (with k up to 5) are considerably lower than those of the ETC_a precision, which is commonly used to evaluate automated process discovery techniques. We also showed on synthetic model-log pairs, that the proposed measure approximates two (unscalable) measures of precision that have been previously advocated as possible ground truths in this field.

Given that our measure abstracts from the model structure and focuses only on its behavior, though in chunks, the only limitation to its usage is scalability, which indirectly affects also the quality of the results. Even if MAP^k is scalable for acyclic process models, for cyclic real-life models, MAP^k showed to be scalable only for low values of k. Despite the evaluation highlights that low k-orders are sufficient to compare (rank) different models discovered from the same log, higher values of k may return more accurate results.

Possible avenues for future work include the design of more efficient and formally grounded instances of this family of precision measures by exploring alternative behavioral abstractions (besides Markovian ones) and alternative comparison operators.

Acknowledgements. This research is partly funded by the Australian Research Council (DP180102839) and the Estonian Research Council (IUT20-55).

References

1. Adriansyah, A., Munoz-Gama, J., Carmona, J., van Dongen, B., van der Aalst, W.: Measuring precision of modeled behavior. IseB **13**(1), 37–67 (2015)
2. Augusto, A., Conforti, R., Dumas, M., La Rosa, M.: Automated discovery of structured process models from event logs: the discover-and-structure approach. DKE (2017)
3. Augusto, A., et al.: Automated discovery of process models from event logs: review and benchmark. TKDE (2018, to appear)
4. Augusto, A., Conforti, R., Dumas, M., La Rosa, M., Polyvyanyy, A.: Split miner: automated discovery of accurate and simple business process models from event logs. KAIS (2018)
5. Augusto, A., Conforti, R., Dumas, M., La Rosa, M.: Split miner: discovering accurate and simple business process models from event logs. In: IEEE ICDM. IEEE (2017)
6. Conforti, R., La Rosa, M., ter Hofstede, A.: Filtering out infrequent behavior from business process event logs. IEEE TKDE **29**(2), 300–314 (2017)
7. De Weerdt, J., De Backer, M., Vanthienen, J., Baesens, B.: A robust F-measure for evaluating discovered process models. In: IEEE Symposium on CIDM. IEEE (2011)
8. Greco, G., Guzzo, A., Pontieri, L., Sacca, D.: Discovering expressive process models by clustering log traces. IEEE TKDE **18**(8), 1010–1027 (2006)
9. Kuhn, H.W.: The Hungarian method for the assignment problem. NRL **2**(1–2), 83–97 (1955)
10. Leemans, S.J.J., Fahland, D., van der Aalst, W.M.P.: Discovering block-structured process models from event logs - a constructive approach. In: Colom, J.-M., Desel, J. (eds.) PETRI NETS 2013. LNCS, vol. 7927, pp. 311–329. Springer, Heidelberg (2013). https://doi.org/10.1007/978-3-642-38697-8_17
11. Leemans, S.J.J., Fahland, D., van der Aalst, W.M.P.: Discovering block-structured process models from event logs containing infrequent behaviour. In: Lohmann, N., Song, M., Wohed, P. (eds.) BPM 2013. LNBIP, vol. 171, pp. 66–78. Springer, Cham (2014). https://doi.org/10.1007/978-3-319-06257-0_6
12. Leemans, S., Fahland, D., van der Aalst, W.: Scalable process discovery and conformance checking. Softw. Syst. Model. (2016)
13. Muñoz-Gama, J., Carmona, J.: A fresh look at precision in process conformance. In: Hull, R., Mendling, J., Tai, S. (eds.) BPM 2010. LNCS, vol. 6336, pp. 211–226. Springer, Heidelberg (2010). https://doi.org/10.1007/978-3-642-15618-2_16
14. Rozinat, A., van der Aalst, W.: Conformance checking of processes based on monitoring real behavior. ISJ **33**(1), 64–95 (2008)
15. Tax, N., Lu, X., Sidorova, N., Fahland, D., van der Aalst, W.: The imprecisions of precision measures in process mining. Inf. Process. Lett. **135**, 1–8 (2018)
16. van Dongen, B.F., Carmona, J., Chatain, T.: A unified approach for measuring precision and generalization based on anti-alignments. In: La Rosa, M., Loos, P., Pastor, O. (eds.) BPM 2016. LNCS, vol. 9850, pp. 39–56. Springer, Cham (2016). https://doi.org/10.1007/978-3-319-45348-4_3
17. vanden Broucke, S., De Weerdt, J.: Fodina: a robust and flexible heuristic process discovery. DSS **100**, 109–118 (2017)
18. Weijters, A., Ribeiro, J.: Flexible heuristics miner (FHM). In: CIDM. IEEE (2011)

Correlating Activation and Target Conditions in Data-Aware Declarative Process Discovery

Volodymyr Leno[1,2], Marlon Dumas[1], and Fabrizio Maria Maggi[1(✉)]

[1] University of Tartu, Tartu, Estonia
{leno,marlon.dumas,f.m.maggi}@ut.ee
[2] University of Melbourne, Melbourne, Australia

Abstract. Automated process discovery is a branch of process mining that allows users to extract process models from event logs. Traditional automated process discovery techniques are designed to produce procedural process models as output (e.g., in the BPMN notation). However, when confronted to complex event logs, automatically discovered process models can become too complex to be practically usable. An alternative approach is to discover declarative process models, which represent the behavior of the process in terms of a set of business constraints. These approaches have been shown to produce simpler process models, especially in the context of processes with high levels of variability. However, the bulk of approaches for automated discovery of declarative process models are focused on the control-flow perspective of business processes and do not cover other perspectives, e.g., the data, time, and resource perspectives. In this paper, we present an approach for the automated discovery of multi-perspective declarative process models able to discover conditions involving arbitrary (categorical or numeric) data attributes, which relate the occurrence of pairs of events in the log. To discover such correlated conditions, we use clustering techniques in conjunction with interpretable classifiers. The approach has been implemented as a proof-of-concept prototype and tested on both synthetic and real-life logs.

1 Introduction

Process mining is a family of techniques for analyzing business processes starting from their executions as recorded in event logs [20]. Process discovery is the most prominent process mining technique. A process discovery technique takes an event log as input and produces a model without using any a-priori information. The dichotomy procedural versus declarative when choosing the most suitable language to represent the output of a process discovery technique has been largely discussed [15,17]: procedural languages can be used for predictable processes working in stable environments, whereas declarative languages can be

Work supported by the Estonian Research Council (IUT20-55).

M. Weske et al. (Eds.): BPM 2018, LNCS 11080, pp. 176–193, 2018.
https://doi.org/10.1007/978-3-319-98648-7_11

used for unpredictable, variable processes working in highly unstable environments. A still open challenge in the discovery of declarative process models is to develop techniques taking into consideration not only the control flow perspective of a business process but also other perspectives like the data, time, and resource perspectives.

In the current contribution, we present an approach that tries to address this challenge. We base our approach on DECLARE a declarative language to represent business processes [14]. In particular, we use the multi-perspective extension of DECLARE, MP-DECLARE, presented in [5]. The proposed approach can be seen as a step forward with respect to the one presented in [12]. In this preliminary work, the discovered models include data conditions to discriminate between cases in which a constraint is satisfied and cases in which the constraint is violated. For example, in a loan application process, we could discover a constraint telling that the submission of an application is eventually followed by a medical history check when the submission has an amount higher than 100 000 euros. Otherwise, the medical check is not performed. In the example above, we have that a response DECLARE constraint (the submission of an application is eventually followed by a medical history check) is satisfied only when a certain condition on the payload of the activation (on the amount associated to the submission of the application) is satisfied.

In this paper, we present an approach to infer two correlated conditions on the payloads of the activation (activation condition) and of the target (target condition) of a constraint. For example, we can discover behaviors like: when an applicant having a salary lower than 24 000 euros per year submits a loan application, eventually an assessment of the application will be carried out, and the type of the assessment is complex. The approach starts with the discovery of a set of *frequent constraints*. A frequent constraint is a constraint having a high number of *constraint instances*, i.e., pairs of events (one activation and one target) satisfying it. Starting from the constraint instances of a frequent constraint, the algorithm clusters the target payloads to find groups of targets with similar payloads. Then, these groups are used as labels for a classification problem. These labels together with the features extracted from the activation payloads are used to train an interpretable classifier (a decision tree). This procedure allows for finding correlations between the activation payloads and the target payloads. The proposed technique is agnostic on how the input set of frequent constraints is derived. In the context of this paper, we identify these constraints using the semantics of MP-DECLARE. The approach has been validated with several synthetic logs to show its ability to rediscover behaviors artificially injected in the logs and its scalability. In addition, the approach has been applied to 6 real-life logs in the healthcare and public administration domains.

The paper is structured as follows. Section 2 provides the necessary background to understand the rest of the paper. Section 3 presents an exemplifying MP-DECLARE model. Section 4 illustrates the proposed discovery approach and Sect. 6 its evaluation. Finally, Sect. 7 provides some related work, and Sect. 8 concludes the paper and spells out directions for future work.

Table 1. Semantics for DECLARE templates

Template	LTL semantics	Activation
Responded existence	$\mathbf{G}(A \to (\mathbf{O}B \vee \mathbf{F}B))$	A
Response	$\mathbf{G}(A \to \mathbf{F}B)$	A
Alternate response	$\mathbf{G}(A \to \mathbf{X}(\neg A\mathbf{U}B))$	A
Chain response	$\mathbf{G}(A \to \mathbf{X}B)$	A
Precedence	$\mathbf{G}(B \to \mathbf{O}A)$	B
Alternate precedence	$\mathbf{G}(B \to \mathbf{Y}(\neg B\mathbf{S}A))$	B
Chain precedence	$\mathbf{G}(B \to \mathbf{Y}A)$	B
Not responded existence	$\mathbf{G}(A \to \neg(\mathbf{O}B \vee \mathbf{F}B))$	A
Not response	$\mathbf{G}(A \to \neg\mathbf{F}B)$	A
Not precedence	$\mathbf{G}(B \to \neg\mathbf{O}A)$	B
Not chain response	$\mathbf{G}(A \to \neg\mathbf{X}B)$	A
Not chain precedence	$\mathbf{G}(B \to \neg\mathbf{Y}A)$	B

2   Preliminaries

In this section, we first introduce the XES standard (Sect. 2.1), then we give some background knowledge about DECLARE (Sect. 2.2) and MP-DECLARE (Sect. 2.3).

2.1   The XES Standard

The starting point for process mining is an event log. XES (eXtensible Event Stream) [1,22] has been developed as the standard for storing, exchanging and analyzing event logs. Each event in a log refers to an *activity* (i.e., a well-defined step in some process) and is related to a particular *case* (i.e., a *process instance*). The events belonging to a case are *ordered* and can be seen as one "run" of the process (often referred to as a *trace* of events). Event logs may store additional information about events such as the *resource* (i.e., person or device) executing or initiating the activity, the *timestamp* of the event, or *data elements* recorded with the event. In XES, data elements can be event attributes, i.e., data produced by the activities of a business process and case attributes, namely data that are associated to a whole process instance. In this paper, we assume that all attributes are globally visible and can be accessed/manipulated by all activity instances executed inside the case.

Let Σ be the set of unique activities in the log. Let $t \in \Sigma^*$ be a trace over Σ, i.e., a sequence of activities performed for one process case. An event log E is a multi-set over Σ^*, i.e., a trace can appear multiple times.

2.2   Declare

DECLARE is a declarative process modeling language originally introduced by Pesic and van der Aalst in [14]. Instead of explicitly specifying the flow of the interactions among process activities, DECLARE describes a set of constraints that must be satisfied throughout the process execution. The possible orderings of activities are implicitly specified by constraints and anything that does not violate them is possible during execution. In comparison with procedural approaches that produce "closed" models, i.e., all that is not explicitly specified is forbidden, DECLARE models are "open" and tend to offer more possibilities for the execution. In this way, DECLARE enjoys flexibility and is very suitable for highly dynamic processes characterized by high complexity and variability due to the changeability of their execution environments.

A DECLARE model consists of a set of constraints applied to activities. Constraints, in turn, are based on templates. Templates are patterns that define parameterized classes of properties, and constraints are their concrete instantiations (we indicate template parameters with capital letters and concrete activities in their instantiations with lower case letters). They have a graphical representation understandable to the user and their semantics can be formalized using different logics [13], the main one being LTL over finite traces, making them verifiable and executable. Each constraint inherits the graphical representation and semantics from its template. Table 1 summarizes some DECLARE templates (the reader can refer to [21] for a full description of the language). Here, the $\mathbf{F}$, $\mathbf{X}$, $\mathbf{G}$, and $\mathbf{U}$ LTL (future) operators have the following intuitive meaning: formula $\mathbf{F}\phi_1$ means that ϕ_1 holds sometime in the future, $\mathbf{X}\phi_1$ means that ϕ_1 holds in the next position, $\mathbf{G}\phi_1$ says that ϕ_1 holds forever in the future, and, lastly, $\phi_1\mathbf{U}\phi_2$ means that sometime in the future ϕ_2 will hold and until that moment ϕ_1 holds (with ϕ_1 and ϕ_2 LTL formulas). The $\mathbf{O}$, and $\mathbf{Y}$ LTL (past) operators have the following meaning: $\mathbf{O}\phi_1$ means that ϕ_1 holds sometime in the past, and $\mathbf{Y}\phi_1$ means that ϕ_1 holds in the previous position.

The major benefit of using templates is that analysts do not have to be aware of the underlying logic-based formalization to understand the models. They work with the graphical representation of templates, while the underlying formulas remain hidden. Consider, for example, the *response* constraint $\mathbf{G}(a \rightarrow \mathbf{F}b)$. This constraint indicates that if a occurs, b must eventually *follow*. Therefore, this constraint is satisfied for traces such as $\mathbf{t}_1 = \langle a, a, b, c \rangle$, $\mathbf{t}_2 = \langle b, b, c, d \rangle$, and $\mathbf{t}_3 = \langle a, b, c, b \rangle$, but not for $\mathbf{t}_4 = \langle a, b, a, c \rangle$ because, in this case, the second instance of a is not followed by a b. Note that, in $\mathbf{t}_2$, the considered response constraint is satisfied in a trivial way because a never occurs. In this case, we say that the constraint is *vacuously satisfied* [10]. In [6], the authors introduce the notion of *behavioral vacuity detection* according to which a constraint is non-vacuously satisfied in a trace when it is activated in that trace. An *activation* of a constraint in a trace is an event whose occurrence imposes, because of that constraint, some obligations on other events (targets) in the same trace. For example, a is an activation for the *response* constraint $\mathbf{G}(a \rightarrow \mathbf{F}b)$ and b is a target, because the execution of a forces b to be executed, eventually. In Table 1, for each template the corresponding activation is specified.

An activation of a constraint can be a *fulfillment* or a *violation* for that constraint. When a trace is perfectly compliant with respect to a constraint, every

Table 2. Semantics for multi-perspective DECLARE constraints

Template	MFOTL semantics
Responded existence	$\mathbf{G}(\forall x.((A \wedge \varphi_a(x)) \rightarrow (\mathbf{O}_I(B \wedge \exists y.\varphi_c(x,y)) \vee \mathbf{F}_I(B \wedge \exists y.\varphi_c(x,y)))))$
Response	$\mathbf{G}(\forall x.((A \wedge \varphi_a(x)) \rightarrow \mathbf{F}_I(B \wedge \exists y.\varphi_c(x,y))))$
Alternate response	$\mathbf{G}(\forall x.((A \wedge \varphi_a(x)) \rightarrow \mathbf{X}(\neg(A \wedge \varphi_a(x))\mathbf{U}_I(B \wedge \exists y.\varphi_c(x,y)))))$
Chain response	$\mathbf{G}(\forall x.((A \wedge \varphi_a(x)) \rightarrow \mathbf{X}_I(B \wedge \exists y.\varphi_c(x,y)))$
Precedence	$\mathbf{G}(\forall x.((B \wedge \varphi_a(x)) \rightarrow \mathbf{O}_I(A \wedge \exists y.\varphi_c(x,y)))$
Alternate precedence	$\mathbf{G}(\forall x.((B \wedge \varphi_a(x)) \rightarrow \mathbf{Y}(\neg(B \wedge \varphi_a(x))\mathbf{S}_I(A \wedge \exists y.\varphi_c(x,y))))$
Chain precedence	$\mathbf{G}(\forall x.((B \wedge \varphi_a(x)) \rightarrow \mathbf{Y}_I(A \wedge \exists y.\varphi_c(x,y)))$
Not responded existence	$\mathbf{G}(\forall x.((A \wedge \varphi_a(x)) \rightarrow \neg(\mathbf{O}_I(B \wedge \exists y.\varphi_c(x,y)) \vee \mathbf{F}_I(B \wedge \exists y.\varphi_c(x,y)))))$
Not response	$\mathbf{G}(\forall x.((A \wedge \varphi_a(x)) \rightarrow \neg\mathbf{F}_I(B \wedge \exists y.\varphi_c(x,y))))$
Not precedence	$\mathbf{G}(\forall x.((B \wedge \varphi_a(x)) \rightarrow \neg\mathbf{O}_I(A \wedge \exists y.\varphi_c(x,y)))$
Not chain response	$\mathbf{G}(\forall x.((A \wedge \varphi_a(x)) \rightarrow \neg\mathbf{X}_I(B \wedge \exists y.\varphi_c(x,y)))$
Not chain precedence	$\mathbf{G}(\forall x.((B \wedge \varphi_a(x)) \rightarrow \neg\mathbf{Y}_I(A \wedge \exists y.\varphi_c(x,y)))$

activation of the constraint in the trace leads to a fulfillment. Consider, again, the response constraint $\mathbf{G}(a \rightarrow \mathbf{F}b)$. In trace $\mathbf{t}_1$, the constraint is activated and fulfilled twice, whereas, in trace $\mathbf{t}_3$, the same constraint is activated and fulfilled only once. On the other hand, when a trace is not compliant with respect to a constraint, an activation of the constraint in the trace can lead to a fulfillment but also to a violation (at least one activation leads to a violation). In trace $\mathbf{t}_4$, for example, the response constraint $\mathbf{G}(a \rightarrow \mathbf{F}b)$ is activated twice, but the first activation leads to a fulfillment (eventually b occurs) and the second activation leads to a violation (b does not occur subsequently). An algorithm to discriminate between fulfillments and violations for a constraint in a trace is presented in [6].

Tools implementing process mining approaches based on DECLARE are presented in [11]. The tools are implemented as plug-ins of the process mining framework ProM.

2.3 Multi-perspective Declare

In this section, we illustrate a multi-perspective version of DECLARE (MP-DECLARE) introduced in [5]. This semantics is expressed in Metric First-Order Linear Temporal Logic (MFOTL) and is shown in Table 2. We describe here the semantics informally and we refer the interested reader to [5] for more details. To explain the semantics, we have to introduce some preliminary notions.

The first concept we use here is the one of *payload* of an event. Consider, for example, that the execution of an activity SUBMIT LOAN APPLICATION (S) is recorded in an event log and, after the execution of S at timestamp τ_S, the attributes *Salary* and *Amount* have values 12 500 and 55 000. In this case, we say that, when S occurs, two special relations are valid $event(\mathrm{S})$ and $p_S(12\,500, 55\,000)$. In the following, we identify $event(\mathrm{S})$ with the event itself S and we call $(12\,500, 55\,000)$, the *payload* of S.

Note that all the templates in MP-DECLARE in Table 2 have two parameters, an activation and a target (see also Table 1). The standard semantics of DECLARE is extended by requiring two additional conditions on data, i.e., the *activation condition* φ_a and the *correlation condition* φ_c, and a time condition. As an example, we consider the response constraint "activity SUBMIT LOAN APPLICATION is always eventually followed by activity ASSESS APPLICATION" having SUBMIT LOAN APPLICATION as activation and ASSESS APPLICATION as target. The activation condition is a relation (over the variables corresponding to the global attributes in the event log) that must be valid when the activation occurs. If the activation condition does not hold the constraint is not activated. The activation condition has the form $p_A(x) \wedge r_a(x)$, meaning that when A occurs with payload x, the relation r_a over x must hold. For example, we can say that whenever SUBMIT LOAN APPLICATION occurs, and the amount of the loan is higher than 50 000 euros and the applicant has a salary lower than 24 000 euros per year, eventually an assessment of the application must follow. In case SUBMIT LOAN APPLICATION occurs but the amount is lower than 50 000 euros or the applicant has a salary higher than 24 000 euros per year, the constraint is not activated.

The correlation condition is a relation that must be valid when the target occurs. It has the form $p_B(y) \wedge r_c(x, y)$, where r_c is a relation involving, again, variables corresponding to the (global) attributes in the event log but, in this case, relating the payload of A and the payload of B. A special type of correlation condition has the form $p_B(y) \wedge r_c(y)$, which we call *target condition*, since it does not involve attributes of the activation.

In this paper, we aim at discovering constraints that correlate an activation and a target condition. For example, we can find that whenever SUBMIT LOAN APPLICATION occurs, and the amount of the loan is higher than 50 000 euros and the applicant has a salary lower than 24 000 euros per year, then eventually ASSESS APPLICATION must follow, and the assessment type will be *Complex* and the cost of the assessment higher than 100 euros.

Finally, in MP-DECLARE, also a time condition can be specified through an interval $(I = [\tau_0, \tau_1))$ indicating the minimum and the maximum temporal distance allowed between the occurrence of the activation and the occurrence of the corresponding target.

3   Running Example

Figure 1 shows a fictive MP-DECLARE model that we will use as a running example throughout this paper. This example models a process for loan applications in a bank. When an applicant submits a loan application with an amount higher than 50 000 euros and she has a salary lower than 24 000 euros per year, eventually an assessment of the application will be carried out. The assessment will be complex and the cost of the assessment higher than 100 euros. This behavior is described by response constraint C_1 in Fig. 1. In case the applicant submits a loan application with an amount lower than 50 000 euros or she has a salary

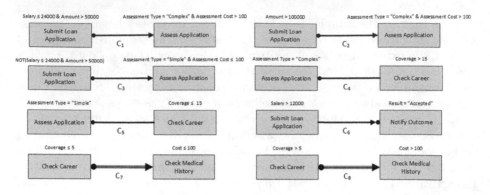

Fig. 1. Running example.

higher than 24 000 euros per year, eventually a simple assessment will be carried out and the cost of the assessment will be lower than or equal to 100 euros. This behavior is described by response constraint C_3. When an applicant submits a loan application with an amount higher than 100 000 euros, eventually a complex assessment with cost higher than 100 euros is performed. This behavior is described by response constraint C_2 in Fig. 1. If the outcome of an application assessment is notified and the result of the outcome is accepted, then this event is always preceded by an application submission whose applicant has a salary higher than 12 000 euros per year. This behavior is described by precedence constraint C_6. Outside the application assessment there are 2 additional checks that can be performed before or after the assessment: the career check and the medical history check. A career check with a coverage lower than 15 years is required if the application assessment is simple (responded existence constraint C_5). The career of the applicant should be checked with a coverage higher than 15 years if the application assessment is complex (responded existence constraint C_4). If the career check covers less than 5 years, a medical history check should be performed immediately after and its cost is lower than 100 euros (chain response constraint C_7). If the career check covers more than 5 years, the medical history check is more complex and more expensive (its cost is higher than 100 euros). This behavior is described by chain response constraint C_8 in Fig. 1.

4   Discovery Approach

The proposed approach is shown in Fig. 2. The approach starts with the discovery of a set of *frequent constraints*. A frequent constraint is a constraint having a high number of *constraint instances*, i.e., pairs of events (one activation and one target) satisfying it. In addition, for each frequent constraint, also activations that cannot be associated to any target (representing a violation for the constraint) are identified. Feature vectors are extracted from the payloads of these activations and associated with a label indicating that they correspond to violations of the constraint (*violation feature vectors*). (Unlabeled) feature vectors are

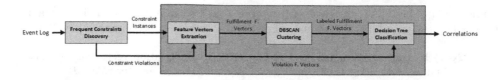

Fig. 2. Proposed approach.

also extracted by using the payloads of targets of the constraint instances identified in the first phase. These feature vectors are then clustered using DBSCAN clustering [8] to find groups of targets with similar payloads. Then, these clusters are used as labels for a classification problem. These labels together with the features extracted from the activation payloads are used to generate a set of *fulfillment feature vectors*. Violation and fulfillment feature vectors are used to train a decision tree. This procedure allows for finding correlations between the activation payloads and the target payloads. Note that the core part of our approach (highlighted with a blue rectangle in Fig. 2) is independent of the procedure used to identify frequent constraints and can be used in combination with other techniques for frequent constraint mining (also based on semantics that go beyond MP-DECLARE).

4.1 Frequent Constraints Discovery

The first step of our discovery algorithm is to identify a set of frequent constraints. In particular, the user specifies a DECLARE template (e.g., response) and, starting from the input template and the input log, a set of candidate constraints is instantiated. To generate the set of candidate constraints, we use the idea behind the well-known Apriori algorithm presented in [4]. In particular, the algorithm first searches for frequent individual activities in the input log. We call the absolute frequency of an activity in the log *activity support*. Individual activities with an activity support lower than an input threshold $supp_{min}$ are filtered out.

The input template is instantiated with all the possible combinations of frequent activities thus identifying a set of candidate constraints. Note that, unlike the classical Apriori algorithm that works with unordered itemsets, the order inside a candidate constraint plays an important role (i.e., RESPONSE(S,A) is not the same as RESPONSE(A,S)). In particular, we work with pairs where the first element is the activation of the constraint and the second element is the target. At this point, the algorithm evaluates, for each candidate constraint, the number of its constraint instances in the log. We call this measure *constraint support*. In particular, this measure is calculated by creating two vectors idx_1 and idx_2 that represent the activation and target occurrences. Then, the algorithm processes each trace in the input log to find the events corresponding either to the activation or the target of the candidate constraint, and their indexes are collected in the corresponding vector. For example, for trace $t = SSSASASSA$

and RESPONSE(S,A), we have $idx_1 = (1,2,3,5,7,8)$ and $idx_2 = (4,6,9)$. Then, based on the template, the number of constraint instances is computed as follows:

- **(Not) Response.** For each element idx_{1i} from the activation vector idx_1 we take the first element idx_{2j} from the target vector idx_2 that is greater than idx_{1i}.
- **(Not) Chain Response.** Here, we check the existence of pairs (i,j) from idx_1 and idx_2 where $j - i = 1$.
- **Alternate Response.** In this case, for each element idx_{1i} from idx_1, we take the first element idx_{2j} from idx_2 that is greater than idx_{1i}. However, we identify a constraint instance only if there are no elements from idx_1 that lie between idx_{1i} and idx_{2j}.
- **Precedence.** For precedence, chain precedence and alternate precedence, the logic is almost the same as for their response counterparts. However, for precedence rules, the idx_1 is considered as target vector and idx_2 as activation vector. In addition, the trace has to be reversed.
- **(Not) Responded Existence.** We associate each element idx_{1i} from the activation vector idx_1, with the first element from the target vector idx_2.

If we enumerate the occurrences of S and A in t, we have $t = S_1 S_2 S_3 A_1 S_4 A_2 S_5 S_6 A_3$. The constraint instances of the standard DECLARE templates instantiated with activities (A, S) and (S, A) are listed in Tables 3 and 4. Note that, these procedures can also be used to identify constraint violations (i.e., activations that cannot be associated to any target). We filter out candidate constraint with a support that is lower than $supp_{min}$ thus obtaining a set of frequent constraints. The constraint instances of the frequent constraints are used for creating fulfillment feature vectors. Activations that do not have a target are used to generate violation feature vectors. We stress again that these procedures only provide an example of how to identify temporal patterns in a log. Any semantics (also beyond standard DECLARE) can be used to identify frequent constraints.

Table 3. Constraint instances of type (S,A) in trace t

Candidate constraint	Constraint instances
RESPONSE(S,A)	$\{S_1 A_1\}, \{S_2 A_1\}, \{S_3 A_1\}, \{S_4 A_2\}, \{S_5 A_3\}, \{S_6 A_3\}$
CHAIN RESPONSE(S,A)	$\{S_3 A_1\}, \{S_4 A_2\}, \{S_6 A_3\}$
ALTERNATE RESPONSE(S,A)	$\{S_3 A_1\}, \{S_4 A_2\}, \{S_6 A_3\}$
PRECEDENCE(S,A)	$\{S_3 A_1\}, \{S_4 A_2\}, \{S_6 A_3\}$
CHAIN PRECEDENCE(S,A)	$\{S_3 A_1\}, \{S_4 A_2\}, \{S_6 A_3\}$
ALTERNATE PRECEDENCE(S,A)	$\{S_3 A_1\}, \{S_4 A_2\}, \{S_6 A_3\}$
RESPONDED EXISTENCE(S,A)	$\{S_1 A_1\}, \{S_2 A_1\}, \{S_3 A_1\}, \{S_4 A_1\}, \{S_5 A_1\}, \{S_6 A_1\}$

Table 4. Constraint instances of type (A,S) in trace t

Candidate constraint	Constraint instances
RESPONSE(A,S)	$\{A_1 S_4\}$, $\{A_2 S_5\}$
CHAIN RESPONSE(A,S)	$\{A_1 S_4\}$, $\{A_2 S_5\}$
ALTERNATE RESPONSE(A,S)	$\{A_1 S_4\}$, $\{A_2 S_5\}$
PRECEDENCE(A,S)	$\{A_1 S_4\}$, $\{A_2 S_5\}$ $\{A_2 S_6\}$
CHAIN PRECEDENCE(A,S)	$\{A_1 S_4\}$, $\{A_2 S_5\}$
ALTERNATE PRECEDENCE(A,S)	$\{A_1 S_4\}$, $\{A_2 S_5\}$
RESPONDED EXISTENCE(A,S)	$\{A_1 S_1\}$, $\{A_2 S_1\}$, $\{A_3 S_1\}$

4.2 Feature Vectors Extraction

Violation feature vectors consist of the payloads of activations of frequent constraints that do not have a corresponding target and are labeled as "violated". Assume to have a constraint instance where the activation SUBMIT LOAN APPLICATION has a payload $(12\,500, 55\,000)$ (see Sect. 2.3). If this activation cannot be associated to any target, we generate the violation feature vector:

$$V_{viol} = [12\,500; 55\,000; violated].\qquad(1)$$

If the same activation is part of a constraint instance of a frequent constraint with target ASSESS APPLICATION and payload $(Complex, 140)$, we generate the (unlabeled) fulfillment feature vector:

$$V_{ful} = [Complex; 140].\qquad(2)$$

The violation feature vectors are used for interpretable classification (see Sect. 4.4). Fulfillment feature vectors are used for clustering with DBSCAN.

4.3 Clustering with DBSCAN

Starting from the fulfillment feature vectors, we use the Density Based Spatial Clustering of Application with Noise (DBSCAN) [8] to find groups of payloads that are similar.

Given a set of points in some space DBSCAN groups together vectors that are closely packed, marking as outliers vectors that lie in low-density regions. To do this, we use the Gower distance

$$d(i, j) = \frac{1}{n} \sum_{f=1}^{n} d_{i,j}^{(f)},\qquad(3)$$

where n denotes the number of features, while $d_{i,j}^{(f)}$ is a distance between data points i and j when considering only feature f. $d_{i,j}^{(f)}$ is a normalized distance. For nominal attribute values, we calculate the normalized Edit Levenshtein distance [9]. Edit distance is a way of quantifying how dissimilar two strings are by counting the minimum number of operations (i.e., removal, insertion, substitution of a character) required to transform one string into other. The normalized edit distance is calculated as:

$$d_{i,j}^{(f)} = \frac{EditDistance(x_i^{(f)}, x_j^{(f)})}{maxEditDistance(x_i^{(f)}, x_j^{(f)})}. \tag{4}$$

For interval scaled attribute values, we use the distance:

$$d_{i,j}^{(f)} = \frac{|x_i^{(f)} - x_j^{(f)}|}{max - min}, \tag{5}$$

where min and max are the maximum and minimum observed values of attribute f. For boolean attribute values, we use the distance:

$$d_{i,j}^{(f)} = \begin{cases} 0, & \text{if } x_i^{(f)} = x_j^{(f)} \\ 1, & \text{if } x_i^{(f)} \neq x_j^{(f)}. \end{cases} \tag{6}$$

When obtained the clusters, we project the target payload attributes in order to describe the characteristics of the elements of the clusters. For numerical attributes, the projection results in the range [min-max], where min and max are the minimum and maximum values of the attribute. When projecting onto categorical attributes, we take the most frequent value. For example, Fig. 3a shows two clusters associated to a frequent constraint with target ASSESS APPLI-CATION. One of them is characterized by the condition $Assessment\ Cost = [10 - 100]$ & $Assessment\ Type = Simple$, while the second one by the condition $Assessment\ Cost = [101 - 198]$ & $Assessment\ Type = Complex$. These clusters/conditions are used as labels to build labeled fulfillment feature vectors. Assume again to have a constraint instance where the activation SUBMIT LOAN APPLICATION has a payload $(12\,500, 55\,000)$. If this activation is part of a constraint instance with target ASSESS APPLICATION and payload $(Complex, 140)$, we generate the labeled fulfillment feature vector:

$$V'_{ful} = [12\,500; 55\,000; Cluster2]. \tag{7}$$

Labeled fulfillment feature vectors and violation feature vectors are used for interpretable classification using decision trees.

4.4   Interpretable Classification

After having created labeled fulfillment feature vectors and violation feature vectors, we use them to train a decision tree. The C4.5 algorithm is used to perform the classification [16]. The data is split in a way that the resulting

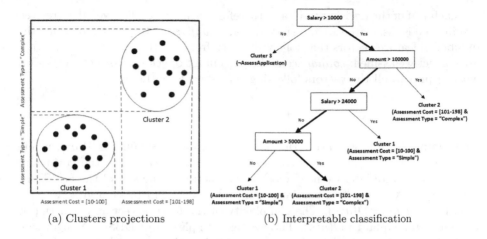

(a) Clusters projections   (b) Interpretable classification

Fig. 3. Core steps of the proposed approach.

parts are more homogenous. The algorithm uses entropy and information gain to choose the split. We can express the overall entropy as:

$$H(Y) = - \sum_{a \in Dom(Y)} Pr[Y = a] * log_2(Pr[Y = a]). \qquad (8)$$

The information gain can be calculated as:

$$Gain = H_S(Y) - \sum_{i=1}^{k} \frac{|S_i|}{|S|} * H_{S_i}(Y), \qquad (9)$$

where $|S|$ the size of the dataset and $|S_i|$ is the size of the split i. $H_S(Y)$ denotes the entropy computed over S and $H_{S_i}(Y)$ denotes the entropy computed over S_i. The information gain measures how much the split makes the target value distribution more homogenous. We select the split with maximal information gain. The splitting is stopped either when information gain is 0 (we will not improve the results when splitting) or when the size of the split is smaller than an input support threshold.

Figure 3b shows a decision tree generated from our running example. Correlations in the activation and target payloads are found by correlating the activation conditions derived from paths from the root to the leaves of the decision tree with the target conditions labeling each leaf of the decision tree. For example, from the path highlighted with thicker arcs in Fig. 3b, we can extract the activation condition $Salary \leq 24000$ & $Amount > 50000$ & $Amount \leq 100000$. If the frequent constraint was a response constraint, the discovery algorithm produces the MP-DECLARE constraint RESPONSE(SUBMIT LOAN APPLICATION, ASSESS APPLICATION) with activation condition $Salary \leq 24000$ & $Amount > 50000$ & $Amount \leq 100000$ and target condition $Assessment\ Cost = [101 - 198]$ & $Assessment\ Type - Complex$.

Each leaf of the decision tree (and therefore each pair of discovered correlated conditions) is associated with a *support* and a *confidence*. Support represents the number of feature vectors that follow the path from the root to the leaf and that are correctly classified; *confidence* is the percentage of vectors correctly classified with respect to all the vectors following that specific path.

5   Algorithm Complexity

The complexity for generating the set of candidate constraints is $O(a^2)$, where a is the number of distinct activities in the log. To calculate the constraint support of a candidate, we need $O(e)$ time, where e is the size of the log. Thus, the complexity of the frequent constraints discovery can be computed as $T = O(ae) + O(a^2 e) = O(a^2 e)$. The complexity of the feature vectors extraction is the same and equal to $O(a^2 e)$. The average complexity of DBSCAN algorithm is $O(nlog(n))$, and $O(n^2)$ in the worst case, when the vectors have one feature [8]. Thus, the total complexity of the DBSCAN clustering is equal to $O(mn^2)$ in the worst case, where m is the number of features (size of the feature vectors) and n is the number of feature vectors. Since DBSCAN is applied to all frequent constraints, its complexity is equal to $O(a^2 mn^2)$. In order to avoid the distances recomputations, a distance matrix of size $(n^2 - n)/2$ can be used. However, this needs $O(n^2)$ memory. In general, the runtime cost to construct a balanced binary tree is $O(mnlog(n))$, where m is the number of features and n is the number of feature vectors. Assuming that the subtrees remain approximately balanced, the cost at each node consists of searching through $O(m)$ to find the feature that offers the largest reduction in entropy. This has a cost of $O(mnlog(n))$ at each node, leading to a total cost over the entire tree of $O(mn^2 log(n))$.[1] Considering the fact that we have a^2 candidate constraints, the complexity is $O(a^2 mn^2 log(n))$. Thus, the total complexity of the algorithm is $T = O(a^2 e) + O(a^2 e) + O(a^2 mn^2) + O(a^2 mn^2 log(n)) = O(a^2 mn^2 log(n))$.

6   Experiments

The evaluation reported in this paper aims at understanding the potential of the proposed discovery approach. In particular, we want to examine the capability of the discovery approach to rediscover behavior artificially injected into a set of synthetic logs. In addition, we want to assess the scalability of the approach and its applicability to real-life datasets. In particular, we investigate the following three research questions:

- **RQ1.** Does the proposed approach allow for rediscovering behavior artificially injected into a set of synthetic logs?
- **RQ2.** What is the time performance of the proposed approach when applied to logs with different characteristics?
- **RQ3.** Is the proposed approach applicable in real-life settings?

---

[1] http://scikit-learn.org/stable/modules/tree.html.

Table 5. Experimental results: rediscovery

Template	Activation/Target	Activation/Target Condition	Support	Confidence
(C_2) Response	Submit Loan Application	$Amount > 100000$	0.17	0.99
	Assess Application	$AssessmentCost = [101 - 198]$ & $Assessment\ Type = Complex$		
(C_3) Response	Submit Loan Application	$Amount \leq 100000$	0.76	0.91
	Assess Application	$Assessment\ Cost = [10 - 100]$ & $Assessment\ Type = Simple$		
(C_3) Response	Submit Loan Application	$Salary > 24000$ & $Amount > 50000$ & $Amount \leq 100000$	0.15	0.95
	Assess Application	$Assessment\ Cost = [10 - 100]$ & $Assessment\ Type = Simple$		
(C_1) Response	Submit Loan Application	$Salary \leq 24000$ & $Amount > 50000$ & $Amount \leq 100000$	0.05	1.0
	Assess Application	$Assessment\ Cost = [101 - 198]$ & $Assessment\ Type = Complex$		
Response	Submit Loan Application	$Amount > 100000$	0.09	0.52
	Check Career	$Coverage = [16 - 25]$		
Response	Submit Loan Application	$Amount \leq 100000$	0.75	0.9
	Check Career	$Coverage = [0 - 20]$		
Response	Submit Loan Application	$Salary \leq 12000$	0.23	1.0
	Notify Outcome	$Result = Rejected$		
Response	Submit Loan Application	$Salary > 12000$	0.39	0.5
	Notify Outcome	$Result = Accepted$		
Response	Assess Application	$Assessment\ Cost > 100$	0.23	1.0
	Check Medical History	$Cost = [101 - 199]$		
Response	Assess Application	$Assessment\ Cost \leq 100$	0.77	0.99
	Check Medical History	$Cost = [10 - 200]$		
Response	Submit Loan Application	$Amount > 156868$	0.048	0.47
	Check Medical History	$Cost = [128 - 199]$		
Response	Submit Loan Application	$Amount \leq 156868$	0.76	0.85
	Check Medical History	$Cost = [10 - 200]$		
(C_4) Responded Existence	Assess Application	$Type = Complex$	0.22	1.0
	Check Career	$Coverage = [16 - 30]$		
(C_5) Responded Existence	Assess Application	$Type = Simple$	0.77	0.99
	Check Career	$Coverage = [0 - 15]$		
Precedence	Submit Loan Application	$Result = Rejected$	0.61	1.0
	Notify Outcome	$Salary = [1022 - 99829]$ & $Amount = [10028 - 249847]$		
(C_6) Precedence	Submit Loan Application	$Result \neq Rejected$	0.39	1.0
	Notify Outcome	$Salary = [12001 - 99229]$ & $Amount = [10267 - 248013]$		
Precedence	Submit Loan Application	$Coverage \leq 15$	0.73	0.95
	Check Career	$Salary = [1022 - 99822]$ & $Amount = [10028 - 100000]$		
Precedence	Submit Loan Application	$Coverage > 15$	0.04	0.2
	Check Career	$Salary = [2551 - 24000]$ & $Amount = [160103 - 246486]$		
Precedence	Assess Application	$Coverage > 15$	0.22	1.0
	Check Career	$AssessmentCost = [101 - 198]$ & $AssessmentType = Complex$		
Precedence	Assess Application	$Coverage \leq 15$	0.77	0.99
	Check Career	$AssessmentCost = [10 - 100]$ & $AssessmentType = Simple$		
Precedence	Check Career	$Cost > 100$	0.46	0.99
	Check Medical History	$Coverage = [6 - 30]$		
Precedence	Check Career	$Cost \leq 100$	0.33	1.0
	Check Medical History	$Coverage = [0 - 5]$		
Precedence	Submit Loan Application	$AssessmentCost > 100$	0.22	0.99
	Assess Application	$Salary = [1145 - 24000]$ & $Amount = [50246 - 249847]$		
Precedence	Submit Loan Application	$AssessmentCost \leq 100$	0.76	0.98
	Assess Application	$Salary = [1022 - 99829]$ & $Amount = [10028 - 100000]$		
(C_8) Chain Response	Check Career	$Coverage > 5$	0.46	0.99
	Check Medical History	$Cost = [101 - 200]$		
(C_7) Chain Response	Check Career	$Coverage \leq 5$	0.52	1.0
	Check Medical History	$Cost = [10 - 100]$		

RQ1 focuses on the evaluation of the quality of the constraints returned by the proposed approach. **RQ2** investigates, instead, the scalability of the approach. Finally, **RQ3** deals with the validation of the discovery approach in real scenarios.

6.1   Rediscovering Injected Behavior (RQ1)

The most standard approach to test a process discovery algorithm is to artificially generate a log by simulating a process model and use the generated log to rediscover the process model. The log used for this experiment was generated by simulating the model in Fig. 1. The log contains 1 000 cases with 5 distinct

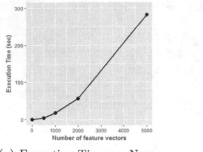
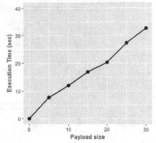

(a) Execution Time vs. Number of Feature Vectors

(b) Execution Time vs. Payload Size

Fig. 4. Experimental results: scalability.

activities (SUBMIT LOAN APPLICATION, ASSESS APPLICATION, NOTIFY OUT-COME, CHECK CAREER, CHECK MEDICAL HISTORY).[2]

The results of the rediscovery are shown in Table 5. The table shows that the approach manages to rediscover mostly all the constraints that generated the log. These constraints are among the ones with the highest confidence (0.99 or 1, highlighted in bold) and are explicitly indicated in the table. The only exception is constraint C_3 (including a disjunctive activation condition) that is included in the behavior described by 2 discovered constraints with a lower confidence.

6.2 Scalability (RQ2)

The scalability of the approach was tested by generating 5 synthetic logs with growing number of constraint instances for one specific MP-DECLARE constraint C (100, 500, 1 000, 2 000 and 5 000 constraint instances), with a default payload size of 10 attributes. In addition, we generated 6 synthetic logs with growing payload sizes (5, 10, 15, 20, 25 and 30 attributes), with a default number of constraint instances of 1 000 for C. Figure 4 shows the execution times (in seconds) needed for the rediscovery of C (averaged over 5 runs). The time required for the discovery from 5 000 feature vectors (with a payload of size 10) is of around 4 min and the time required for the discovery from 1 000 feature vectors (with a payload of size 30) is of around 32 s. Therefore, we can say that the execution times are reasonable when the discovered models are not extremely large.

6.3 Validation in Real Scenarios (RQ3)

For the validation of the proposed approach in real scenarios, we used six datasets provided for the BPI Challenges 2011 [2] and 2015 [3]. The first dataset is an

---

[2] For generating the log, we used the log generator based on MP-DECLARE available at https://github.com/darksoullock/MPDeclareLogGenerator.

event log pertaining to the treatment of patients diagnosed with cancer in a Dutch academic hospital. The remaining five datasets were provided for the BPI Challenge 2015 by five different Dutch Municipalities. The logs contain all building permit applications received over a period of four years. The full results obtained from these logs are not reported here for space limitations and can be downloaded from[3]. In the hospital log, the most important correlations link together medical exams and treatments. In the Municipality logs, there are several correlations between certain actors and the successful completion of a request handling.

7   Related Work

Two contributions are available in the literature that are close to the work presented in this paper. They have been presented in [12,19]. The work in [12] is similar to the one presented in the current contribution but only consider activation conditions that discriminate between positive and negative cases without correlating activation and target conditions.

In [19], the authors present a mining approach that works with RelationalXES, a relational database architecture for storing event log data. The relational event data is queried with conventional SQL. Queries can be customized and cover the semantics of MP-DECLARE. Differently from [19], in the current contribution, we do not check the input log with user-specified queries, but we automatically identify correlations.

8   Conclusion and Future Work

We presented a technique for the automated discovery of multi-perspective business constraints from an event log. The technique allows us to discover conditions that relate the occurrence of pairs of events in the event log. The technique has been implemented as an open-source tool[4] and evaluated using both synthetic and real-life logs. The evaluation shows that the technique is able to rediscover behavior injected into a log and that it is sufficiently scalable to handle real-life datasets.

As future work, we aim at improving the efficiency of the approach. This can be done, for example, by using clustering techniques such as Canopy [18] that can be used when the clustering task is challenging either in terms of dataset size, or in terms of number of features, or in terms of number of clusters. Finally, the current approach works also with pure correlation conditions (involving attributes of activation and target together). However, these correlations should be specified manually as features that can possibly be discovered. Techniques based on Daikon [7] could help to automatically discover this type of conditions.

---

[3] https://bitbucket.org/volodymyrLeno/correlationminer/downloads/results.pdf.
[4] Available at https://github.com/volodymyrLeno/CorrelationMinerForDeclare.

References

1. IEEE Task Force on Process Mining: XES Standard Definition (2013)
2. 4TU Data Center: BPI Challenge 2011 Event Log (2011)
3. 4TU Data Center: BPI Challenge 2015 Event Log (2015)
4. Agrawal, R., Srikant, R.: Fast algorithms for mining association rules in large databases. In: VLDB 1994, pp. 487–499 (1994)
5. Burattin, A., Maggi, F.M., Sperduti, A.: Conformance checking based on multi-perspective declarative process models. Expert Syst. Appl. **65**, 194–211 (2016)
6. Burattin, A., Maggi, F.M., van der Aalst, W.M.P., Sperduti, A.: Techniques for a posteriori analysis of declarative processes. In: EDOC, Beijing, pp. 41–50 (2012)
7. Ernst, M.D., Cockrell, J., Griswold, W.G., Notkin, D.: Dynamically discovering likely program invariants to support program evolution. IEEE Trans. Softw. Eng. **27**(2), 99–123 (2001)
8. Ester, M., Kriegel, H.-P., Sander, J., Xu, X.: A density-based algorithm for discovering clusters in large spatial databases with noise. In: KDD, pp. 226–231 (1996)
9. Gomaa, W.H., Fahmy, A.A.: A survey of text similarity approaches. Int. J. Comput. Appl. **68**(13), 13–18 (2013)
10. Kupferman, O., Vardi, M.Y.: Vacuity detection in temporal model checking. STTT **4**(2), 224–233 (2003)
11. Maggi, F.M.: Declarative process mining with the Declare component of ProM. In: BPM (Demos) (2013)
12. Maggi, F.M., Dumas, M., García-Bañuelos, L., Montali, M.: Discovering data-aware declarative process models from event logs. In: Daniel, F., Wang, J., Weber, B. (eds.) BPM 2013. LNCS, vol. 8094, pp. 81–96. Springer, Heidelberg (2013). https://doi.org/10.1007/978-3-642-40176-3_8
13. Montali, M., Pesic, M., van der Aalst, W.M.P., Chesani, F., Mello, P., Storari, S.: Declarative specification and verification of service choreographies. ACM Trans. Web **4**(1), 3 (2010)
14. Pesic, M., Schonenberg, H., van der Aalst, W.M.P.: DECLARE: full support for loosely-structured processes. In: EDOC, pp. 287–300 (2007)
15. Pichler, P., Weber, B., Zugal, S., Pinggera, J., Mendling, J., Reijers, H.A.: Imperative versus declarative process modeling languages: an empirical investigation. In: Daniel, F., Barkaoui, K., Dustdar, S. (eds.) BPM 2011. LNBIP, vol. 99, pp. 383–394. Springer, Heidelberg (2012). https://doi.org/10.1007/978-3-642-28108-2_37
16. Quinlan, J.R.: C4.5: Programs for Machine Learning. M. Kaufmann Publishers Inc., San Francisco (1993)
17. Reijers, H.A., Slaats, T., Stahl, C.: Declarative modeling–an academic dream or the future for BPM? In: Daniel, F., Wang, J., Weber, B. (eds.) BPM 2013. LNCS, vol. 8094, pp. 307–322. Springer, Heidelberg (2013). https://doi.org/10.1007/978-3-642-40176-3_26
18. Rokach, L., Maimon, O.: Clustering methods. In: Maimon, O., Rokach, L. (eds.) Data Mining and Knowledge Discovery Handbook, pp. 321–352. Springer, Boston (2005). https://doi.org/10.1007/0-387-25465-X_15
19. Schönig, S., Di Ciccio, C., Maggi, F.M., Mendling, J.: Discovery of multi-perspective declarative process models. In: Sheng, Q.Z., Stroulia, E., Tata, S., Bhiri, S. (eds.) ICSOC 2016. LNCS, vol. 9936, pp. 87–103. Springer, Cham (2016). https://doi.org/10.1007/978-3-319-46295-0_6
20. van der Aalst, W.M.P.: Process Mining - Data Science in Action, 2nd edn. Springer, Heidelberg (2016). https://doi.org/10.1007/978-3-662-49851-4

21. van der Aalst, W.M.P., Pesic, M., Schonenberg, H.: Declarative workflows: balancing between flexibility and support. Comput. Sci. - R&D **23**(2), 99–113 (2009)
22. Verbeek, H.M.W., Buijs, J.C.A.M., van Dongen, B.F., van der Aalst, W.M.P.: XES, XESame, and ProM 6. In: Soffer, P., Proper, E. (eds.) CAiSE Forum 2010. LNBIP, vol. 72, pp. 60–75. Springer, Heidelberg (2011). https://doi.org/10.1007/978-3-642-17722-4_5

Track II: Alignments and Conformance Checking

Efficiently Computing Alignments
Using the Extended Marking Equation

Boudewijn F. van Dongen[(✉)]

Eindhoven University of Technology, Eindhoven, The Netherlands
`B.F.v.Dongen@tue.nl`

Abstract. Conformance checking is considered to be anything where observed behaviour needs to be related to already modelled behaviour. Fundamental to conformance checking are alignments which provide a precise relation between a sequence of activities observed in an event log and a execution sequence of a model. However, computing alignments is a complex task, both in time and memory, especially when models contain large amounts of parallelism.

When computing alignments for Petri nets, (Integer) Linear Programming problems based on the marking equation are typically used to guide the search. Solving such problems is the main driver for the time complexity of alignments. In this paper, we adopt existing work in such a way that (a) the extended marking equation is used rather than the marking equation and (b) the number of linear problems that is solved is kept at a minimum.

To do so, we exploit fundamental properties of the Petri nets and we show that we are able to compute optimal alignments for models for which this was previously infeasible. Furthermore, using a large collection of benchmark models, we empirically show that we improve on the state-of-the-art in terms of time and memory complexity.

Keywords: Alignments · Conformance checking · Process Mining

1 Introduction

Conformance checking is considered to be anything where observed behaviour needs to be related to already modelled behaviour. Conformance checking is embedded in the larger contexts of Business Process Management and Process Mining [2], where conformance checking is typically used to compute metrics such as fitness, precision and generalization to quantify the relation between a log and a model.

Fundamental to conformance checking are alignments [3,4]. Alignments provide a precise relation between a sequence of activities observed in an event log and a execution sequence of a model. For each trace, this precise relation is expressed as a sequence of "moves". Such a move is either a "synchronous move" referring to the fact that the observed event in the trace corresponds directly to

© Springer Nature Switzerland AG 2018
M. Weske et al. (Eds.): BPM 2018, LNCS 11080, pp. 197–214, 2018.
https://doi.org/10.1007/978-3-319-98648-7_12

the execution of a transition in the model, a "log move" referring to the fact that the observed event has no corresponding transition in the model, or a "model move" referring to the fact that a transition occurred which was not observed in the trace.

Computing alignments is a complex task, both in time and memory, especially when models contain large amounts of parallelism and traces contain swapped events. Consider the example in Fig. 1. In this example, the model requires the process to finish with transitions A and B, while the trace shows them in the wrong order. The technique of [3,4] will, when reaching the indicated marking, have to investigate all interleavings of the parallel parts of the model inside the upper cloud before reaching the conclusion that there is a swap in the end.

In this paper, we present a technique to efficiently compute alignments using the extended marking equation which, in the example above, would recognize the swapped activities. We present related work in Sect. 2 followed by some preliminaries in Sect. 3. In

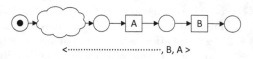

Fig. 1. A model ending in A, B and an example trace ending in B, A.

Sect. 4, we present alignments as well as our incremental technique for computing them. In Sect. 5 we compare our technique to existing approaches and discuss time and memory use before concluding the paper in Sect. 6.

2   Related Work

In [13], Rozinat et al. laid the foundation for conformance checking. They approached the problem by firing transitions in the model, regardless of available tokens and they kept track of missing and remaining tokens after completion of an observed trace. However, these techniques could not handle duplicate labels (identically labelled transition occurring at different locations in the model) or "invisible" transitions, i.e. τ-labelled transitions in the model purely for routing purposes.

As an improvement to token replay, alignments were introduced in [4]. The work proposes to transform a given Petri net and a trace from an event log into a synchronous product net, and solve the shortest path problem using $A^\star$ [6] on the its reachability graph. This graph may be considerable in size as it is worst-case exponential in the size of the synchronous product, in some cases, the proposed algorithm has to investigate the entire graph despite of tweaks identified in [16].

To mitigate these complexity issues, [1,10] proposes decomposition techniques for computing alignments. While this leads to shorter processing times, these techniques cannot always guarantee optimality of the alignment returned. Instead, they focus on giving bounds on the cost of the alignment. In [7], the authors provide techniques to identify misconformance based on sets of labels. The result is not an alignment.

Recently approximation schemes for alignments, i.e. computation of near-optimal alignments, have been proposed in [14]. The techniques use a recursive partitioning scheme, based on the input traces, and solve multiple Integer Linear Programming problems. In this paper, we build on the ideas outlined in [14] and we combine them with the state-of-the-art to compute optimal alignments for large process models.

Several techniques exist to compute alignments. Planner-based techniques [5] are available for safe Petri nets. For safe, acyclic models, constraint satisfaction [8] can be used. When not using a model, but its reachability graph as input, automata matching [12] can be applied. The models in this paper are safe. However, they are cyclic and, due to parallelism, the reachability graphs cannot be computed in reasonable time.

Finally, in [15] the authors provide a pre-processing mechanism to reduce the complexity of the alignment problem for structured input. This can also be applied in the context of this paper.

3 Preliminaries

In the context of conformance checking, we generally assume that there is a global set of activities A. An event log consists of a set of traces where each event in a trace refers to a specific activity execution. For sake of simplicity, we assume a log to be a set of sequences of activities and we also assume all sets to be totally ordered so that, when translating sets to vectors and vice versa, the translation respects the ordering.

Event Logs. An event log $L \subseteq A*$ is a set of sequences over A and $\sigma \in L$ a trace. We use $\sigma = \langle a, b, c \rangle$ to denote the fact that σ is a sequence of three activities $a, b, c \in A$. With $\sigma(i)$ we denote the ith element of σ. **Petri nets.** Petri nets provide a modelling language for modelling business processes. Petri nets consist of places and transitions and exact semantics are provided to describe the behaviour of a Petri net [11]. In this paper, transitions can be labelled with activities or be silent, i.e. their execution would not be recorded in an event log.

We define $PN = (P, T, F, \lambda)$ as a labelled Petri net, where T is a set of transitions, P is a set of places, $F \subseteq P \times T \cup T \times P$ the flow relation and $\lambda : T \rightarrow A \cup \{\tau\}$ the labelling function, labelling each transition with an activity or τ indicating there is no label associated to the transition.

For $t \in T$ we use $\bullet t = \{p \in P \mid (p, t) \in F\}$ to denote the preset of t, i.e. the set of places serving as input of a transition. The postset of t is defined as $t\bullet = \{p \in P \mid (t, p) \in F\}$.

A marking m is a multiset of places, i.e., $m \in \mathcal{B}(P)$. A transition $t \in T$ is *enabled* in a marking m if and only if $\bullet t \leq m$. *Firing* transition t in m results in a new marking $m' = m - \bullet t + t\bullet$, i.e., tokens are removed from $\bullet t$ and added to $t\bullet$. A marking m' is *reachable* from m if there is a sequence of firings $\theta = \langle t_1, t_2 \ldots t_n \rangle$ that transforms m into m', denoted by $m[\theta\rangle m'$. We call θ a firing sequence. For marking m, we use $\vec{m}$ to represent them as column vector, where for each $p \in P$ holds that $\vec{m}(p) = m(p)$, i.e. the number of tokens in place p in marking m. Similarly, for firing sequence θ, we use $\vec{\theta}$ to denote the parikh vector of θ, i.e. a column vector over T with $\vec{\theta}(t) = \#_{0 \leq i < |\theta|}\theta(i) = t$, i.e. a vector counting the occurrences of each transition in θ. The set of all reachable markings of a Petri net given an initial marking is called the statespace.

Marking Equation. The structure of a Petri net can be translated into a so-called incidence matrix.

For Petri net $PN = (P, T, F, \lambda)$, The incidence matrix $\mathbf{C}$ is a matrix with $|P|$ rows and $|T|$ columns, such that for all $t \in T$ and $p \in P$ holds that

$$\mathbf{C}(p, t) = \begin{cases} -1 & \text{if } (p, t) \in F \text{ and } (t, p) \notin F \\ 1 & \text{if } (t, p) \in F \text{ and } (p, t) \notin F \\ 0 & \text{othwerwise} \end{cases}$$

The incidence matrix can be used to mathematically relate markings through the so-called marking equation in the following way.

For a firing sequence $\theta \in T^*$ between two markings m_1 and m_2, i.e. $m_1[\theta\rangle m_2$, the marking equation states that: $\vec{m_1} + \mathbf{C} \cdot \vec{\theta} = \vec{m_2}$.

The marking equation implies that *if* there is a firing sequence θ from marking m_1 to marking m_2, *then* this equation holds. The inverse is not necessarily true, i.e. if $\vec{\theta}$ provides a solution to this equation, then it is not guaranteed that there exists a firing sequence θ corresponding to $\vec{\theta}$.

Where applicable, we mix sets and multisets by assuming a set is a multiset containing each element only once. As Petri nets in this paper are used to model processes, they have a clear initial and final marking.

$PM = (PN, m_i, m_f)$ is a process model consisting of a Petri net (P, T, F, λ) with initial marking m_i and final marking m_f. We assume there exists $\sigma \in T^*$ such that $m_i[\sigma\rangle m_f$, i.e. the final marking is reachable from the initial marking.

4   An Incremental Technique for Computing Alignments

Given a Petri net and its initial and final marking, the behaviour of the model can be seen as the set of possible firing sequences leading from the initial to the final marking. This set is, in many cases, infinite. An event log captures sequences of activity executions observed in practice and an *alignment* is a concept able to express the precise relation between this sequence and the model.

To explain the concept of an alignment, we use a so-called synchronous product net. These synchronous produces are specific types of nets, constructed from the combination of a model and a trace, where the trace is first converted into a trace model.

Definition 1 (Trace model). *Let $\sigma \in A^*$ be a trace. $TN = ((P, T, F, \lambda), m_i, m_f)$ is a process model called the trace model of σ if and only if:*

- $P = \{p_0 \dots p_{|\sigma|}\}$,
- $T = \{t_1 \dots t_{|\sigma|}\}$,
- $F = \{(p_i, t_{i+1}) \in P \times T \mid 0 \leq i < |\sigma|\} \cup \{(t_i, p_i) \in T \times P \mid 0 < i \leq |\sigma|\}$,

- for all $0 < i \leq |\sigma|$ holds $\lambda(t_i) = \sigma(i)$,
- $m_i = [p_0]$, and
- $m_f = [p_{|\sigma|}]$.

A trace model is a straightforward translation of a sequence of activities into a linear Petri net model. Using a process model and a trace model, we define the synchronous product as follows:

Definition 2 (Synchronous Product). Let $PN = (P^m, T^m, F^m, \lambda^m)$ and $PM = (PN, m_i^m, m_f^m)$ a process model. Furthermore, let $\sigma \in A^*$ be a trace with $TN = ((P^l, T^l, F^l, \lambda^l), m_i^l, m_f^l)$ its trace model.

The synchronous product $SN = ((P, T, F, \lambda), m_i, m_f)$ is a Process Model, with:

- $P = P^m \cup P^l$ is the combined set of places,
- $T = \{(t^m, t^l) \in (T^m \cup \{\gg\}) \times (T^l \cup \{\gg\}) \mid t^m \neq\gg \vee t^l \neq\gg \vee \lambda^m(t^m) = \lambda^l(t^l)\}$ is the set of original transitions merged with synchronous ones,
- $F = \{((t^m, t^l), p) \in T \times P \mid (t^m, p) \in F^m \vee (t^l, p) \in F^l\} \cup \{(p, (t^m, t^l)) \in P \times T \mid (p, t^m) \in F^m \vee (p, t^l) \in F^l\}$, is the set of edges,
- for all $(t^m, t^l) \in T$ holds that $\lambda((t^m, t^l)) = \begin{cases} \lambda^m(t^m) & \text{if } t^m \neq\gg \\ \lambda^l(t^l) & \text{if } t^l \neq\gg \end{cases}$, i.e. λ respects the original labelling,
- $m_i = m_i^m \uplus m_i^l$ is the combined initial marking, and
- $m_f = m_f^m \uplus m_f^l$ is the combined final marking.

A synchronous product net is a combination of a process model and a trace model in such a way that for each pair of transitions of which the labels agree, a synchronous transition is added. Any complete firing sequence of the synchronous product represents an alignment in which each transition firing represents either a model move (transitions of the type $(t^m, \gg)$), a log move (transitions of the type $(\gg, t^l)$ or a synchronous move (transitions of the form (t^m, t^l)).

An optimal alignment is an alignment for which a cost function, associating costs to the firing of each transition, is minimized.

Definition 3 (Optimal alignment). Let $SN = ((P, T, F, \lambda), m_i, m_f)$ be a synchronous product model and let $c : T \to \mathbb{R}_{\geq 0}$ be a cost function associating costs to each transition in the synchronous product. An optimal alignment $\gamma \in T^*$ is a firing sequence of SN, such that $m_i[\gamma\rangle m_f$ and there is no $\theta \in T^*$ with $m_i[\theta\rangle m_f$ and $\sum_{t \in \theta} c(t) < \sum_{t \in \gamma} c(t)$. We use $c(\theta)$ as shorthand for $\sum_{t \in \theta} c(t)$.

Note that there may be many optimal alignments in the general case, i.e. alignments with the same minimal costs. Typically, the cost function c is chosen in such a way that 0 costs are only associated to synchronous transitions, i.e. transitions $(t^m, t^L) \in T$ for which $t_M \neq\gg$ and $t_L \neq\gg$. Routing transitions of the form $(\tau, \gg)$ are typically given a small cost $\epsilon > 0$, while transitions of the form $(t, \gg)$ or $(\gg, t)$ with $t \neq \tau$ receive cost 1. For the work in this paper the cost function should be chosen such that there are no infinite sets of markings reachable with the same costs, for which the above cost function provides a sufficient condition. For models that do not contain infinite sets of markings reachable through routing transitions only, $\epsilon = 0$ is a valid assignment, which is what we use in this paper.

4.1   Underestimation Using the Marking Equation

To compute alignments, we build on a technique introduced in [4] with the parameter settings optimized according to [16]. This technique is based on $A^\star$, which is a shortest path search algorithm that can be guided towards the destination using a function that *underestimates* the remaining costs of the optimal alignment.

Definition 4 (Underestimation Function). *Let* $PN = ((P, T, F, \lambda), m_i, m_f)$ *be a Process model and let* $c : T \to \mathbb{R}_{\geq 0}$ *be a cost function. We define* $h : \mathcal{B}(P) \to \mathbb{R}_{\geq 0}$ *to be an underestimation function if and only if for all* $m \in \mathcal{B}(P)$ *and* $\sigma \in T^*$ *with* $m[\sigma\rangle m_f$ *holds that* $h(m) \leq c(\sigma)$.

Several underestimation functions exist in literature. The most trivial one is the function that always returns 0. This function leads to $A^\star$ behaving like Dijkstra's shortest path algorithm, essentially doing a breadth-first search through the reachability graph of the synchronous product.

Using the marking equation of the synchronous product, several other underestimation functions can be defined.

Definition 5 (ILP-based   Underestimation   Function). *Let   PN   =* $((P, T, F, \lambda), m_i, m_f)$ *be a process model with incidence matrix* $\mathbf{C}$ *and let* $c : T \to \mathbb{R}_{\geq 0}$ *be a cost function. We use* $\vec{c}$ *to denote a column vector of the cost values for each transition, i.e.* $\vec{c}(t) = c(t)$.

We define $h^{ILP} : \mathcal{B}(P) \to \mathbb{R}_{\geq 0}$ *as an ILP based underestimation, such that for each marking* $m \in \mathcal{B}(P)$ *holds that* $h^{ILP}(m) = \vec{c}^{\mathsf{T}} \cdot \vec{x}$ *where* $\vec{x}$ *is the solution to:*

$$\begin{aligned} minimize \quad & \vec{c}^{\mathsf{T}} \cdot \vec{x} \\ subject\ to \quad & \vec{m} + \mathbf{C} \cdot \vec{x} = \vec{m}_f \\ & \forall_{t \in T} \vec{x}(t) \in \mathbb{N} \end{aligned}$$

If no solution $\vec{x}$ *to the linear equation system exists, then* $h^{ILP}(m) = +\infty$

The ILP-based estimation function uses the marking equation to underestimate the remaining cost to the final marking from any marking reached. Recall that for any firing sequence, the marking equation has a solution, but not the other way around. As a consequence, if the marking equation has no solution, there is no firing sequence to the final marking. Furthermore, by minimizing, we guarantee to provide a lower bound on the remaining distance.

Theorem 1 (h^{ILP} Provides a lower bound on the costs [4]). *Let* $PN = ((P, T, F, \lambda), m_i, m_f)$ *be a process model with incidence matrix* $\mathbf{C}$ *and let* $c : T \to \mathbb{R}_{\geq 0}$ *be a cost function. Let* $m \in \mathcal{B}(P)$ *be a marking and* $\gamma \in T^*$ *an optimal firing sequence such that* $m[\gamma\rangle m_f$, *i.e. no firing sequence* $\theta \in T^*$ *with* $m[\theta\rangle m_f$ *exists such that* $c(\theta) < c(\gamma)$. *We prove that* $h^{ILP}(m) \leq c(\gamma)$.

Proof. Since γ is a firing sequence, we know that $\vec{m} + \mathbf{C}\vec{\gamma} = \vec{m}_f$ (marking equation). Furthermore, for all $t \in T$ $\vec{\gamma}(t) = \#_{0 \leq i < |\gamma|}\theta(i) = t \in \mathbb{N}$ and $\vec{c}^{\mathsf{T}} \cdot \vec{\gamma} = c(\gamma)$. Assume $\vec{x}$ minimizes the ILP shown in Definition 5 for marking m. It is trivial that either $\vec{c}^{\mathsf{T}} \cdot \vec{x} \leq \vec{c}^{\mathsf{T}} \cdot \vec{\gamma}$, otherwise $\vec{x}$ is not minimizing. Hence $h^{ILP}(m) \leq c(\gamma)$.

The estimation function h^{ILP} can be used to underestimate the remaining cost of reaching the final marking in the synchronous product from any marking reached and is therefore suitable to be used as a heuristic in $A^\star$. Unfortunately, solving the integer linear program for every marking in the model is a complex task. It is well-known that solving integer linear programs is exponential in the rank of the matrix, i.e. in the number of transitions and places in the synchronous product.

Therefore, we relax the constraints a bit and to use a non-integer linear program to underestimate the remaining cost, i.e. we replace the last line of Definition 5 by $\forall_{t \in T}\vec{x}(t) \in \mathbb{R}_{\geq 0}$ to obtain $h^{LP} : \mathcal{B}(P) \to \mathbb{R}_{\geq 0}$.

Since any integer solution is also a real valued solution, for any marking m holds that $h^{LP}(m) \leq h^{ILP}(m)$. Therefore, h^{LP} also provided an underestimate, but the computational complexity is polynomial in the rank of the matrix.

In practice, as h^{LP} provides a worse underestimate than h^{ILP}, more markings need to be expanded when computing alignments, i.e. with a worse underestimation function, $A^\star$ investigates a larger part of the search space. However, due to the fact that most LP solvers use Simplex as a solving technique, it is unlikely that the actual solutions returned are non-integer for real-life examples.

An important property of both functions is the fact that the vectors $\vec{x}$ that provide the minimum can be used to derive a solution in the next marking of the search space.

Theorem 2 (Minimizing solutions can be reused). *Let $PN = ((P, T, F, \lambda), m_i, m_f)$ be a process model with incidence matrix $\mathbf{C}$ and let $c : T \to \mathbb{R}_{\geq 0}$ be a cost function. Let $m, m' \in \mathcal{B}(P)$ be two markings and $\vec{x}$ a vector minimizing the linear program of Definition 5. Furthermore, let $t \in T$ be such that $m[t\rangle m'$ with $\vec{x} \geq 1$. We show that $h^{ILP}(m') = h^{ILP}(m) - c(t)$ with solution vector $\vec{x}' = \vec{x} - \vec{1}_t$ (where $\vec{1}_t$ is a vector with 0 on all rows except the row corresponding to t, which has value 1).*

Proof. We prove by contradiction. Let $\vec{y}$ be a vector satisfying the constraints of Definition 5 such that $h^{ILP}(m') = \vec{c}^{\mathsf{T}} \cdot \vec{y} < h^{ILP}(m) - c(t)$, i.e. $\vec{y}$ provides a better solution than $\vec{x}'$.

Since $m[t\rangle m'$, we know that $\vec{m} + \mathbf{C} \cdot \vec{1}_t = \vec{m}'$ (marking equation). Therefore $\vec{m} + \mathbf{C} \cdot \vec{1}_t + \mathbf{C}\vec{y} = \vec{m}_f$, i.e. $\vec{m} + \mathbf{C} \cdot (\vec{1}_t + \vec{y}) = \vec{m}_f$ and since this provides an integer solution for the linear program, we know $h^{ILP}(m) \leq \vec{c}^{\mathsf{T}} \cdot (\vec{1}_t + \vec{y})$. However, $\vec{c}^{\mathsf{T}} \cdot (\vec{1}_t + \vec{y}) = \vec{c}^{\mathsf{T}} \cdot \vec{1}_t + \vec{c}^{\mathsf{T}} \cdot \vec{y} = c(t) + \vec{c}^{\mathsf{T}} \cdot \vec{y}$. In other words: $h^{ILP}(m) - c(t) \leq \vec{c}^{\mathsf{T}} \cdot \vec{y}$ which contradicts the fact that y provides a better solution than $\vec{x}'$.

It is trivial to see that $\vec{x}'$ is indeed an integer solution to the linear equation system since $\vec{m} = \vec{m}' - \mathbf{C} \cdot \vec{1}_t$, and therefore $\vec{m}' + \mathbf{C} \cdot \vec{x} - \mathbf{C} \cdot \vec{1}_t = \vec{m}_f$, i.e.

$\vec{m'} + \mathbf{C} \cdot (\vec{x} - \vec{1}_t) = \vec{m_f}$ and thus $\vec{m'} + \mathbf{C} \cdot \vec{x'} = \vec{m_f}$. Since $\vec{x}(t) \geq 1, \vec{x}(t) - 1 = \vec{x'}(t) \geq 0$ and hence $\vec{x'}$ provides a solution to the linear equation system for $h^{ILP}(m')$.

Note that the proof is analogous for the h^{LP} case. Also there, a new solution can be derived from a previous one when firing transition t, as long as $\vec{x}(t) \geq 1$. By guiding $A^\star$ to favour markings for which the solution can be reused, the search time can be improved considerably.

Unfortunately, even when re-using solutions, there are scenarios that are common in practical cases where $A^\star$ performs poorly (almost at worst-case, i.e. exploring the full reachability graph of the synchronous product net), for example in the case of Fig. 2.

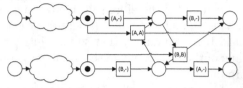

Fig. 2. Synchronous product for the example of Fig. 1.

This is due to the fact that in all reachable markings, the solution to the linear equation system suggests that the two synchronous transitions (A, A) and (B, B) can be executed with cost 0, while in fact, only two sequences are possible with minimal cost of 2, namely $\langle (A, -), (B, B), (-, A) \rangle$ or $\langle (B, -), (A, A), (-, B) \rangle$.

The root cause for the poor performance in this case is the use of the marking equation in the heuristic function. In particular, the fact that the vector $\vec{x}$ minimizing the (integer) linear program does not necessarily correspond to a realizable firing sequence. In the next subsection, we show how the marking equation can be extended to improve the solution vectors of the linear program.

4.2   Underestimation Using the Extended Marking Equation

We first consider the marking equation again, but this time we also include the consumption matrix.

Definition 6 (Consumption Matrix). *Let $PN = (P, T, F, \lambda)$ be a Petri net. The consumption matrix $\mathbf{C}^-$ is a matrix with $|P|$ rows and $|T|$ columns, such that for all $t \in T$ and $p \in P$ holds that* $\mathbf{C}^-(p, t) = \begin{cases} -1 & \text{if } (p, t) \in F \\ 0 & \text{othwerwise} \end{cases}$

The consumption matrix can be used to guarantee that before the firing of a specific transition, the necessary tokens are present in a Petri net by extending the marking equation slightly.

Definition 7 (Extended Marking Equation). *Let $PN = (P, T, F, \lambda)$ be a Petri net, $\mathbf{C}$ its incidence matrix and $\mathbf{C}^-$ its consumption matrix. Let $\theta_1, \theta_2 \in T^*$ be two firing sequences and m_1, m_2, m_3 three markings, such that $m_1[\theta_1\rangle m_2$ and $m_2[\theta_2\rangle m_3$. Furthermore, assume $\theta_2 = \langle t, \ldots \rangle$, i.e. θ_2 starts with transition t. The extended marking equation states that:* $\begin{aligned} \vec{m_1} + \mathbf{C} \cdot (\vec{\theta_1} + \vec{\theta_2}) = \vec{m_3}, \text{ and} \\ \vec{m_1} + \mathbf{C} \cdot \vec{\theta_1} + \mathbf{C}^- \cdot \vec{1}_t \geq \vec{0} \end{aligned}$

The second part of the extended marking equation is essentially a translation of the enabling condition for Petri nets. If a transition is enabled, we know that each input place contains sufficiently many tokens. The marking reached after firing any sequence θ_1 in the net is determined by the first part of the original marking equation. The consumption matrix is then used to express that after *consumption* of all tokens required by transition t (but before production of its output tokens), the net does not have any place with a negative number of tokens.

As with the original marking equation, the reverse does not hold, i.e. not all solutions for $\vec{\theta_1}$ and $\vec{\theta_2}$ that satisfy these conditions correspond to realizable firing sequences. However, as these conditions are more restrictive, we can use them in the estimation function for A^* if we can identify the point where we should split between θ_1 and θ_2. For this, we reason over the original trace.

Theorem 3 (An optimal alignment is at least as long as the trace). *Let $SN = ((P, T, F, \lambda), m_i, m_f)$ be a synchronous product with $PM = ((P^m, T^m, F^m, \lambda^m), m_i^m, m_f^m)$ the process model and $TN = ((P^l, T^l, F^l, \lambda^l), m_i^l, m_f^l)$ the trace model for trace $\sigma \in A^*$ it is constructed from.*

Let $\gamma \in T^*$ be an optimal alignment for some cost function $C : T \to \mathbb{R}_{\geq 0}$. We show that $|\gamma| \geq |\sigma| = |T^l|$, i.e. the length of the optimal alignment is longer than the length of the trace σ.

Proof. This follows trivially from the structure of the synchronous product and the fact that the token from the place initially marked in the trace model needs to be "transported" to the final place of the trace model in any trace reaching this final marking.

We use this property of the optimal alignment to provide a better estimation function. Simply put, we cut the optimal alignment into a number of k predefined pieces (where k is less or equal to the length of the sequence) and we guarantee that after consumption of each transition at the start of each subtrace, the marking is non-negative.

Definition 8 (Underestimation with k subtraces in the initial marking). *Let $SN = ((P, T, F, \lambda), m_i, m_f)$ be a synchronous product model for trace σ. SN has incidence matrix $\mathbf{C}$ and consumption matrix $\mathbf{C}^-$. Let $c : T \to \mathbb{R}_{\geq 0}$ be a cost function and let $\sigma = \sigma_1 \circ \ldots \circ \sigma_k$ be a division of the trace into k non-empty subtraces.*

We define $h_{m_i} = \vec{c}^{\mathsf{T}} \cdot \sum_{0 \leq a \leq k} \vec{x_a} + \vec{c}^{\mathsf{T}} \cdot \sum_{0 \leq a < k} \vec{y_a}$ where $\vec{x_a}$ and $\vec{y_a}$ provide the solution to:

$$\text{minimize} \quad \vec{c}^{\mathsf{T}} \cdot \sum_{0 \leq i \leq k} \vec{x_i} + \vec{c}^{\mathsf{T}} \cdot \sum_{1 \leq i \leq k} \vec{y_i}$$

$$\text{subject to} \quad \vec{m_i} + \mathbf{C} \cdot \vec{x_0} + \mathbf{C} \cdot \sum_{1 \leq a \leq k}(\vec{x_a} + \vec{y_a}) = \vec{m_f} \quad (1)$$

$$\forall_{1 \leq a \leq k} \quad \vec{m_i} + \mathbf{C} \cdot \vec{x_0} + \mathbf{C} \cdot \sum_{1 \leq b < a}(\vec{x_b} + \vec{y_b}) + \mathbf{C}^- \cdot \vec{y_a} \geq \vec{0} \quad (2)$$

$$\forall_{0 \leq a \leq k} \; \forall_{t \in T} \quad \vec{x_a}(t) \in \mathbb{N} \quad (3)$$

$$\forall_{1 \leq a \leq k} \; \forall_{t \in T} \quad \vec{y_a}(t) \in \{0, 1\} \quad (4)$$

$$\forall_{1 \leq a \leq k} \; \forall_{(t^m, t^l) \in T} \text{ with } t^l \neq t_{(1 + \sum_{1 \leq b < a} |\sigma_b|)} \; \vec{y_a}((t^m, t^l)) = 0 \quad (5)$$

$$\forall_{1 \leq a \leq k} \quad \vec{1}^{\mathsf{T}} \cdot \vec{y_a} = 1 \quad (6)$$

Note that the assumption is that there is at least one alignment, hence this equation system is guaranteed to have a solution.

The equation system in Definition 8 appears complicated, but is a fairly straightforward translation of the extended marking equation. Consider an optimal alignment $\gamma \in T^*$ for a trace σ of length $|\sigma|$. As we know, we can split this optimal alignment into $k \leq |\sigma|$ subtraces, each of which starts with one transition corresponding to an event from the trace, i.e. $\gamma = \gamma_0 \circ \ldots \circ \gamma_k$, where for $0 < i \leq k$ holds that $\gamma_i = \langle (t, t^l), \ldots \rangle$ with $t^l \in T^l$, i.e. γ_i starts with one transition that moves the token in the original trace model. γ_0 is a (possibly empty) prefix of transitions of the form $(t^m, \gg)$.

In Definition 8, variables $\vec{y_a}$ refer to the first transition in each γ_i, i.e. these vectors encode the firing of a single transition at the start of each subtrace. Variables $\vec{x_a}$ (bound by rule 3) correspond to any other transitions firing (in any order). These vectors may be empty. Rule 4 guarantees that every element of $\vec{y_a}$ is 0 or 1 and rule 5 and 6 ensure that only one element of $\vec{y_a}$ equals 1 and that that element corresponds to a transition which is a transition of the synchronous product corresponding to the start of σ_a.

Rule 1 is a translation of the original marking equation. It simply states that when combining the firing of all transitions in all $\vec{x_a}$ and $\vec{y_a}$, the final marking is reached from the initial marking. Finally, rule 2 uses the extended marking equation to guarantee that after firing a prefix of transitions γ_0 through γ_{a-1}, sufficient tokens are available to fire the first transition in γ_a (expressed by $\vec{y_a}$).

Definition 8 provides again an underestimate for the total cost of an optimal alignment in the initial marking. It can be generalized to any arbitrary reachable marking by assuming that we know how many events of the original trace remain to be explained, i.e. by adjusting both m_i and σ. This information is generally available by considering the marking of the places in the trace model. Furthermore, like before, we can relax constraints 3 and 4 from integers to real valued numbers to reduce the complexity of minimizing the linear inequation system.

Definition 9 (Underestimation with k subtraces). *Let $SN = ((P, T, F, \lambda), m_i, m_f)$ be a synchronous product model for trace σ. Let $m \in \mathcal{B}(P)$ be a marking in which l events of trace σ are explained and let $\sigma = \sigma_0 \circ \sigma_1 \circ \ldots \circ \sigma_k$ be a division of the trace into $k + 1$ non-empty subtraces, such that the first l events are in σ_0, i.e. $k \leq |\sigma| - l$.*

We define $h^{ILP,k} : \mathcal{B}(P) \to \mathbb{R}_{\geq 0}$ as an underestimation function, such that $h^{ILP,k}(m) = h_m$ following Definition 8.

Like before, we prove that $h^{ILP,k}$ indeed provides an underestimation function, provided that we know, for each marking, how many events have been explained by the path leading up to this marking. Recall that this information is trivially derived from the location of the token in the places corresponding to the trace model.

Theorem 4 ($h^{ILP,k}$ is an underestimation function). *Let $SN = ((P, T, F, \lambda), m_i, m_f)$ be a synchronous product model for trace σ and let $\sigma = \sigma_0 \circ \sigma_1 \circ \ldots \circ \sigma_k$ be a division of the trace into $k+1$ non-empty subtraces, such that the first l events are in σ_0.*

Let $m \in \mathcal{B}(P)$ be a marking in which l events of trace σ are explained. We prove that for each $\theta \in T^*$ with $m[\theta\rangle m_f$ holds that $h^{ILP,k}(m) \leq c(\theta)$.

Proof. Let $\gamma \in T^*$ be a firing sequence with $m[\gamma\rangle m_f$, such that for each $\theta \in T^*$ with $m[\theta\rangle m_f$ holds that $c(\gamma) \leq c(\theta)$, i.e. γ is optimal.

We know that $\gamma = \gamma_0 \circ \ldots \circ \gamma_k$ such that for $1 \leq a \leq k$ holds that $\gamma_a = \langle t_a, \ldots \rangle$ with $t_a = (t^m, t^l)$ and $t^l = t_{(1 + \sum_{0 \leq b < a} |\sigma_b|)} \in T^l$, i.e. γ can be split into $k + 1$ subtraces such that all but the first subtrace start with a transition related to an event in the original trace. We show that γ provides a solution for the inequation system.

Let $\vec{x_0} = \vec{\gamma_0}$ and for all $1 \leq a \leq k$, let $\vec{y_a} = \vec{1_{t_a}}$ and let $\vec{x_a} = \vec{\gamma_a} - \vec{y_a}$, i.e. we translate the subtraces of γ into Parikh vectors, where we separate the first transitions into vectors $\vec{y}$ and the remainder into vectors $\vec{x}$.

It is trivial to see that by definition, conditions 3, 4, and 6 are met by this translation. Condition 5 is also met, since the only element with value 1 is the element that satisfies the given condition. Furthermore, Condition 1 is met, since $\vec{m} + \mathbf{C} \cdot \vec{x_0} + \mathbf{C} \cdot \sum_{1 \leq a \leq k} (\vec{x_a} + \vec{y_a}) = \vec{m} + \mathbf{C} \cdot (\vec{x_0} + \sum_{1 \leq a \leq k} (\vec{x_a} + \vec{y_a})) = \vec{m} + \mathbf{C} \cdot \vec{\gamma} = \vec{m_f}$. This is again the marking equation which holds for any sequence.

Condition 2 is more complicated. Let $1 \leq a \leq k$ and let $m' \in \mathcal{B}(P)$ be the marking, such that $m[\gamma_0, \ldots, \gamma_{a-1}\rangle m'$, i.e. m' is the marking reached after executing the first a subtraces. We know that $\vec{m} + \mathbf{C} \cdot \vec{x_0} + \mathbf{C} \cdot (\sum_{1 \leq b < a} (\vec{x_b} + \vec{y_b})) = \vec{m'}$ (again from the marking equation). Furthermore, we know that firing transition t_a is possible in marking m' as this is the first transition in γ_a. Hence $\vec{m'} + \mathbf{C}^- \cdot \vec{y_a} \geq \vec{0}$ which follows from the extended marking equation. Combining the two yields $\vec{m} + \mathbf{C} \cdot \vec{x_0} + \mathbf{C} \cdot (\sum_{1 \leq b < a} (\vec{x_b} + \vec{y_b})) \geq -\mathbf{C}^- \cdot \vec{y_a}$. Therefore $\vec{m} + \mathbf{C} \cdot \vec{x_0} + \mathbf{C} \cdot (\sum_{1 \leq b < a} (\vec{x_b} + \vec{y_b})) + \mathbf{C}^- \cdot \vec{y_a} \geq \vec{0}$.

Since γ provides a solution to the inequation system, we know that any minimal solution has less or equal costs. Hence $h^{ILP,k}$ is an underestimation function.

It is fairly easy to see that any solution in terms of $\vec{x}$ and $\vec{y}$ to the inequation system of Definition 8 can be translated into a solution to the equation systems of Definition 5 in terms of $\vec{z}$ by defining $\vec{z} = \vec{x_0} + \sum_{1 \leq a \leq k} (\vec{x_a} + \vec{y_a})$. Hence for any marking m holds that $h^{LP}(m) \leq h^{ILP}(m) \leq h^{ILP,k}(m)$. Furthermore, the k-variant can also easily be extended to the real domain in which case $h^{LP}(m) \leq h^{LP,k}(m) \leq h^{ILP,k}(m)$.

This implies that $h^{ILP,k}$ provides a better underestimation function than the previous ones, since it guarantees that *if* the first vector $\vec{x_0}$ corresponds to a realizable firing sequence, *then* it reaches a marking that enables the transition indicated by $\vec{y_1}$. So for each splitpoint, the extended marking equations comes with an additional "guarantee" on a transition being enabled.

However, this comes at a cost of computational complexity, since the inequation system has more variables (and constraints). Where Definition 5 has $|T|$ variables, Definition 8 has $(2 * k + 1) * |T|$ variables (of which many are bound to 0 by constraint 5).

The observation that there are many more variables involved in the k variants of the underestimation function leads to the question which value for k should be chosen. Furthermore, the re-use of solutions of the inequation system to derive new solutions is not as trivial as it is for the traditional underestimation functions. Therefore, we propose an incremental search technique to find optimal alignments which combines Theorem 2 with Definition 9.

4.3   Incrementally Extending the Heuristic Function

Using the k-based underestimation function, we propose a special version of the $A^\star$ search algorithm which incrementally increases the value of k when needed and maximizes the reuse of previously computes solution vectors. The basic principle of the search is simple:

We start in the initial marking m_i by computing an underestimate for $k = 1$. After solving the inequation system, we remember the solution vector $\vec{z_{m_i}} = \vec{x_0} + \sum_{1 \leq a \leq k}(\vec{x_a} + \vec{y_a})$ as well as the estimate h.

Then, we follow the classical $A^\star$ algorithm to find the optimal alignment. For each marking m' reached by firing transition t in marking m, the new underestimate value is computed either exactly if for marking m holds that $\vec{z_m}(t) \geq 1$, in which case $h(m') = h(m) - c(t)$ and $\vec{z_{m'}} = \vec{z_m} - \vec{1_t}$, (cf. Theorem 2). If $\vec{z_m}(t) < 1$, then $h(m') = \max(0, h(m) - c(t))$, but $\vec{z_{m'}} =\perp$, i.e. the underestimate is decreased, but the solution vector is unknown.

During the $A^\star$ search, we prioritize markings for which the solution vector is known and since we monotonously decrease the underestimate with the actual costs, we know that if we ever have to expand a marking for which the solution vector is unknown, there are no markings with known solution vectors in the open set of the search algorithm.

If we do visit a marking with an unknown solution vector, we have reached a point where the original estimate computed in the initial marking corresponds to a firing sequence that is guaranteed not to be realizable. In that case, we try to improve the underestimate function by increasing k and choosing a new way to split the trace. Here, we consider the maximum number of events already explained by any marking reached in the search space so-far and we split there, i.e. if a marking explains a events for $k = 1$, we restart the procedure from scratch with $k = 2$ and $\sigma = \sigma_1 \circ \sigma_2$ with $|\sigma_1| = a$.

Suppose that in the next iteration, we encounter an unknown solution vector again, while we have found a marking explaining b events with $b > a$, we then restart with $k = 3$ and $\sigma = \sigma_1 \circ \sigma_2 \circ \sigma_3$ with $|\sigma_1| = a$ and $|\sigma_2| = b - a$. If $b < a$, then we restart with $k = 3$ and $\sigma = \sigma_1 \circ \sigma_2 \circ \sigma_3$ with $|\sigma_1| = b$ and $|\sigma_2| = a - b$.

If we encounter again an unknown solution vector with $b = a$ events explained, we follow the normal steps in the search, i.e. we compute an exact solution for the state and we requeue the state if the new estimate is higher than the previous estimate.

This procedure incrementally improves the underestimation function used in the $A^\star$. Worst case, this procedure leads to the entire alignment problem being encoded as a linear program and especially for larger process models, this implies that the computation time for the initial marking might increase substantially. It is therefore advised not to use the ILP variant, but the LP variant. In practical examples, k does not grow so big and in the experiment section, we show that this procedure can considerably reduce the computation time of finding alignments.

5   Experiments

The procedure outlined in this paper was implemented in ProM as part of the normal alignment plugins. It requires the user to install the "alignment" package in which case the incremental algorithm shows up as a advanced variant in the conformance checking plugin. The input objects as well as the returned objects are identical to the classical conformance checking, i.e. optimal alignments are returned.

In this section, we compare the performance of the algorithm with two existing alignment techniques. First, the classical $A^\star$ using optimal parameter settings [16] and second the planner-based approach [5]. Note that, for the former, we use a completely new codebase which is optimized for memory use and also part of the aforementioned alignment package in ProM. For the latter, we use the ProM package "PlanningBasedAlignment"[1].

We report the CPU times of all three techniques. For the planner, we show both the time reported by the software, as well as the wallclock time. The latter is considerably larger due to the overhead of reading and writing files and instantiating the external planner processed from Python. As the planner can only be run in a single thread, we conducted all experiments in single-threaded mode on a 2.8 GHz Intel Xeon W3530 CPU. The tools were given maximum 10 GB of RAM and for the planner, file IO was done on a 2 GB RAM disk in physical RAM, eliminating slow disk access. The $A^\star$ variants were bound to 60 s per trace. The planner does not support a time limit. It was given as much time as needed and was only terminated after running out of memory (10 GB). The $A^\star$ approaches never needed more than 10 MB of memory to store the internal data structures for a single trace (including the memory used by the LP Solver LpSolve). The Java VM however needed about 1 GB to store the event logs and Petri net objects in memory. The total wallclock time to conduct all experiments (not including the logs for which the planner ran out of memory) was 9 h for the planner, 5 h for $A^\star$ and 35 min for incremental $A^\star$. The remaining logs took 32 h in $A^\star$ (however, 45% of the traces timed out after 60 s) and 2 h using our incremental version.

---

[1] Due to the new codebase, the time performance of $A^\star$ cannot directly be compared to the results presented in [5]. Furthermore, the CPU times reported in [5] were obtained using a proprietary implementation which is not available for download.

Table 1. Time and memory to align benchmark event logs.

Log	Cases	Classical $A^\star$ [16]			Incremental $A^\star$		Planner [5]			cost
		time (s)	(timeout)	solved LPs	time (s)	solved LPs	pre (s)	src (s)	clock (s)	
[18] road fines	10,247	0.2		458	**0.2**	176	4.3	1.6	2,362.4	7,284
[17] bpi12	12,136	7.9		29,089	**1.3**	759	10.8	43.4	3,111.0	57,727
[19] sepsis	1,051	120.0		440,825	**10.7**	5,553	1.8	59.6	323.4	2,448
[9] Fitting logs	4,004	4.6		2,935	**4.5**	2,935	15.2	13.7	1,123.9	53
[9] Noisy logs	16,016	52.6		100,529	**21.7**	14,155	55.3	237.8	4,671.8	12,111
[20] Fitting log (prBm6)	1,201	5.0		1,126	**4.3**	1,126	72.6	24.5	679.4	14
[20] Noisy logs	6,506	132,824.2	2,182	24,016,461	**8,190.7**	47,270	1,465.0	Out of Mem		98,708
[21] Fitting logs	38,019	53.2		25,620	**51.5**	25,620	326.2	185.1	12,022.5	253
[21] Noisy logs	28,014	15,503.4		11,756,515	**488.5**	76,857	288.1	575.5	9,618.0	91,603
[20] prAm6	1,201	**55.0**		29,496	99.2	2,752	64.3	798.3	1,475.1	4,969
[20] prCm6	501	1,780.4		804,518	1,302.7	7,683	31.7	166.1	**445.1**	12,552
[20] prEm6	1,201	**29.1**		7,200	60.8	4,800	148.5	Out of Mem		4,880
[20] prFm6	1,201	30,847.8	514	2,702,736	**1,354.8**	7,878	396.1	Out of Mem		21,425
[20] prGm6	1,201	71,941.9	1,200	18,643,107	**1,526.6**	13,794	253.4	Out of Mem		26,411
[21] pr1151_l4	2,001	2,467.9		1,457,592	**68.9**	7,486	26.9	61.4	715.8	9,044
[21] pr191_l4	2,001	3,978.8		2,702,250	**126.5**	11,707	30.5	94.1	763.7	16,044

Table 1 shows the results for three real life logs, three collections divided in fitting and non-fitting logs[2] and selected individual model/log combinations from these collections. We show the number of cases and per algorithm, the time needed to align all traces, the number of traces that reached the timeout and the number of linear programs solved. The time is the total time needed to align all traces in the log (including the timed-out ones). If the planner ran out of memory, we show the pre-processing time (excluding IO) for translating the traces to PDDL files. For each row, the fastest technique (wallclock time) is highlighted.

The table shows that for the top three real-life cases, the incremental $A^\star$ outperforms the others, mainly due to the significant reduction in number of LP's that are solved. For the entire collections, the totals are shown over the full collection, divided into fitting and noisy event logs. For fitting event logs, the performance of both $A^\star$ variants is equal. For noisy logs, the incremental version is orders of magnitude faster. The planner's preprocessing time is already higher than the computation time for both $A^\star$ variants.

The models from [20] present the worst-case scenario for incremental $A^\star$. These logs are created using swapping of events in various parts of the model. The incremental $A^\star$ needs to reach the swapped event, after which a splitpoint will be introduced. For prAm6 and prEm6, we see that $A^\star$ is faster despite the reduction in number of LPs solved, i.e. the CPU can investigate enough states per second to eliminate the need for incremental $A^\star$. This is due to the location of the swaps in these logs. For prEm6 the planner runs out of memory, but the pre-processing time (excluding IO) is already higher than the $A^\star$ time.

For prCm6, a model which does not have a large amount of parallelism, the planner is fastest in the reported time, but not in wallclock time.

---

[2] Note that the cost are never 0, due to the empty trace that is always included in the computation. Furthermore, for the largest of these logs, no optimal alignments could be computed before.

For prFm6 and prGm6, regular $A^\star$ runs out of time for many (if not all) of the traces. This is due to the fact that these models contain vast amounts of parallelism and swapped events towards the end of the traces. The full reachability graph is therefore expanded by $A^\star$. Our incremental version however correctly identifies the swaps and returns optimal alignments, whereas the planner runs out of memory, probably also due to the parallelism.

Figure 3 shows the relation between the cost of the alignment, the time to compute the alignment and the memory use for the two $A^\star$ variants for model prCm6. The planner is not included here as it does not report the time per trace, but only the total time for the entire event log. We see that the times per trace are comparable for the two algorithms (time is plotted on the lefthand axis, with logarithmic scale). As the costs of the alignments increase, so does the computation time. The time complexity for this model shows a polynomial trend in the cost of the alignment, as does the memory use. For regular $A^\star$ the memory use is rather low, as markings are not explicitly stored (they are computed when needed by following the firing sequence back to the root). For incremental $A^\star$,

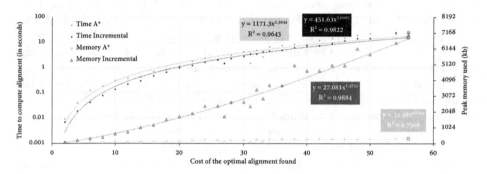

Fig. 3. Time and memory use of both approaches vs. the cost of the optimal alignment found for prCm6 of [20]. The trendlines show polynomial time and memory complexity.

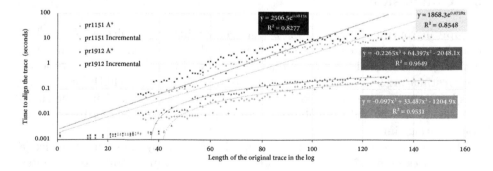

Fig. 4. Time to compute optimal alignments for the two most complex models in [21]. Trendlines show exponential time complexity for $A^\star$ and cubic complexity for the incremental approach.

the memory is dominated by the size of the linear program. For example, the maximum number of linear programs solved was 31 for model prCm6 with trace "instance_124" (highlighted). About 6 MB of memory is needed there, mainly to store the non-zero coefficients of a linear program of 11,688 rows and 10,262 columns in the solver. The total time to compute the optimal alignment was 22 s for incremental $A^\star$, of which 16 s was spent in the LP solver (for all 31 LPs together) and the remaining 6 s on investigating 7,575 markings. For the same trace, the classical $A^\star$ searched 9,792 markings and solved 8,706 LPs in 25 s.

In Fig. 4, we show the computation time per trace for the two most complex event logs in Table 1. For each trace, we plot the time needed to align it in the model using classic $A^\star$ (on a logarithmic axis) as well as our incremental version (again the planner is omitted as it does not report times per trace). These models contain considerably more parallelism which results in an exponential time trendline for $A^\star$. More interestingly though, the time complexity for the incremental version shows a cubic trend in both cases. This difference is explained by the parallelism in the model, which leads to considerable differences in the percentage of the time that is spent in the LP solver, as well as the number of solved LPs. The incremental version spends 72% of the computation time in the solver for pr1151 (70% for pr1912), vs. 99% (99%) for the classical one. Per trace however, the incremental version solves only 7.0 (4.9) LPs per trace rather than 953 (1612). The time per solve call is roughly 4.5 times higher for the incremental version as the linear programs are larger.

The main difference between the various datasets is the length of the traces compared to the size of the model. As our technique explicitly exploits the trace to identify split-points, it benefits from long traces in relatively small models. In [20], the models do not contain loops, so the traces are relatively short (up to 271 events) compared to the model size (up to 429 transitions). In the other datasets, the trace lengths are up to three times the number of transitions in the models.

6   Conclusion

Computing optimal alignments is a time consuming task which is essential in the context of conformance checking. Traditional algorithms for computing optimal alignments use the marking equation borrowed from Petri net theory as an underestimating heuristic function in the context of an $A^\star$ search. Unfortunately, this heuristic is proven to perform poorly in certain cases which cause the $A^\star$ to expand nearly the full reachability graph of the Petri net, leading to excessive computation times.

In this paper, we reconsider the heuristic function by exploiting knowledge of the traces being aligned. Essentially, we use the original trace to guarantee progress in the depth of the $A^\star$ search. We do this by splitting the marking equation into a number of sub-problems which together provide a more accurate under estimation of the remaining cost. Rather than starting the search with the fully split marking equation, we use $A^\star$ itself to decide when to split the marking equation and we show this leads to a considerable reduction in number of computed linear programs.

The work is implemented in ProM and we use publicly available benchmark datasets to compare our work to existing techniques, showing significant improvements in computation time. For future work, we develop techniques select appropriate splitpoints in advance.

References

1. van der Aalst, W.M.P.: Decomposing Petri nets for process mining: a generic approach. Distrib. Parallel Databases **31**(4), 471–507 (2013)
2. van der Aalst, W.M.P.: Process Mining - Data Science in Action, 2nd edn. Springer, Heidelberg (2016). https://doi.org/10.1007/978-3-662-49851-4
3. van der Aalst, W.M.P., Adriansyah, A., van Dongen, B.F.: Replaying history on process models for conformance checking and performance analysis. Wiley Interdiscip. Rev.: Data Min. Knowl. Discov. **2**(2), 182–192 (2012)
4. Adriansyah, A.: Aligning observed and modeled behavior. Ph.D. thesis, Department of Mathematics and Computer Science. Eindhoven University of Technology, July 2014
5. de Leoni, M., Marrella, A.: Aligning real process executions and prescriptive process models through automated planning. Expert Syst. Appl. **82**, 162–183 (2017)
6. Hart, P.E., Nilsson, N.J., Raphael, B.: A formal basis for the heuristic determination of minimum cost paths. IEEE Trans. Syst. Sci. Cybern. **4**(2), 100–107 (1968)
7. Leemans, S.J.J., Fahland, D., van der Aalst, W.M.P.: Scalable process discovery and conformance checking. Softw. Syst. Model. **17**(2), 599–631 (2018)
8. López, M.T.G., Borrego, D., Carmona, J., Gasca, R.M.: Computing alignments with constraint programming: the acyclic case. In: Proceedings of ATAED, Torun, Poland, CEUR Workshop Proceedings, vol. 1592, pp. 96–110. CEUR-WS.org (2016)
9. Măruşter, L., Weijters, A.J., van der Aalst, W.M., van den Bosch, A.: A rule-based approach for process discovery: dealing with noise and imbalance in process logs. Data Min. Knowl. Discov. **13**(1), 67–87 (2006)
10. Munoz-Gama, J., Carmona, J., van der Aalst, W.M.P.: Single-entry single-exit decomposed conformance checking. Inf. Syst. **46**, 102–122 (2014)
11. Murata, T.: Petri nets: properties, analysis and applications. Proc. IEEE **77**(4), 541–580 (1989)
12. Reißner, D., Conforti, R., Dumas, M., La Rosa, M., Armas-Cervantes, A.: Scalable conformance checking of business processes. In: Panetto, H., et al. (eds.) OTM 2017. LNCS, vol. 10573, pp. 607–627. Springer, Cham (2017). https://doi.org/10.1007/978-3-319-69462-7_38
13. Rozinat, A., van der Aalst, W.M.P.: Conformance checking of processes based on monitoring real behavior. Inf. Syst. **33**(1), 64–95 (2008)
14. Taymouri, F., Carmona, J.: A recursive paradigm for aligning observed behavior of large structured process models. In: La Rosa, M., Loos, P., Pastor, O. (eds.) BPM 2016. LNCS, vol. 9850, pp. 197–214. Springer, Cham (2016). https://doi.org/10.1007/978-3-319-45348-4_12
15. Taymouri, F., Carmona, J.: Model and event log reductions to boost the computation of alignments. In: Ceravolo, P., Guetl, C., Rinderle-Ma, S. (eds.) SIMPDA 2016. LNBIP, vol. 307, pp. 1–21. Springer, Cham (2018). https://doi.org/10.1007/978-3-319-74161-1_1

16. van Zelst, S.J., Bolt, A., van Dongen, B.F.: Tuning alingment computation: an experimental evaluation. In: Proceedings of ATAED, 25–30 June 2017, Zaragoza, Spain, pp. 1–15 (2017)
17. van Dongen, B.F.: BPI challenge dataset, 2012, in 4TU Center for Research Data. https://doi.org/10.4121/uuid:3926db30-f712-4394-aebc-75976070e91f
18. de Leoni, M., Mannhardt, F.: Road Fines dataset, 2015, in 4TU Center for Research Data. https://doi.org/10.4121/uuid:270fd440-1057-4fb9-89a9-b699b47990f5
19. Mannhardt, F.: Sepsis dataset, 2016, in 4TU Center for Research Data. https://doi.org/10.4121/uuid:915d2bfb-7e84-49ad-a286-dc35f063a460
20. Munoz-Gama, J.: Synthetic dataset, 2013, in 4TU Center for Research Data. https://doi.org/10.4121/uuid:44c32783-15d0-4dbd-af8a-78b97be3de49
21. Munoz-Gama, J.: Synthetic dataset, 2014, in 4TU Center for Research Data. https://doi.org/10.4121/uuid:b8c59ccb-6e14-4fab-976d-dd76707bcb8a

An Evolutionary Technique to Approximate Multiple Optimal Alignments

Farbod Taymouri and Josep Carmona[✉]

Universitat Politècnica de Catalunya, Barcelona, Spain
{taymouri,jcarmona}@cs.upc.edu

Abstract. The alignment of observed and modeled behavior is an essential aid for organizations, since it opens the door for root-cause analysis and enhancement of processes. The state-of-the-art technique for computing alignments has exponential time and space complexity, hindering its applicability for medium and large instances. Moreover, the fact that there may be multiple optimal alignments is perceived as a negative situation, while in reality it may provide a more comprehensive picture of the model's explanation of observed behavior, from which other techniques may benefit. This paper presents a novel evolutionary technique for approximating multiple optimal alignments. Remarkably, the memory footprint of the proposed technique is bounded, representing an unprecedented guarantee with respect to the state-of-the-art methods for the same task. The technique is implemented into a tool, and experiments on several benchmarks are provided.

1 Introduction

Current conformance checking techniques strongly rely on *alignments*: given an observed trace representing a process instance, to find the best model trace that resembles it [1]. This way, the best model explanation of the reality is reported, so that one sees the reality through the model' perspective. This opens the door for further techniques, including root-cause analysis, model enhancement and predictive monitoring.

Since the reality can be explained in many ways, costs need to be defined on the deviations so that certain explanations are rendered less interesting, coining the notion of *optimal alignment*. In spite of this, many different optimal explanations may exist for a given trace, a concept denoted *all-optimal alignments* in [1]. The derivation of more than one explanation may provide a better, more global, analysis: for instance, to estimate the *precision* of a process model in describing an event log, different metrics are defined in [2] when considering all or just one optimal alignment.

Due to the existence of concurrency and iteration, the behaviour of underlying process models can be exponential, a fact that hampers the application of the state-of-the-art technique for computing alignments, which is based on

© Springer Nature Switzerland AG 2018
M. Weske et al. (Eds.): BPM 2018, LNCS 11080, pp. 215–232, 2018.
https://doi.org/10.1007/978-3-319-98648-7_13

exploring the model state space using A^* search [1]. The situation becomes significantly worse in case multiple optimal alignments need to be computed, since none of the heuristics to speed-up the search is applicable, and therefore the full exploration of the model' search space is then unavoidable. It is well-known (e.g., [3]) that the memory requirements of the A^*-based alignment technique are the most limiting factor to apply it on the large.

In this paper we propose an evolutionary technique to approximate multiple optimal alignments. We trade-off computation time for memory, i.e., assume that in some contexts, it is acceptable to spend more time in the computation, provided that the memory footprint is guaranteed to not exceed a given bound. To accomplish this, we encode the computation of alignments as a *genetic algorithm* (GA), where tailored *crossover* and *mutation* operators are applied to an initial population of candidate model explanations. This way, the derivation of a set of alignments is the result of genetic evolution.

The technique proposed has some weakness that should be reported: first, it can only provide optimal alignments when certain conditions are satisfied (variability in the population and genetic convergence). In practice, however, the number of iterations may be decided a priori, which may be insufficient for genetic convergence, and the initial population may not contribute to reach optimal solutions. Second, the number of optimal alignments obtained is in practice inferior to the real number of all optimal alignments, due to the dependence to the initial population and genetic convergence. Hence, the proposed technique only approximates several optimal alignments.

In spite of the approximation nature of the technique proposed, we still see a clear value for several reasons: first, to obtain more than one model explanation of an observed trace may open the door to apply a posteriori root-cause analysis to identify the most likely explanation, as has been described in [4]. Second, the technique proposed represents the first algorithmic alternative to search for multiple optimal alignments, which can be applied on large instances under bounded memory. In the same way as GA provided an interesting perspective for process discovery [5–7], this work contributes to open a research direction for computing alignments on the large. Third, in contrast to the A^*-based alignment technique, our technique is non-deterministic in providing alignments, so that two runs of the method may obtain different result. This may be very useful in *multi-perspective alignments* [8]: since control-flow is aligned before other perspectives, randomness in the generation of the control-flow alignment will enable the exploration of a broader solution space in the rest of perspectives.

The paper is organized as follows: in Sect. 2 we provide related work. In Sect. 3, the necessary ingredients to understand the contents of this paper are presented. Then in Sect. 4 we describe the encoding as a GA of the problem of searching several best model explanations. The general framework for approximating multiple optimal alignments is described in Sect. 5. Tool support and experiments with various benchmarks are reported in Sect. 6. Finally, conclusions and pointers to future work are reported in Sect. 7.

2   Related Work

The work in [1] proposed the notion of alignment, and developed a technique to compute optimal alignments for a particular class of process models. For each trace σ, the approach consists on exploring the synchronous product of model's state space and σ. In the exploration, the shortest path is computed using the A^* algorithm, once costs for model and log moves are defined. The approach represents the state-of-the-art technique for computing alignments, and can be adapted (at the expense of increasing significantly the memory footprint) to provide all optimal alignments.

Alternatives to the A^* have appeared very recently: in the approach presented in [9], the alignment problem is mapped as an *automated planning* instance. Unlike the A^*, the aforementioned work is only able to produce one optimal alignment (not all optimal), but it is expected to consume considerably less memory. Automata-based techniques have also appeared [10, 11]. In particular, the technique in [10] can compute all optimal alignments. The technique in [10] relies on state space exploration and determinization of automata, whilst the technique in [11] is based on computing several subsets of activities and projecting the alignment instances accordingly.

The work in [12], presented the notion of *approximate* alignment to alleviate the computational demands of the current challenge by proposing a recursive paradigm on the basis of structural theory of Petri nets. In spite of resource efficiency, the solution is not guaranteed to be executable. A follow-up work of [12] is presented in [13], which proposes a trade-off between complexity and optimality of solutions, and guarantees executable properties of results. The technique in [3], presents a framework to reduce a process model and the event log accordingly, with the goal to alleviate the computation of alignments. The obtained alignment, which is called *macro-alignment* since some of the positions are high-level elements, is expanded based on the gathered information during the initial reduction. Decompositional techniques have been presented [14], [15] that instead of computing optimal alignments, they focus on the *decisional problem* of whereas a given trace fits or not a process model.

3   Preliminaries

3.1   Petri Nets, Event Logs and Parikh Vector Representations of Traces

A *Petri Net* [16] is a 3-tuple $N = \langle P, T, \mathcal{F} \rangle$, where P is the set of places, T is the set of transitions, $P \cap T = \emptyset$, $\mathcal{F} : (P \times T) \cup (T \times P) \rightarrow \{0, 1\}$ is the flow relation. A *labeled Petri net (LPN)* is a 3-tuple $\langle N, \Sigma, \ell \rangle$, where N is a Petri net, Σ is an alphabet (a set of labels) and $\ell : T \rightarrow \Sigma \cup \{\tau\}$ is a *labeling function* that assigns to each transition $t \in T$ either a symbol from Σ or the empty symbol τ.

Given an alphabet of events $\Sigma = \{a_1, \ldots, a_n\}$, a trace is a word $\sigma \in \Sigma^*$ that represents a finite sequence of events. An *event log* $L \in \mathcal{B}(\Sigma^*)$ is a multiset of traces[1]. $|\sigma|_a$ represents the number of occurrences of a in σ. The *Parikh vector* of a sequence of events σ is a function $\hat{\ }: \Sigma^* \to \mathbb{N}^n$ defined as $\hat{\sigma} = (|\sigma|_{a_1}, \ldots, |\sigma|_{a_n})$.

Workflow processes can be represented in a simple way by using Workflow Nets (WF-nets). A WF-net is a Petri net where there is a place *start* (denoting the initial state of the system) with no incoming arcs and a place *end* (denoting the final state of the system) with no outgoing arcs, and every other node is within a path between *start* and *end*. For the sake of simplicity, in this paper we assume WF-nets.

3.2   Alignment of Observed Behavior

The notion of aligning event log and process model was introduced by [1]. To achieve an alignment between a process model and a observed trace, we need to relate *moves* in the trace to *moves* in the model. When some

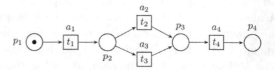

Fig. 1. Process model M_1.

of the moves in the observed trace can not be mimicked by the model or vice versa, there is an *asynchronous move*. When both model and trace agree in the performed label a *synchronous move* arises. For instance, consider the model M_1 in Fig. 1, with the following labels, $\ell(t_1) = a_1, \ell(t_2) = a_2, \ell(t_3) = a_3$ and $\ell(t_4) = a_4$, and trace $\sigma = a_1 a_1 a_4 a_2$; two possible alignments between M_1 and σ are:

$$\alpha_1 = \frac{\left| a_1 \right| a_1 \left| \perp \right| a_4 \left| a_2 \right|}{\left| t_1 \right| \perp \left| t_3 \right| t_4 \left| \perp \right|} \qquad \alpha_2 = \frac{\left| a_1 \right| a_1 \left| \perp \right| a_4 \left| a_2 \right|}{\left| \perp \right| t_1 \left| t_2 \right| t_4 \left| \perp \right|}$$

The moves are represented in tabular form, where moves in the observed trace are at the top and moves by model are at the bottom. For example the first move in α_2 is asynchronous: $(a_1, \perp)$, and it means that the observed trace performs a_1 while the model does not make any move, i.e., a_1 is an *inserted* transition. In contrast, the fourth move in α_2, (a_4, t_4), is a synchronous move. Cost can be associated to alignments, with asynchronous moves having greater cost than synchronous ones [1]. Given assigned cost values, an alignment with optimal cost is preferred. Alignments open the door to compute metrics, report diagnosis, enhance the model, among others.

4   GA for Computing Several Explanations of Observed Behavior

GA starts by creating an initial population, and then combining the best solutions through operators, to create a new generation of solutions which should be

---

[1] $\mathcal{B}(\Sigma)$ denotes the set of all multisets of the set Σ.

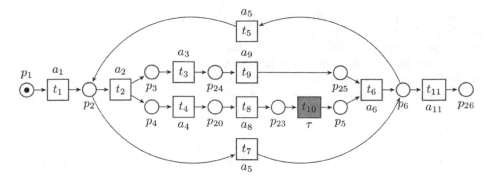

Fig. 2. Process model M_2.

better than the previous generation. As it will be noticed bellow, in some cases the evaluation of solutions will be adapted depending on the operator applied, so that the search for solutions can be better guided. A GA approach to a problem usually starts by encoding a solution which is called a *chromosome*, and define functions to evaluate how good it is. Next, generating the initial population of chromosomes and defining corresponding operators, i.e., *crossover* and *mutation*. In our setting, chromosomes will be potential model traces, which are combined through tailored crossover and mutation operators.

Given an observed trace σ and WF-net N, a random population of chromosomes is first generated (Sect. 4.1). Then, it evaluates each chromosome based on a specific fitness function[2], which considers both the initial model (for measuring replayability), and the observed trace (for measuring similarity) (see Sect. 4.2). It then applies traditional crossover and mutation operators, as well as novel ones defined for this problem, to speed up the process of evolving chromosomes and convergence (see Sect. 4.3). This process continues until reaching satisfactory results, or will be stopped by a predefined number of iterations. The detailed descriptions will be presented in the next sections.

4.1   Generation of the Initial Population

Given an observed trace σ, and WF-net N, the objective of this part is to generate an initial population. The population size is an important decision, which often affects the final solution in terms of accuracy and convergence [17]. Also, diversity in the population will help reaching different parts of the solution space. We rely on previous work for obtaining different model explanations [3,12]: these methods are based on finding a maximal set of transitions that the model can reproduce to mimic the observed trace and deriving a Parikh vector thereof. More in detail, the *marking equation* of Petri nets is used to solve an Integer Linear Programming (ILP) model for obtaining the corresponding Parikh vector.

---

[2] As the reader will soon realize, we refer to the term fitness in the genetic algorithms context.

The ILP model has some additional constraints and a tailored cost function, that jointly guide the search for a maximal set with respect to its similarity (in terms of support) to the observed trace σ. To obtain the traces from the Parikh vector, we perform linearizations of the Parikh vector, obtained by either replaying it in the Petri net, or by arbitrary (possibly non-replayable) linearizations that do not consider the Petri net. Those are the seeds for generating new chromosomes.

For example consider the model in Fig. 2, and the observed trace $\sigma = a_1a_3a_8a_4a_2a_9a_5a_6$. Some chromosomes with respect to σ could be $\chi_1 = t_1t_7t_{11}$, $\chi_2 = t_1t_2t_3t_9t_4t_8t_{10}t_6t_{11}$, $\chi_3 = t_2t_1t_3t_8t_4t_9t_{10}t_6t_{11}$ and $\chi_4 = t_1t_7t_5t_{11}$. It is worth mentioning that some chromosomes may not be replayable at this stage (e.g., χ_4 above).

4.2   Evaluation Criteria

In GA's jargon a fitness function is a particular type of objective function that prescribes the optimality of a solution (that is, a chromosome) in the corresponding population. Elevated chromosomes, which are the best ones at the corresponding time are allowed to breed and mix their datasets by any of several techniques, producing a new generation that will (hopefully) be even better. An ideal fitness function correlates closely with the algorithm's goal, and yet may be computed quickly. Speed of execution is very important, as a typical GA must be iterated many times in order to produce a usable result for a non-trivial problem.

In this paper a chromosome χ is evaluated based on two metrics. The fitness value of a chromosome χ is summed over the following terms:

$$f^t(\chi) = \lambda_1 \cdot f^m(\chi) + \lambda_2 \cdot f^{ed}(\chi) \tag{1}$$

where $f^m(\chi)$ denotes the ratio of *missed tokens*[3] to the total tokens while χ is being replayed in the model, and $f^{ed}(\chi)$ shows the normalized *Edit Distance* between χ, and observed trace σ. Both λ_1 and λ_2 denote the penalization terms which will be adjusted individually for each genetic operator, as will be discussed in the next sections. It is clear that the lower value of $f^t(\chi)$ represents a better chromosome, i.e., modeled trace, by which the observed trace is mimicked. It should be pointed out that always having a chromosome χ with small $f^t(\chi)$ does not represent a desired or good solution if it is not replayable (i.e., $f^m(\chi) \neq 0$).

To get the idea of evaluation criteria consider chromosome $\chi = t_1t_2t_9t_3t_8$ $t_4t_{10}t_{11}t_6$, the model in Fig. 2 and observed trace $\sigma = a_2a_1a_3a_9a_8a_4a_{26}a_6$; the number of missed and total tokens while χ is replayed equals to 3 and 23 respectively, thus $f^m(\chi) = \frac{3}{23}$. Additionally, unreplayable transitions t_9, t_8 and t_{11}, are likely to be considered through genetic operators, in the next step of the proposed approach. Also, the corresponding edit distance, i.e., $f^{ed}(\chi)$, between χ ans σ[4] is 5. Thereby by selecting $\lambda_1, \lambda_2 = 1$, the corresponding fitness value is $f^t(\chi) = 1 * \frac{3}{23} + 1 * 5 \approx 5.130$.

---

[3] In Petri net terms, missed tokens represent tokens that hamper the firing of a transition.

[4] Note that indeed the edit distance is computed between σ and $\ell(\chi)$.

4.3   Genetic Operators

Genetic operators used in GA are analogous to those which occur in the natural world: survival of the fittest, or selection; reproduction (crossover); and mutation. When GA proceeds, both the search direction to optimal solution and the search speed should be considered as important factors, in order to keep a balance between exploration and exploitation in search space. In general, the exploitation of the accumulated information resulting from GA search is done by the selection mechanism, while the exploration to new regions of the search space is accounted for by genetic operators. In the remainder of this section, several genetic operators will be proposed. Some of them are inspired from ones found in analogous problems, whilst new ones are proposed that tend to improve the evaluation criteria described in the previous section.

Crossover Operators. Crossover is the main genetic operator. It operates on two chromosomes at a time and generates two new chromosomes by combining both chromosomes' features. A standard way to achieve crossover is to choose a random segment at both chromosomes, and generate two new chromosomes as the result of interchanging the two segments among the original two chromosomes. We apply an adaptation of this standard crossover operator, denoted *Modified Partially-Mapped Crossover (MPMX)*, for chromosomes having the same Parikh vector representation (see Sect. 3.1)[5]. The intuitive idea for operating over chromosomes with identical Parikh vector is due to the fact that the search space is reduced, and in particular the generation of the initial population is oriented towards satisfying this property. In order to keep Parikh vector representation of the original chromosomes, some modifications are done after the segments are interchanged. Let us look at the example in Fig. 3 to illustrate the MPMX operator; the initial chromosomes χ_1 and χ_2 are mixed with this operator, generating the new chromosomes χ_3 and χ_4, choosing a segment between positions 4 and 6.

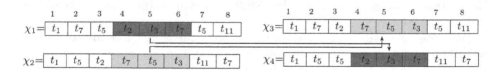

Fig. 3. The MPMX operator.

To keep the Parikh vector representation of the original chromosomes, some modifications are performed circularly starting from the first position after the segment (in the example, position 7). For instance, in χ_3 (that arised from χ_2

---

[5] The restriction on having the same Parikh vector is for the sake of simplicity of application.

inserting the segment from χ_1), in the third position t_5 is removed since $|\chi_1|_{t_5} = 2$ and when we reach this position we already have 2 occurrences of t_5.

The next crossover operator, denoted *Cross-Insert Crossover (CIX)*, tries to guide the search towards chromosomes that are replayable in the model. Still, the CIX operator works under the assumption of both initial chromosomes have the same Parikh vector representation. To induce replayability, it focuses on the parts of a chromosome that are not replayable, and uses the other chromosome in order to find candidate positions where it may be possible to reply the set of unreplayable transitions. This is done for each unreplayable transition in each one of the chromosomes that are merged. For each candidate position, the transition is moved to that position and the chromosome is shifted accordingly to fill the space left. For instance, let us look at the two chromosomes χ_1 and χ_2 in Fig. 4, and model M_2.

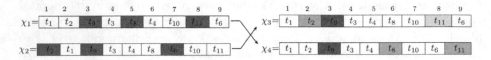

Fig. 4. The CIX operator: in the figure, red background means unreplayable positions of the trace, while green denotes positions in the new chromosomes where unreplayable transitions have been fixed. Yellow denotes transitions that, in spite of being initially replayable, due to other moves, they became unreplayable. (Color figure online)

In χ_1 the transitions in the third, fifth and the eighth position cannot be replayed, namely t_9, t_8 and t_{11}. Transition t_9 cannot be moved, since in both chromosomes it is unreplayable (so there is no candidate position in this case). However, for transition t_8 in χ_1 (which is at position 5) there is a candidate position (position 6, extracted from χ_2) to move. Moving t_8 to position 6 and shifting once from position 6 will leave the space in position 6 to put t_8 in χ_4. A similar situation happens with t_{11} position from χ_1 to the new position in χ_4. Notice that, as denoted in χ_3 in yellow, shifting may introduce new unreplayable transitions: see t_{11}.

We stress that, since this operator tries to generate more executable offspring regardless of the corresponding edit distance, in our experiments we assign small values to λ_2 of Eq. 1, to retain the new generated chromosomes in next generations even if the edit distance has been degraded at the expense of improving replayability.

Mutation Operators. Mutation applies to a single chromosome, generating a new chromosome as a modification of the initial one. It is viewed as a background operator to maintain genetic diversity in the population. Mutation helps escaping from local minima's trap and maintains diversity in the population. This part presents both generic and specific mutation operators related to the problem considered in this paper.

As with the crossover operators, we start by adapting a generic one. The *Scramble Mutation (SM)* operator simply chooses a segment in the chromosome, and randomly shuffles it. For instance, in Fig. 5 we show how the operator works.

Fig. 5. Scramble mutation operator (SM).

In contrast, the *Mimic Mutation (MM)* operator is a specific operator proposed exclusively for the problem at hand. It tries to mimic the observed trace, by repositioning a transition t as close as possible into the position that $\ell(t)$ was observed in σ. Hence, this operator tends to reduce the edit distance to σ for the mutated chromosome. To implement this idea, we need to reflect on the observation that, due to the Central Limit Theorem, in the limit the position of labels in observed traces follows a Normal distribution (with the corresponding parameters)[6]. Since we are assuming a considerable amount of observed traces in the event log, we can compute these distributions for each one of the distinct labels, using statistical methods like *Regression Splines* [18][7]. This is done only once, before of applying the genetic algorithm.

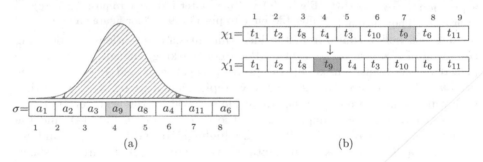

Fig. 6. (a) A probability distribution of locations of a_9 in σ, (b) Mimic mutation.

Figure 6 shows an example for the observed trace (Fig. 6(a)). In there, the probability of the position of a_9 follows a Normal distribution $N(4, 4)$. It implies that in practice a_9 should be in positions near by position 4.

---

[6] In case we have a limited number of observations, the real distribution can be estimated by traditional methods, like Kernel Smoothing [18].

[7] In case of duplicate labels in the traces, the normality assumption may be violated and therefore the estimation may be less accurate. In spite of this, the distribution used is only an oracle for generating new locations and does not limit the applicability or our approach.

Once the density function of a certain label is known, a random number following the function will be generated (in the previous example, it should be around position 4). Once this is obtained, the current position of the event in the chromosome and the new position are swapped. This explains the mutation from χ_1 to χ'_1 in Fig. 6(b).

For a chromosome χ, and its offspring χ', since the goal is to mimic the observed trace, if $f^m(\chi) < f^m(\chi')$ and $f^{ed}(\chi) > f^{ed}(\chi')$ then λ_1 and λ_2 in Eq. 1 are adjusted so that sometimes χ' survives in the next iteration. In other words replayability is overshadowed for this operator. Playing with these parameters would decrease the risk of getting stuck in a local optimum.

Launch Mutation (LM). Up to this point, none of the mutation operators above try to improve the replayable property of chromosomes. The intuitive idea is, for a given chromosome with an unreplayable position i, the transition in i will be relocated forward (i.e., in a position $j > i$) by this operator. The rationale

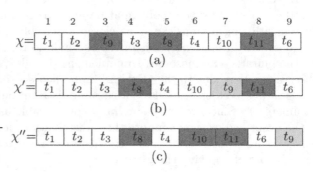

Fig. 7. (a) χ, (b) χ' after LM on χ to position 7, (c) χ'' after LM on χ to position 9. (Color figure online)

behind this policy comes from the idea that in some situations, by delaying the firing of a transition to a future Petri net marking, enough tokens will be placed by the transitions occupying positions between i and $j - 1$. Since the overall goal of this operator is to improve replayability, we set $\lambda_1 < \lambda_2$ so that replayability has more importance to decide survival for the next iteration.

To give a concrete example consider the chromosome in Fig. 7(a), and the model M_2. Unreplayable transitions are highlighted (positions 3, 5 and 8). Assume that t_9 is selected to be mutated with is operator. Figure 7(b) shows one possible launch mutation from position 3 to position 7. One can see that transition t_9 can now be replayable, as highlighted in green. Unfortunately, this operator can sometimes introduce new unreplayable transitions, as demonstrated in Fig. 7(c) for t_{10}.

5   General Framework for Obtaining Multiple Alignments

Given a process model represented as a WF-net, N, and a trace σ, the schema of the proposed framework is depicted in Fig. 8. Explanations of each part are provided below:

– *Genetic Algorithm Framework:* In the initial stage, the genetic approach described in the previous section is performed. Once finished it generates a

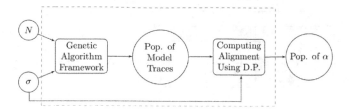

Fig. 8. Overall description of the general approach to compute alignments.

final population of model traces. Among them, we choose those chromosomes χ having both $f^m(\chi) = 0$ (so, replayable), and minimal $f^{ed}(\chi)$.

- *Computing Alignment Using Dynamic Programming:* This part concerns the computation of alignments between the chromosomes of final population and σ. The adopted method in this section is a dynamic programming approach inspired from aligning two sequence of genes [19,20]. The alignments computed are called *best alignments*, which are not necessarily optimal: this is due to the lack of guarantees that the model explanations provided in the previous stage correspond to the optimal model explanation for σ.

5.1   Computing an Alignment Using Dynamic Programming

To compute an alignment between a chromosome like χ and observed trace σ, the technique presented in this paper is inspired from [20]. This technique was already applied in [3] for the same task, so we informally describe it here. Consider an oversimplified example, $\chi = t_3 t_{11} t_{17}$ with $\ell(\chi) = a_3 a_{11} a_{17}$ and $\sigma = a_3 a_{11}$. To obtain an alignment α between these two sequences, a two-dimensional table is created, where the first row and first column are filled with the observed trace and chromosome, respectively, as depicted in Fig. 9(a). The second row and second column are initialized with numbers starting from $0, -1, -2, ...$, they are depicted in yellow color. The task then is to fill the remaining cells with the recurrence Eq. (2), in which δ represents the *gap penalty*[8] and $s(t_i, a_j)$ represents both the match and mismatch cost between two elements t_i and a_j which are modeled and observed trace elements, respectively.

$$SIM(t_i, a_j) = MAX \begin{cases} SIM(t_{i-1}, a_{j-1}) + s(t_i, a_j) \\ SIM(t_{i-1}, a_j) - \delta \\ SIM(t_i, a_{j-1}) - \delta \end{cases} \qquad s(t_i, a_j) = \begin{cases} \beta & \text{If } \ell(t_i) = a_j \\ -\beta & \text{If } \ell(t_i) \neq a_j \end{cases}$$

$$(2)$$

$SIM(t_i, a_j)$ represents the similarity score between t_i and a_j.

After filling the matrix, to compute the alignment we start from the bottom right entry, and compare the value with three possible sources, i.e., top, left and diagonal to identify from which one of them it came from. If it was fed by a diagonal entry, it represents a synchronous move between corresponding elements and if it was fed by a top or left entries then it represents an asynchronous move or a gap. For the mentioned chromosome and observed trace the computed alignment is shown in Fig. 9(b).

---

[8] The gap penalty represents asynchronous move in our setting.

		a_3	a_{11}
	0	-1	-2
a_3	-1	0	-1
a_{11}	-2	-1	0
a_{17}	-3	-2	-1

(a)

$$\alpha = \begin{array}{|c|c|c|} a_3 & a_{11} & \perp \\ \hline a_3 & a_{11} & a_{17} \end{array}$$

(b)

Fig. 9. (a) Computing alignment using dynamic programming (b) Obtained alignment. (Color figure online)

6   Experiments

The approach of this paper has been incorporated into the tool ALI [21]. This section evaluates the method proposed over the following perspectives: What is its sensitivity on the number of evolutionary iterations? How does it compares to [1,10] for the memory and execution time? How does it compares to [1,10] for the quality and quantity of alignments obtained? What is the impact on the fitness calculation?

The tool has been evaluated over different family of examples from artificial to realistic, containing transitions with duplicate labels and from well-structured to completely Spaghetti[9]. The number of transitions varies between models, i.e., minimum 15 and maximum 429. We also included a real-life benchmark from [9], where a model was discovered using the Inductive Miner by sampling 10000 observed traces of a *Road Traffic* process dataset[10], and using the rest of the log for alignment computation. Also, to examine the proposed approach in dealing with models that contain duplicate labels, a set of models $ML_1, \ldots, ML_5$ consisting of duplicate transitions, with the corresponding logs obtained by injecting different degree of noise (25%, 35%, 50% and 75%), were generated by PLG2 [22][11]. The specification of benchmark datasets can be found in [3], [14], [12], [23]. The results on these benchmarks are compared with the state-of-the-art technique for computing one optimal alignments [1], since the version for computing all optimal alignments ran out of memory for all models considered in this paper. We also compare ALI with a recent technique to compute all-optimal alignments [10].

Configuration of the Genetic Algorithm. Since the decision of the initial population and the probabilities of operators strongly influences the genetic algorithm, we have chosen to customize it so that diversity is kept in early stages of the genetic evolution. It should be stressed that in the general application of genetic algorithms, populations are significantly larger than the ones considered for this paper, so that the probabilities for operators have been set accordingly, to avoid that few high-ranked chromosomes dominate the rest in early stages of the genetic evolution. In the implementation, the application of crossover had a high probability (80%, and then tuned with the individual probabilities set by parameters λ_1 and λ_2), whilst mutation operators were

---

[9] The experiments have been done on Intel Core i7-2.20 GHz computer with 8 GB of RAM.

[10] https://data.4tu.nl/repository/uuid:270fd440-1057-4fb9-89a9-b699b47990f5.

[11] At the time of generating models for the experiments, PLG2 in fact was unable to produce models containing duplicate labels from scratch, therefore the generated models and logs were modified in order to have transitions with duplicate labels.

applied with a low probability (5%). In any case, the chosen probabilities are common when applying genetic algorithms in other scenarios. As we commented in the previous section, best alignments in our setting correspond to replayable chromosomes with minimal edit distance to the observed trace among the final population. For each observed trace a population of 700 chromosomes was generated, and the quality and quantity of them were compared with state of the art approach at iterations 10,20,30 and 100. We also experimented with population sizes significantly smaller (e.g., 100 chromosomes), and the results obtained were proportional in the main perspectives considered in this paper: less memory footprint and execution time, but slightly worst quality.

Execution Times. Figure 10 shows violin plots of execution time (in seconds) for each model per iteration given an observed trace. Obviously the required execution time varies from different observed traces and this is why the corresponding distributions via violin plots are presented. One can see that for big models with large traces (*prDm6, prEm6, prFm6*), models with many deviations in observed traces (*prCm6*) and models with many duplicate transitions (*ML5*), the corresponding distributions are wider due to more operations made by the proposed operators at any iteration. An important point should be done: although the computation time per trace (corresponding to multiplying the execution time per iteration shown in the plot by the number of iterations performed) is significantly higher with respect to [1], our evaluation is done with a simple, unoptimized implementation of the technique of this paper.

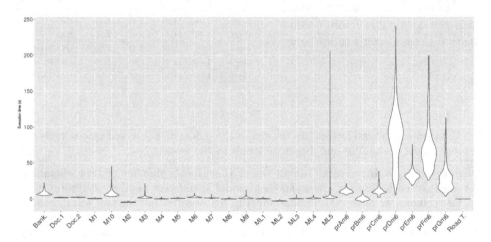

Fig. 10. Distribution of execution time for an observed trace per iteration.

Fitness Comparison. Table 1 represents the *mean square error* (MSE) of fitness values based on metric presented in [1,12], between the best alignments provided by our technique and the approach in [1] as optimal solutions, respectively. One can see that the quality of alignments is improved from 10 iterations to 100 iterations for all models. For models *prAm6*, *prBm6* and *prEm6* optimal alignments were found (for some of them, at iterations 2 and 3 respectively). Comparisons were done only for those benchmark datasets whenever the approach in [1] could provide solutions. Overall, one can see that the approach of this paper is very close to the optimal solutions computed

Table 1. MSE comparison of fitness values of chromosomes at different iterations and A^* as the optimal one

Model	It.(10)	It.(20)	It.(30)	It.(100)
prAm6	0.0106	0.0055	0.0035	0
prBm6	0.0009	0	0	0
prCm6	0.0776	0.0044	0.0031	0.0012
prEm6	0.0436	0.0353	0	0
M_1	0.0571	0.0089	0.0084	0.0046
M_2	0.0475	0.0116	0.0094	0.0070
M_3	0.0351	0.0341	0.0327	0.0219
M_4	0.1980	0.0512	0.0508	0.0398
M_5	0.0958	0.0289	0.0226	0.0155
M_8	0.0836	0.0384	0.0379	0.0357
M_9	0.0725	0.0220	0.0214	0.0203
ML_1	0.0775	0.0400	0.0146	0.0091
ML_3	0.1864	0.0991	0.0159	0.0142
ML_4	0.3113	0.1740	0.0330	0.0251
Bank	0.0013	0.0008	0.0005	0.0002
Doc. 1	0.2912	0.1971	0.0702	0.0698
Doc. 2	0.2581	0.1701	0.0579	0.0570
Road T	0.0036	0.0022	0.0018	0.0014

Table 2. Number of difference alignments for the best solution found in average

Model	It.(10)	It.(20)	It.(30)	It.(100)	ATM. A_*
prAm6	1.11	1.21	1.25	1.45	NA
prBm6	1.00	1.15	1.15	1.34	NA
prCm6	1.16	1.52	1.79	3.46	NA
prDm6	1.11	1.33	1.57	1.71	NA
prEm6	1.16	1.43	1.52	1.56	NA
prFm6	1.03	1.17	1.36	1.46	NA
prGm6	1.08	1.30	1.49	1.78	NA
M_1	1.94	2.91	3.32	4.32	62.12 (92%)
M_2	2.98	4.97	5.89	7.13	320.1 (53%)
M_3	1.30	1.98	2.41	2.79	NA
M_4	1.00	1.01	1.21	1.62	7.40 (39%)
M_5	1.77	2.62	3.44	6.01	114.78 (10%)
M_6	1.68	2.34	2.87	4.37	NA
M_7	2.05	3.38	4.27	7.12	NA
M_8	1.36	1.55	1.71	2.14	7.81 (69%)
M_9	1.01	1.02	1.31	1.46	8.32 (30%)
M_{10}	1.02	2.56	3.54	5.23	NA
ML_1	1.04	1.15	1.27	1.29	11.78 (34%)
ML_2	1.85	2.49	3.38	4.85	NA
ML_3	1.75	2.71	3.25	3.60	7.94 (21%)
ML_4	1.72	2.98	3.39	5.80	NA
ML_5	1.05	1.84	2.42	3.42	NA
Bank	1.08	1.44	1.83	2.66	NA
Doc1	1.00	1.08	2.21	2.70	I/O Error
Doc2	1.00	1.01	1.68	1.98	I/O Error
Road T	1.00	1.01	1.03	1.21	1.41 (100%)

by [1], in spite of several factors like the size of the model and observed traces, presence of loops, silent transitions and duplicate labels in the model.

Quantity of Best Alignments. Table 2 shows the average number of best alignment obtained per each observed trace and each model at different number of iterations. In the last column, we report this number for [10]: NA denotes that the tool was unable to provide the result due to memory problems. When it can, we also provide in parenthesis the percentage of the log traces where [10] can find solutions; for instance, for M_4, only 39% of the traces have a solution. One sees that these average numbers are improved from 10 to 100 iterations and this improvements are usually more tangible in models containing loops, i.e., M_1, M_2, M_3, M_7. The approach from [10] usually obtains more alignments than our method, but that only holds for small or medium instances.

Also, Figs. 11 and 12 show for each model, the violin plots or distribution of number of best alignment for 30 and 100 iterations, respectively (the corresponding average values are shown in the fourth and fifth column of Table 2, respectively). When focusing in the experiment for 100 iterations (Fig. 12), It can be seen from the plot that, for some models like M_1, M_2 M_5, M_7 and ML_2, the number of distinct best solutions are close to 30 for some cases and for *Documentflow2* the best solutions are unique.

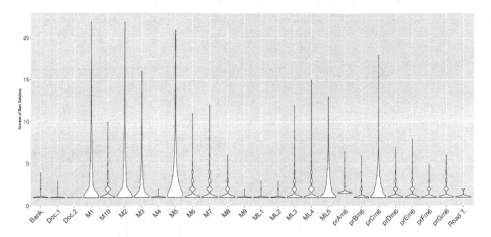

Fig. 11. Distribution of number of best solutions for 30 iterations.

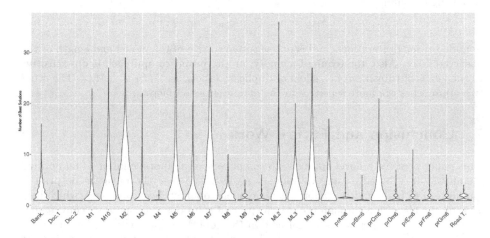

Fig. 12. Distribution of number of best solutions for 100 iterations.

Memory Consumption. The memory footprint of the proposed technique and the ones in [1,10], for all the benchmarks of this paper are represented in Fig. 13, using black, gray and brown colors, respectively. It must be stressed that the comparison reported in Fig. 13 provides just an indication of the huge difference in terms of memory footprint between the technique of this paper and the other techniques: for [1], experiments were only done for computing one optimal alignment inevitably, since the implementation for all optimal alignments ran out of memory. In contrast, in Fig. 13 we provide the results of our technique and the technique in [10] to compute multiple alignments.

One can see that the proposed technique requires considerable less memory than the other two techniques. Obviously for small and medium models, the memory footprints are similar. For large models the tendency is inversed: as an example, for *prDm6* the proposed method required around 1.5 GB whereas [1,10] need more than 5.5 GB Notice that the memory footprint of the proposed approach for computing best alignments is

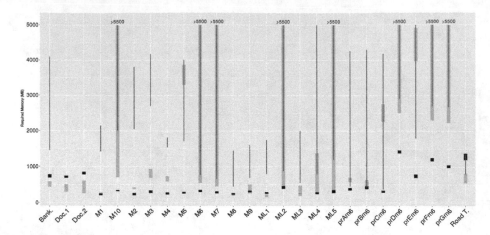

Fig. 13. Mem. footprint of our approach (black), [1] (gray), and [10] (brown, thin line). (Color figure online)

bounded through iterations, and is not sensitive to size of the model and length of the observed trace. Also, the required memory for the proposed approach is not sensitive to the labels of transitions i.e., silent or duplicate labels, see $ML_1, \ldots, ML_5$. The other two approaches are more sensitive to the aforementioned factors.

7   Conclusion and Future Work

This paper presents a novel approach to compute several approximation of an optimal alignment. It is based on an evolutionary algorithm, where the memory footprint is guaranteed to be bounded. Tailored genetic operators have been proposed, which help guiding the algorithm through the search space of solutions, and speed up convergence accordingly. The experiments performed on the tool developed witness the quality of obtained alignments, deriving solutions that are close to optimal ones, and which can be improved iteratively. Moreover, the quantity of alignments improves considerably as more genetic iterations are performed. In spite of not having theoretical guarantees on optimality or replayability, the results show that in practice it is always the case that replayable, quasi-optimal or optimal solutions are produced. For the future work there are many possibilities to explore, like introducing more efficient operators, exploring the parameter setting of the genetic algorithm to boost its application, or devising mechanism to alleviate the mentioned drawbacks.

Acknowledgments. We would like to thank B. van Dongen for interesting discussions. This work has been supported by MINECO and FEDER funds under grant TIN2017-86727-C2-1-R.

References

1. Adriansyah, A.: Aligning observed and modeled behavior. Ph.D. thesis, Technische Universiteit Eindhoven (2014)
2. Adriansyah, A., Munoz-Gama, J., Carmona, J., van Dongen, B.F., van der Aalst, W.M.P.: Measuring precision of modeled behavior. Inf. Syst. E-Bus. Man. **13**(1), 37–67 (2015)
3. Taymouri, F., Carmona, J.: Model and event log reductions to boost the computation of alignments. In: Ceravolo, P., Guetl, C., Rinderle-Ma, S. (eds.) SIMPDA 2016. LNBIP, vol. 307, pp. 1–21. Springer, Cham (2018). https://doi.org/10.1007/978-3-319-74161-1_1
4. Koorneef, M., Solti, A., Leopold, H., Reijers, H.A.: Automatic root cause identification using most probable alignments. In: Teniente, E., Weidlich, M. (eds.) BPM 2017. LNBIP, vol. 308, pp. 204–215. Springer, Cham (2018). https://doi.org/10.1007/978-3-319-74030-0_15
5. van der Aalst, W.M.P., de Medeiros, A.K.A., Weijters, A.J.M.M.: Genetic process mining. In: Ciardo, G., Darondeau, P. (eds.) ICATPN 2005. LNCS, vol. 3536, pp. 48–69. Springer, Heidelberg (2005). https://doi.org/10.1007/11494744_5
6. Buijs, J.C.A.M., van Dongen, B.F., van der Aalst, W.M.P.: A genetic algorithm for discovering process trees. In: Proceedings of the IEEE Congress on Evolutionary Computation, CEC 2012, Brisbane, Australia, 10–15 June 2012, pp. 1–8 (2012)
7. Vázquez-Barreiros, B., Mucientes, M., Lama, M.: ProDiGen: mining complete, precise and minimal structure process models with a genetic algorithm. Inf. Sci. **294**, 315–333 (2015)
8. Mannhardt, F., de Leoni, M., Reijers, H.A., van der Aalst, W.M.P.: Balanced multi-perspective checking of process conformance. Computing **98**(4), 407–437 (2016)
9. de Leoni, M., Marrella, A.: Aligning real process executions and prescriptive process models through automated planning. Expert Syst. Appl. **82**, 162–183 (2017)
10. Reißner, D., Conforti, R., Dumas, M., Rosa, M.L., Armas-Cervantes, A.: Scalable conformance checking of business processes. In: Panetto, H., et al. (eds.) OTM 2017. LNCS, vol. 10573, pp. 607–627. Springer, Cham (2017). https://doi.org/10.1007/978-3-319-69462-7_38
11. Leemans, S.J.J., Fahland, D., van der Aalst, W.M.P.: Scalable process discovery and conformance checking. Softw. Syst. Model. **17**(2), 599–631 (2018)
12. Taymouri, F., Carmona, J.: A recursive paradigm for aligning observed behavior of large structured process models. In: La Rosa, M., Loos, P., Pastor, O. (eds.) BPM 2016. LNCS, vol. 9850, pp. 197–214. Springer, Cham (2016). https://doi.org/10.1007/978-3-319-45348-4_12
13. van Dongen, B., Carmona, J., Chatain, T., Taymouri, F.: Aligning modeled and observed behavior: a compromise between computation complexity and quality. In: Dubois, E., Pohl, K. (eds.) CAiSE 2017. LNCS, vol. 10253, pp. 94–109. Springer, Cham (2017). https://doi.org/10.1007/978-3-319-59536-8_7
14. Munoz-Gama, J., Carmona, J., Van Der Aalst, W.M.P.: Single-entry single-exit decomposed conformance checking. Inf. Syst. **46**, 102–122 (2014)
15. van der Aalst, W.M.P.: Decomposing Petri nets for process mining: a generic approach. Distrib. Parallel Databases **31**(4), 471–507 (2013)
16. Murata, T.: Petri nets: properties, analysis and applications. Proc. IEEE **77**(4), 541–574 (1989)
17. Piszcz, A., Soule, T.: Genetic programming: optimal population sizes for varying complexity problems. In: Conference on Genetic and Evolutionary Computation, pp. 953–954 (2006)

18. Ruppert, D., Wand, M.P., Carroll, R.J.: Scatterplot Smoothing. Cambridge Series in Statistical and Probabilistic Mathematics, pp. 57–90. Cambridge University Press, Cambridge (2003)
19. Neapolitan, R.: Foundations of Algorithms, 5th edn, pp. 138–146. Jones and Bartlett Publishers Inc., Burlington (2014)
20. Needleman, S.B., Wunsch, C.D.: A general method applicable to the search for similarities in the amino acid sequence of two proteins. J. Mol. Biol. **48**(3), 443–453 (1970)
21. Taymouri, F.: ALI: Alignment for Large Instances (2017). https://www.cs.upc.edu/~taymouri/tool.html
22. Burattin, A.: PLG2: multiperspective process randomization with online and offline simulations. In: BPM Demo Track 2016, pp. 1–6 (2016)
23. Taymouri, F.: Conformance datasets (2017). https://www.cs.upc.edu/~taymouri/dataset.html

Maximizing Synchronization for Aligning Observed and Modelled Behaviour

Vincent Bloemen[1(✉)], Sebastiaan J. van Zelst[2], Wil M. P. van der Aalst[3], Boudewijn F. van Dongen[2], and Jaco van de Pol[1]

[1] University of Twente, Enschede, The Netherlands
v.bloemen@utwente.nl
[2] Eindhoven University of Technology, Eindhoven, The Netherlands
[3] RWTH Aachen University, Aachen, Germany

Abstract. Conformance checking is a branch of process mining that aims to assess to what degree event data originating from the execution of a (business) process and a corresponding reference model conform to each other. Alignments have been recently introduced as a solution for conformance checking and have since rapidly developed into becoming the de facto standard.

The state-of-the-art method to compute alignments is based on solving a shortest path problem derived from the reference model and the event data. Within such a shortest path problem, a cost function is used to guide the search to an optimal solution. The standard cost-function treats mismatches in the model and log as equal. In this paper, we consider a variant of this standard cost function which maximizes the number of correct matches instead. We study the effects of using this cost-function compared to the standard cost function on both small and large models using over a thousand generated and industrial case studies.

We further show that the alignment computation process can be sped up significantly in specific instances. Finally, we present a new algorithm for the computation of alignments on models with many log traces that is an order of magnitude faster (in maximizing synchronous moves) compared to the state-of-the-art A* based solution method, as a result of a preprocessing step on the model.

1 Introduction

Process mining [1] is a field of study involved with the *discovery, conformance checking*, and *enhancement* of processes, using event data recorded during process execution. In process discovery, we aim to discover process models based on traces of executed event data. In conformance checking, we assess to what degree a process model (potentially discovered) is in line with recorded event data. Finally, in process enhancement, we aim at improving or extending the process based on facts derived from event data.

V. Bloemen—This work is supported by the 3TU.BSR project.

M. Weske et al. (Eds.): BPM 2018, LNCS 11080, pp. 233–249, 2018.
https://doi.org/10.1007/978-3-319-98648-7_14

Modern information systems allow us to track, often in great detail, the behaviour of the process it supports. Moreover, instrumentation and/or program tracing tools allow us to track the behavioural profile of the execution of enterprise-level software systems [2,3]. Such behavioural data is often referred to as an event log, which can be seen as a multiset of log traces, i.e. sequences of observed events in the system. However, it is often the case, due to noise or under/over-specification, that the observed behaviour does not conform to a valid process instance, i.e., it deviates from its intended behaviour as specified by its reference model.

Conformance checking assesses to what degree the event log and model conform to each other. Early conformance checking techniques [4] are based on simple heuristics and therefore, may yield ambiguous/unpredictable results.

Alignments [5,6] were introduced to overcome the limitations of early conformance checking techniques. Alignments map observed behaviour onto behaviour described by the process model. As such, we identify four types of relations between the model and event log in an alignment:

1. A *log move*, in which we are unable to map an observed event, recorded in the event log, onto the reference model.
2. A *model move*, in which an action is described by the reference model, yet this is not reflected in the event log.
3. A *synchronous move*, in which we are able to map an event, observed in the event log, to a corresponding action described by the reference model.
4. A *silent move*, in which the model performs a silent or invisible action (denoted with τ).

Consider the example model of a simple file reading system given in Fig. 1 and the trace $\sigma = \langle A, D, B, D \rangle$. An alignment for the model and σ is given by γ^0 (top right in Fig. 1). Here, the upper-part depicts the trace and the bottom-part depicts an execution path described by the model, starting at state p_0 and ending at state p_5. The first pair, $|\frac{A}{A}|$, represents a synchronous move, in which both the log and the path in the model describe the execution of an A activity. The next pair, $|\frac{D}{\gg}|$, is a log move where the log trace describes the execution of a D activity that is not mapped to a model move. The *skip* ($\gg$) symbol is used to represent such a mismatch. Observe that the model remains in the same state. This is continued by a model move in which the model executes a C activity, which is not recorded in the trace, i.e., $|\frac{\gg}{C}|$. Finally, the alignment ends with two synchronous moves.

An optimal alignment is an alignment that minimizes a given cost function. Typically, each type of move gets a value assigned $\mathbb{R}_{\geq 0}$. The cost of an alignment is simply the sum of the costs of its individual moves. The most common way to do this is to assign a cost of 1 to both model and log moves and 0 to synchronous and silent moves. In practice, the A* shortest path algorithm [7] is often used for computing optimal alignments.

We argue that the standard cost function is not always the best-suited function for optimal alignments. Consider the model from Fig. 1 again, with the trace $\sigma' = \langle B \rangle$. An optimal alignment using the standard cost function would result

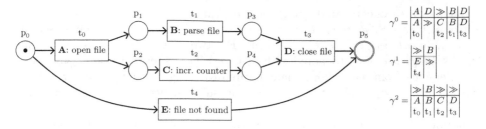

Fig. 1. Example process model (in Petri net formalism) for a simple file reading system and an alignment for the trace $\sigma = \langle A, D, B, D \rangle$ (γ^0). For the trace $\sigma = \langle B \rangle$, two optimal alignments are given using the standard- (γ^1) and variant (γ^2) cost functions.

in γ^1. Considering that event B is observed behaviour, i.e., the system logged "parse file", it seems illogical to map this behaviour with a path in the model indicating that the file was not found. In case we set up the cost function such that the number of synchronous moves are maximized, an optimal alignment would result in γ^2. Arguably, a more likely scenario is that not all parts of the program produced log output and γ^2 would be preferred.

Motivated by the example shown in Fig. 1, we consider the applicability of a cost function that maximizes the number of synchronous moves in a more general setting and study its effects. Our contributions are as follows.

- We formalise the relation between the event log and the reference model to distinguish different cases of alignment problems. We show how the cost functions affect the resulting alignments for these cases. We further show that when the reference model is an abstraction of the event log, the alignment computation process can be significantly improved.
- We study the differences in alignments and their computation times on over a thousand large instances that exhibit various characteristics. We also compare the results from the A* algorithm with a recent symbolic algorithm [8].
- We present a new algorithm for computing alignments that exploits our new cost function in a preprocessing step. Using a set of industrial models, we show that it performs an order of magnitude faster than the A* algorithm.

The remainder of this paper is structured as follows. Section 2 introduces preliminaries. Then, in Sects. 3 and 4 we introduce the synchronous cost function and formalise the relation between the event log and the reference model. We discuss existing algorithms for computing alignments in Sect. 5. In Sect. 6 we present the new algorithm that preprocesses the model to improve the alignment computation process. Experiments are presented in Sect. 7. Section 8 discusses related work. Section 9 concludes the paper.

2 Preliminaries

We assume that the reader is familiar with the basics of automata theory and Petri nets. We denote a trace or sequence by $\sigma = \langle \sigma_0, \sigma_1, \ldots, \sigma_{|\sigma|-1} \rangle$, two sequences are concatenated using the · operation. Given a sequence σ and a set of elements S, we refer to $\sigma \setminus S$ as the sequence without any elements from S, e.g., $\langle a, b, b, c, a, f \rangle \setminus \{b, f\} = \langle a, c, a \rangle$. For two sequences σ_1 and σ_2, we call σ_1 a *subsequence* of σ_2 (denoted with $\sigma_1 \sqsubseteq \sigma_2$) if σ_1 is formed from σ_2 by deleting elements from σ_2 without changing its order, e.g., $\langle c, a, t \rangle \sqsubseteq \langle a, c, r, a, t, e \rangle$. Similarly, $\sigma_1 \sqsubset \sigma_2$ implies that σ_1 is a strict subsequence of σ_2, thus $\sigma_1 \neq \sigma_2$.

Traces are sequences $\sigma \in \Sigma^*$, for which each element is called an *event* and is contained in the alphabet Σ, also called the set of events. We globally define the alphabet Σ, which does not contain the skip event ($\gg$) nor the invisible action or silent event (τ). Given a set S, we denote the set of all possible multisets as $\mathcal{B}(S)$, and its power-set by 2^S. An *event log* E is a multiset of traces, i.e., $E \subseteq \mathcal{B}(\Sigma^*)$.

2.1 Preliminaries on Petri Nets

Petri nets are a mathematical formalism that allow us to describe processes, typically containing parallel behaviour, in a compact manner. Consider Fig. 1 which is a simple example of a Petri net. The Petri net consists of *places*, visualized as circles, that allow us to express the state (or *marking*) of the Petri net. Furthermore, it consists of *transitions*, visualized as boxes, that allow us to manipulate the state of the Petri net. We are never able to connect a place with another place nor a transition with another transition. Thus, from a graph-theoretical perspective, a Petri net is a bipartite graph.

Definition 1 (Petri net, marking). *A Petri net is defined as a tuple* $N = (P, T, F, \Sigma_\tau, \lambda, m_0, m_F)$ *such that:*

- *P is a finite set of* places,
- *T is a finite set of* transitions *such that* $P \cap T = \emptyset$,
- $F \subseteq (P \times T) \cup (T \times P)$ *is a set of directed arcs, called the* flow relation,
- Σ_τ *is a set of activity events, with* $\Sigma_\tau = \Sigma \cup \{\tau\}$,
- $\lambda : T \to \Sigma_\tau$ *is a labelling function for each transition,*
- $m_0 \in \mathcal{B}(P)$ *is the initial* marking *of the Petri net,*
- $m_F \in \mathcal{B}(P)$ *is the final marking of the Petri net.*

A marking is defined as a multiset of places, denoting where tokens reside in the Petri net. A transition $t \in T$ *can be fired if, according to the flow relation, all places directing to t contain a token. After firing a transition, the tokens are removed from these places and all places having an incoming arc from t receive a token. It may be possible for a place to contain more than one token.*

Definition 2 (Marking graph). *For a Petri net* $N = (P, T, F, \Sigma_\tau, \lambda, m_0, m_F)$, *the corresponding* marking graph *or* state-space $MG = (Q, \Sigma_\tau, \delta, q_0, q_F)$ *is a non-deterministic automaton such that:*

- $Q \subseteq \mathcal{B}(P)$ *is the (possibly infinite) set of vertices in MG, which corresponds to the set of* reachable *markings from* m_0 *(obtained from firing transitions),*

- $\delta \subseteq (Q \times T \times Q)$ *is the set of edges in MG, i.e.,* $(m, t, m') \in \delta$ *iff there is a* $t \in T$ *such that* m' *is obtained from firing transition t from marking m.*
- $q_0 = m_0$ *is the initial state of the graph,*
- $q_F = m_F$ *is the final state of the graph.*

For an edge $e = (m, t, m') \in \delta$, *we write $\lambda(e)$ to denote $\lambda(t)$ and use the notation* $m \xrightarrow{a} m'$ *to represent the edge e for which $\lambda(e) = a$ (we assume that for two edges* $(m, t_1, m') \in \delta$ *and* $(m, t_2, m') \in \delta$, *if $\lambda(t_1) = \lambda(t_2)$ then $t_1 = t_2$). The source and target markings of edge e are respectively denoted by $src(e)$ and $tar(e)$.*

Definition 3 (Path, language). *Given a Petri net N and corresponding marking graph* $MG = (Q, \Sigma_\tau, \delta, q_0, q_F)$, *a sequence of edges* $P = \langle P_0, P_1, \ldots, P_n \rangle \in \delta^*$ *is called a* path *in N if it forms a path on the marking graph of N:* $src(P_0) = m_0 \wedge tar(P_n) = m_F \wedge \forall_{0 \le i < n} : tar(P_i) = src(P_{i+1})$. *The set of all paths in N is denoted by $Paths(N)$. With $\lambda(P)$ we refer to the sequence of labels visited in P, i.e.,* $\lambda(P) = \langle \lambda(P_0), \lambda(P_1), \ldots, \lambda(P_n) \rangle$ *(there may be different paths P and P' such that $\lambda(P) = \lambda(P')$). We define the* language $\mathcal{L}$ *of a Petri net N by* $\mathcal{L}(N) = \{\lambda(P) \mid P \in Paths(N)\}$.

Definition 4 (Trace to Petri net). *Given a trace* $\sigma = \langle \sigma_1, \sigma_2, \ldots, \sigma_n \rangle \in \Sigma^*$, *its corresponding Petri net is defined as* $N_\sigma = (P, T, F, \Sigma_\tau, \lambda, m_0, m_F)$ *with* $P = \{p_0, p_1, \ldots, p_n, p_{n+1}\}$, $T = \{t_0, t_1, \ldots, t_n\}$, $F = \{(p_0, t_0), (p_1, t_1), \ldots, (p_n, t_n)\} \cup \{(t_0, p_1), (t_1, p_2), \ldots, (t_n, p_{n+1})\}$, $\Sigma_\tau = \bigcup_{0 \le i < n} \{\sigma_i\}$, $\forall_{0 \le i < n} : \lambda(t_i) = \sigma_i$, $m_0 = p_0$, *and* $m_F = p_{n+1}$.

2.2 Preliminaries on Alignments

Definition 5 (Alignment). *Let $\sigma \in \Sigma^*$ be a log trace and let N be a Petri net model, for which we obtain the marking graph* $MG = (Q, \Sigma_\tau, \delta, q_0, q_F)$. *We refer to $\Sigma_\gg$ as the alphabet containing skips:* $\Sigma_\gg = \Sigma \cup \{\gg\}$ *and $\Sigma_{\tau\gg}$ as the alphabet that also contains the silent event:* $\Sigma_{\tau\gg} = \Sigma \cup \{\gg, \tau\}$. *Let $\gamma \in (\Sigma_\gg \times \Sigma_{\tau\gg})^*$ be a sequence of log-model pairs (note that τ steps are only possible in the model). For* $\gamma = \langle (\gamma_0^0, \gamma_0^1), (\gamma_1^0, \gamma_1^1), \ldots, (\gamma_{|\gamma|-1}^0, \gamma_{|\gamma|-1}^1) \rangle$, *we define γ^ℓ as* $\gamma^\ell = \langle \gamma_0^0, \gamma_1^0, \ldots, \gamma_{|\gamma|-1}^0 \rangle \setminus \{\gg\}$ *and γ^m by* $\gamma^m = \langle \gamma_0^1, \gamma_1^1, \ldots, \gamma_{|\gamma|-1}^1 \rangle \setminus \{\gg\}$. *We call γ an* alignment *if the following conditions hold:*

1. $\gamma^\ell = \sigma$ *(the activities of the log-part, equals to σ),*
2. $\gamma^m \in \mathcal{L}(N)$ *(γ^m forms a path in N),*
3. $\forall a, b \in \Sigma : a \ne b \Rightarrow (a, b) \notin \gamma$ *(illegal moves),*
4. $(\gg, \gg) \notin \gamma$, *(the 'empty' move may not exist in γ).*

Definition 6 (Alignment cost). *Let $\gamma \in (\Sigma_\gg \times \Sigma_{\tau\gg})^*$ be an alignment for* $\sigma \in \Sigma^*$ *and the Petri net N. The cost function c for pairs of γ is given as follows;* $c : (\Sigma_\gg \times \Sigma_{\tau\gg}) \to \mathbb{R}_{\ge 0}$, *and we overload c for alignments;* $c : (\Sigma_\gg \times \Sigma_{\tau\gg})^* \to \mathbb{R}_{\ge 0}$, *for which we have* $c(\gamma) = \sum_{i=0}^{|\gamma|-1} c(\gamma_i)$.

We call an alignment γ under cost function c optimal iff $\nexists \gamma' : c(\gamma') < c(\gamma)$, *i.e., there does not exist an alignment γ' with a smaller cost.*

Definition 7 (Standard cost function). *The* standard cost function c_{st} *is defined for an alignment pair* $(\ell, m) \in (\Sigma_\gg \times \Sigma_{\tau\gg})$ *as follows:*

$$
c_{\mathrm{st}}(\ell, m) = \begin{cases} 0 & \ell = \gg \text{ and } m = \tau \text{ (silent move, e.g., } (\gg, \tau)) \\ 0 & \ell \in \Sigma \text{ and } m \in \Sigma \text{ and } \ell = m \text{ (e.g., synchronous move } (a, a)) \\ 1 & \ell \in \Sigma \text{ and } m = \gg \text{ (e.g., log move } (a, \gg)) \\ 1 & \ell = \gg \text{ and } m \in \Sigma \text{ (e.g., model move } (\gg, a)) \end{cases}
$$

3   Maximizing Synchronous Moves

We gather that the standard cost function from Definition 7 is the most commonly used cost function in literature [1,7,9,10], though note that any cost function could be used. The standard cost function may, however, lead to undesired results, as illustrated by the example from Fig. 1. We consider a new cost function that maximizes the number of synchronous moves, since it explains as many log moves as possible. We propose the alternative cost function as follows.

Definition 8 (max-sync cost function). *We define the* max-sync cost function c_{sync} *for an alignment pair as follows (for small $\varepsilon > 0$):*

$$
c_{\mathrm{sync}}(\ell, m) = \begin{cases} 0 & \ell = \gg \text{ and } m = \tau \text{ (silent move, e.g., } (\gg, \tau)) \\ 0 & \ell \in \Sigma \text{ and } m \in \Sigma \text{ and } \ell = m \text{ (e.g., synchronous move } (a, a)) \\ 1 & \ell \in \Sigma \text{ and } m = \gg \text{ (e.g., log move } (a, \gg)) \\ \varepsilon & \ell = \gg \text{ and } m \in \Sigma \text{ (e.g., model move } (\gg, a)) \end{cases}
$$

This cost function only penalizes log moves, which as a consequence causes an optimal alignment to minimize the number of log moves and thus maximize the number of synchronous moves. The ε cost for model moves further filters optimal alignments to only include shortest paths through the model that maximize synchronous moves.

An advantage of the max-sync cost function over the standard one is that synchronized behaviour is not sacrificed for shorter paths through the model (as Fig. 1 illustrates). A disadvantage is that in order to maximize the number of synchronous moves, it may be possible that many model moves are required.

4   Relating the Model and Event Log

Given a Petri net model N and an event log $E \subseteq \mathcal{B}(\Sigma^*)$, we can distinguish four cases based on the languages that they describe. By distinguishing the relative granularities of N and E we define cases of alignment problems as follows.

C1: $\forall \sigma_1 \in E : (\exists \sigma_2 \in \mathcal{L}(N) : \sigma_1 = \sigma_2)$; all log traces correspond to paths in the model. Then, every log trace can be mapped onto the model by only using synchronous and silent moves, which is optimal for c_{st} and c_{sync}.

C2: $\forall \sigma_1 \in E : (\exists \sigma_2 \in \mathcal{L}(N) : \sigma_1 \sqsubseteq \sigma_2)$; all log traces correspond to subsequences of paths in the model. Then, every log trace can be mapped onto the model without using any log moves. The example from Fig. 1 for $\sigma = \langle B \rangle$ is such an instance. We hypothesize that c_{sync} provides better alignments in such instances as c_{st} may avoid synchronization in favour of shorter paths through the model.

C3: $\forall \sigma_1 \in E : (\exists \sigma_2 \in \mathcal{L}(N) : \sigma_2 \sqsubseteq \sigma_1)$; for every log trace there is a path that forms a subsequence of the log trace. Then, every log trace can be mapped onto the model without using any model moves. Here, c_{sync} and to some extent c_{st} can arguably lead to bad results as model moves may be taken to synchronize with 'undesired' behaviour.

C4: None of the properties hold. All move types may be necessary for alignments. We regard this as a standard scenario. Depending on the use case, either c_{st} or c_{sync} could be preferred.

Aside from C4, we consider cases C2 and C3 as common instances in practice, as logging software often causes either too many or too little events to be logged or in case the model is over/underspecified. Discrepancies then show whether the model is of the right granularity. We note that it is also possible to hide certain activities in the model or log before alignment. This is however not trivial, especially if there are (slight) deviations in the log such that the alignment problem does not fit C2 or C3 exactly anymore.

When considering instances that exactly fit case C2 or C3, we can construct alignments by respectively removing all log or model moves from the product of the model and log. We define the cost functions c_{add} and c_{rem} to be variants of c_{st} such that model and log moves respectively have a cost of ∞. We argue that this results in a better 'alignment quality' and reduces the time for its construction.

5   Algorithms for Computing Alignments

We consider two algorithms for computing alignments, which we discuss as follows. Both algorithms take the product Petri net as input.

A*. The A* algorithm [7] computes the shortest path from the initial marking to the final marking on the marking graph for a given cost function. The heuristic function for A* exploits the Petri net marking equation, which can be achieved using Integer Linear Programming (ILP), to prune the search space.

Symbolic Algorithm. The symbolic algorithm [8] was recently developed as an improvement over A* for large state spaces. It exploits symbolic reachability to search for an alignment, i.e., considering sets of markings instead of single ones. By restricting the cost function to only allow 0 or 1-cost moves, optimal alignments can be computed by only taking a 1-cost move after exploring all markings reachable via 0-cost steps. We refer to this algorithm by Sym.

6   Preprocessing Reference Models for Large Event Logs

When constructing an alignment under the c_{sync} cost function, we can disregard the cost for model moves to a certain extent. The goal is to find a path through the model that maximizes the number of synchronous moves. We can achieve this by searching for a subsequence in the log trace that is also included in the language of the reference model. By computing the transitive closure of the model's marking graph, we find all paths and subsequences of paths through the model. For every log trace we can use dynamic programming to search for the maximum-length subsequence in the log trace that can be replayed in the transitive closure graph (TCG), from which we can construct a path through the marking graph and obtain an optimal alignment.

We construct a TCG as described in Definition 9. Here, τ-edges are added to the marking graph such that every marking is reachable via τ-steps. After determinization, for every path P in the original marking graph the TCG contains all paths P' such that $\lambda(P') \sqsubseteq \lambda(P)$.

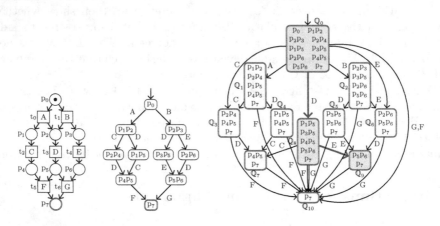

Fig. 2. Example Petri net model (left), its corresponding marking graph (middle) and transitive closure graph (right) with the sequence $\langle D, E \rangle$ highlighted.

We can use this property to search for a subsequence of the log trace that can fully synchronize with the model. For instance in the example of Fig. 2, consider a log trace $\sigma = \langle F, D, E, B \rangle$. The F event can be fired from $\mathbb{Q}_0$, after which the TCG is in state $\mathbb{Q}_{10}$. From this state, it is not possible to perform any other event from log trace. A better choice would be to skip the F event (which would then be a log move) and form the subsequence $\langle D, E \rangle$, as highlighted[1]. We call the maximum-length subsequence $\hat{\sigma}$ from the log trace a *maximum fitting subsequence* if $\hat{\sigma}$ also forms a path through the TCG, as defined in Definition 10.

---

[1] It might be interesting to note that after performing the D action in the TCG, in the Petri net we have not yet made the choice to fire either an A or a B transition; we implicitly make the decision to fire the B transition after choosing the E event.

Definition 9 (Transitive closure graph). *Given a marking graph $MG =$ $(Q, \Sigma_\tau, \delta, q_0, q_F)$, we first construct an extended marking graph $MG' =$ $(Q, \Sigma_\tau, \delta', q_0, q_F)$ with $\delta' = \delta \cup \{(src(e), \tau, tar(e)) \mid e \in \delta\}$. A transitive closure graph (TCG), $TCG = (\mathbb{Q}, \Sigma, \Delta, \mathbb{Q}_0, \mathbb{Q}_F)$ is defined as the result of determinizing MG' (by using a standard determinization algorithm [11]) and by then removing all non-final states from the TCG such that $\mathbb{Q} \subseteq 2^Q$, $\Sigma = \Sigma_\tau \setminus \{\tau\}$, $\Delta \subseteq (\mathbb{Q} \times \Sigma \times \mathbb{Q})$, $\mathbb{Q}_0 = Q$, and $\mathbb{Q}_F = \mathbb{Q}$.*

For an edge $e \in \Delta$ we also use the notation $src(e)$ and $tar(e)$ to respectively refer to the source and target marking sets in the TCG. Paths over the TCG are defined analogously to paths over marking graphs (Definition 3) and we use $Paths(TCG)$ and $\mathcal{L}(TCG)$ to respectively denote the set of all paths in the TCG and the language of the TCG.

Definition 10 (Maximum fitting subsequence). *Given a sequence (log trace) $\sigma \in \Sigma^*$ and $TCG = (\mathbb{Q}, \Sigma, \Delta, \mathbb{Q}_0, \mathbb{Q}_F)$, then $\hat{\sigma} \sqsubseteq \sigma$ is a maximum fitting subsequence if and only if $\hat{\sigma} \in \mathcal{L}(TCG) \wedge \forall \hat{\sigma}' \sqsubseteq \sigma : \hat{\sigma}' \in \mathcal{L}(TCG) \Rightarrow |\hat{\sigma}| \geq |\hat{\sigma}'|$. We construct $\hat{\sigma}$ by using dynamic programming to search for a subsequence of σ that is a maximum-length path in the TCG.*

Algorithm 1. Path construction from a maximum fitting subsequence $\hat{\sigma}$

---

1 **func** $PC(TCG = (\mathbb{Q}, \Sigma, \Delta, \mathbb{Q}_0, \mathbb{Q}_F), MG = (Q, \Sigma_\tau, \delta, q_0, q_F), \hat{\sigma} = \langle \hat{\sigma}_0, \hat{\sigma}_1, \ldots, \hat{\sigma}_n \rangle)$
2   // Construct path MFP on TCG such that $\lambda(\text{MFP}) = \hat{\sigma}$
3      $\text{MFP} := \langle (\mathbb{Q}_0, \hat{\sigma}_0, S), (S, \hat{\sigma}_1, S'), \ldots, (S'', \hat{\sigma}_n, S''') \rangle$ s.t. $\forall_{0 \leq i \leq n} : \text{MFP}_i \in \Delta$
4      $P := \text{BWD}(MG, q_F, \hat{\sigma}_n, tar(\text{MFP}_n))$ // Path $\hat{\sigma}_n$ to q_F on MG
5   **for** $i := n - 1;\ i \geq 0;\ i := i - 1$ **do** // Add paths from $\hat{\sigma}_i$ to $\hat{\sigma}_{i+1}$
6          $P := \text{BWD}(MG, src(P_0), \hat{\sigma}_i, tar(\text{MFP}_i)) \cdot P$
7   **return** $\text{BWD}(MG, src(P_0), \bot, \mathbb{Q}_0) \cdot P$ // Add path from q_0 to $\hat{\sigma}_0$

8 **func** $BWD(MG = (Q, \Sigma_\tau, \delta, q_0, q_F), m \in Q, a \in (\Sigma \cup \bot), S \subseteq Q)$
9      $W := \langle m \rangle$ // Sequence of unvisited markings in the backward search
10      $\forall m \in S : F[m] := \text{Null}$ // Mapping from markings to edges ($F : Q \rightarrow \delta$)
11   **for** $i := 0;\ i < |W|;\ i := i + 1$ **do** // Continue for all markings in W
12   **if** $\exists m' \in Q, a' \in \Sigma : (m', a', W_i) \in \delta \wedge (a' = a \vee (a = \bot \wedge m' = q_0))$ **then**
13              $P := \langle (m', a', W_i) \rangle$ // Found path from a (or initial marking)
14   **while** $tar(P_{|P|-1}) \neq m$ **do** $P := P \cdot F[tar(P_{|P|-1})]$
15   **return** P // Shortest path from a (or q_0) to m
16   **forall the** $e \in \delta : src(e) \in (S \setminus W) \wedge tar(e) = W_i$ **do**
17              $W := W \cdot \langle src(e) \rangle$ // Add predecessor markings of m to W
18              $F[src(e)] := e$ // Direct the source markings towards m
19   **return** $\langle \rangle$ // No path from a (or q_0) is found (should never occur)

Once we have found the maximum fitting subsequence $\hat{\sigma}$ for a given model and log trace, we still have to determine which model moves should be applied to form a path through the original model. This can be achieved by using the TCG and traversing $\hat{\sigma}$ in a backwards fashion as we show in Algorithm 1.

We first construct a path MFP from the subsequence $\hat{\sigma}$ (line 3), in the example from Fig. 2 with $\hat{\sigma} = \langle D, E \rangle$ (see also Fig. 3 for an illustration of the path

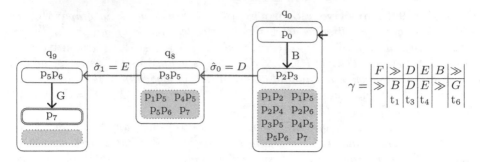

Fig. 3. Path construction using Algorithm 1 on the example from Fig. 2 for a maximum fitting subsequence $\hat{\sigma} = \langle D, E \rangle \sqsubseteq \langle F, D, E, B \rangle$. Markings in the grey region are not part of the path. The resulting alignment γ is shown on the right.

construction process) this would be $\text{MFP} = \langle (Q_0, D, Q_8), (Q_8, E, Q_9) \rangle$. Then in line 4, a backward search procedure (BWD) is called to search for a path P in the marking graph from an E-edge to the final marking (p_7).

The BWD procedure takes a target marking m, label a and search space S as arguments. A sequence W is maintained to process unvisited markings from S and a mapping $F : Q \rightarrow \delta$ is used for reconstructing the path. Starting from the target marking m (which is W_0), the procedure searches for edges e directing towards m in line 16–18 such that $\text{src}(e)$ is in S and not already visited. For every such edge e, its source is appended to W (to be considered in a future iteration) and $\text{src}(e)$ is mapped to e for later path reconstruction.

Following iterations of the for loop in line 11–18 consider a predecessor W_i of m and search for edges directing to W_i. This way, the search space is traversed backwards in a breadth-first manner, resulting in shortest paths to m.

In line 12–15 the BWD procedure checks whether there is an edge $m' \xrightarrow{a} W_i$ for some m' (or an edge $q_0 \xrightarrow{a'} W_i$ for arbitrary a' in case $a = \perp$) and if so, constructs a path towards m in line 14 which is then returned. In the example, the path $\langle (p_3 p_5, E, p_5 p_6), (p_5 p_6, G, p_7) \rangle$ will be returned for the first BWD call.

After the first BWD call, the main function iterates backwards over all remaining edges from MFP (line 5–6) to create paths between $\hat{\sigma}_i$ and $\hat{\sigma}_{i+1}$, which are inserted in the path before P. Finally, in line 7 a path from the initial marking q_0 towards the first label $\hat{\sigma}_0$ is inserted before P to complete the path (here the label is set to $\perp$ to search for q_0 in the BWD procedure).

In the example we first compute the path $\langle (p_3 p_5, E, p_5 p_6), (p_5 p_6, G, p_7) \rangle$ in line 4, then after line 5–6 we insert the path $\langle (p_2 p_3, D, p_3 p_5) \rangle$, and in line 7 we insert the path from the initial state $q_0 = p_0$, $\langle (p_0, B, p_2 p_3) \rangle$ to create the complete minimal-length path P in the marking graph such that $\hat{\sigma} \sqsubseteq \lambda(P)$.

The alignment can be reconstructed by marking all events in the maximum fitting subsequence as synchronous moves, by marking the remaining labels in the log trace as log moves, and inserting the model and silent moves (as computed by Algorithm 1) at the appropriate places.

Note that the TCG algorithm does not exactly compute an alignment for the cost function c_{sync}. The backwards BFS does ensure a shortest path through the model from the initial to the final marking while synchronizing with the maximum fitting subsequence. However, there might exist a different maximum fitting subsequence that leads to a different path through the model with a lower total cost (fewer model moves). This can be repaired by computing the alignments for all maximum fitting subsequences. If the marking graph contains cycles, the corresponding markings get contracted to a single state in the TCG with a self-loop for each activity in the cycle. Also, the TCG may in theory contain exponentially more states than there are markings in the marking graph. However, in industrial models (Sect. 7.3), we found that in many cases the number of states in the TCG is at most two times more than the number of markings in the marking graph.

7 Experiments

For the experiments, we considered two types of alignment problems. On the one hand, a large reference model accompanied by an event log consisting of a single log trace, and on the other hand a smaller reference model accompanied by an event log of many traces. All experiments were performed on an Intel® Core™ i7-4710MQ processor with 2.50 GHz and 7.4 GiB memory. For all experiments, we have set a timeout of 60 s. When computing averages, a timeout also counts as 60 s.

We investigate differences between the alignments resulting from using the standard- and max-sync cost functions, and compare alignment computation times for A* (with ILP, using the implementation from RapidProM [12]) and the symbolic algorithm (implemented in the LTSMIN model checker [13]). We further investigate specific alignment problems, cases C2 and C3 as discussed in Sect. 4. Finally, we also look at models accompanied by many log traces to compare the performance of the TCG algorithm (implemented in ProM [14]) with the other algorithms. For all large models with singleton log traces we used 8 threads for computing alignments, and for smaller models with many log traces we only used a single thread per alignment computation[2]. All results are available online at https://github.com/utwente-fmt/MaxSync-BPM2018.

7.1 Experiments Using Large Models and Singleton Event Logs

Model Generation. Using the PTandLogGenerator [15] we generated Petri net models with process operators and additional features set to their defaults; where the respective probabilities for sequence, XOR, parallel, loop, OR are set to 45%, 20%, 20%, 10%, and 5%. The additional features for the occurrence of silent and duplicate activities, and long-term dependencies were all set to 20%.

---

[2] We consider multi-threaded experiments not as useful in this scenario, as the problem can be parallelized by dividing the log traces over the different threads and computing the alignments independently.

To examine scalability we ranged the average number of activities from 25, 50, and 75, resulting in respectively 110, 271, and 370 transitions on average. For these settings, we generated 30 models (thus 90 in total) and generated a single log trace per model. For this log trace we added 10%, 30%, 50%, and 70% noise in three different ways (thus 12 noisy singleton logs are created); by (1) adding, removing and swapping events (resembling case C4), (2), by only adding events (resembling case C3), and (3) by only removing events (resembling case C2). In total there are 1,080 noisy singleton logs. We first consider noise of type 1.

Alignment Differences. In Table 1 we compare the resulting alignments, produced by Sym, for the different cost functions. When comparing the overall results of c_{st} and c_{sync} (rightmost column), we observe that c_{sync} uses about 43% fewer log moves, which are added as synchronous moves. However in doing so, more than six times as many model moves are required.

When looking at an increase in the amount of noise, the relative difference between the number of log moves remains the same, while this difference in model moves slightly drops. When increasing the number of activities from 25 to 75, We observe an increase in the number of model moves for c_{sync} from 3.2 times to 9.3 times as many compared to c_{st}. As a corresponding result from this effect, the difference between log moves from c_{sync} and c_{st} stays relatively the same for increasing activities.

We conclude that for c_{sync} the relative reduction in log moves stays mostly the same, when fluctuating the amount of noise or size of the model. The size of the model seems to greatly affect the number of model moves for c_{sync}, making alignments from c_{st} and c_{sync} more diverse for larger models.

Table 1. Comparison between alignments generated using the c_{st} and c_{sync} cost functions. The numbers show averages, e.g., the value of 2.3 in the top-left corner denotes the average number of log moves for all computed alignments for which 10% noise is added, using the c_{st} cost function.

| | Noise added (add, remove, swap) | | | | | | | | Number of activities | | | | | | Average | |
| | 10% | | 30% | | 50% | | 70% | | 25 | | 50 | | 75 | | | |
	c_{st}	c_{sync}	c_{st}	c_{sync}	c_{st}	c_{sync}	c_{st}	c_{sync}	c_{st}	c_{sync}	c_{st}	c_{sync}	c_{st}	c_{sync}	c_{st}	c_{sync}
Log	2.3	1.3	6.5	3.6	9.4	5.4	10.9	6.3	4.7	3.2	8.9	4.6	8.4	4.5	7.0	4.0
Model	2.0	15.7	4.6	30.9	5.8	35.3	6.2	38.1	3.3	10.7	5.6	39.1	5.4	50.2	4.5	29.4
Sync	28.5	29.6	20.9	23.7	16.8	20.8	14.5	19.1	13.8	15.4	23.2	27.5	29.4	33.3	20.6	23.6
Silent	17.3	24.4	14.7	30.4	13.6	35.3	12.8	35.1	10.0	13.3	16.2	39.6	21.6	51.6	14.7	31.0

Performance Results. We observed that while Sym is faster in computing alignments than the A* algorithm on c_{st} (it takes on average 15.8 s for computing an alignment using A* and 10.5 s for Sym), for the c_{sync} cost function A* is outperforming the symbolic algorithm (13.7 s for A* and 16.5 s for Sym). This has to do with the effect that the symbolic algorithm will explore the entire model before attempting a single log move whereas A* does not.

7.2   Alignment Problems that only Add or Remove Events

Alignment Differences. In Table 2 we compare the resulting alignments for adding or removing events. When inspecting the Add case, we find that the c_{st} already avoids model moves for the most part as we would expect. Moreover, there are only small differences between alignments from c_{st} and c_{add}. For c_{sync}, many model moves may be chosen to increase the number of synchronous moves. These additional synchronous moves are arguably not part of the 'desired' alignment since they require a large detour through the model.

When removing events from the log trace, the c_{st} cost function is only partly able to describe the removal of events as it still chooses log moves. The c_{sync} cost function does not take any log moves as this maximizes the number of synchronous moves, making it equal to c_{rem}. When comparing c_{st} and c_{sync}, we could argue that for the Add case, the c_{st} cost function better represents a 'correct' alignment and for the Rem case c_{sync} is better suited.

Performance Results. We observed that for c_{st}, A* performs relatively bad for the Rem case (14.1 s on average), but significantly better for c_{sync} (2.3 s on average). We argue that A* for c_{st} tries to perform many log moves, that results in a lot of backtracking, while for c_{sync} the algorithm avoids log moves entirely. The symbolic algorithm uses 6.6s and 7.3s on average for c_{st} and c_{sync} respectively. For the Rem case, we do not observe a significant difference in the performance times when considering c_{rem}, i.e., removing the log moves. This is because both algorithms already avoid log moves for the c_{sync} cost function.

For the Add case, both A* and Sym require more time for computing alignments for c_{sync} than for c_{st}. When removing model moves (c_{add}), A* and Sym perform in respectively 36% and 77% of the time required for c_{st} (thus 3.4 s and 9.3 s). By removing the model moves, both algorithms no longer have to explore a large part of the state-space and only have to decide on which log moves, synchronous and silent actions to chose, which is especially beneficial for A*.

Table 2. Comparison between alignments generated using the c_{st} and c_{sync} cost functions for alignment problems, where noise only consist of adding (Add) or removing (Rem) events. The cost functions c_{add} and c_{rem} are variations on c_{st} such that model and log moves respectively have a cost of ∞

| | Log events added (Add) | | | | | | | | | Log events removed (Rem) | | | | | | | | |
| | 10% | | | 30% | | | 50% | | | 10% | | | 30% | | | 50% | | |
	c_{st}	c_{sync}	c_{add}	c_{st}	c_{sync}	c_{add}	c_{st}	c_{sync}	c_{add}	c_{st}	c_{sync}	c_{rem}	c_{st}	c_{sync}	c_{rem}	c_{st}	c_{sync}	c_{rem}
Log	3.1	2.0	3.1	7.5	5.1	7.6	10.6	7.4	10.8	0.3	0.0	0.0	1.0	0.0	0.0	2.5	0.0	0.0
Model	0.0	13.1	0.0	0.1	21.6	0.0	0.2	23.1	0.0	3.0	3.3	3.3	6.3	7.7	7.7	8.0	11.9	11.9
Sync	29.4	30.5	29.4	26.5	28.9	26.4	24.1	27.4	23.9	30.3	30.7	30.7	21.0	22.0	22.0	13.9	16.4	16.4
Silent	16.3	23.6	16.2	15.5	31.0	15.4	14.0	30.0	13.8	18.4	18.5	18.5	16.0	16.7	16.7	13.2	16.0	16.0

7.3   Experiments Using Event Logs with More Traces

We now consider smaller models that have to align many log traces. For our experiments, we selected 9 instances from the 735 industrial business process Petri net models from financial services, telecommunications and other domains, obtained from the data sets presented in Fahland et al. [16].

For our selection, we computed the transitive closure graph (TCG) and considered the instances for which we were able to compute TCG within 60 s. From this set, we selected the 9 most interesting cases, e.g., the models with the largest Petri net models, largest marking graphs, largest TCG graph, and largest TCG construction time. On average the marking graph contains 108 markings and the TCG 134 states. In the worst case, the number of states in the TCG was 200, which doubled the number of markings in the marking graph. We did not find a large difference between the performance results of the individual experiments.

For each model, we generated a set of 10, 100, 1,000, and 10,000 log traces for 10%, 30%, 50%, and 70% noise added by adding, removing, and swapping events. Thus in total, we have 16 event logs per model. We compared the performance of the TCG algorithm with that of A* using a single thread. We also experimented with the symbolic algorithm, but its setup time per alignment computation is too large to provide meaningful results. Note that in our experiments, we only consider the c_{sync} cost function. The TCG algorithm is not applicable to the c_{st} cost function.

Table 3. Alignment computation time (in milliseconds) for models with many log traces. TCG-comp, TCG-align, and TCG respectively denote the time for computing the TCG, the time for aligning all log traces, and the sum of the two.

Log size	TCG-comp	TCG-align	TCG	A*	Noise	TCG	A*
10	272	9	281	426	10%	727	9,320
100	269	20	289	3,539	30%	729	13,919
1,000	265	161	426	13,247	50%	750	14,199
10,000	274	1,542	1,936	33,906	70%	727	13,679

Results. The results are summarized in Table 3. On average, the TCG algorithm used 270 ms for computing the transitive closure graph. When increasing the number of log traces (left table), we see that the preprocessing step of the TCG algorithm remains a significant part of its total time for up to 1,000 log traces. The A* algorithm has to create a synchronous product of the model and log trace for each instance, and expectedly takes more time in total. For 10,000 log traces, A* is 17 times slower than the TCG algorithm. But even for 10 log traces, the TCG algorithm outperforms A* by almost a factor of two.

When comparing the results for different amounts of noise (right table), we see practically no difference in the computation times for the TCG algorithm. The A* algorithm does require significantly more time for 30%, 50%, and 70% noise compared to the 10% case. We argue that from 30% noise onwards, A* has to visit most of the state-space to construct an optimal alignment. In the TCG algorithm, noise does not seem to affect its performance.

8   Related Work

One of the earliest works in conformance checking was from Cook and Wolf [17]. They compared log traces with paths generated from the model.

One technique to check for conformance is *token-based replay* [4]. The idea is to 'replay' the event logs by trying to fire the corresponding transitions, while keeping track of possible missing and remaining tokens in the model. However, this technique does not provide a path through the model. When traces in the event log deviate a lot, the Petri net may get flooded with tokens and the tokens do not provide good insights anymore.

Alignments were introduced [5,7] to overcome the limitations of the token-based replay technique. Alignments formulate conformance checking as an optimization problem, i.e., minimizing the alignment cost-function. Since its introduction, alignments have quickly become the standard technique for conformance checking along with the A* algorithm for computing alignments [9]. In previous work [8] we presented the symbolic algorithm for alignments and we analysed how different model characteristics influence the computation times for c_{st}.

For larger models, techniques have been developed to decompose the Petri net in smaller subprocesses [18]. For instance, fragments that have a single-entry and single-exit node (SESE) represent an isolated part of the model. This way, localizing conformance problems becomes easier in large models. It would be interesting to combine the TCG algorithm with such decomposed models.

A sub-field of alignments is to compute a *prefix-alignment* for an incomplete log trace. This is useful for analysing processes in real-time instead of a-posteriori. Several techniques exist for computing prefix-alignments [7,19]. The TCG approach that we introduced in this paper could also be suitable for computing prefix-alignments. Recently, Burattin and Carmona [20] introduced a technique similar to the TCG approach, in which the marking graph is extended with additional edges to allow for deviations. However, it cannot guarantee optimality as a *single* successor marking is chosen per event, while instead we consider all possible successors and can, therefore, better adapt for future events.

In a more general setting, conformance checking is related to finding a *longest common subsequence*, computing a *diff*, or computing minimal *edit distances*. Here, the problem is translated to searching for a string B from a regular language L such that the edit distance of B and an input word α is minimal [21].

9   Conclusion

In this paper, we considered a max-sync cost function that instead of minimizing discrepancies between the log trace and the model, maximizes the number of synchronous moves. We empirically evaluated the differences with the standard cost function, compared the alignment computation times. The max-sync cost function also lead to a new algorithm for computing alignments.

We observed that in general, a considerable amount of model moves may be required to add a few additional synchronous moves, when comparing max-sync with the standard cost function. However, when alignment problems are structured such that log moves are on a lower granularity than the model, a max-sync cost function may be better suited. We also observed a significant performance improvement in alignment construction if alignments can be formed without taking any model moves or without any log moves.

On industrial models with many log traces, we showed that our new algorithm, which uses a preprocessing step on the model, is an order of magnitude faster in computing alignments on many log traces for the max-sync cost function.

We conclude that the max-sync cost function is complementary to the standard one as it provides an alternative view that may be preferable in some contexts, and it may also significantly reduce the alignment construction time.

References

1. van der Aalst, W.M.P.: Process Mining: Data Science in Action. Springer, Heidelberg (2016). https://doi.org/10.1007/978-3-662-49851-4
2. Liu, C., van Dongen, B.F., Assy, N., van der Aalst, W.M.P.: Component behavior discovery from software execution data. In: 2016 IEEE Symposium Series on Computational Intelligence, SSCI 2016, 6–9 December 2016, pp. 1–8 (2016)
3. Leemans, M., van der Aalst, W.M.P.: Process mining in software systems: discovering real-life business transactions and process models from distributed systems. In: 18th ACM/IEEE International Conference on Model Driven Engineering Languages and Systems, MoDELS 2015, 30 September–2 October 2015, pp. 44–53 (2015)
4. Rozinat, A., van der Aalst, W.M.P.: Conformance checking of processes based on monitoring real behavior. Inf. Syst. **33**(1), 64–95 (2008)
5. van der Aalst, W.M.P., Adriansyah, A., van Dongen, B.F.: Replaying history on process models for conformance checking and performance analysis. Wiley Interdiscip. Rev.: Data Min. Knowl. Discov. **2**(2), 182–192 (2012)
6. Adriansyah, A., Sidorova, N., van Dongen, B.F.: Cost-based fitness in conformance checking. In: 11th International Conference on Application of Concurrency to System Design, ACSD 2011, 20–24 June 2011, pp. 57–66 (2011)
7. Adriansyah, A.: Aligning observed and modeled behavior. Ph.D. thesis, Eindhoven University of Technology, The Netherlands (2014)
8. Bloemen, V., van de Pol, J., van der Aalst, W.M.P.: Symbolically aligning observed and modelled behaviour. In: 18th International Conference on Application of Concurrency to System Design, ACSD 2018, 24–29 June 2018 (2018)
9. van Zelst, S.J., Bolt, A., van Dongen, B.F.: Tuning alignment computation: an experimental evaluation. In: Proceedings of the International Workshop on Algorithms and Theories for the Analysis of Event Data, ATAED 2017, 25–30 June 2017, pp. 1–15 (2017)
10. Adriansyah, A., van Dongen, B.F., van der Aalst, W.M.P.: Memory-efficient alignment of observed and modeled behavior. Technical report (2013)

11. Sudkamp, T.A.: Languages and Machines: An Introduction to the Theory of Computer Science. Addison-Wesley Longman Publishing Co., Inc., Boston (1988)
12. van der Aalst, W.M.P., Bolt, A., van Zelst, S.J.: RapidProM: mine your processes and not just your data. CoRR abs/1703.03740 (2017)
13. Kant, G., Laarman, A., Meijer, J., van de Pol, J., Blom, S., van Dijk, T.: LTSmin: high-performance language-independent model checking. In: Baier, C., Tinelli, C. (eds.) TACAS 2015. LNCS, vol. 9035, pp. 692–707. Springer, Heidelberg (2015). https://doi.org/10.1007/978-3-662-46681-0_61
14. Verbeek, H.M.W., Buijs, J.C.A.M., van Dongen, B.F., van der Aalst, W.M.P.: XES, XESame, and ProM 6. In: Soffer, P., Proper, E. (eds.) CAiSE Forum 2010. LNBIP, vol. 72, pp. 60–75. Springer, Heidelberg (2011). https://doi.org/10.1007/978-3-642-17722-4_5
15. Jouck, T., Depaire, B.: PTandLogGenerator: a generator for artificial event data. In: Proceedings of the BPM Demo Track 2016 Co-located with the 14th International Conference on Business Process Management (BPM 2016), 21 September 2016, pp. 23–27 (2016)
16. Fahland, D., Favre, C., Koehler, J., Lohmann, N., Völzer, H., Wolf, K.: Analysis on demand: instantaneous soundness checking of industrial business process models. Data Knowl. Eng. 70(5), 448–466 (2011)
17. Cook, J.E., Wolf, A.L.: Software process validation: quantitatively measuring the correspondence of a process to a model. ACM Trans. Softw. Eng. Methodol. 8(2), 147–176 (1999)
18. Polyvyanyy, A., Vanhatalo, J., Völzer, H.: Simplified computation and generalization of the refined process structure tree. In: Bravetti, M., Bultan, T. (eds.) WS-FM 2010. LNCS, vol. 6551, pp. 25–41. Springer, Heidelberg (2011). https://doi.org/10.1007/978-3-642-19589-1_2
19. van Zelst, S.J., Bolt, A., Hassani, M., van Dongen, B.F., van der Aalst, W.M.P.: Online conformance checking: relating event streams to process models using prefix-alignments. Int. J. Data Sci. Anal. (2017)
20. Burattin, A., Carmona, J.: A framework for online conformance checking. In: Teniente, E., Weidlich, M. (eds.) BPM 2017. LNBIP, vol. 308, pp. 165–177. Springer, Cham (2018). https://doi.org/10.1007/978-3-319-74030-0_12
21. Wagner, R.A.: Order-n correction for regular languages. Commun. ACM 17(5), 265–268 (1974)

Online Conformance Checking Using Behavioural Patterns

Andrea Burattin[1]([envelope]), Sebastiaan J. van Zelst[2], Abel Armas-Cervantes[3], Boudewijn F. van Dongen[2], and Josep Carmona[4]

[1] Technical University of Denmark, Kgs. Lyngby, Denmark
andbur@dtu.dk
[2] Eindhoven University of Technology, Eindhoven, The Netherlands
[3] The University of Melbourne, Melbourne, Australia
[4] Universitat Politècnica de Catalunya, Barcelona, Spain

Abstract. New and compelling regulations (e.g., the GDPR in Europe) impose tremendous pressure on organizations, in order to adhere to standard procedures, processes, and practices. The field of conformance checking aims to quantify the extent to which the execution of a process, captured within recorded corresponding event data, conforms to a given reference process model. Existing techniques assume a *post-mortem* scenario, i.e. they detect deviations based on *complete* executions of the process. This limits their applicability in an online setting. In such context, we aim to detect deviations online (i.e., *in-vivo*), in order to provide recovery possibilities before the execution of a process instance is completed. Also, current techniques assume cases to start from the initial stage of the process, whereas this assumption is not feasible in online settings. In this paper, we present a generic framework for online conformance checking, in which the underlying process is represented in terms of behavioural patterns and no assumption on the starting point of cases is needed. We instantiate the framework on the basis of Petri nets, with an accompanying new unfolding technique. The approach is implemented in the process mining tool ProM, and evaluated by means of several experiments including a stress-test and a comparison with a similar technique.

Keywords: Conformance checking · Online processing
Behavioural patterns · Stream processing · Petri nets · Unfoldings

1 Introduction

Organizations are facing challenges that arisie by digital transformation. Important concerns to face are the way processes are managed, their strategic alignment w.r.t. the organization's goals and their compliance with respect to applicable regulations. An example of these challenges is the compliance with the new regulations on the protection of data in Europe, i.e. *GDPR*[1], where unprecedented requirements on the use of data of EU citizens by organizations will

---

[1] See http://eur-lex.europa.eu/eli/reg/2016/679/oj.

© Springer Nature Switzerland AG 2018
M. Weske et al. (Eds.): BPM 2018, LNCS 11080, pp. 250–267, 2018.
https://doi.org/10.1007/978-3-319-98648-7_15

be applicable from May 2018. Are current business processes in organizations aligned with these new regulations?

Conformance checking is acknowledged as one of the key enabling technologies for verifying compliance monitoring of regulations [11]. It compares (prescriptive) process models to the actual execution of a process, and allows us to pinpoint deviations. The detection of compliance problems can be narrowed to the set of detected deviations [1]. In spite of being a powerful aid, a rigid exploration of conformance checking techniques has only been performed relatively recently [2,4,8,14,17,18,20–22].

A widespread application mode of conformance in literature is *post-mortem*: the relation between the model and the observed behaviour is computed, assuming that traces of observed process behaviour are complete. Such analysis, though meaningful and accurate, only allows us to detect deviations *after they occurred*, which, in some contexts, is too late. For example, consider the case where a trace of process behaviour represents the treatment of a patient during her life, and the model encompasses the clinical guidelines to follow for a given disease.

In contrast, *online conformance checking* techniques consider a live, real-time stream of events as input, where every event belongs to a particular case, i.e. process instance. As such, several different unfinished (running) cases at any position in the stream need to be considered [4,22]. Moreover, in real scenarios cases may start at different points in the process, not necessarily in its initial stage, e.g. a patient process being monitored in the middle of her clinical life. Such *warm start* mode of online conformance checking allows us to not only analyze cases from which the full history is available, but also those cases that lack historical process information.

In this paper, we present a novel framework, accompanied with a corresponding instantiation that builds on top of the notion of Petri net unfoldings [13], that enables the application of online conformance checking in warm start settings. To the best of our knowledge, this is the first solution for this important problem. We present a framework that relies on the notion of *behavioural patterns*, i.e., relations between process activities. In particular, for each possible behavioural pattern, the number of different behavioural patterns preceding/following it for a case is assumed to be known. Subsequently, the approach assesses *compliance* by checking whether the expected behavioural patterns are either observed or violated. Additionally, *completeness* (is the running case expected to be complete?) and *confidence* (is the compliance metric reliable?) values provide a more holistic view on the compliance of running cases.

We provide an instance of the framework based on *weak order relations*, accompanied by an implementation in the process mining framework ProM [19]. We validate the approach by means of a synthetic data set containing models and traces of varying sizes and a data set containing cases that start in different stages of the process (warm start). Furthermore we assess the applicability of the approach on a real data set. We also asses the correlation of the technique w.r.t. the technique presented in [22], which confirms that our framework provides a good estimation of conformance.

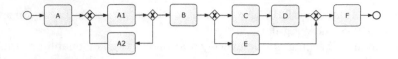

Fig. 1. Running example considered throughout this paper.

Table 1. Comparison of offline [2] and online conformance values (as proposed in this paper) based on the process model in Fig. 1.

Trace	Offline	Online		
	Conformance	Conformance	Completeness	Confidence
$t_1 = \langle A, A1, B, E, F \rangle$	1.00	1.00	1.00	1.00
$t_2 = \langle B, C, D, F \rangle$	0.78	1.00	0.60	1.00
$t_3 = \langle A, A1, A2, A1, B \rangle$	0.80	1.00	1.00	0.50
$t_4 = \langle B, C, D \rangle$	0.62	1.00	0.50	0.75

The remainder of the paper is structured as follows. In Sect. 2, we motivate the need for an online conformance checking technique capable of handling the warm start scenario. In Sect. 3, we present related work. In Sect. 4, we briefly present background terminology. In Sect. 5, the general framework is described, which is instantiated in Sect. 6 for weak order relations. We evaluate the instantiation in Sect. 7. In Sect. 8 we discuss limitations of the work, whereas Sect. 9 concludes this paper.

2   Motivation

Consider the process model reported in Fig. 1. Furthermore, consider some possible executions of such process and their corresponding conformance values as reported in Table 1. Trace t_1 conforms w.r.t. the model: it represents a possible complete execution of the process. This information is properly captured by both the offline technique [2] and our online approach. Execution t_2, on the other hand, is compliant with the process but just from activity B onward, i.e. assuming that the initial activity A was executed yet not observed. Such case is known as a warm start scenario: we start monitoring ongoing process instances rather than processes started after monitoring. Our approach is explicitly designed to deal with this problem by additionally quantifying the *completeness of the execution*. Note that, as Table 1 reports, offline approaches do not capture the notion of completeness, and thus, in case of warm start, the final conformance value is simply decreased. Trace t_3 suffers from the opposite problem: it conforms to the process model only up-until activity B, i.e., we expect to observe future behaviour. If we do not assume to be in a *post-mortem* scenario, this trace has no conformance problem, but is simply *partial*. Our approach is designed to explicitly handle this problem by quantifying the *confidence* of the execution, i.e. the degree of reliability of the reported conformance metric. Again, Table 1 shows that offline techniques cannot handle this situation. The combination of the last two problems is present in trace t_4, i.e. the trace captures an intermediate execution of the process which conforms the model but lacks initial and final parts of the execution. The offline approach reports a conformance of 0.62, whereas the online approach indicates that, subject to incompleteness and a little lack of confidence, the behaviour as seen conforms to the model.

3 Related Work

Until recently, conformance checking has only focused on relating modeled and observed behaviour in a post-mortem fashion. Techniques for this task have been proposed, with different assumptions and guarantees. Among existing techniques, we observe: rule-based [14,20], token replay-based [14], and alignment-based techniques [2,3,8,17,18,21]. The work presented in this paper can be seen as an evolution of the rule-based approaches, where important new features, i.e. from offline to online and the warm start capability, have been properly incorporated.

For online conformance checking, we identify two research lines. In [22] the authors propose to compute *prefix-alignments*, i.e. providing explanations for prefixes of complete behaviour. Unfortunately the complexity requirements are high and the technique is unable to handle the warm start scenario. An alternative approach is presented in [4], where all the possible deviations are precomputed on top of the model behaviour, which is used to walk through the input stream.

4 Background

4.1 Process Models and Behavioural Patterns

We do not assume a specific process modelling formalism, yet we do assume process models to be defined in context of collections of activities. As such, we assume a process model to constrain the relative ordering of its activities, e.g. reconsider the BPMN diagram in Fig. 1, which specifies that we are able to execute activity A prior to activity $A1$, yet the reverse is not the case. We furthermore assume the execution process activities to be atomic. A model M is potentially an imperative model, e.g. BPMN, Petri net or EPC. The only requirement we impose on the considered model(s) is the fact that we are able to deduce a language in terms of the activities it is defined upon.

Given a process model, with a corresponding language and relative ordering on its activities, we assume that we are able to derive more advanced behavioural relations, i.e. *behavioural patterns*, such as weak ordering, parallelism, causality and conflict. Given two activities part of a process model, formally, we define a behavioural pattern as a relation that the process imposes on them. As an example, consider the model in Fig. 1, which dictates that activity A is *always followed* by activity $A1$.

Definition 1 (Behavioural Pattern). *Given a set of activities $\mathcal{A}$ and a set of possible control-flow relations $\mathcal{R}$,* a behavioural pattern *is defined as $b(a_1, a_2)$ where $a_1, a_2 \in \mathcal{A}$ are activities and $b \in \mathcal{R}$ represents a control-flow relation. An alternative writing of $b(a_1, a_2)$ is $a_1\ b\ a_2$.*

Using the notion of behavioural patterns, we formalize process models as follows.

Definition 2 (Process Model). *A* process model *B is the set of all behavioural patterns prescribed by the process, such that $B \subseteq \mathcal{R} \times \mathcal{A} \times \mathcal{A}$, where $\mathcal{A}$ is the set of activities and $\mathcal{R}$ is the set of possible control-flow relations.*

In context of this paper, we are primarily interested in behavioural patterns induced by the possible sequential ordering of activities, i.e. we take a *control-flow perspective*. As such we assume the existence of a universe of *control-flow relations* $\mathcal{R}$ that allow us to induce behavioural patterns. Examples of control-flow relations present in $\mathcal{R}$ are defined in [15]. Consider for example the *weak order relation*. Let's assume the existence of two activities a_1 and a_2. They are in weak order relation, expressed as $a_1 \prec a_2$, if there exists an execution of the process where a_1 occurs before a_2. Such relations are used not only for the formal definition of the process, but also for the definition of our observations: instances of these relations represent the observable units against which we want to compute the conformance. For example, consider the BPMN model in Fig. 1. Based on the semantics of BPMN, we deduce, for the control-flow relation $\prec$ (weak order relation), to have $\{(\prec, A, A1), (\prec, A1, B), \ldots, (\prec, D, F), (\prec, E, F)\}$.

4.2   Data Streams

A data stream is typically defined as an infinite sequence of data items. As such, we define a sequence over set X of length n as a function $\sigma \colon \{1, \ldots, n\} \to X$, and an infinite sequence as $\sigma \colon \mathbb{N}^+ \to X$. We also refer to a sequence using string representation: $\sigma = \langle x_1, x_2, \ldots, x_n \rangle$ where $x_i = \sigma(i) \in X$. In context of this paper, the streams we observe refer to executions of a certain behavioural pattern. Therefore, we define an *observable unit* as a behavioural pattern which is observed in a process instance.

Definition 3 (Observable Unit). *Let $\mathcal{C}$ denote the set of case ids, let $\mathcal{R}$ denote the set of control-flow relations and let $\mathcal{A}$ denote the set of activities. Let $b \in \mathcal{R} \times \mathcal{A} \times \mathcal{A}$ denote a behavioural pattern. An* observable unit *$o = (c, b) \in \mathcal{C} \times \mathcal{R} \times \mathcal{A} \times \mathcal{A}$ is a tuple describing a behavioural pattern $b \in B$ that is observed in context of case id c.*

The universe of all possible observable units is defined as $\mathcal{O} = \mathcal{C} \times \mathcal{R} \times \mathcal{A} \times \mathcal{A}$.

For each observable unit we assume to have projection operators to extract the case id and the pattern i.e. given $o = (c, b)$, $\pi_c(o) = c$ and $\pi_b(o) = b$.

Definition 4 (Stream of Behavioural Patterns). *Given the universe of observable units $\mathcal{O} = \mathcal{C} \times \mathcal{R} \times \mathcal{A} \times \mathcal{A}$, a* stream of behavioural patterns *is defined as an infinite sequence of observable units: $S : \mathbb{N}^+ \to \mathcal{O}$.*

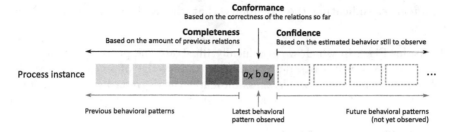

Fig. 2. General idea of the 3 conformance measures computed based on a partially observed process instance: *conformance*, *completeness*, and *confidence*.

A stream of behavioural patterns can be seen as an unbounded sequence of observable units where their ordering complies with the time order of the observable units, as defined by the underlying execution time of the corresponding activities. Note that, a stream of behavioural patterns refers to information at a high level of abstraction, i.e. when compared to the commonly used stream of executed process events [5]. However, under specific circumstances, e.g. the behavioural pattern considered in Sect. 6 (we consider a stream of direct follows relations), a stream of behavioural patterns is easily extracted from a stream of simple events. We refer to [5,6], where techniques to convert a stream of events to a stream of behavioural patterns are described.

5 Online Conformance Checking Using Behavioural Patterns

In this section we present conformance checking in terms of behavioural patterns. We first present the envisioned requirements for an online conformance checking approach after which we propose a generic framework that fulfills these requirements.

5.1 Problem Statement

Existing conformance checking techniques quantify conformance using one specific metric, typically in terms of compliance or deviation costs. In online settings however, we suffer both from the fact that we perform in-vivo analysis, i.e. new event data is likely to be observed in the future, as well as the warm start scenario. Using only one metric to express conformance, therefore, leads to misleading results, i.e. cases that already started and/or that are not finished yet get falsely penalized for this. To solve these issues, we propose a breakdown of conformance in:

1. *Conformance*: Indicating the amount of correct behaviour observed thus-far;
2. *Completeness*: Indicating whether the entire trace is observed since the beginning.

3. *Confidence*: Indicating the possibility that the conformance score remains stable.

Consider Fig. 2 in which we graphically illustrate the proposed conformance metrics. *Conformance* is based on the current knowledge of a case, witnessed by the observed behaviour. *Completeness* indicates the degree to what behaviour is potentially missed for a case. *Confidence* signifies to what degree we are able to trust the conformance metric, i.e. if more behaviour is expected in the future, deviations may occur later as well.

5.2   Process Representation

The foundation of our online conformance checking technique is the notion of behavioural pattern. Hence, we need a model capturing the following information:

1. The set of behavioural patterns prescribed by the model;
2. For each behavioural pattern, the minimum and maximum number of distinct prescribed patterns that must be *observed before*, since the beginning of the case;
3. For each behavioural pattern, the minimum number of distinct patterns *still to observe* in order to reach the end of the process (as prescribed by the reference model).

We formalize such (process) model as follows.

Definition 5 (Process Model for Online Conformance (PMOC)). *A process model for online conformance (PMOC) $M = (B, P, F)$ is defined as a triplet containing the set of prescribed behavioural patterns B. Each pattern is defined according to Definition 1. P contains, for each behavioural pattern $b \in B$, the pair of minimum and maximum number distinct prescribed patterns (i.e., B) to be seen before b. We refer to these values as $P_{\min}(b)$ and $P_{\max}(b)$. Finally, for each pattern $b \in B$, $F(b)$ refers to the minimum number of distinct patterns (i.e., B) required to reach the end of the process from b.*

5.3   Computing Online Conformance Metrics

The procedure for the online computation of the conformance checking is reported in Algorithm 1. The algorithm requires a stream of behavioural patterns (cf. Definition 4) and a PMOC (cf. Definition 5) as input. The algorithm initializes two maps/functions: obs and inc (lines 1–2). Given a case id as key, these maps store the set of observed prescribed behavioural patterns and the number of observed patterns not prescribed. Note that, for each case, the amount of data to store is bounded by the model, and thus, constant w.r.t. the stream.

The online conformance procedure has an infinite loop to process the unbounded stream of behavioural relations (lines 3 and 4). The procedure is then split into 3 steps: *(i)* updating the maps; *(ii)* computing the conformance;

Algorithm 1: Online conformance computation

Input: S: stream of behavioural patterns
$M = (B, P, F)$: process model for online conformance

1   Initialize map obs   // Maps case ids to (finite) set of observed prescribed patterns
2   Initialize map inc   // Maps case ids to integers
3   **forever do**
4      $(c, b, t) \leftarrow observe(S)$ // New observable unit from the stream

       // Step 1: update internal data structures
5   **if** $b \in B$ **then**
6 | $obs(c) \leftarrow obs(c) \cup \{b\}$ // If b already in obs(c), then no effect
7   **else**
8 | $inc(c) \leftarrow inc(c) + 1$

       // Step 2: compute online conformance values
9      $conformance(c) \leftarrow \dfrac{|obs(c)|}{|obs(c)| + inc(c)}$
10   Notify new value of conformance(c)
11   **if** $b \in B$ **then**
12   **if** $P_{\min}(b) \leq |obs(c)| \leq P_{\max}(b)$ **then**
13 | $completeness(c) \leftarrow 1$
14   **else**
15         $completeness(c) \leftarrow \min\left\{1, \dfrac{|obs(c)|}{P_{\min}(b) + 1}\right\}$
16       $confidence(c) \leftarrow 1 - \dfrac{F(b)}{\max_{b' \in B} F(b')}$
17   Notify new values of completeness(c) and confidence(c)

       // Step 3: cleanup
18   **if** *size of* obs *and* inc *is close to max capacity* **then**
19 | Remove oldest entries from obs and inc

---

and *(iii)* housekeeping. In the first step (lines 5–8) the obs and inc data structures are updated with the new observation: if the pattern refers to prescribed relation, then it is added to the obs(c) set[2]. Otherwise, the value of incorrect observations is incremented.

The second step of the algorithm (lines 9–7) computes the actual conformance. The *conformance* for a (partial) process instance c is calculated in line 9: the number of distinct observed prescribed patterns in c (i.e., $|obs(c)|$) divided by the sum of the number of prescribed observed patterns and the incorrect patterns (i.e., $|obs(c)| + inc(c)$). We quantify, in the interval $[0,1]$, the correct behaviour observed, where 1 indicates full conformance (i.e., no incorrect behaviour) and 0 indicates no conformance at all (i.e., only incorrect behaviour). Completeness and confidence are updated only when a prescribed behavioural pattern is observed (line 11) since they require to locate the pattern itself in the process. Specifically, the *completeness* of process instance c is calculated in lines 12–15. It depends on whether the number of distinct behavioural patterns observed so far is within the expected interval for current pattern b (i.e.,

---

[2] If obs has no key c, obs(c) returns the empty set. If inc has no key c then inc(c) returns 0.

$P_{\min}(b) \leq |\mathrm{obs}(c)| \leq P_{\max}(b)^3)$ or not. In the former case, we assume completeness is perfect (therefore value 1). In the latter case, the problem could be due to two reasons: we observe less patterns than expected ($|\mathrm{obs}(c)| < P_{\min}(b)$) and in this case we have the ratio of observed pattern over the minimum expected. Alternatively we observe more behavioural patterns than expected ($|\mathrm{obs}(c)| > P_{\max}(b)$) and in this case we assume a completeness value of 1. Note that this last case could represent a "false positive": we count the number of observed correct patterns without checking which exact patters we are dealing with. This approximation is imposed by online processing constraints. Finally, the *confidence* of case c is calculated in line 16 as 1 minus the ratio of patterns still to observe (i.e., $F(b)$) and the overall maximum number of future patterns (i.e., $\max_{b' \in B} F(b')$). Confidence also ranges in $[0, 1]$: 1 indicates strong confidence (i.e., the execution reached the end of the process), 0 means low confidence (i.e., the execution is still far from completion, therefore there is room for changes). Observe that, the metrics computed by the algorithm implement the metrics described in the problem statement section (cf. Subsect. 5.1).

The third step of the algorithm (lines 18, 19) consists of cleanup operations. Specifically, only a finite amount of memory is available: we can store only some process instances. This step of the algorithm takes care of that: once the size of obs and inc reaches the memory limit, oldest entries are removed. For the sake of readability, we do not focus on the actual procedures to achieve that (cf. [5,6] for possible solutions).

Suitability of the Algorithm for Online Settings. The computational complexity of the main loop of the algorithm is constant for each event (given the reference model as input). Specifically, step 1 (lines 5–8) updates hash maps in constant time. All computations in step 2 (lines 9–17) require constant time complexity (note that $\max_{b' \in B} F(b')$ depends just on the model and can be pre-computed in advance). Finally, step 3 (lines 18, 19), can be realized to require constant time complexity (e.g., using LinkedHashMaps). The space required by the procedure is bounded by an imposed maximum number of keys in obs and inc. Then, since obs stores sets of prescribed behavioural patterns (which are finite) and inc stores just one integer, the whole memory can not grow above the imposed threshold. Since processing a single event takes a constant amount of time and fixed amount of space, the procedure is suitable for online processing.

6   Online Conformance Checking Using Weak Ordering Relations

In this section, we present an instantiation of the framework proposed in this paper. We do so by computing three matrices out of the original model before

---

[3] $P_{\min}(b)$ and $P_{\max}(b)$ refer to the min./max. number of distinct patterns to be seen before b.

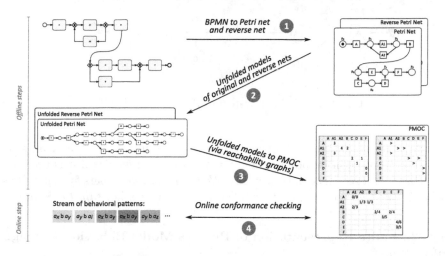

Fig. 3. General idea of the approach presented in this paper. Steps 1–3 are performed once, offline. Step 4 is the only online activity.

the actual online analysis. These matrices contain the information about the possible relations between pairs of activities in the process model (behavioral patterns) that is needed by PMOC (cf. Definition 5). In particular, for each possible behavioural pattern, we compute the (min. and max.) number of different behavioural patterns preceding/following it for each case in the model. The computation of these matrices allows us to retrieve information online in constant time. The roadmap for the computation of the three matrices out of a process model is shown in Fig. 3, while each of the steps is described in more detail in the remaining of the section.

Step ①: Input Process Models

As mentioned in Sect. 4, we do not assume a specific process modelling formalism. However, in the context of this particular instantiation, we assume that the model can be represented as a Petri net, possibly through a transformation from other process modelling languages (e.g., transforming BPMN into Petri nets [7]). For instance, Fig. 4 shows the Petri net system representation of the BPMN process in Fig. 1, where transitions, places, arcs and tokens are represented as squares, circles, directed black arrows and black dots, respectively.

Given a (transformed) Petri net, an additional *reverse* net is computed. The reverse net is a net with the same set of places and transitions as the original one, but where the direction of the edges is inverted. The use of this additional net is made clear in Step ③. Some notions used later in this section relate to the execution semantics of Petri nets, which we briefly/informally introduce here. A transition t is enabled iff there is at least one token in each place in the preset of t. An enabled transition t can be fired and, as a consequence, modifies the distribution of tokens over the net, thus producing a new marking. The firing

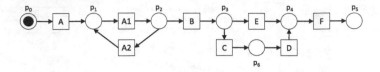

Fig. 4. Labeled net system of the model in Fig. 1.

of a transition t removes one token from each place in its preset and puts one token in each place in its poset. Finally, a marking is *reachable* if it is produced by the firing of a sequence of transitions. We restrict to Petri nets systems whose reachable markings contain up to 1 token in every place, i.e. *safe Petri nets*.

Step ②: Finite Representation of Process Model Behavior Through Unfoldings

The information about the behavioral patterns required by our framework can be extracted by analyzing the state space (markings) of the Petri net. Specifically, at each marking, the number of behavioral patterns are computed and counted, and the number of different behavioural patterns preceding/following the last observed pattern is stored. Nevertheless, if a net is cyclic then the number of behavioural patterns it can produce is infinite. Several authors have proposed techniques for computing finite Petri net representations of the behavior of a net known as *complete prefix of an unfolding*. For instance, [13] introduces a way to truncate the unfolding of a net at a finite level, while keeping a representation of any reachable marking. Then, a framework for constructing a *canonical unfolding prefix*, complete with respect to a suitable property, not limited to reachability, was proposed in [10]. Our own work relies on such a framework, i.e. we compute a finite fragment of the unfolding capturing enough information about the distinct behavioral patterns in a net.

The new unfolding, specially developed for this instantiation, analyses each reachable marking at every possible case and computes the set of behavioral patterns between the transitions (activities) that were fired to reach such marking. The idea of this new unfolding is to keep firing transitions in the original net and create new instances of places and transitions whenever they are fired, in the case of transitions, or visited by a token, in the case of places. Then, the unfolding stops once it finds information that has been observed before. As a concrete example, consider the *weak order relation* between activities. Figure 5 shows the complete prefix unfolding for the running example (unfolding of the net shown in Fig. 4). Observe that p_2', p_2'' and p_2''' are instances of the place p_2. However, the unfolding stopped at p_2''' because the weak order relations are the same as those captured at the marking in p_2''. In [10], the necessary conditions that a notion of equivalence between execution states shall satisfy to guarantee that the complete prefix unfolding is canonical and finite are defined. In our case, a pair of markings are equivalent if they have *(i)* the same places, *(ii)* the same relations (i.e., weak order) between activities executed to reach such marking,

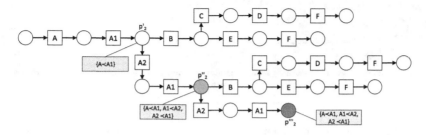

Fig. 5. Weak order relation preserving unfolding computed by our new unfolding technique. The unfolding stops when reaching p_2''' since the induced weak order relations are the same as those observed in p_2''.

and *(iii)* the same set of activities that were lastly executed for reaching such markings. These conditions allow to prove canonicity and finiteness of the new complete prefix unfolding.

Step ③: Computation of PMOC's Data Structures via Reachability Graphs

Given the complete prefix unfolding described above, different ways to compute the weak order relations can be envisioned. For simplicity, in our implementation we construct the corresponding reachability graph $TS = (S, TR, s_0)$. Such graph, which is always finite, is used to derive the set of allowed weak order relations: for each state $s \in S$ it is possible to compute the set $t_{in}^s \subset TR$ with all non-silent transitions immediately leading to s, and the set $t_{out}^s \subset TR$ with all non-silent transitions immediately leaving s. In case there is a silent transition connected to s it is necessary to recursively follow it and retrieve all incoming/outgoing transitions which will be part of t_{in}/t_{out}. The set $\bigcup_{s \in S}\{x \prec y \mid x \in t_{in}^s, y \in t_{out}^s\}$ represents all weak order relations that can be extracted from TS. A weak order relation $x \prec y$, defined as x entering $s \in S$ and y leaving s, might appear several time in TS. By finding the longest and shortest paths from s_0 to all occurrences of s, and converting these paths into distinct weak order relations, it is possible to identify the minimum and maximum number of weak order relations preceding $x \prec y$.

For the purpose of this paper we do not only require the minimum and maximum number of relations preceding a given one, but also the minimum number of relations required to reach the end of the model. Thus, we use the reverse net for computing such information by computing the complete prefix of the reverse net (reusing the methodology in Step ②), and then counting the distinct relations over the corresponding reachability graph.[4] Observe that

---

[4] In general, not all Petri nets can be reversed for computing the minimal number of relations to reach the *end*. Hence, for computing confidence, we assume in the realization of the framework presented in Sect. 6 a proper subclass, i.e., sound workflow nets.

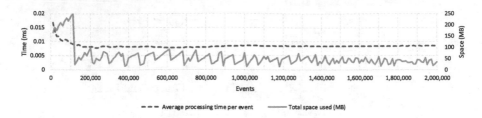

Fig. 6. Performance of the system during a stress test involving 2 million events.

by inverting the direction of the weak order relations in the reverse net, we obtain information referring to the end of the model: the distances now refer to the minimum/maximum number of relations to reach the *end*. The techniques described in this section allow the computation of the information needed to have a proper process model abstraction for online conformance checking (cf. Definition 5).

7   Experimental Evaluation

In this section, we present an experimental evaluation of the proposed techniques in terms of performance, as well as its indicative power of conformance. We additionally compare our technique against an alternative, state-of-the-art online conformance checking technique. The proposed technique is available as ProM plugin.[5]

7.1   Stress Test

We performed a stress test of our prototype. We randomly generated a BPMN model containing 64 activities and 26 gateways. The model was then used to simulate an event stream of 2 million events[6]. The test was performed on a standard machine, equipped with Java 1.8(TM) SE Runtime Environment on Windows 10 64 bit, an Intel Core i7-7500U 2.70 GHz CPU and 16 GB of RAM. Results of the test are reported in Fig. 6. After an initial phase, when the constructed data structures were still in memory, the Java Virtual Machine was able to remove these unreferenced objects. This explains the drop in the memory and the stabilization of the processing time, after about 100k event. From that moment on, the memory used remained permanently around 100 MB and the average processing time persisted below 0.009 ms/event.

This test shows that the implemented prototype is capable of sustaining a high load of events on a standard laptop machine. Moreover, we observe that both the processing time and memory usage show a relatively stable, non-increasing trend. This aligns well with our expectations and the general requirements of data stream analysis.

---

[5] See https://svn.win.tue.nl/repos/prom/Packages/StreamConformance/.

[6] Models and streams available at https://doi.org/10.5281/zenodo.1194057.

7.2   Correlation with Alternative Conformance Metrics

In this section, we examine the correlation of the proposed metrics with the alternative described in [22], which reports a potential deviation in terms of *costs*, rather than a conformance metric. Hence, the higher the cost of deviation, the less conformance. As the metric in [22] is a more informed technique (at the expense of using more memory) than the one proposed in this paper, a correlation between both metrics shows that our technique reflects online conformance well.

We generated 12 random process models [9] with number of activities according to a triangular distribution with lower bound 10, mode 20, and upper bound 30. We did not include duplicate labels, a probability of 0.2 for addition of silent activities, moreover, the probability of control-flow operator insertion was: 0.45 for sequence, 0.2 for parallel and xor-split operators, 0.05 for an inclusive-or operator and 0.1 for loop constructs. From these models a collection of event logs has been created (each log contains 1000 traces), subsequently treated as streams by both techniques. Incremental noise levels (both on a trace- and event-level) were introduced in the logs. Probability of trace- and event-level noises ranged from 0.1 to 0.5 with steps of 0.1. In order to compute the conformance, the technique presented in this paper needs, at least, two events. Hence for a fair comparison, we only consider conformance values from the second event onward, yielding a total of 2,977,744 analyzed events (See footnote 6).

In Fig. 7 we present a scatter-plot of the conformance metric (this paper) versus the incremental alignment-based costs (alternative approach). Figure 7a plots all results, i.e. all events, where the size of the dot indicates the number of instances for the specific value combination. *Spearman's rank correlation coefficient* for the whole data set (ρ-value) is -0.9538502. As the chart reports, coordinate $(0, 1.0)$ dominates the data (in 73.4% of cases both techniques agree on no deviation). Hence, the data is extremely skewed (vast majority of results at coordinate $(0, 1.0)$) which explains the strong negative correlation. Nonetheless, the result shows that the two metrics generally agree when no deviations occur. In Fig. 7b, we present the same results but only for combinations in which at least one of the techniques identifies a deviation. In general, when alignment costs increases, the conformance metric decreases. However, we observe that the conformance values are spread around, i.e. we do not observe a clear linear trend. This is supported by the corresponding ρ-value of -0.2951334, presented in Table 2, which shows a correlation matrix for non-conforming results (cf. Fig. 7b) of the conformance metrics presented in the paper and the costs as defined in [22]. Correlations among the metrics presented in this paper are the strongest. The fact that completeness and conformance depict the strongest correlation is explained by the fact that the data set in general contains complete cases. For confidence, weaker correlation is found. Based on the data used, it is expected that once a case matures, relatively more correct behaviour is observed than incorrect behaviour. The correlation between costs and completeness is negligible. For costs and confidence we observe a weak positive correlation: towards the end of a trace, the likelihood of having observed noise, and thus costs, goes up.

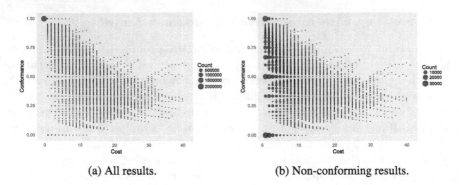

(a) All results.   (b) Non-conforming results.

Fig. 7. Scatter plots of conformance metric versus incremental alignment costs [22].

Table 2. Correlation matrix (ρ-values, Spearman) for non-conforming results (cf. Fig. 7b), showing the conformance metrics of this paper and costs as defined in [22].

		Metrics from this paper			Cost [22]
		Conformance	Completeness	Confidence	
Metrics from this paper	Conformance		0.52282662	0.3862707	−0.29513342
	Completeness			0.1851850	−0.02546182
	Confidence				0.25104526

We conclude that the two metrics largely agree when no noise is present (with a minor number of outliers). When both methods observe deviations, corresponding quantifications do not clearly correlate. This is partly due to the fact that the alignment based approach always explains observations in terms of the model, whereas the approach in this paper does not. Secondly, the use of weak order relation as a behavioural pattern leads to the use of a strong abstraction of the model: this representational bias seems not in-line with the deviation approximation of the alternative approach.

7.3   Real-World Event Data Test

Finally, we investigated the real event log of an information system managing road traffic fines for the Italian police [12]. This log has a reference model, designed with the help of domain experts and regulators [12]. To avoid the state explosion problem during the computation of the matrices, we removed self-loops from the model[7]. Additionally, to focus on most relevant traces, we discarded all process instances with just one or two activities. The resulting log contains 316 868 events, over 83 614 cases. The processing of the log, (excluded the offline computations, and with support for up to 10 000 process instance in

---

[7] This limitation only affects Sect. 6: it is possible to manually define the behavioural patterns.

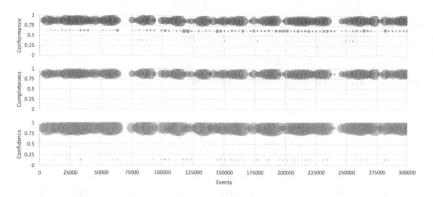

Fig. 8. Online conformance on the road traffic fines log.

parallel) took 44 967 ms (0.14 ms/event). Figure 8 contains the binned results of
the analysis. The x axis reports the different events (by time, grouped in bins
of 3 000 events). The y axes of the charts report conformance, completeness and
confidence levels (grouped in bins of 0.25). Each point represents several events
(bubble size proportional to number of events) but different process instances
can be intertwined. Therefore, two consecutive events could refer to cases with
very different conformance levels (this explains fluctuations). We can see that
the conformance values are mostly at 1: only few events deviated from the reference model (93.5% of the events have conformance 1 and 99.6% of events have
conformance ≥ 0.5). Average conformance value is 0.97, suggesting very high conformance in general. 99.8% of the events have completeness of 1: most executions
actually started from the beginning with just sporadic warm starts. Finally, confidence has mostly value 1 (99.4% of events). This is due to the specific behaviour
of the process which allows immediate termination of the execution right after
the execution of the first activity.

8 Discussion

The approach presented in this paper can be used to monitor any set of
behavioural patterns, i.e. we represent processes models as sets of prescribed
behavioural patterns and streams as infinite sequences of behavioural patterns.
Because of this, the framework is rather abstract and allows us to monitor any
possible set of relations. Note that we could also use the *organizational perspective*, rather than the control-flow perspective. An example relation which might
be relevant for monitoring purposes is whenever pairs of activities have to be performed collaboratively and simultaneously (i.e., *cooperation* [16]). The provided
instantiation automatically extract instances of weak order behavioural patterns
out of a Petri net and an event stream. We focus on weak order relations since
they are widely used and relatively easy to deduce. Clearly, using more advanced
behavioural patterns such as causality, parallism and/or different perspectives,

i.e. organizational, requires a corresponding algorithmic design to deduce such patterns from the process model and/or stream under study.

9   Conclusions and Future Work

In this paper we present a generic approach to compute the conformance of data streams against a reference process model. In order to cope with all possible scenarios, the approach decomposes the actual conformance into 3 metrics: the actual *conformance*, the *completeness* and the *confidence*. Thus, the technique can be used on partial executions and on traces already running (i.e., warm start). Moreover, we provide an instantiation of the generic approach for the case of weak order relations, which is based on a new unfolding technique. This instantiation is implemented and available in ProM and it has been verified on large dataset for stress test, on a real dataset, and it has also been compared against a prefix-alignment based approach. As future work we plan to investigate further realizations of the framework, including declarative models, to understand which behavioural patterns are useful in order to converge towards optimal approaches bearing in mind that, being online, approximations must be in place.

Acknowledgments. This work has been partially supported by MINECO and FEDER funds under grant TIN2017-86727-C2-1-R.

References

1. van der Aalst, W.M.P.: Process Mining: Data Science in Action. Springer, Heidelberg (2016). https://doi.org/10.1007/978-3-662-49851-4
2. Adriansyah, A.: Aligning observed and modeled behavior. Ph.D. thesis, Technische Universiteit Eindhoven (2014)
3. vanden Broucke, S.K.L.M., Munoz-Gama, J., Carmona, J., Baesens, B., Vanthienen, J.: Event-based real-time decomposed conformance analysis. In: Meersman, R., et al. (eds.) OTM 2014. LNCS, vol. 8841, pp. 345–363. Springer, Heidelberg (2014). https://doi.org/10.1007/978-3-662-45563-0_20
4. Burattin, A., Carmona, J.: A framework for online conformance checking. In: Teniente, E., Weidlich, M. (eds.) BPM 2017. LNBIP, vol. 308, pp. 165–177. Springer, Cham (2018). https://doi.org/10.1007/978-3-319-74030-0_12
5. Burattin, A., Cimitile, M., Maggi, F.M., Sperduti, A.: Online discovery of declarative process models from event streams. IEEE TSC **8**(6), 833–846 (2015)
6. Burattin, A., Sperduti, A., van der Aalst, W.M.: Control-flow discovery from event streams. In: Proceedings of the IEEE CEC, pp. 2420–2427 (2014)
7. Dijkman, R., Dumas, M., Ouyang, C.: Semantics and analysis of business process models in BPMN. Inf. Softw. Technol. **50**(12), 1281–1294 (2008)
8. van Dongen, B., Carmona, J., Chatain, T., Taymouri, F.: Aligning modeled and observed behavior: a compromise between computation complexity and quality. In: Dubois, E., Pohl, K. (eds.) CAiSE 2017. LNCS, vol. 10253, pp. 94–109. Springer, Cham (2017). https://doi.org/10.1007/978-3-319-59536-8_7

9. Jouck, T., Depaire, B.: PTandLogGenerator: a generator for artificial event data. In: Proceedings of the BPM Demo Track, pp. 23–27 (2016)
10. Khomenko, V., Koutny, M., Vogler, W.: Canonical prefixes of Petri net unfoldings. Acta Informatica **40**(2), 95–118 (2003)
11. Ly, L.T., Maggi, F.M., Montali, M., Rinderle-Ma, S., van der Aalst, W.M.P.: Compliance monitoring in business processes: functionalities, application, and tool-support. Inf. Syst. **54**, 209–234 (2015)
12. Mannhardt, F., de Leoni, M., Reijers, H.A., van der Aalst, W.M.P.: Balanced multi-perspective checking of process conformance. Computing **98**(4), 407–437 (2016)
13. McMillan, K.L., Probst, D.K.: A technique of state space search based on unfolding. Formal Methods Syst. Des. **6**(1), 45–65 (1995)
14. Rozinat, A., van der Aalst, W.M.P.: Conformance checking of processes based on monitoring real behavior. Inf. Syst. **33**(1), 64–95 (2008)
15. Smirnov, S., Weidlich, M., Mendling, J.: Business process model abstraction based on behavioral profiles. In: Maglio, P.P., Weske, M., Yang, J., Fantinato, M. (eds.) ICSOC 2010. LNCS, vol. 6470, pp. 1–16. Springer, Heidelberg (2010). https://doi.org/10.1007/978-3-642-17358-5_1
16. Song, M.: Organizational mining in business process management. Ph.D. thesis, Pohang University of Science and Technology, Pohang, South Korea (2006)
17. Taymouri, F., Carmona, J.: A recursive paradigm for aligning observed behavior of large structured process models. In: La Rosa, M., Loos, P., Pastor, O. (eds.) BPM 2016. LNCS, vol. 9850, pp. 197–214. Springer, Cham (2016). https://doi.org/10.1007/978-3-319-45348-4_12
18. Taymouri, F., Carmona, J.: Model and event log reductions to boost the computation of alignments. In: Ceravolo, P., Guetl, C., Rinderle-Ma, S. (eds.) SIMPDA 2016. LNBIP, vol. 307, pp. 1–21. Springer, Cham (2018). https://doi.org/10.1007/978-3-319-74161-1_1
19. Verbeek, H.M.W., Buijs, J.C.A.M., van Dongen, B.F., van der Aalst, W.M.P.: XES, XESame, and ProM 6. In: Soffer, P., Proper, E. (eds.) CAiSE Forum 2010. LNBIP, vol. 72, pp. 60–75. Springer, Heidelberg (2011). https://doi.org/10.1007/978-3-642-17722-4_5
20. Weidlich, M., Polyvyanyy, A., Desai, N., Mendling, J., Weske, M.: Process compliance analysis based on behavioural profiles. Inf. Syst. **36**(7), 1009–1025 (2011)
21. van Zelst, S.J., Bolt, A., van Dongen, B.F.: Tuning alignment computation: an experimental evaluation. In: Proceedings of ATAED, pp. 6–20 (2017)
22. van Zelst, S.J., Bolt, A., Hassani, M., van Dongen, B.F., van der Aalst, W.M.P.: Online conformance checking: relating event streams to process models using prefix-alignments. Int. J. Data Sci. Anal. (2017). https://doi.org/10.1007/s41060-017-0078-6

Track II: Process Model Analysis and Machine Learning

BINet: Multivariate Business Process Anomaly Detection Using Deep Learning

Timo Nolle[✉], Alexander Seeliger, and Max Mühlhäuser

Telecooperation Lab, Technische Universität Darmstadt, Darmstadt, Germany
{nolle,seeliger,max}@tk.tu-darmstadt.de

Abstract. In this paper, we propose BINet, a neural network architecture for real-time multivariate anomaly detection in business process event logs. BINet has been designed to handle both the control flow and the data perspective of a business process. Additionally, we propose a heuristic for setting the threshold of an anomaly detection algorithm automatically. We demonstrate that BINet can be used to detect anomalies in event logs not only on a case level, but also on event attribute level. We compare BINet to 6 other state-of-the-art anomaly detection algorithms and evaluate their performance on an elaborate data corpus of 60 synthetic and 21 real life event logs using artificial anomalies. BINet reached an average F_1 score over all detection levels of 0.83, whereas the next best approach, a denoising autoencoder, reached only 0.74. This F_1 score is calculated over two different levels of detection, namely case and attribute level. BINet reached 0.84 on case and 0.82 on attribute level, whereas the next best approach reached 0.78 and 0.71 respectively.

Keywords: Business process management · Anomaly detection
Artificial process intelligence · Deep learning
Recurrent neural networks

1 Introduction

Anomaly detection is an important topic for today's businesses because its application areas are so manifold. Fraud detection, intrusion detection, and outlier detection are only a few examples. However, anomaly detection can also be applied to business process executions, for example to clean up datasets for more robust predictive analytics and robotic process automation (RPA). Especially in RPA, anomaly detection is an integral part because the robotic agents must recognize tasks they are unable to execute to not halt the process. Naturally, businesses are interested in anomalies within their processes, as these can be indicators for inefficiencies, insufficiently trained employees, or even fraudulent activities. Consequently, being able to detect such anomalies is of great value, for they can have an enormous impact on the economic well-being of the business.

© Springer Nature Switzerland AG 2018
M. Weske et al. (Eds.): BPM 2018, LNCS 11080, pp. 271–287, 2018.
https://doi.org/10.1007/978-3-319-98648-7_16

In today's digital world, companies rely more and more on process-aware information systems (PAISs) to accelerate their processes. A byproduct of such PAISs is an enormous data base that often remains unused. The log files these systems are storing can be used to extract valuable information about a process. One key data structure is an event log, which contains information about what activities have been executed in a process, who executed it, at which time, etc. These event logs are a great source of information and are frequently used for different data mining techniques, such as process mining.

In this paper, we propose BINet (Business Intelligence Network), a novel neural network architecture that allows to detect anomalies on attribute level. Often, the actual cause of an anomaly is only captured by the value of a single attribute. For example, a user has executed an activity without permission. This anomaly is only represented by the user attribute of exactly this event. Anomaly detection algorithms must work on the lowest (attribute) level, to provide the greatest benefit. BINet has been designed to process both the control flow (sequence of activities) and the data flow (see [1]).

Due to the nature of the architecture of BINet it can be used for ex-post analysis, but can also be deployed in a real-time setting to detect anomalies at runtime. Being able to detect anomalies at runtime is important because otherwise no counter measures can be undertaken in time. BINet can be trained during the execution of the process and therefore can adapt to concept drift. If unseen attribute values occur during the training, the network can be altered and retrained on the historic data to include the new attribute value in the future. Dealing with concept drift is also important as most business processes are flexible systems. BINet is a recurrent neural network architecture and therefore can detect point anomalies as well as contextual anomalies (see [12]). BINet works under the following assumptions.

- No domain knowledge about the process
- No clean dataset (i.e., dataset contains anomalous examples)
- No reference model
- No labels (i.e., no knowledge about anomalies)

In the context of business processes an anomaly is defined as a deviation from a defined behavior, i.e., the business process. An anomaly is an event that does not typically occur as a consequence of preceding events, specifically their order and combination of attributes. Anomalies that are attributed to the order of activities (e.g., two activities are executed in the wrong order) are called control flow anomalies. Anomalies that are attributed to the attributes (e.g., a user that is not part of a certain security group has illicitly executed an event) of events are called data flow anomalies.

Many anomaly detection algorithms rely on the manual setting of a threshold value to determine anomalies. We propose an unsupervised method for automatically setting the threshold using a heuristic.

We compare BINet to 6 state-of-the art anomaly detection methods and evaluate on a comprehensive dataset of 60 synthetic logs and 20 real-life logs, using artificial anomalies. This work contains four main contributions.

1. BINet neural network architecture[1]
2. Automatic threshold heuristic
3. Comprehensive evaluation of state-of-the-art methods.

2 Related Work

In the field of process mining [1], it is popular to use discovery algorithms to mine a process model from an event log and then use conformance checking to detect anomalous behavior [2,4,27]. However, the proposed methods do not utilize the event attributes, and therefore cannot be used to detect anomalies on attribute level.

A more recent publication proposes the use of likelihood graphs to analyze business process behavior [5]. Specifically, the authors describe a method to extend the likelihood graph to include event attributes. This method works on noisy event logs and includes important characteristics of the process itself by including the event attributes. A drawback of this method is that the attributes are checked in a specific order, thereby introducing a bias towards certain attributes.

A review of classic anomaly detection methodology can be found in [22]. Here, the authors describe and compare many methods that have been proposed over the last decades. Another elaborate summary on anomaly detection in discrete sequences is given by Chandola in [7]. The authors differentiate between five different basic methods for novelty detection: probabilistic, distance-based, reconstruction-based, domain-based, and information-theoretic novelty detection.

Probabilistic approaches estimate the probability distribution of the normal class, and thus can detect anomalies as they come from a different distribution. An important probabilistic technique is the sliding window approach [26]. In window-based anomaly detection, an anomaly score is assigned to each window in a sequence. Then the anomaly score of the sequence can be inferred by aggregating the window anomaly scores. Recently, Wressnegger et al. used this approach for intrusion detection and gave an elaborate evaluation in [28]. While being inexpensive and easy to implement, sliding window approaches show a robust performance in finding anomalies in sequential data, especially within short regions [7].

Distance-based novelty detection does not require a clean dataset, yet it is only partly applicable for process cases, as anomalous cases are usually very similar to normal ones. A popular distance-based approach is the one-class support vector machine (OC-SVM). Schölkopf et al. [23] first used support vector machines [9] for anomaly detection.

Reconstruction-based novelty detection (e.g., neural networks) is based on the idea to train a model that can reconstruct normal behavior but will fail to do so with anomalous behavior. Therefore, the reconstruction error can be

---

[1] https://github.com/tnolle/binet.

used to detect anomalies [15]. This approach has successfully been used for the detection of control flow anomalies [21] as well as data flow anomalies [20] in event logs of PAISs.

Domain-based novelty detection requires domain knowledge, which violates our assumption of no domain knowledge about the process. Information-theoretic novelty detection defines anomalies as the examples that influence an information measure (e.g., entropy) on the whole dataset the most. Iteratively removing the data with the highest impact will yield a cleaned dataset, and thus a set of anomalies.

The core of BINet is a recurrent neural network, trained to predict the next event and its attributes. The architecture is influenced by the works of Evermann [10,11] and Tax [24], who utilized long short-term memory [13] (LSTM) networks for next event prediction, demonstrating their utility. LSTMs have been used for anomaly detection in different contexts like acoustic novelty detection [18] and predictive maintenance [17]. These applications mainly focus on the detection of anomalies in time series and not, like BINet, on multivariate anomaly detection in discrete sequences of events.

The novelty of BINet lies in the tailored architecture for business processes, including the control and data flow, the scoring system to assign anomaly scores, and the automatic threshold heuristic.

3   Datasets

As a basis for the understanding of the following sections, we first need to define the terms case, event, log, and attribute. A log consists of cases, each of which consists of events executed within a process. Each event is defined by an activity name and its attributes, e.g., a user who executed the event. We use a nomenclature adapted from [1].

Definition 1. *Case, Event, Log, Attribute. Let C be the set of all cases and $\mathcal{E}$ be the set of all events. The event sequence of a case $c \in C$, denoted by $\hat{c}$, is defined as $\hat{c} \in \mathcal{E}^*$, where $\mathcal{E}^*$ is the set of all sequences over $\mathcal{E}$. An event log is a set of cases $\mathcal{L} \subseteq C$. Let $\mathcal{A}$ be a set of attributes and $\mathcal{V}$ be a set of attribute values, where $\mathcal{V}_a$ is the set of possible values for the attribute $a \in \mathcal{A}$. Note that $|\hat{c}|$ is the number of events in case c, $|\mathcal{L}|$ is the number of cases in $\mathcal{L}$, and $|\mathcal{A}|$ is the number of event attributes.*

To evaluate our method, we generated synthetic event logs from random process models of different complexities. We used PLG2 [6] to generate five process models: Small, Medium, Large, Huge, and Wide. The complexity of the models varies in number of activities, breadth, and width; for Small to Huge, activities, breadth, and width increase uniformly, whereas Wide features a much larger breadth than width. Wide was designed as a challenge because it features a high branching factor, thereby making it hard to predict the next activity or attribute. We also use a handmade procurement process model called P2P for demonstrative purposes because it features human readable activity names.

Now, we randomly generate logs from these process models, following the control flow and generating attributes for each event. Each possible sequence of activities was assigned a random probability sampled from a normal distribution with $\mu = 1$ and $\sigma = 0.2$, so that not all sequences appear equally likely.

To generate the attributes, we first create a set of possible values $\mathcal{V}_a$ for each attribute a, with set sizes ranging from 20 to 100. Then we assign random subsets of $\mathcal{V}_a$ to each activity, ranging from 5 to 40 in size. When generating a sequence from the process model, we also sample one possible value for each attribute in each event. While sampling we enforce long term dependencies between attributes. For example, 2 of the 10 attribute values of one activity always occur when 1 of the attribute values for a different event occurred earlier in the sequence; hence, we model causal relationships between attributes within sequences.

Table 1. Overview showing dataset information

Name.	#Logs	#Activities	#Cases	#Events	#Attributes
P2P	10	12	12.5K	102K	0–5
Small	10	20	12.5K	111K	0–5
Medium	10	32	12.5K	73K	0–5
Large	10	42	12.5K	138K	0–5
Huge	10	54	12.5K	100K	0–5
Wide	10	34	12.5K	75K	0–5
BPIC12	1	36	13K	262K	0
BPIC13	3	5–13	0.8K–7.5K	2.4K–66K	2–4
BPIC15	5	355–410	0.8K–1.4K	44K–60K	2–3
BPIC17	2	8–26	31K–43K	194K–1.2M	1
Comp	10	7–18	0.9K–56K	4K–180K	1

In addition to the synthetic logs we also use the event logs from the Business Process Intelligence Challenge (BPIC): BPIC12[2], BPIC13[3], BPIC15[4] and BPIC17[5]. Furthermore, we evaluate our method on 10 real-life event logs of procurement processes, made available to us by a consulting company. Refer to Table 1 for information about the datasets.

Like Bezerra [3] and Böhmer [5], we apply artificial anomalies to the event logs, altering 30 percent of all cases. In addition to the three anomaly types used in [3,5], we introduced a new, attribute-based anomaly. The anomalies are defined as follows: *Skip*, a necessary activity has not been executed; *Switch*, two

---

[2] http://www.win.tue.nl/bpi/doku.php?id=2012:challenge.
[3] http://www.win.tue.nl/bpi/doku.php?id=2013:challenge.
[4] http://www.win.tue.nl/bpi/doku.php?id=2015:challenge.
[5] http://www.win.tue.nl/bpi/doku.php?id=2017:challenge.

events have been executed in the wrong order; *Rework*, an activity has been executed too many times; *Attribute*, an incorrect attribute value is set (e.g., a user does not have the necessary security level).

Notice that we do apply the artificial anomalies to the real-life event logs as well, which very likely already contain natural anomalies. Thereby, we can measure the performance of the algorithms on the real-life logs to demonstrate feasibility while using the synthetic logs to evaluate accuracy.

When applying the artificial anomalies, we also gather a ground truth dataset. Whenever the anomaly is of type *Skip*, *Rework*, or *Switch*, it is a control flow anomaly, and hence the activity name attribute is marked as anomalous for affected events. If the anomaly is of type *Attribute*, the affected attribute is marked as anomalous. Hence, we obtain ground truth data on attribute level. The ground truth data can easily be adapted to case level by the following rule: A case is anomalous if any of the attributes in its events are anomalous.

We generated 10 flavors of each synthetic process model with different numbers of attributes (0, 1, 2, 3, and 5) and different sizes for $\mathcal{V}_a$, resulting in 60 synthetic logs. Together with BPIC12 (1 log), BPIC13 (3 logs), BPIC15 (5 logs), and BPIC17 (2 logs), and the 10 procurement event logs (Comp), the corpus consists of 81 event logs.

4   Method

In this section we will describe the BINet architecture and all necessary steps for the implementation.

4.1   Preprocessing

Due to the mathematical nature of neural networks, we must transform the logs into a numerical representation. To accomplish this, we encode all string attribute values. Multiple options are available, such as integer encoding or one-hot encoding. We chose to use an integer encoding, which is a mapping $\mathcal{I}_a : \mathcal{V}_a \rightarrow \mathbb{N}$, mapping all possible attribute values for an attribute a to a unique positive integer. The integer encoding is applied to all attributes of the log, including the activity name.

We will represent the event logs as third-order tensors. Each event e is a first-order tensor $\boldsymbol{e} \in \mathbb{R}^A$, with $A = |\mathcal{A}|$, the first attribute always being the activity name, representing the control flow. Hence, an event is defined by its activity name and the event attributes. Each case is then represented as a second-order tensor $\boldsymbol{C} \in \mathbb{R}^{E \times A}$, with $E = \max_{c \in \mathcal{L}} |\hat{c}|$, being the maximum case length of all cases in the log $\mathcal{L}$. To force all cases to have the same size, we pad all shorter cases with event tensors only containing zeros, which we call padding events (these will be ignored by the neural network). The log $\mathcal{L}$ can now be represented as a third-order tensor $\boldsymbol{L} \in \mathbb{R}^{C \times E \times A}$, with $C = |\mathcal{L}|$, the number of cases in log $\mathcal{L}$. Using matrix

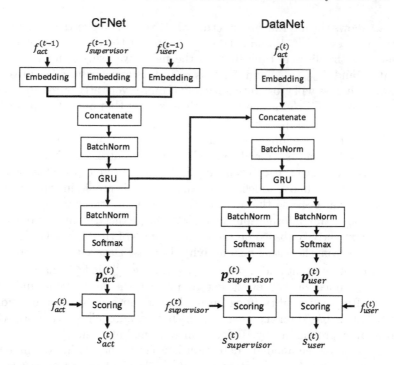

Fig. 1. BINet architecture for a log with two event attributes, *supervisor* and *user*

index notation, we can now obtain the second attribute of the third event in the ninth case with $L_{9,3,2}$. Now we can define a preprocessor as follows:

Definition 2. *Preprocessor Let C, E, and A be defined as above, then a preprocessor is a mapping $\mathcal{P} : \mathcal{L} \to \mathbb{R}^{C \times E \times A}$.*

The BINet preprocessor $\mathcal{P}$ will encode all attribute values and then transform the log $\mathcal{L}$ into its tensor representation. In the following, we will refer to the preprocessed log $\mathcal{L}$ by F (features), with $F = \mathcal{P}(\mathcal{L})$.

4.2   BINet Architecture

BINet is based on a neural network architecture that is trained to predict the next event, including all its attributes. To model the sequential nature of event log data, the core of BINet is a recurrent neural network, using a Gated Recurrent Unit (GRU) [8], an alternative for the popular long short-term memory (LSTM) [13].

We must distinguish between the control flow and the data flow aspect of event log data. Therefore, BINet is composed of two parts: CFNet (control flow) and DataNet (data flow). CFNet is responsible for predicting the next activity name of the next event, while DataNet is responsible for predicting all attributes of the next event.

Figure 1 shows the internal architecture of BINet. CFNet retrieves as input all attributes of an event $e^{(t-1)}$ and is trained to predict the activity name of event $e^{(t)}$, where t is the discrete time step. In the figure we use f_{act} for the activity name feature and $f_{supervisor}$ and f_{user} as examples for attribute features. Note that the architecture will grow automatically if more event attributes are present. The output layer of CFNet is one single softmax layer that outputs a probability distribution $p_{act}^{(t)}$ over all possible activity names.

DataNet retrieves as input the activity name of $e^{(t)}$ and the internal state of the CFNet GRU and is trained to predict all attributes. DataNet will have a separate softmax layer for each event attribute. Note that DataNet is being fed the activity name at time t, which is the same time it is predicting the attributes for. This is crucial because the attributes of an event strongly depend on the activity. Without this information, DataNet will predict the attributes for the most likely next activity (based on the internal state of the CFNet GRU), which can be different from the actual next activity. Because DataNet is only predicting the attributes of an event and not its activity, using $f_{act}^{(t)}$ is legitimate.

BINet is trained on the event log to predict the next event and all its attributes. After the initial training phase, BINet can now be used for anomaly detection. This is based on the assumption that an anomalous attribute will be assigned a lower probability by BINet than a normal attribute.

The last step of the anomaly detection process is the scoring of the events. Therefore, we use a scoring function in the last layer of the architecture. This scoring function receives as input the probability distribution $p_a^{(t)}$ for an attribute a and the actual value of the attribute $f_a^{(t)}$.

A softmax layer outputs a probability distribution over all possible values. When an event log features 5 different activity names, $p_{act}^{(t)}$ will be a first-order tensor of size 5. Each of the dimensions of $p_{act}^{(t)}$ holds the probability that the softmax layer assigns to one of the 5 possible activity values.

We can now define the scoring function σ as the difference between the probability of the most likely attribute (according to BINet) and the probability of the attribute value encountered $f_a^{(t)}$, where p_a is the probability tensor for attribute a and an event e, and i is the corresponding index of $f_a^{(t)}$ in p_a.

$$\sigma(p_a, i) = \max_{p \in p_a} p - p_{a_i}$$

The effect of normalization is demonstrated by Fig. 2, which shows three example cases of a P2P dataset with 2 attributes with and without normalization. Without normalization, the scoring function is defined as $\sigma(p_a, i) = 1 - p_{a_i}$, i.e., the inverse probability. We can observe that BINet is able to accurately predict the activities after the first activity, however, the anomaly scores for the user attribute are close to 1 because multiple users are permitted to execute an activity. We can counteract this effect by applying the normalization by confidence as demonstrated in Fig. 2.

The complete BINet architecture is then comprised of CFNet, DataNet and the scoring function σ applied to each softmax layer. We can obtain the anomaly scores tensor S by applying BINet to the feature tensor $\boldsymbol{F}$

$$S = (s_{ijk}) \in \mathbb{R}^{C \times E \times A} = \text{BINet}(\boldsymbol{F}),$$

mapping an anomaly score to each attribute in each event in each trace. The anomaly score for attributes of padding events will always be 0.

4.3 Training

BINet is trained without the scoring function. The GRU units are trained in sequence to sequence fashion. With each event that is fed in, the network is trained to predict the attributes of the next event. We train BINet with a GRU size of $2E$ (two times the maximum case length), on mini batches of size 100 for 50 epochs using the Adam [16] optimizer using the parameters stated in the original paper. We use batch normalization [14] between all layers to counteract overfitting. Every feature is passed through a separate embedding layer (see [19]) to reduce the input dimension.

	1 act	1 supervisor	1 user	2 act	2 supervisor	2 user	3 act	3 supervisor	3 user	4 act	4 supervisor	4 user
Normal	Create SC 0.54	Amanda 0.88	Iluminada 0.94	Purchase SC 0.03	Roy 0.84	Velda 0.65	Approve SC 0.05	Roy 0.39	Marilyn 0.94	Create PO 0.06	Alyce 0.87	Amanda 0.82
normalized	Create SC 0.05	Amanda 0.04	Iluminada 0.02	Purchase SC 0.01	Roy 0.00	Velda 0.00	Approve SC 0.00	Roy 0.00	Marilyn 0.01	Create PO 0.00	Alyce 0.06	Amanda 0.06
Switch 2 and 3	Create SC 0.54	Amanda 0.88	Hannah 0.94	Approve SC 0.97 X	Melany 0.91	Amanda 0.95	Purchase SC 0.92 X	Tiffiny 0.78	Velda 0.88	Create PO 0.10	Melany 0.86	Lucy 0.94
normalized	Create SC 0.05	Amanda 0.04	Hannah 0.02	Approve SC 0.94 X	Melany 0.55	Amanda 0.09	Purchase SC 0.72 X	Tiffiny 0.00	Velda 0.08	Create PO 0.00	Melany 0.05	Lucy 0.27
Wrong user at 1	Create PR 0.49	Clayton 0.94	Alyce 1.00 X	Release PR 0.05	Melany 0.93	Rossie 0.64	Create PO 0.02	Lucy 0.88	Alyce 0.84	Decrease PO 0.70	Hannah 0.89	Roy 0.91
normalized	Create PR 0.00	Clayton 0.02	Alyce 0.52 X	Release PR 0.00	Melany 0.01	Rossie 0.00	Create PO 0.00	Lucy 0.07	Alyce 0.11	Decrease PO 0.05	Hannah 0.03	Roy 0.04

Fig. 2. Effect of confidence normalization on BINet anomaly scores (high scores indicate anomalies); anomalies are marked with X

4.4 Detection

An anomaly detector only outputs anomaly scores. We need to define a function that maps anomaly scores to a label $l \in \{0, 1\}$, 0 indicating normal and 1 indicating anomalous, by applying a threshold t. Whenever an anomaly score for an attribute is greater than t, this attribute is flagged as anomalous. To obtain a separate threshold for each anomaly score in $\boldsymbol{S}$, we define a threshold tensor $\boldsymbol{T} = \alpha \cdot \boldsymbol{\tau}$, where $\boldsymbol{\tau} \in \mathbb{R}^{C \times E \times A}$ is a baseline threshold tensor and $\alpha \in \mathbb{R}$ is a scaling factor. Now we can define the function θ, with inputs $\boldsymbol{S}$, α, and $\boldsymbol{\tau} = (\tau_{ijk})$, as

$$\theta(\boldsymbol{S}, \alpha, \boldsymbol{\tau}) = (p_{ijk}) = \begin{cases} 1 & \text{if } (s_{ijk}) > \alpha \cdot (\tau_{ijk}) \\ 0 & \text{otherwise} \end{cases}.$$

To obtain a baseline threshold $\boldsymbol{\tau} \in \mathbb{R}^{C \times E \times A}$, we propose four different strategies τ_0, τ_e, τ_a, and τ_{ea}. In the following we will use $N = \sum_{c \in \mathcal{L}} |\hat{c}|$ to denote the number of non-padding events. The first baseline threshold function, τ_0, is defined by the average anomaly score over all cases, events, and attributes.

$$\tau_0(\boldsymbol{S}) = (\tau_{ijk}) = \frac{1}{NA} \sum_a^C \sum_b^E \sum_c^A (s_{abc})$$

It is sensible to use a separate threshold for each event position in a case because the branching factor can vary for different points in a process model. Therefore, we define the second baseline threshold function, τ_e, based on the position of an event e in a case c.

$$\tau_e(\boldsymbol{S}) = (\tau_{ijk}) = \frac{1}{CA} \sum_a^C \sum_c^A (s_{ajc})$$

It is also sensible to use a separate threshold for each attribute because V_a has a different size for each event e and attribute a. Thus, we define the third baseline threshold function, τ_a, to output a separate threshold for each event attribute.

$$\tau_a(\boldsymbol{S}) = (\tau_{ijk}) = \frac{1}{N} \sum_a^C \sum_b^E (s_{abk})$$

The fourth baseline threshold function, τ_{ea} is a combination of τ_e and τ_a, and outputs a threshold for each event position and attribute separately.

$$\tau_{ea}(\boldsymbol{S}) = (\tau_{ijk}) = \frac{1}{C} \sum_a^C (s_{ajk})$$

Note that we use matrix index notation to broadcast $\boldsymbol{\tau}$ to the right dimensionality to conform with the definition of θ from before. Utilizing the automatic broadcasting functionality in modern numerical computing libraries, such as TensorFlow[6] or NumPy[7], this can be implemented very efficiently. Remember that anomaly scores for padding events are set to 0, and hence they do not influence the sums. By normalizing with N we calculate the average based only on non-padding events.

4.5   Threshold Heuristic

Most anomaly detection algorithms rely on a manual setting for the threshold. We have proposed four different methods of obtaining a baseline threshold tensor $\boldsymbol{\tau}$ from the anomaly scores tensor $\boldsymbol{S}$. We still need to set the scaling factor α manually. To overcome this, we need to introduce a heuristic to set α automatically.

---

[6] https://tensorflow.org.
[7] http://numpy.org.

Because anomaly detection is an unsupervised task and no labels are available at runtime, we cannot optimize α based on the detection F_1 score. However, using the F_1 score function we can define the heuristic h_{best} that computes the best possible α for a given anomaly score tensor S, a baseline threshold τ, and the ground truth label tensor L as

$$h_{best}(L, S, \tau) = \arg \max_{\alpha} F_1(L, \theta(S, \alpha, \tau)).$$

As we do not have access to L at runtime, we cannot use h_{best}. We propose a new heuristic that works like the elbow method, commonly used to find an optimal number of clusters for clustering algorithms (see [25]). As we cannot rely on the F_1 score as our metric, we propose the use the anomaly ratio r, which can be defined as, with $\theta(S, \alpha, \tau) = (p_{ijk})$.

$$r(S, \alpha, \tau) = \frac{1}{CEA} \sum_{i}^{C} \sum_{j}^{E} \sum_{k}^{A} (p_{ijk})$$

The optimal α must lie between α_{low} and α_{high}, where $r(S, \alpha_{low}, \tau) = 1$ and $r(S, \alpha_{high}, \tau) = 0$. We can reduce our search space to this interval. Now we span a grid G of size s between α_{low} and α_{high} to define our candidates for α.

$$G = \left\{ \alpha_{low} + \frac{1}{s} (\alpha_{high} - \alpha_{low}), \ldots, \alpha_{low} + \frac{s}{s} (\alpha_{high} - \alpha_{low}) \right\}$$

In our experiments we found that $s = 20$ is a good choice for s, however, any reasonable choice of $s \in \{5, \ldots, 100\}$ generally works.

Because r is a discrete function, we use the central difference approximation to obtain the second order derivative r'' of r.

$$r''(S, \alpha, \tau) \approx \frac{r(S, \alpha - s, \tau) - 2r(S, \alpha, \tau) + r(S, \alpha + s, \tau)}{s^2}$$

Now we can define the elbow heuristic

$$h_{elbow}(S, \tau) = \arg \max_{\alpha \in G} r''(S, \alpha, \tau).$$

h_{elbow} mimics the way a human would set α manually. When given a user interface with a heatmap visualization (like in Fig. 2) for $\theta(S, \alpha, \tau)$ and control over the value of α, a human would start with a value of α where all attributes are marked as anomalous (i.e., the heatmap shows only blue and no white), and then gradually decrease α until the point where the heatmap switches from mostly showing blue to mostly showing white.

Figure 3 shows the F_1 score and the corresponding anomaly ratio r for a model of BINet, trained on a P2P dataset and using τ_a as the baseline threshold. We can see that the highest possible F_1 score correlates with the maximum of r'', and hence with the "elbow" of r. Interestingly, we found that h_{elbow} works just as well for other anomaly detection algorithms, such as t-STIDE [26], Naive [3], and DAE [21].

Evaluating h_{elbow} for BINet over all synthetic datasets and all baseline threshold strategies, we can see in Fig. 4 that the best baseline threshold is τ_a. We also find, that the h_{elbow} works remarkably well over all strategies, for the performance of h_{elbow} is very close to h_{best}.

5   Evaluation

We evaluated BINet on all 81 event logs and compared it to two methods from [7]: a sliding window approach (t-STIDE+) [26]; and the one-class SVM (OC-SVM). Additionally, we compared BINet to two approaches from [3]: the Naive algorithm and the Sampling algorithm. Furthermore, we provide the results of the denoising autoencoder (DAE) approach from [20]. Lastly, we compared BINet to the approach from [5], which utilizes an extended likelihood graph (Likelihood). As a baseline, we provide the results of a random classifier.

For the OC-SVM, we relied on the implementation of scikit-learn[8] using an RBF kernel of degree 3 and $\nu = 0.5$. The Naive, Sampling, Likelihood, and DAE methods were implemented as described in the original papers. t-STIDE+ is an

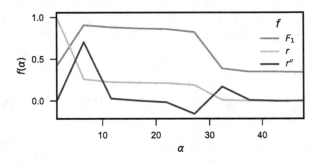

Fig. 3. F_1 score, anomaly ratio r, and second order derivative r'' (scaled for clarity) by α for BINet on a dataset with 5 attributes using τ_a as the baseline threshold

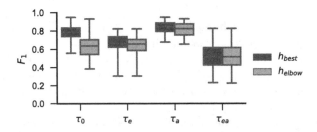

Fig. 4. F_1 score by strategy and heuristic for BINet on the P2P dataset

[8] http://scikit-learn.org.

implementation of the t-STIDE method from [26], which we adapted to work with event attributes (see [20]).

Sampling, Likelihood, Baseline, and the OC-SVM do not rely on a manual setting of the threshold and were unaltered. For the remaining algorithms we used h_{elbow} and chose the following baseline threshold strategies following a grid search: τ_0 for Naive, τ_{ea} for t-STIDE+, τ_a for DAE, and τ_a for BINet.

Figure 5 shows the F_1 score distribution for all methods over all datasets and for the two detection levels. F_1 score is calculated as the macro average F_1 score over the normal and the anomalous class. BINet outperforms all other methods on both detection levels. The more important detection level is the attribute level, as this measure demonstrates how accurately an anomaly detection algorithm can detect the actual attribute that caused an anomaly.

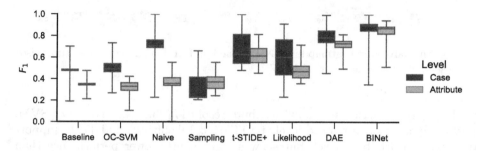

Fig. 5. F_1 score by method and detection level using h_{elbow} where applicable

Table 2. Results showing F_1 score over all datasets by detection level and method; best results are shown in bold typeface

Level	Method	P2P	Small	Medium	Large	Huge	Wide	BPIC12	BPIC13	BPIC15	BPIC17	Comp
Case	Baseline	0.47	0.50	0.47	0.48	0.48	0.50	0.55	0.50	0.49	0.47	0.45
	OC-SVM [23]	0.49	0.50	0.51	0.53	0.52	0.52	0.42	0.52	0.47	0.60	0.51
	Naive [3]	**0.91**	0.80	0.71	0.71	0.76	0.72	0.58	0.47	0.24	0.53	0.63
	Sampling [3]	0.23	0.23	0.44	0.34	0.23	0.23	0.45	0.26	0.22	0.22	0.57
	t-STIDE+ [26]	0.72	0.71	0.73	0.68	0.69	0.67	**0.81**	0.57	**0.51**	0.68	0.68
	Likelihood [5]	0.64	0.65	0.62	0.61	0.65	0.60	0.65	0.29	0.30	0.53	**0.73**
	DAE [20]	0.86	0.81	0.78	0.89	0.80	0.75	0.76	0.52	0.45	**0.75**	0.68
	BINet	**0.91**	**0.92**	**0.92**	**0.92**	**0.92**	**0.92**	0.58	**0.58**	0.49	0.58	0.66
Attribute	Baseline	0.34	0.36	0.35	0.35	0.35	0.36	0.35	0.35	0.34	0.35	0.36
	OC-SVM [23]	0.33	0.34	0.32	0.35	0.34	0.32	0.11	0.36	0.32	0.37	0.26
	Naive [3]	0.50	0.41	0.35	0.35	0.39	0.37	0.09	0.14	0.00	0.24	0.36
	Sampling [3]	0.28	0.32	0.40	0.46	0.40	0.36	0.37	0.30	0.29	0.32	0.48
	t-STIDE+ [26]	0.66	0.66	0.68	0.64	0.64	0.65	0.67	0.56	0.51	0.62	0.49
	Likelihood [5]	0.50	0.50	0.51	0.48	0.50	0.50	0.47	0.40	0.36	0.47	0.52
	DAE [20]	0.77	0.72	0.72	0.75	0.75	0.71	**0.72**	0.52	0.51	**0.73**	0.61
	BINet	**0.85**	**0.89**	**0.87**	**0.89**	**0.88**	**0.88**	0.59	**0.57**	**0.54**	0.64	**0.68**

	1 act	1 supervisor	1 user	2 act	2 supervisor	2 user	3 act	3 supervisor	3 user	4 act	4 supervisor	4 user
Normal	Create SC 0.05	Amanda 0.04	Iluminada 0.02	Purchase SC 0.00	Roy 0.01	Velda 0.00	Approve SC 0.00	Roy 0.00	Marilyn 0.01	Create PO 0.00	Alyce 0.06	Amanda 0.06
Normal	Create PR 0.00	Roy 0.02	Marilyn 0.06	Release PR 0.00	Rossana 0.03	Rossie 0.00	Create PO 0.00	Rossie 0.00	Lucy 0.07	Increase PO 0.00	Hannah 0.04	Iluminada 0.00
Switch 2 and 3	Create SC 0.05	Amanda 0.04	Hannah 0.02	Approve SC 0.94 X	Melany 0.55	Amanda 0.09	Purchase SC 0.72 X	Tiffiny 0.00	Velda 0.08	Create PO 0.00	Melany 0.05	Lucy 0.27
Switch 1 and 2	Release PR 0.50 X	Clayton 0.09	Rossie 0.17	Create PR 0.21 X	Rossie 0.43	Steve 0.31	Create PO 0.00	Rossana 0.12	Brant 0.00	Release PO 0.24	Alyce 0.00	Roy 0.07
Skip at 3	Create SC 0.05	Lourdes 0.04	Amanda 0.02	Purchase SC 0.00	Steve 0.01	Jack 0.06	Create PO 0.92 X	Rossie 0.19	Brant 0.08	Release PO 0.00	Alyce 0.43	Roy 0.20
Skip at 1	Release PR 0.50 X	Jack 0.16	Brant 0.29	Create PR 0.08	Melany 0.10	Rossie 0.04	Decrease PO 0.18	Jack 0.36	Tiffiny 0.20	Release PO 0.00	Alyce 0.36	Velda 0.00
Rework at 3	Create SC 0.05	Alyce 0.04	Marilyn 0.03	Purchase SC 0.00	Steve 0.00	Jack 0.07	Approve SC 0.00	Roy 0.01	Hannah 0.02	Approve SC 0.93 X	Roy 0.32	Hannah 0.45
Rework at 2	Create PR 0.00	Sharee 0.01	Steve 0.00	Release PR 0.00	Clayton 0.04	Brant 0.07	Release PR 0.93 X	Clayton 0.25	Brant 0.17	Create PO 0.00	Melany 0.15	Amanda 0.08
Wrong user at 2	Create SC 0.05	Iluminada 0.03	Marilyn 0.03	Purchase SC 0.00	Ryan 0.05	Ryan 0.37 X	Approve SC 0.00	Melany 0.11	Velda 0.02	Create PO 0.00	Amanda 0.05	Alyce 0.08
Wrong user at 1	Create PR 0.00	Clayton 0.02	Alyce 0.52 X	Release PR 0.00	Melany 0.01	Rossie 0.00	Create PO 0.00	Lucy 0.07	Alyce 0.11	Decrease PO 0.05	Hannah 0.03	Roy 0.04

Fig. 6. Anomaly score heatmap for BINet trained on P2P with 2 attributes (supervisor and user); anomalies are marked by X

Expectedly, methods without attribute resolution, like Naive and OC-SVM, perform poorly on attribute level. t-STIDE+, Likelihood, and DAE support detection on attribute level, but show a significantly lower performance than BINet. DAE and BINet are both neural network based. The main advantage of BINet over DAE is that BINet makes use of the time dimension in the sequential data, whereas DAE does not.

Table 2 contains the detailed results for each method by dataset and detection level. BINet performs best across levels on the synthetic logs. On the real event logs, BINet performs best on attribute level, whereas on case level, the field is mixed. An accurate prediction on attribute level is to be favored over an accurate prediction on case level because only the attribute level allows to identify the exact cause of an anomaly.

Figure 6 shows a heatmap of BINet anomaly scores for a P2P dataset with 2 attributes. Two example cases are chosen for each type of anomaly and the normal class to demonstrate how BINet detects anomalies based on attribute level. No threshold has been applied to the anomaly scores. This way, the severity of an anomaly can be illustrated by the colors in the heatmap. Overall, we find that BINet detects control flow anomalies very effectively. For example, a shopping cart (SC) cannot be approved before it has been purchased (first Switch). Similarly, a purchase requisition (PR) cannot be released before it has been created (second Skip). Rework and Attribute anomalies are also detected accurately. In the case of Attribute, only the incorrect attribute is assigned a significantly high anomaly score. In the two examples, Ryan cannot be his own supervisor and Alyce is not permitted to create a PR.

6   Conclusion

In this paper we presented BINet, a neural network architecture for multivariate anomaly detection in business process event logs. Additionally, we proposed a heuristic for setting the threshold of an anomaly detection algorithm automatically, based on the anomaly ratio function.

BINet is a recurrent neural network, and can therefore be used for real-time anomaly detection, since it does not require a completed case for detection. BINet does not rely on any information about the process modeled by an event log, nor does it depend on a clean dataset. Utilizing the elbow heuristic, BINet's internal threshold can be set automatically, reducing manual workload. It can be used to find point anomalies as well as contextual anomalies because it models the time dimension in event sequences and utilizes both the control flow and the data flow information. Furthermore, BINet can cope with concept drift, as it can be setup to continuously train on new cases in real-time.

Based on the empirical evidence obtained in the evaluation, BINet is a promising method for anomaly detection, especially in business process event logs. BINet outperformed the opposition on all detection levels (case, event, and attribute level). Specifically, on the synthetic datasets BINet's performance surpasses those of other methods by an order of magnitude.

For an accurate detection of an anomaly it is essential that an anomaly detection algorithm processes event logs on attribute level; otherwise, a control flow anomaly cannot be distinguished from a data flow anomaly. To allow easy analysis of an event log, the attribute level is the most important detection level. On attribute level, BINet performs significantly better than the other methods.

Overall, the results presented in this paper suggest that BINet is a reliable and versatile method for detecting attribute anomalies in business process logs.

Acknowledgments. This project [522/17-04] is funded in the framework of Hessen ModellProjekte, financed with funds of LOEWE, Förderlinie 3: KMU-Verbundvorhaben (State Offensive for the Development of Scientific and Economic Excellence), and by the German Federal Ministry of Education and Research (BMBF) Software Campus project "AI-PM" [01IS17050].

References

1. van der Aalst, W.M.P.: Process Mining: Data Science in Action. Springer, Heidelberg (2016). https://doi.org/10.1007/978-3-662-49851-4
2. Bezerra, F., Wainer, J.: Anomaly detection algorithms in logs of process aware systems. In: Proceedings of the 2008 ACM Symposium on Applied Computing, pp. 951–952. ACM (2008)
3. Bezerra, F., Wainer, J.: Algorithms for anomaly detection of traces in logs of process aware information systems. Inf. Syst. **38**(1), 33–44 (2013)
4. Bezerra, F., Wainer, J., van der Aalst, W.M.P.: Anomaly detection using process mining. In: Halpin, T., et al. (eds.) BPMDS/EMMSAD -2009. LNBIP, vol. 29, pp. 149–161. Springer, Heidelberg (2009). https://doi.org/10.1007/978-3-642-01862-6_13

5. Böhmer, K., Rinderle-Ma, S.: Multi-perspective anomaly detection in business process execution events. In: Debruyne, C., et al. (eds.) Move to Meaningful Internet Systems. LNCS, pp. 80–98. Springer, Cham (2016). https://doi.org/10.1007/978-3-319-48472-3_5

6. Burattin, A.: PLG2: multiperspective processes randomization and simulation for online and offline settings. arXiv:1506.08415 (2015)

7. Chandola, V., Banerjee, A., Kumar, V.: Anomaly detection for discrete sequences: a survey. IEEE Trans. Knowl. Data Eng. **24**(5), 823–839 (2012)

8. Cho, K., et al.: Learning phrase representations using RNN encoder-decoder for statistical machine translation. arXiv:1406.1078 (2014)

9. Cortes, C., Vapnik, V.: Support-vector networks. Mach. Learn. **20**(3), 273–297 (1995)

10. Evermann, J., Rehse, J.-R., Fettke, P.: A deep learning approach for predicting process behaviour at runtime. In: Dumas, M., Fantinato, M. (eds.) BPM 2016. LNBIP, vol. 281, pp. 327–338. Springer, Cham (2017). https://doi.org/10.1007/978-3-319-58457-7_24

11. Evermann, J., Rehse, J.R., Fettke, P.: Predicting process behaviour using deep learning. Decis. Support Syst. **100**, 129–140 (2017)

12. Han, J., Pei, J., Kamber, M.: Data Mining: Concepts and Techniques. Elsevier, New York City (2011)

13. Hochreiter, S., Schmidhuber, J.: Long short-term memory. Neural Comput. **9**(8), 1735–1780 (1997)

14. Ioffe, S., Szegedy, C.: Batch normalization: accelerating deep network training by reducing internal covariate shift. In: International Conference on Machine Learning, pp. 448–456 (2015)

15. Japkowicz, N.: Supervised versus unsupervised binary-learning by feedforward neural networks. Mach. Learn. **42**(1), 97–122 (2001)

16. Kingma, D., Ba, J.: Adam: a method for stochastic optimization. arXiv:1412.6980 (2014)

17. Malhotra, P., Ramakrishnan, A., Anand, G., Vig, L., Agarwal, P., Shroff, G.: LSTM-based encoder-decoder for multi-sensor anomaly detection. arXiv:1607.00148 (2016)

18. Marchi, E., Vesperini, F., Eyben, F., Squartini, S., Schuller, B.: A novel approach for automatic acoustic novelty detection using a denoising autoencoder with bidirectional LSTM neural networks, April 2015

19. Mikolov, T., Chen, K., Corrado, G., Dean, J.: Efficient estimation of word representations in vector space. arXiv:1301.3781 (2013)

20. Nolle, T., Luettgen, S., Seeliger, A., Mühlhäuser, M.: Analyzing business process anomalies using autoencoders. arXiv:1803.01092 (2018)

21. Nolle, T., Seeliger, A., Mühlhäuser, M.: Unsupervised anomaly detection in noisy business process event logs using denoising autoencoders. In: Calders, T., Ceci, M., Malerba, D. (eds.) DS 2016. LNCS (LNAI), vol. 9956, pp. 442–456. Springer, Cham (2016). https://doi.org/10.1007/978-3-319-46307-0_28

22. Pimentel, M.A.F., Clifton, D.A., Clifton, L., Tarassenko, L.: A review of novelty detection. Sig. Process. **99**, 215–249 (2014)

23. Schölkopf, B., et al.: Support vector method for novelty detection. In: NIPS. vol. 12, pp. 582–588 (1999)

24. Tax, N., Verenich, I., La Rosa, M., Dumas, M.: Predictive business process monitoring with LSTM neural networks. In: Dubois, E., Pohl, K. (eds.) CAiSE 2017. LNCS, vol. 10253, pp. 477–492. Springer, Cham (2017). https://doi.org/10.1007/978-3-319-59536-8_30

25. Tibshirani, R., Walther, G., Hastie, T.: Estimating the number of clusters in a data set via the gap statistic. J. R. Stat. Soc.: Ser. B (Stat. Methodol.) **63**(2), 411–423 (2001)
26. Warrender, C., Forrest, S., Pearlmutter, B.: Detecting intrusions using system calls: alternative data models. In: Proceedings of the 1999 IEEE Symposium on Security and Privacy, pp. 133–145. IEEE (1999)
27. Wen, L., van der Aalst, W.M.P., Wang, J., Sun, J.: Mining process models with non-free-choice constructs. Data Min. Knowl. Disc. **15**(2), 145–180 (2007)
28. Wressnegger, C., Schwenk, G., Arp, D., Rieck, K.: A close look on n-grams in intrusion detection: Anomaly detection vs. classification. In: Proceedings of the 2013 ACM Workshop on Artificial Intelligence and Security, pp. 67–76. AISec 2013. ACM (2013)

Finding Structure in the Unstructured: Hybrid Feature Set Clustering for Process Discovery

Alexander Seeliger(✉) ⓘ, Timo Nolle ⓘ, and Max Mühlhäuser

Telecooperation Lab, Technische Universität Darmstadt, Darmstadt, Germany
{seeliger,nolle,max}@tk.tu-darmstadt.de

Abstract. Process discovery is widely used in business process intelligence to reconstruct process models from event logs recorded by information systems. With the increase of complexity and flexibility of processes, it is getting more and more challenging for discovery algorithms to generate accurate and comprehensive models. Trace clustering aims to overcome this issue by splitting event logs into smaller behavioral similar sub-logs. From these sub-logs more accurate and comprehensive process models can be reconstructed. In this paper, we propose a novel clustering approach that uses frequent itemset mining on the case attributes to also reveal relationships on the data perspective. Our approach includes this additional knowledge as well as optimizes the fitness of the underlying process models of each cluster to generate accurate clustering results. We compare our method with six other clustering methods and evaluate our approach using synthetic and real-life event logs.

Keywords: Knowledge discovery · Process discovery
Trace clustering · Process mining · Business process intelligence

1 Introduction

Business process intelligence supports organizations to optimize and improve their business processes. In particular, *process mining* [1] helps to understand the actual use of information systems in various environments. The basis for process mining are event logs, recorded by process-aware information systems (PAISs), industrial machines or sensors. Event logs reflect the activities performed by employees or machines, allowing the analysis of the relationships between activities. Additionally, the event log may also store much more information, such as the executor of an activity and the context of the case, such as the vendor, used material or the customer.

An essential part of process mining is *process discovery* which is an unsupervised method for reconstructing a process model from an event log. The challenge is to find a model that accurately matches the recorded observations, but is also human interpretable. Many business processes in the real world are

M. Weske et al. (Eds.): BPM 2018, LNCS 11080, pp. 288–304, 2018.
https://doi.org/10.1007/978-3-319-98648-7_17

often executed in highly flexible environments, such as health care or product development. Here a dense distribution of cases with a high variety of complex behavior can be found. In such scenarios, process discovery often produces a spaghetti-like model which suffers from inaccuracy and high complexity. Furthermore, behavior on other process perspectives (e.g., data attributes) is not considered which may also be interesting for process analysts. For example, in an hospital the same admission process may be applied to emergency and non-emergency patients. Although the sequence of activities may be the same for both kinds of patients, the underlying process may be different with respect to resource assignments or activity durations.

Trace clustering tries to overcome these issues by splitting the different observed behaviors into multiple sub-logs of similar behavior. For each sub-log, process discovery is then applied separately to retrieve more accurate and human interpretable process models. However, existing methods either only rely on the control-flow perspective or ignore the quality of the discovered models. In particular, including different process perspectives and providing accurate process models is challenging.

In this paper, we introduce a new clustering method that uses a hybrid feature set to consider multiple process perspectives and optimizes the fitness of the underlying process models. Inspired by our previous work [14] in which we clustered documents together into meaningful groups by combining the document content and the user behavior, we transfer the idea to trace clustering in process mining. Our method considers the control-flow and the data perspective to extract process behaviors on both views to split the event log into multiple sub-logs. The basic idea is that different process behaviors often depend on the context of the case, e.g., the product category or the customer. These differences may only be small on the trace level, but large on the data perspective. Our clustering approach additionally uses the case attributes to distinguish between the different process behaviors. We extract frequent itemsets using frequent pattern mining [10] to find common relationships between case attributes and use them for clustering. Studies have shown that approaches relying solely on the control-flow are unable to identify different process behaviors adequately [18]. Furthermore, our method automatically optimizes the fitness of the sub-log process models, ensuring that the underlying models sufficiently represent the clustered traces.

In summary, the contributions of this paper are as follows:

1. We provide a hybrid feature set clustering approach that splits event logs into sub-logs containing similar cases based on multiple perspectives.
2. Our approach automatically adapts to the given event logs and determines the optimal clustering parameters by applying the particle swarm optimization algorithm to optimize model fitness.
3. We provide a comprehensive evaluation of six other trace clustering methods in the domain of business process intelligence.

The paper is structured as follows. First, we introduce related work. Second, we introduce our method that combines the data and the control-flow perspective to generate clusters. Third, we evaluate our method and discuss the results. Finally, we conclude with a discussion and a short summary.

2   Related Work

Our work is related to trace clustering in process mining, aiming to improve accuracy and interpretability of reconstructed process models by separating different behaviors on different process perspectives into multiple sub-logs. A summary and an evaluation framework of trace clustering methods are provided in [18]. The authors elaborate a systematic empirical analysis of different techniques and evaluate their applicability in two scenarios: the identification of different processes and the improvement of the understandability of the mined models. We classify the related work into *distance-based* and *model-based* trace clustering.

Distance-based trace clustering such as [4,9,16] use a vector space model on the event traces to segment the event log into smaller sub-logs. Greco et al. [9] use significant subsequences of activities and activity transitions to generate clusters of traces. Similar, Bose et al. [4] propose the use of different sequence similarity measures to find traces with similar behavior with respect to the order of activities. Alignment-based approaches can also be used for specifying the difference between traces, as for example presented in [7]. In [16] the use of log profiles is proposed, allowing to incorporate different perspectives into the clustering. Each profile describes its own vector space which can be separately used for clustering. A co-training strategy for multiple view clustering was presented by Appice et al. [2]. The authors combine multiple log profiles using unsupervised co-training. Clustering of one log profile is iteratively constrained by the similarities of the other profiles, leading to a unique clustering pattern.

To overcome the issue of heterogeneous scaled similarity criteria, occurring when multiple criteria are included, Delias et al. [6] propose an outranking approach. The authors build an overall metric using a non-compensatory methodology to overcome this issue. Song et al. [15] presented a comparative study to improve trace clustering using dimensionality reduction methods. The authors show the effect of applying three different reduction methods on the performance of trace clustering. To further improve the clustering result, De Konick et al. [12] incorporate expert knowledge into the clustering to produce results that are more consistent with the expert's expectations. However, distance-based trace clustering do not consider the model evaluation bias, neglecting the accuracy of the reconstructed process models from the sub-logs.

Model-based trace clustering techniques such as [17,21] combine the clustering bias and the model bias into an integrated view. ActiTraC [21] directly optimizes the fitness of the underlying process models to produce accurate results. In [17] a similar method is proposed which overcomes the stability issues of ActiTraC by first optimizing the average complexity of the models and then improve the accuracy of each model separately. While model-based approaches are good to produce accurate process models, they neglect the other process perspectives such as the data perspective.

The method presented in this paper aims at addressing both mentioned issues. It is inspired by our prior work [14] in which we cluster documents based on usage behavior and content into activity-centric groups. In this paper, we combine distance-based and model-based trace clustering methods to improve clustering results. We transfer the same idea to trace clustering in process mining, combining two distance measures to calculate the similarity between cases, including the control-flow and the data perspective. To retrieve accurate process models, we use an optimization algorithm to adjust the weighting of the distance measures and clustering parameters.

3   Hybrid Feature Set Clustering

In this section, we introduce our hybrid feature set clustering method. We extend existing trace clustering by optimizing the clusters using model fitness as a quality measure and incorporating case attributes.

3.1   Notation

First, we introduce the notations that are used throughout this paper. They were derived from [1].

Definition 1 *(Event, Attribute). Let $\mathcal{E}$ be the set of all possible event identifiers. Events may be described by attributes, such as the timestamp. Let $\mathcal{A}$ be the set of attributes and $\mathcal{V}_a$ the set of all possible values of attribute $a \in \mathcal{A}$. For an event $e \in \mathcal{E}$ and an attribute $a \in \mathcal{A}$: $\#_a(e)$ is the value of attribute a for event e.*

Definition 2 *(Case, Trace, Event Log). Let $\mathcal{C}$ be the set of all possible case identifiers. Cases can also have attributes. For a case $c \in \mathcal{C}$ and an attribute $a \in \mathcal{A}$: $\#_a(c)$ is the value of attribute a for case c. Each case contains a mandatory attribute trace: $\#_{trace}(c) \in \mathcal{E}^*$, also denoted as $\hat{c} = \#_{trace}(c)$.*

A trace is a finite sequence of events $\sigma \in \mathcal{E}^*$ such that each event only occurs once: $1 \leq i < j \leq |\sigma| : \sigma(i) \neq \sigma(j)$.

An event log is a set of cases $L \subseteq \mathcal{C}$ such that each event only occurs at most once in the log.

Definition 3 *(Classifier). For an event $e \in \mathcal{E}$, $\underline{e} = \#_{activity}(e)$ the activity name of the event e. The classifier can also be applied to sequences $\langle e_1, e_2, ..., e_n \rangle = \langle \underline{e_1}, \underline{e_2}, ..., \underline{e_n} \rangle$.*

Table 1 shows an example event log of a procurement process. The log consists of the cases $L = \{1, 2\}$, events $E = \{11, 12, 13, 14, 21, 22\}$ and the attributes $\mathcal{A} = \{$Case id, Event id, Vendor, Category, Timestamp, Activity, Resource$\}$. The table also shows the values of the attributes, for example, $\#_{category}(1) =$ Office supplies or $\#_{activity}(13) =$ PO created.

Table 1. Simplified example event log of a procurement process.

Case id	Event id	Vendor	Category	Timestamp	Activity	Resource
1	11	B. Trug	Office supplies	2017-04-17 10:11	PR created	John
1	12			2017-04-18 14:55	PR released	Maria
1	13			2017-04-18 17:12	PO created	Roy
1	14			2017-04-29 09:06	Goods receipt	Ryan
2	21	Company	Computer	2017-04-19 17:45	PO created	Emily
2	22			...	...	...

3.2 Our Approach

Most trace clustering methods define a similarity function between the event sequence of cases (e.g., Levenshtein distance or bag-of-activities) and then apply a clustering algorithm (e.g., k-means, hierarchical, or partition methods) to segment the event log. As a result, such methods provide a set of sub-logs which contain maximized intra-cluster and minimized inter-cluster similarity, neglecting the quality of the underlying process model [4,21]. This may lead to unsatisfactory results of the reconstructed models. Another issue of most existing trace clustering approaches is that they do not explore the relationships between case attributes to identify different behaviors on the data perspective. However, cases might be influenced by their data attributes. For process analysts, it might also be interesting to separate similar traces and put them into a different cluster if their corresponding data attributes are inconsistent.

Our approach addresses both issues. Instead of solely relying on the similarity function between traces, we additionally use the fitness [20] of the underlying process model as a criterion for quality of sub-logs. The fitness of a model is a normalized measure reflecting how many behaviors featured in the event log are also contained in the discovered process model. Our goal is to find an optimal separation of the event log such that the different process behaviors on both perspectives are clustered separately, while optimizing the fitness of the underlying process models. Additionally, the number of clusters should be kept reasonably small. Including additional data attributes uncovers certain behavior, for example, different levels of product quality checks, which would normally be clustered together despite being completely different regarding the data attributes. By adding this additional perspective, we are able to separate such behaviors, despite the cases being similar with respect to the control-flow. We combine both perspectives to extract more valuable knowledge about the execution of processes allowing us to distinguish between the different process behaviors more accurately.

In the following, we will describe our hybrid feature set clustering approach in detail.

Algorithm 1. Algorithm to retrieve the clusters

---

1 Let L be the event log, and let $\hat{L} = \{(\hat{c}) : c \in L\}$ be the set of distinct event traces of L.

2 Define $lev(x, y)$ to be the edit distance of the event traces $x, y \in \hat{L}$.

3 Define $sim^*_{lev}(X, Y)$ to be the edit distance between two sets of event traces $X, Y \subseteq \hat{L}$:

$$sim^*_{lev}(X, Y) = \sum_{x \in X} \sum_{y \in Y} lev(x, y) \; / \; (|X| \cdot |Y|)$$

4 Define $cases(t) = \{c : c \in L \land (\hat{c}) = t\}$ as the cases following event trace $t \in \hat{L}$.

5 Define $encode(c) = \{\mathcal{I}_a(\#_a(c)) : a \in \mathcal{A}\}$ with $c \in L$ and $\mathcal{I}$ as an integer index function; further define $encodes(C) = \{encode(c) : c \in C\}$.

6 Let $\mathcal{S}$ be the universe of all possible itemsets, $S \subseteq \mathcal{S}$ and s_i being the i-th itemset in S.

7 Define $itemsets : \hat{L} \to \mathcal{P}(S)$ as the function that returns the frequent itemsets using the $FPclose$ algorithm with θ being the *minimum support threshold*:

$$itemsets(t) = FPclose(encodes(cases(t)), \theta)$$

8 Define $sim_{itemsets}(S_a, S_b)$ to be the similarity function of the itemsets with $S_a, S_b \subseteq S$:

$$sim_{itemsets}(S_a, S_b) = \frac{2 \cdot |S_a \cap S_b|}{|S_a| + |S_b|}$$

9 Define $traces(s) = \{t : t \in \hat{L} \land s \in itemsets(t)\}$ with $s \in S$ to be the inverse function of $itemsets$ which returns the traces for a given itemset.

10 Define $sim(s_a, s_b)$ to be the combined similarity function with $s_a, s_b \in S$, $w \in \mathbb{R}$ and $0 \leq w \leq 1$ to be the weighting factor:

$$sim(s_a, s_b) = w \cdot sim^*_{lev}(traces(s_a), traces(s_b)) + (1 - w) \cdot sim_{itemsets}(s_a, s_b)$$

11 Define $M : S \times S \to \mathbb{R}^{|S| \times |S|}$ to be the itemset distance matrix

$$M = (m_{ij}) = sim(s_i, s_j)$$

12 Let $cluster(M, n)$ be the hierarchical clustering function that returns the cluster index of each itemset as a vector of size $|S|$ with $n \in \mathbb{N}$ as the number of clusters to generate.

13 Define $C(k)$ to be the set of the traces in cluster k

$$C(k) = \{cases(traces(s_j)) \mid s_j \in S \land c_j \in cluster(M) \land c_j = k\}$$

Candidate Clusters. The first step of our approach is to generate a candidate set of clusters. Algorithm 1 shows the generation of the clusters using the combined similarity measure. From the event log L we extract all distinct event traces $\hat{L}$ (Line 1). For comparing the traces, we use the Levenshtein edit distance. It is defined as the minimum number of edit operations that are required to transform one sequence into another. The edit operations are insertion, deletion, or substitution of an element in the sequence. Each operation has a cost of 1. We denote the normalized Levenshtein edit distance between two traces $x, y \in \hat{L}$ as $lev(x, y)$ (line 2). In line 3, we additionally define a function sim^*_{lev} which calculates the pairwise normalized Levenshtein distance between the traces in two sets. It is noteworthy that the Levenshtein distance can be replaced by a more advanced measure (e.g., one that also recognizes concurrency).

As a second similarity measure, we incorporate the attributes and their values of a case by extracting further knowledge about the underlying case relations. The idea is to use the case attributes in the event log to extract dependencies between attributes in certain process behaviors. Consider the procurement of supplies. Usually, there are different order approval steps involved depending, for example, on the material type of the purchased item. So, for office supplies there might be only one approval step whereas for the spare part of an expensive machine multiple approval steps of different departments are required. Such variations are usually deployed to reduce the amount of process steps. With the use of the case attributes, our approach is able to distinguish such behaviors even if the behavior on the control-flow perspective is very similar.

To extract such knowledge patterns from case attributes, we use frequent itemset mining (line 5–7). Specifically, we use the FPclose algorithm [8] to extract closed frequent itemsets to limit the number of itemsets. An itemset is closed if there exists no suitable superset which has the same support. We calculate the frequent itemsets for all cases that follow the same trace $t \in \hat{L}$. To retrieve all cases that follow a specific trace t, we define a function $cases(t)$ (line 5) which maps a given event trace t to their respective cases based solely on the event sequence. Note that while $cases(t)$ yields a set of cases with identical behavior, their case attributes might be quite different for which we mine frequent itemsets. We calculate the frequent itemsets for a given *minimum support threshold* θ. So, attribute-value pairs that occur in a certain amount of cases are extracted as frequent itemsets, directly taking the frequency of cases following the same trace into account. Case attributes are transformed using integer encoding, which is a mapping $\mathcal{I}_a : \mathcal{V}_a \rightarrow \mathbb{N}$ (line 6), assigning each attribute-value pair a unique positive integer. $encode(c)$ is the encoding function that encodes all attributes of a case $c \in L$. A similarity function between the two itemsets $S_1, S_2 \subseteq S$ is defined in line 8. It compares the two itemsets, in particular, the attribute-value pairs, and returns the proportion of items which are contained in both sets.

In our approach, we do not cluster the cases itself but the itemsets of all cases that follow the same trace. We define a similarity function $sim(s_a, s_b)$ that, on the one hand, calculates the similarity between itemsets and, on the other hand, compares the traces that share the same itemsets (line 10). With the weighting factor w we can control the balance between itemset similarity and trace similarity. It is noteworthy that even if $w = 1$ the itemset similarity is indirectly incorporated because traces that share the same itemsets are merged together. Proceeding further, we generate a distance matrix M (line 11) and use the *Agglomerative Hierarchical Clustering* algorithm to build a vector that contains the cluster index for each itemset (line 12). In line 13 the result is generated. $C(k)$ contains a set of traces that are clustered into cluster k.

Generating Non-overlapping Clusters. Due to the construction of the Algorithm 1, generated clusters are overlapping. This is because we create the clusters based on the itemsets and not based on the cases. For all cases that follow a specific trace, multiple frequent itemsets can be mined which are not necessarily clustered together, for example if the distance to other itemsets is lower. Whenever this occurs, a trace and their corresponding cases are part of multiple clusters. Even if it might also be interesting to analyze overlapping clusters, in this paper we aim for non-overlapping clusters to reconstruct process models using a discovery algorithm. To resolve the overlapping clusters, we assign traces that are assigned to multiple clusters to the one with the minimum distance with respect to sequence similarity.

We denote *buildCluster*(θ, n, w) to be the function which executes Algorithm 1 to generate the candidate clusters and the algorithm to resolve the overlapping.

Determine Optimal Parameters. In the last step, we optimize the minimum support threshold θ, the number of clusters N, the weighting factor w such that the fitness of the underlying model is maximized. The goal is to find an optimal separation of cases such that the different behaviors are separated into clusters while still being able to reconstruct accurate process models. We use the *Flexible HeuristicsMiner* [22] to reconstruct the models for each cluster because of its low computational costs and high accuracy in real-life scenarios. From these models we calculate the weighted average improved continuous semantics fitness measure (ICS) over all models.

Definition 4 *(Weighted ICS Fitness). Let ics_k the ICS-Fitness of a model k and n_k the number of cases in k, then the weighted ICS Fitness is defined as:*

$$ICS - Fitness = \frac{\sum_{k=1}^{N}(n_k \cdot ics_k)}{|L|}$$

Besides the weighted fitness of the models, we also optimize the number of cases assigned to a cluster, the number of clusters and the cluster silhouette coefficient. It might occur that θ is chosen too high such that a small amount of frequent patterns were extracted which should be avoided.

We use the *Particle Swarm Optimization* (PSO) [11] algorithm to maximize the fitness of the mined models of each cluster, finding optimal values for θ, n, and w. PSO is an evolutionary optimization algorithm which was initially inspired by bird flocking, specifically, the group dynamics of the bird behavior. PSO maintains a swarm of n particles p, the candidate solutions, which move around in the search-space. Initially, particles p_0 are randomly distributed in the search space and assigned an initial movement velocity v_0. Each particle maintains, the inertia ω, the best known position p_{best}, the global best position over all particles g_{best}, a cognitive weighting factor c_k and a social weighting factor c_s.

Definition 5. *For each iteration, a new velocity vector v_{n+1} is calculated with r_1, r_2 being random factors for each iteration:*

$$v_{n+1} = \omega \cdot v_n + c_k \cdot r_1 \cdot (p_{best} - p_n) + c_s \cdot r_2 \cdot (g_{best} - p_n)$$

The movement of the particles is determined by their current position and velocity as well as the local and global best known positions. Particles are moved until the maximum number of iterations is reached. PSO executes the *buildCluster*(θ, N, w) function and optimizes the model fitness as well as the proportion of assigned traces. In our experiments, we found that 10 iterations with 5 particles are appropriate. Although PSO does not guarantee a global optimum, our evaluation results suggest that even local optima yield good results.

4   Evaluation

In this section, we evaluate our proposed approach in two different evaluation settings. First, synthetic event logs are used to evaluate the quality of the generated clusters based on well-known evaluation measures for cluster analysis as well as process mining related measures. We compare our approach with six other clustering methods of the related work. Secondly, we use real-life event logs to show the applicability of our approach. Here, we focus on the quality of the generated models with respect to comprehensibility and accuracy. Our proposed approach is implemented as the *HybridCluster plugin*[1] in ProM.

4.1   Synthetic Event Logs Evaluation

We use synthetic event logs to evaluate and compare the performance of our clustering approach with respect to the quality of the generated clusters.

Datasets. Currently, there exists no comprehensive benchmark for the evaluation of trace clustering in the related work that focuses on the different behaviors on both the control-flow and data perspective. We generated synthetic event logs from random process models of different complexity (varying number of activities, maximum depth and branching factor). Five process models (see Table 2) are generated using PLG2 [5]: Small, Medium, Large, Huge, Wide and custom designed model with human readable activity names, all derived from [13]. For a representative data perspective, we generate sets of possible attribute values V_a, of size 20, for each of the case attributes $a \in \mathcal{A}$. Then we assign case attribute values to all cases by sampling from V_a, for each attribute a. To introduce some causal relationships, we force certain combinations of attribute values

---

[1] Source code available at: https://github.com/alexsee/HybridClusterer.

Table 2. Process models used for generating the event log: Number of activity types (# at), number of transitions (# tr), number of variants (# dpi), maximal trace length and out-degree.

Model	# at	# tr	# dpi	max length	out-degree
P2P	14	16	6	9	1.14
Small	22	26	6	10	1.18
Medium	34	48	25	8	1.41
Large	44	56	28	12	1.27
Wide	56	75	39	11	1.34
Huge	36	53	19	7	1.47

to occur more frequently than others depending on the event sequence of a case. Hence, each sequence of events will have certain attributes and attribute values that represent causalities for this sequence. Note that it is possible that multiple event sequences feature the same attribute value patterns. Consequently, we obtain patterns that correlate both with the data perspective and the control-flow, which will then represent our ground truth for the clusters. In the evaluation we generate five clusters consisting of the different generated patterns.

To increase the complexity of the task we also incorporate some level of noise (ranging from 0.0 to 0.2) into the control-flow [13]. For example by perturbating the order of events, skipping of events or executing multiple events. In summary, we generated event logs with different sizes (1 000, 2 000, 5 000 and 10 000), varying number of attributes (5, 10, 15, 20) and 3 different noise levels (0.0, 0.1, 0.2), resulting in 288 event logs[2].

Accuracy Results. We compare our hybrid feature set clustering approach (HC) with six other trace clustering methods: bag-of-activities (BOA) [4], Levenshtein edit distance (LED) [4], Context-Aware-Clustering (CAC) [3] and Acti-TraC (ATC) [21]. For BOA and LED agglomerative hierarchical clustering with ward linkage is used. For BOA+ and LED+ we filter out the event sequences that occur less than 2 times. BOA, BOA+, LED, LED+ and CAC do not provide any optimization to find the optimal number of clusters, thus we vary the number of clusters from 2 to the number of distinct traces and show the best results for the same or less clusters as HC. For ATC we use standard setting, 80% stopping criterion for the frequency-based and MRA distance-based selective sampling. As a baseline without clustering (FHM) and for generating the process models of each cluster, we use the Flexible Heuristics Miner [22].

Weighted Fitness, Precision and Generalization. To evaluate the accuracy of the discovered models, we report the ICS fitness of the FHM which is in the range of $(-\infty, 1]$. For precision and generalization, we first use the *Heuristics*

---

[2] Models are openly available: https://doi.org/10.7910/DVN/QBL1K0.

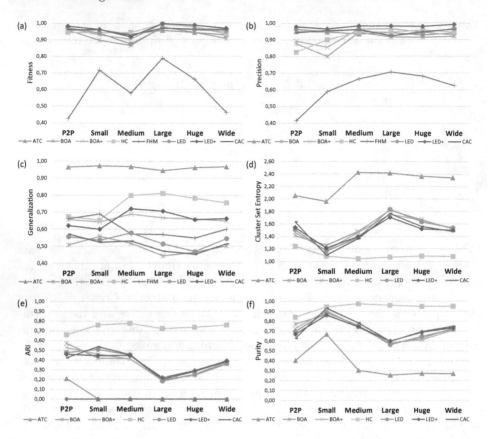

Fig. 1. Evaluation results of the synthetic event logs: (a) weighted fitness of the models; (b) weighted precision of the process models; (c) weighted generalization of the models; (d) cluster set entropy (lower is better); (e) adjusted rand index; (f) purity.

Net to Petri Net plugin in ProM and then calculate the measures introduced in [19]. Note that precision and generalization are sensitive measures that strongly depend on the petri net (e.g., if the petri net contains silent or duplicate transitions). While the weighted fitness (see Fig. 1(a)) of the process model without any clustering is quite low for all evaluated event logs, all trace clustering method produced good results within the range of 0.969 and 0.925. The best ICS fitness was achieved by LED+, followed by CAC and our HC method. Comparing the precision (see Fig. 1(b)) of the models reveals that LED+ slightly outperformed all other methods, followed by LED, ATC and BOA. Our HC method achieved an average precision of 0.920 compared to 0.980 of the best. However, it is still significantly better than the FHM. With respect to generalization (see Fig. 1(c))) ATC outperformed all methods significantly, whereas our HC method is the second best.

Table 3. Performance of the related work and our hybrid cluster approach with respect to process model and clustering evaluation; best values in bold typeface.

	Process model			Clustering			
	Fitness	Precision	Generalization	Purity	ARI	CSE	$|C|$
FHM [22]	0.606	0.612	0.607	-	-	-	1.0
ActiTraC [21]	0.946	0.940	**0.964**	0.363	0.036	2.258	1.8
Bag-of-activities [4]	0.925	0.895	0.499	0.719	0.372	1.526	17.5
Bag-of-activities+	0.952	0.925	0.661	0.713	0.367	1.507	20.1
Levensthein [4]	0.945	0.951	0.533	0.711	0.374	1.509	21.9
Levensthein+	**0.969**	**0.980**	0.661	0.716	0.376	1.480	18.6
CAC [3]	0.957	0.948	0.510	0.729	0.381	1.484	22.6
Hybrid clusterer	0.956	0.920	0.746	**0.937**	**0.736**	**1.100**	26.7

Cluster Set Entropy, Adjusted Rand Index and Purity. While fitness, precision and generalization evaluate the accuracy of the process models, cluster set entropy (CSE), adjusted rand index (ARI) and purity (see Fig. 1 (d)–(f) and Table 3) evaluate the calculated clustering against the ground truth. Our approach outperforms all other methods over all event logs with an average CSE of 1.100, an average ARI of 0.736 and an average purity of 0.937. A clear ranking of the other methods cannot be made, as the field is mixed here. For smaller models, the performance difference between our HC and the related work is less significant.

Discussion. In our experiments, the HC approach is the only method that incorporates case attributes. Hence, the existence of case attributes that are somehow related to the traces is essential for our method. While we used synthetic event logs that contain such relationships, other event logs may not have these relations. Here, other trace clustering methods may perform better. Still, the aim of our method is to combine both perspectives, the control-flow and the case attributes, to generate behavioral similar sub-logs. Our method tries to overcome the issue of unrelated case attributes by optimizing the balance between both perspectives. In cases where no relationship is found, our method will prefer giving the control-flow similarity more contribution.

However, when such a relationship exists, as presented in our evaluation, our method provides a solid separation of the event log. With respect to CSE, ARI and purity, our method works better than control-flow only based methods. Our method is also able to generate process models with a good fitness, precision and generalization, although being outperformed by other methods. The variance of the traces within a cluster might be high which negatively influences the fitness, precision and generalization. Another observation is that our approach generates more clusters than other methods (see Table 1). This is mainly caused by the separation of the different behaviors on the different perspectives.

Table 4. Overview of real-life logs: The number of process instances (# pi), number of events (# ev), number of activity types (# at) and the number of variants (# dpi).

Event log	Event log properties				Description
	# pi	# ev	# at	# dpi	
P2P	33 277	255 427	37	7 026	Procurement process
EV	1 434	8 577	27	116	Case handling system
HOSBILL	100 000	451 359	18	1 020	Hospital invoice billing
HOSLOG	1 143	150 291	624	981	Case handling in hospital
ROAD	150 370	561 470	11	231	Road traffic fine process

4.2 Real-Life Event Logs Evaluation

The second part of the evaluation applies our approach to real-life event logs. In Table 4 we show some basic statistics of the used event logs, originated from different environments to show the applicability of our approach in various scenarios. All event logs except for P2P are openly available[3]. Because we use real-life event logs for which we do not know their real behavior, we focus on the evaluation of the following measures: First, we measure the weighted fitness, precision and generalization of models discovered by the FHM of each cluster. Again, the heuristics nets are converted using the ProM plugin as mentioned before. Second, we measure the complexity of the resulting models. For the five real-life experiments we use the same settings as used in the synthetic evaluation. Again, for BOA, BOA+, LED and LED+ we only report the best fitness for the same or less number of clusters as HC.

Accuracy Results. The weighted fitness, precision and generalization results are presented in Table 5. In all cases the FHM was outperformed by all clustering techniques, concluding that a single model is not sufficient to accurately model the observed behavior. The vector-based clustering approaches, i.e., BOA, BOA+, LED and LED+, provide a relatively good fitness of the models. Acti-TraC in general provides better fitness values as the vector-based approaches. HC provides better or similar results as other clustering methods, except for HOSLOG. This is due to the fact that HOSLOG contains many unique traces. The HOSLOG event log origins from an hospital where the examination of each patient is recorded and contains cases that usually do not follow a strictly defined sequence. With respect to the precision and generalization, the field is mixed.

---

[3] http://data.4tu.nl/repository/collection:event_logs_real.

Table 5. Results of the real-life logs showing average weighed fitness. Precision and generalization in parenthesis. Missing values due to canceled calculation after 12 h.

	P2P	EV	HOSBILL	HOSLOG	ROAD
FHM [22]	−0.799	0.649	−0.781	0.554	0.434
	(0.15/0.27)	(0.66/0.41)	(0.63/0.50)	-	(0.98/0.39)
ActiTraC [21]	**0.729**	0.893	0.651	**0.719**	0.973
	(0.70/0.81)	(0.72/0.91)	-	-	(0.99/0.93)
Bag-of-activities [4]	0.146	0.793	−0.354	0.346	0.755
	(0.29/0.48)	(0.70/0.37)	(0.86/0.37)	-	(0.95/0.50)
Bag-of-activities+	0.519	0.839	−0.134	0.685	0.813
	(0.50/0.77)	(0.99/0.36)	(0.98/0.73)	(0.42/0.88)	(0.90/0.65)
Levensthein [4]	0.196	0.871	−0.291	0.406	0.837
	(0.42/0.55)	(0.70/0.47)	(0.77/0.62)	-	(0.99/0.72)
Levensthein+	0.725	0.857	−0.017	0.001	0.979
	(0.68/0.79)	(0.84/0.73)	(0.93/0.64)	(0.68/0.68)	(0.99/0.98)
CAC [3]	0.121	0.739	0.839	0.198	0.887
	(0.40/0.53)	(0.77/0.39)	(0.74/0.72)	-	(1.00/0.56)
Hybrid clusterer	0.723	**0.975**	**0.958**	0.515	**0.993**
	(0.61/0.77)	(0.96/0.55)	(0.97/0.81)	(0.88/0.86)	(0.99/0.99)

Complexity Results. We calculate four complexity measures based on related work [6,18]. The graph density GD is defined as $GD = \frac{|E|}{|N| \cdot (|N|-1)}$ where $|N|$ are the number of nodes in the discovered heuristics net and $|E|$ the number of edges. The cyclomatic number CN is defined as $CN = |E| - |N| + 1$. The coefficient of connectivity CNC is defined as $CNC = |E|/|N|$ and the coefficient of network complexity $CNCK$ is defined as $CNCK = |E|^2/|N|$.

Table 6 shows the average and maximum of each complexity measure. We can see that our approach creates more clusters than other methods. This might be due to the fact that even similar traces are split into multiple clusters when they differ from the data perspective. The high number of clusters for ActiTraC is caused by the explosion in clusters for the P2P and the HOSLOG event log. When comparing the graph density, our method produces slightly denser process models than other methods. For cyclomatic number, coefficient of connectivity and coefficient of network complexity our hybrid clustering approach performs better than all compared methods. Comparing the average and the maximum numbers concludes that generated clusters of our method do not vary heavily.

Table 6. Complexity measures for each approach.

	$|C|$	GD		CN		CNC		CNCK	
		avg	max	avg	max	avg	max	avg	max
ActiTraC [21]	198.4	0.15	0.66	11.10	116.60	1.38	2.22	46.37	477.28
Bag-of-activities [4]	18.2	0.12	0.20	82.33	114.20	1.15	1.20	329.05	456.40
Bag-of-activities+	17.6	0.24	0.51	2.84	5.40	0.78	1.00	12.06	20.00
FHM [22]	1	**0.10**	0.18	185.00	803.00	1.80	3.00	754.40	3258.00
Levensthein [4]	29.2	0.11	0.32	37.96	120.80	1.00	1.40	154.28	484.60
Levensthein+	22	0.20	0.42	3.61	7.60	0.76	1.20	15.10	28.80
CAC [3]	16.6	**0.10**	0.18	41.22	112.80	1.06	1.40	166.76	451.60
Hybrid clusterer	31.2	0.23	0.63	**2.15**	7.80	**0.67**	1.20	**10.62**	31.60

5   Conclusion

In this paper, we proposed a novel hybrid-feature set clustering approach in the area of process mining. While other trace clustering methods mainly rely on the control-flow, we use frequent pattern mining to extract further knowledge from the case attributes. To produce accurate process models, our method uses the particle swarm optimization to optimize the fitness of the underlying process models by automatically finding appropriate parameters.

We implemented our approach as an openly available ProM plugin. We evaluated our approach by conducting a comprehensive evaluation using synthetic and real-life event logs. We compared our approach to six other methods and showed that our approach is able to separate process behaviors on the control-flow as well as on the data perspective. For the synthetic event logs, our method reaches an ARI of 0.736 in average over all models evaluated, whereas the second best approach CAC reaches an ARI of 0.381. Besides using synthetic event logs, we used 5 real-life event logs to show the applicability of our method.

Something that we did not inspect is the question, if the identified clusters are relevant for process analysts during their analysis. Because this question is hard to answer without an extended user study, we would like to address this in future work. A first interview with a process mining consultant showed the interest in the idea of incorporating both perspective into the clustering. Additionally, it might also be interesting to incorporate expert knowledge to adjust the cluster quality. We also want to extend the approach to support numeric attributes and optimize runtime performance, because currently the algorithm has to perform certain operations multiple times which can be precalculated or cached.

Overall, we can conclude that our hybrid feature set clustering approach is a promising method for segmenting the event log into behavioral similar clusters.

Acknowledgements. This project [522/17-04] is funded in the framework of Hessen ModellProjekte, financed with funds of LOEWE, Förderline 3: KMU-Verbundvorhaben (State Offensive for the Development of Scientific and Economic Excellence), and by the German Federal Ministry of Education and Research (BMBF) Software Campus project "AI-PM" [01IS17050].

References

1. van der Aalst, W.M.P.: Process Mining: Discovery, Conformance and Enhancement of Business Processes, 2nd edn. Springer, Heidelberg (2011). https://doi.org/10.1007/978-3-642-19345-3
2. Appice, A., Malerba, D.: A co-training strategy for multiple view clustering in process mining. IEEE Trans. Serv. Comput. **9**(6), 832–845 (2016). https://doi.org/10.1109/tsc.2015.2430327
3. Bose, R.P.J.C., van der Aalst, W.M.P.: Context aware trace clustering: towards improving process mining results. In: Proceedings of the 2009 SIAM International Conference on Data Mining, pp. 401–412. Society for Industrial and Applied Mathematics (2009). https://doi.org/10.1137/1.9781611972795.35
4. Bose, R.P.J.C., van der Aalst, W.M.P.: Trace clustering based on conserved patterns: towards achieving better process models. In: Rinderle-Ma, S., Sadiq, S., Leymann, F. (eds.) BPM 2009 Workshops. LNBIP, vol. 43, pp. 170–181. Springer, Heidelberg (2010). https://doi.org/10.1007/978-3-642-12186-9_16
5. Burattin, A.: PLG2: multiperspective process randomization with online and offline simulations. In: CEUR Workshop Proceedings, vol. 1789, pp. 1–6 (2016)
6. Delias, P., Doumpos, M., Grigoroudis, E., Matsatsinis, N.: A non-compensatory approach for trace clustering. Int. Trans. Oper. Res. (2017). https://doi.org/10.1111/itor.12395
7. Evermann, J., Thaler, T., Fettke, P.: Clustering traces using sequence alignment. In: Reichert, M., Reijers, H.A. (eds.) BPM 2015 Workshops. LNBIP, vol. 256, pp. 179–190. Springer, Cham (2016). https://doi.org/10.1007/978-3-319-42887-1_15
8. Grahne, G., Zhu, J.: Fast algorithms for frequent itemset mining using FP-trees. IEEE Trans. Knowl. Data Eng. **17**(10), 1347–1362 (2005). https://doi.org/10.1109/tkde.2005.166
9. Greco, G., Guzzo, A., Pontieri, L., Sacca, D.: Discovering expressive process models by clustering log traces. IEEE Trans. Knowl. Data Eng. **18**(8), 1010–1027 (2006). https://doi.org/10.1109/tkde.2006.123
10. Han, J., Cheng, H., Xin, D., Yan, X.: Frequent pattern mining: current status and future directions. Data Min. Knowl. Disc. **15**(1), 55–86 (2007). https://doi.org/10.1007/s10618-006-0059-1
11. Kennedy, J., Eberhart, R.: Particle swarm optimization. In: Proceedings of ICNN 1995 - International Conference on Neural Networks. IEEE. https://doi.org/10.1109/icnn.1995.488968
12. De Koninck, P., Nelissen, K., Baesens, B., vanden Broucke, S., Snoeck, M., De Weerdt, J.: An approach for incorporating expert knowledge in trace clustering. In: Dubois, E., Pohl, K. (eds.) CAiSE 2017. LNCS, vol. 10253, pp. 561–576. Springer, Cham (2017). https://doi.org/10.1007/978-3-319-59536-8_35
13. Nolle, T., Seeliger, A., Mühlhäuser, M.: Unsupervised anomaly detection in noisy business process event logs using denoising autoencoders. In: Calders, T., Ceci, M., Malerba, D. (eds.) DS 2016. LNCS (LNAI), vol. 9956, pp. 442–456. Springer, Cham (2016). https://doi.org/10.1007/978-3-319-46307-0_28

14. Seeliger, A., Schmidt, B., Schweizer, I., Mühlhäuser, M.: What belongs together comes together. Activity-centric document clustering for information work. In: Proceedings of the 21st International Conference on Intelligent User Interfaces - IUI 2016. ACM Press (2016). https://doi.org/10.1145/2856767.2856777
15. Song, M., Yang, H., Siadat, S.H., Pechenizkiy, M.: A comparative study of dimensionality reduction techniques to enhance trace clustering performances. Expert Syst. Appl. **40**(9), 3722–3737 (2013). https://doi.org/10.1016/j.eswa.2012.12.078
16. Song, M., Günther, C.W., van der Aalst, W.M.P.: Trace clustering in process mining. In: Ardagna, D., Mecella, M., Yang, J. (eds.) BPM 2008 Workshops. LNBIP, vol. 17, pp. 109–120. Springer, Heidelberg (2009). https://doi.org/10.1007/978-3-642-00328-8_11
17. Sun, Y., Bauer, B., Weidlich, M.: Compound trace clustering to generate accurate and simple sub-process models. In: Maximilien, M., Vallecillo, A., Wang, J., Oriol, M. (eds.) ICSOC 2017. LNCS, vol. 10601, pp. 175–190. Springer, Cham (2017). https://doi.org/10.1007/978-3-319-69035-3_12
18. Thaler, T., Ternis, S., Fettke, P., Loos, P.: A comparative analysis of process instance cluster techniques. In: Proceedings der 12. Internationalen Tagung Wirtschaftsinformatik, WI 2015, August, pp. 423–437 (2015)
19. Vanden Broucke, S.K., De Weerdt, J., Vanthienen, J., Baesens, B.: Determining process model precision and generalization with weighted artificial negative events. IEEE Trans. Knowl. Data Eng. **26**(8), 1877–1889 (2014). https://doi.org/10.1109/TKDE.2013.130
20. Weerdt, J.D., Backer, M.D., Vanthienen, J., Baesens, B.: A multi-dimensional quality assessment of state-of-the-art process discovery algorithms using real-life event logs. Inf. Syst. **37**(7), 654–676 (2012). https://doi.org/10.1016/j.is.2012.02.004
21. Weerdt, J.D., vanden Broucke, S., Vanthienen, J., Baesens, B.: Active trace clustering for improved process discovery. IEEE Trans. Knowl. Data Eng. **25**(12), 2708–2720 (2013). https://doi.org/10.1109/tkde.2013.64
22. Weijters, A.J.M.M., Ribeiro, J.T.S.: Flexible heuristics miner (FHM). In: 2011 IEEE Symposium on Computational Intelligence and Data Mining (CIDM). IEEE (2011). https://doi.org/10.1109/cidm.2011.5949453

act2vec, trace2vec, log2vec, and model2vec: Representation Learning for Business Processes

Pieter De Koninck$^{(\boxtimes)}$, Seppe vanden Broucke, and Jochen De Weerdt

Department of Decision Sciences and Information Management,
Faculty of Economics and Business, KU Leuven, Leuven, Belgium
{pieter.dekoninck,seppe.vandenbroucke,jochen.deweerdt}@kuleuven.be

Abstract. In process mining, the challenge is typically to turn raw event data into meaningful models, insights, or actions. One of the key problems of a data-driven analysis of processes, is the high dimensionality of the data. In this paper, we address this problem by developing representation learning techniques for business processes. More specifically, the representation learning paradigm is applied to activities, traces, logs, and models in order to learn highly informative but low-dimensional vectors, often referred to as embeddings, based on a neural network architecture. Subsequently, these vectors can be used for automated inference tasks such as trace clustering, process comparison, predictive process monitoring, anomaly detection, etc. Accordingly, the main contribution of this paper is the proposal of representation learning architectures at the level of activities, traces, logs, and models that can produce a distributed representation of these objects and a thorough analysis of potential applications. In an experimental evaluation, we show the power of such derived representations in the context of trace clustering and process model comparison.

Keywords: Representation learning · Process mining
Word embedding

1 Introduction

Process mining is a set of techniques which are both data-driven and process-centric [2], often grouped into process discovery, conformance checking and extension categories. One of the key challenges in process mining is dealing with the high dimensionality of the data, given that real-life event logs present a large number of cases, potentially representing a highly varied set of distinct event sequences, and usually also containing information on resources and a diverse set of other event or case-related attributes. This makes that featurizing event data (i.e. attempting to extract structured instances from raw event logs) for learning tasks such as trace clustering [5,10] or predictive process monitoring [3,11] suffer from a dimensionality problem and usually resort to an ad-hoc definition of input features.

© Springer Nature Switzerland AG 2018
M. Weske et al. (Eds.): BPM 2018, LNCS 11080, pp. 305–321, 2018.
https://doi.org/10.1007/978-3-319-98648-7_18

Naturally, this dimensionality problem poses itself in other disciplines as well, e.g. in natural language processing (NLP) [14], image recognition [21] and social network analytics [16]. Accordingly, despite being originally developed in the NLP domain [25], representation learning is gaining a lot of traction outside NLP. The key novelty of representation learning is the use of neural network-based architectures to automatically learn high quality distributed vector representations of a concept of interest (an image, word, document, or node in a social graph). Such dense yet informative vectors (e.g. in text mining applications, vectors of length 128 are common) representing input objects are often referred to as "embeddings". Despite showing excellent potential in NLP, image recognition, and social network analytics, representation learning has so far not been leveraged in the field of business process management.

Accordingly, the main contribution of this paper is the development of representation learning architectures for deriving distributed vector representations of activities (*act2vec*), traces (*trace2vec*), logs (*log2vec*) and process models (*model2vec*). In addition, we discuss different use cases of such embeddings. Hereto, the remainder of this paper is structured as follows: Sect. 2 gives a detailed overview of embedding architectures for representing activities, traces, event logs, and process models. Section 3 investigates different BPM uses cases of these distributed representations. Next, we experimentally illustrate that representation learning techniques can be applied successfully in the context of trace clustering and process model comparison in Sect. 4. Related work from the representation learning field as well as other applications of neural network-based learning techniques in process mining are discussed in Sect. 5. The paper is concluded with a discussion and outlook towards future research in Sect. 6.

2   Representation Learning Architectures

Representation learning relies on neural networks to construct vectors representing objects. In the natural language processing domain, many different models and architectures have been proposed for obtaining distributed representations of words, ranging from Latent Dirichlet Allocation (LDA) to the neural network language model (NNLM) [4]. More recently, Mikolov et al. [24] proposed the Continuous Bag of Words (CBOW) and SkipGram architectures, mainly improving previous works in terms of scalability, making it possible to learn representations of words from huge datasets. In this section, we propose representation learning architectures for deriving distributed vector representations of activities, traces, event logs, and process models.

2.1   *act2vec*: Obtaining Representations of Activities

The first representation learning architecture relates to deriving representations of activities. For doing so, we assume that we have input data in the form of an event log. In line with the *word2vec* approach in natural language processing, we can learn representations for activities by considering activities as words in a corpus, with the corpus being the event log in our case. Observe that our default architecture for *act2vec*, as visualized in Fig. 1, is similar to the CBOW model for constructing word representations [25]. The CBOW neural network architecture is based on the principle that a word can be predicted from its context (i.e. the words appearing before and after the focus word). In this work, we consider traces as sentences and events as words. As such, we learn general purpose representations of activities which is not tailored towards for instance predicting the next activity or predicting the remaining time of instances. In our experimental evaluations, we show by example that such general purpose representations could be powerful, nonetheless, it is important to note that it remains to be seen whether the contextual similarity based on co-occurrence on which the representations are based is useful in the full range of potential applications.

Observe that, despite the fact that traces in an event log are sequential in nature, we opt to define the context of an activity based on the unordered set of preceding and following activities. As such, we obtain a learning architecture as depicted in Fig. 1 that can be used to understand the context of two different activities. Depending on their context, their representation will either be similar or very different. While activity representations could be used for particular BPM-related tasks (an overview is provided in the next section), the application potential can be considered less significant than the learning architectures discussed below. Nevertheless, the architecture is considered foundational for the subsequent representation learning models, as discussed below.

To speed up the training process of the neural network, we apply negative sampling as described in [25]. The idea of negative sampling is that in each network update, only a small percentage of its parameters (i.e. its weights) are updated, instead of all of them, by updating the weights based on sampling the output vector (a one-hot encoded vector). For more information, we refer to [30].

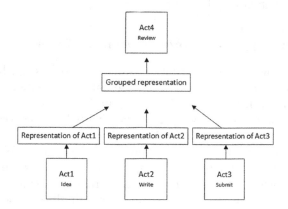

Fig. 1. The *act2vec*-architecture for learning vector representations of activities. The context consisting of activities "idea", "write", and "submit" is used to predict activity "review".

Architectural Extensions. The *act2vec* architecture as represented in Fig. 1 can be extended in several ways. Most importantly, other related attributes could be leveraged in order to include other data available in the event log in the definition of an activity's context or in the concept to predict. For instance, resource information can be included in the neural network architecture in order to also incorporate information regarding who executed a particular activity or data attributes can be taken into account as well. These additional data dimensions can be considered as additional inputs of the architecture proposed above, but more complex hierarchical or multilayer designs could be explored as well. However, for now, we keep the neural network architectures simple given that these simple models tend to work very well in other domains.

Embeddings Through Recurrent Neural Networks. Long short term memory networks (LSTM) and other recurrent neural network (RNN) architectures allow for keeping track of the sequential nature of the input data. That is, they explicitly incorporate the dependent aspect of the input data in the architecture of the neural network, and can be applied in the setting of representation learning as well, see e.g. [9]. This is not the case for the standard CBOW and SkipGram models, which assume the contextual inputs to be independent (typically, the grouping of representations of the contextual inputs is performed by averaging or summing them, though a concatenation approach can be used here as well, which partially resolves the independence assumption). In process analysis, it seems logical to assume that order is important to take into account. Nevertheless, also in speech recognition or text mining, word order seems to be important at first sight, though it has been demonstrated that neural network embedding architectures neglecting order provide solid results using default hyperparametrization. Although this argument should be further investigated for application of embedding techniques for business processes, please observe that in this paper, we explicitly avoid looking into a comparison between ordered and unordered neural network architectures.

2.2   *trace2vec*: Obtaining Representations of Traces

Following the analogy between activities and words, traces can be regarded as sentences as well. Learning distributed representations of sentences, paragraphs, or documents has been introduced in the natural language processing domain in the form of the *doc2vec* approach [21]. The idea behind *doc2vec* is simple yet clever as the authors extend the CBOW architecture with a paragraph vector, resulting in the so-called Distributed Memory Model of Paragraph Vectors (PV-DM). This idea can be adopted for traces as well, giving rise to the *trace2vec* architecture as shown in Fig. 2. Given that this architecture includes a representation of traces (based on the trace identifier), it will allow for joint learning of representations of activities and traces. Obviously, for further analysis, we are mostly interested in the trace representations.

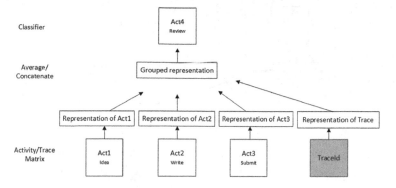

Fig. 2. The *trace2vec*-architecture for learning vector representations of traces. The context consisting of activities "idea", "write", and "submit", as well as the trace-id, is used to predict activity "review".

Alternative Architectures. Contrasting the Distributed Memory Model of Paragraph Vectors (PV-DM) approach, which can be regarded as an extension of CBOW, an alternative option called Distributed Bag of Words model for Paragraph Vectors (PB-DBOW) exists as well, which is comparable to the SkipGram model. Here, the idea is to only use a paragraph vector on the input side to predict a small window of words in the paragraph on the output side. Following the reasoning to opt for CBOW above SkipGram as outlined above, we have chosen to investigate PV-DM first as an approach towards *trace2vec* in this work. Apart from using these *doc2vec*-based architectures, it would also be possible to derive trace-level representations by making use of aggregator architectures. Such an aggregator architecture would simply infer trace-level representations by aggregating activity representations. A wide variety of potential aggregators exist, ranging from a simple mean operator, i.e. compute the element-wise mean of the representation vectors up to much more complex aggregator architectures such as LSTM aggregators or pooling aggregators [19]. Observe that, in order to manage the scope of this paper, aggregator architectures are not further investigated. Note also that, in line with *act2vec*, additional data dimensions can be included here as well (e.g. resource, timing, etc.).

2.3 *log2vec*: Obtaining Representations of Logs

Several data-driven learning tasks within the BPM domain will rely on representations of an entire process (either represented by an event log or a model). As such, in line with *trace2vec*, an architectural design can be devised to learn distributed representations of logs. A simple method could be to replace the representation of a trace with a representation of a log in Fig. 2. Though given that such a setup architecture would be unable to incorporate trace-level information, and thus will consider an event log as a set of ungrouped activities, it makes sense from a business process perspective to extend the architecture to

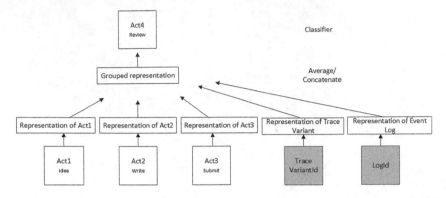

Fig. 3. The *log2vec*-architecture for learning vector representations of logs. The context consisting of activities "idea", "write", and "submit", an identifier for each trace variant, as well as the log-id from which the words are sampled, is used to predict activity "review".

also include trace information. To do so, we could simply use the trace identifier as before, however, given that business processes could share similar execution variants, we propose to include an artificial identifier relating to distinct process instances in the architecture (i.e. a trace "variant" identifier). More specifically, all traces from the different event logs under consideration are joined into one event log based on that input, a distinct process instance identifier is computed. The *log2vec* architecture is then illustrated in Fig. 3.

2.4 *model2vec*: Obtaining Representations of Process Models

The final architecture we propose here is *model2vec*. Here, the goal is to represent process models as low-dimensional vectors. Given the architectures discussed above, a trivial extension could be to obtain model representations by first simulating the models, keeping track of the event data produced while simulating, and subsequently applying *log2vec* to learn a representation of the process model. Nonetheless, given that it can be argued that process models are not solely represented by the execution variants they are able to produce, opportunities open up to apply different representation learning techniques of models. In this paper, we propose to treat a process model as an undirected graph, and learn representations by generating input data through performing random walks within this undirected graph. This is in line with graph representation learning techniques in the social network analytics domain [16,26,27]. Obviously, this second approach is only indirectly considering the actual behavior of the model, but is putting more emphasis on the graphical layout and relationships between modeling elements, including those which are not directly related to an activity's representation (such as gateways, places, or other constructs). Through the random walking procedure, the neural network architecture itself is then represented by Fig. 4.

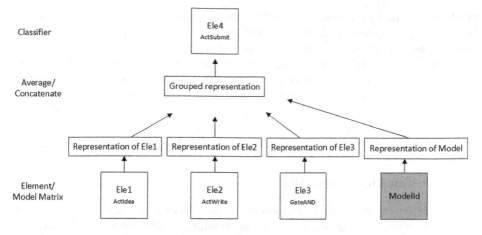

Fig. 4. The *model2vec*-architecture for learning vector representations of logs. The context of elements resulting from either simulation or random walk ("idea", "write", "And-gateway"), as well as the model-id for the model on which the simulation is performed or the walk is sampled, are used to predict activity "submit".

3   Applications of Process Representation Learning

In this section, we discuss potential use cases of *act2vec*, *trace2vec*, *log2vec*, and *model2vec*, relying on the classification in [1].

3.1   Model Discovery Preprocessing and Trace Clustering

Regarding process discovery, our representation learning architectures can be used in the context of preprocessing event data before providing it to a discovery algorithm. First of all, *act2vec* can be used to "vertically" partition event logs, which is often necessary when event logs contain too much fine-granular events. As such, several researchers have looked into abstraction techniques, e.g. [7,17]. More recently, in [33] event abstraction is performed based on conditional random fields. Nevertheless, the input features are, in contrast to *act2vec*, user-defined and the setup is supervised. In comparison to existing works, *act2vec*-based event abstraction will focus more heavily on the "similarity by co-occurrence"-principle inherent to CBOW, while existing techniques are more strongly focused on the specific, ordered control-flow relations. Whether this difference can actually result in better or different abstractions is not further considered in this work.

Secondly, trace clustering is a natural application of *trace2vec*. Trace clustering deals with splitting an event log horizontally instead of vertically. It is important to point out that a key advantage of *trace2vec* over existing trace clustering solutions relates to the fact that the input features are not defined in an ad-hoc and expert driven way. More specifically, the usage of typical features such as activity profiles and n-grams is subject to user preference. Whether a clustering technique is using the right types of features is typically not known by the user. As such, in line with performance benefits reported in the NLP domain, we expect that learning a dense but high-quality representation of features in an unsupervised way, will allow to automatically take all relevant differences between traces into account. Note that *trace2vec*-based trace clustering is more in line with existing similarity-based clustering techniques, thus it can be expected that similar advantages and disadvantages can be observed when compared to behaviorally-oriented clustering techniques such as [10]. Finally, observe that this is the first application of embeddings for trace clustering.

3.2   Process Model Selection

Our representation learning architectures also have application potential for model selection related use cases. In particular, *model2vec* (and also *log2vec* when combined with simulation) would allow for an alternative approach to process model comparison. Current methods such as behavioural profiles [32] and untanglings [28] focus heavily on control-flow relationships, while techniques such as the GED-based graph matching [13] are more graph-based. Regardless the unfavourable but also immature experimental results in Sect. 4.2, on the theoretical level, *model2vec*-based process model comparison could leverage the structural and graphical dimension of process models, with existing approaches strongly focusing on the behavioral similarity. In case this structural/graphical dimension should be de-emphasized, we expect that *log2vec* or a combination of both, might yield an opportunity to approach process model comparison differently. Observe that this is the first research to proposes an embedding-based process model comparison technique. In [31], process querying based on Latent Semantic Analysis is proposed, also relying on an NLP technique to develop a process model similarity technique. In contrast to our proposal, the technique of [31] does not rely on the local context of activities as it only takes into account the presence of "words" extracted from the process model. As such, this technique is accounting for only semantic similarity. As mentioned in [31], representation learning techniques have shown to outperform probabilistic methods in the computational linguistics field. As such, and although text is different from event logs and the random walks in a process model, we expect similar performance gains for our methods, especially given that a balance between semantical and control-flow based similarity seems the right avenue for future work.

3.3   Process Monitoring

Finally, predictive process monitoring is considered as another potential application. We are witnessing a strong uptake of predictive process monitoring research [11,22], presenting techniques and methodologies to make business process execution systems smarter by predicting outcomes of ongoing process instances. As with trace clustering, for predictive process monitoring, a wide set of potential input features can be important. Nevertheless, most of the presented techniques again rely on ad-hoc featurization. As such, *act2vec* or *trace2vec*, applied to partial traces, can be considered as powerful alternative featurization methods with the advantage being that the number of input features can be kept low, but at the same time taking into account a wide range of potentially discriminating information in the event log. For this particular application, a more profound analysis regarding the use of an ordered vs. unordered architecture should be performed, as well as regarding the exact definition of the context (e.g. by only looking at preceding activities). A more detailed comparison with existing works can be found in Sect. 5.

4   Implementation and Experimental Evaluation

In this section, two of the proposed architectures for learning representations are evaluated: on the one hand, *trace2vec* is applied in a trace clustering context, and on the other hand, *model2vec* is used to calculate process model similarity[1].

4.1   Trace Clustering with *trace2vec*

To evaluate the usefulness of *trace2vec*, it is applied to an event log for which a ground truth is known. This event log is a pre-processed version of the dataset used in the BPI Challenge of 2015, a collection of event data from the permit processes of five Dutch municipalities [35]. The log contains 5649 traces and 29 types of activities. The five distinct municipalities can be considered the "true" clusters of traces.

As described in Sect. 2.2, *trace2vec* produces representations for each trace in an event log. The neural network is trained using a window size of 3. Four different feature vector lengths are tested: 16, 32, 64, and 128. To cluster the traces using the learned representations, a general-purpose clustering algorithm can be used. In this case, we opt to include k-means and hierarchical clustering based on Ward's minimum variance method, on each of the four learned representations.

We compare the results obtained to the following existing trace clustering techniques: frequency-based (*ActFreq*), and distance-based (*ActMRA*) ActiTraC [10]. These clustering techniques are active in the sense that each cluster is represented by a discovered process model: a clustering is created based on process model quality. Furthermore, we compare with two hierarchical clustering approaches based on defined features: *3-grams* and *MRA*. These are well known approaches in the area of text mining and were ported to the domain of process mining by [6]. Finally, a hierarchical clustering based on the Generic Edit Distance (*GED* [5]) between traces is included as well. All techniques are applied with a cluster size of 5.

---

[1] Implementations of *act2vec*, *trace2vec*, *log2vec*, and *model2vec* are available alongside the data and models used for the experimental evaluation and full color figures on http://processmining.be/replearn. The implementations are based on the Gensim-library for unsupervised semantic modelling from plain text [29].

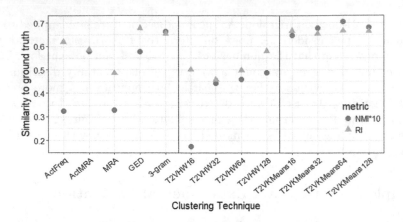

Fig. 5. The similarity of each of the clustering solutions to the ground truth, as measured by Rand Index (RI) and Normalized Mutual Information (NMI).

The results of each of these clustering approaches is represented in Fig. 5. Each point visualizes the similarity of the resulting clustering to the ground truth. In terms of notation, *T2VHW16* is the result of applying Hierarchical clustering with Ward's method onto the features of length 16 learned by *trace2vec*. The similarity is calculated using two metrics: Rand Index (RI) and Normalized Mutual Information (NMI), both measuring the extent to which the resulting cluster solution captures the same information as the ground truth. The NMI is scaled by a factor 10 to increase legibility.

A couple of observations can be made: first, for the Rand Index, the best results are obtained by *GED* and *3-gram*. The results of *trace2vec* combined with k-means is similar to those values across all vector sizes. When looking at Normalized Mutual Information, *trace2vec* combined with k-means scores better than the best alternative, *3-gram*, for all vector sizes larger than 16. Furthermore, observe that for both the Rand Index and NMI, *trace2vec* combined with Ward's method performs worse than with k-means.

Apart from their similarity to the ground truth, we are also interested in the extent to which *trace2vec* leads to unique solutions. In Fig. 6, the pairwise similarity of each of the clustering solutions is plotted, calculated using the Rand Index. From this figure, it is clear that varying the vector sizes influences the final solution, but the chosen clustering technique that is applied to the learned representation influences the clustering more than the vector sizes. All other solutions are reasonably unique, apart from *GED* and *3-gram*. There is also some slight correspondence between the clusters found by *3-gram* and those found by *trace2vec* combined with k-means.

In summary, *trace2vec* shows a lot of potential for trace clustering, given that even with a very simple architecture and without any significant hyperparameterization, it already performs on par or better than existing techniques, with an additional benefit of yielding unique solutions.

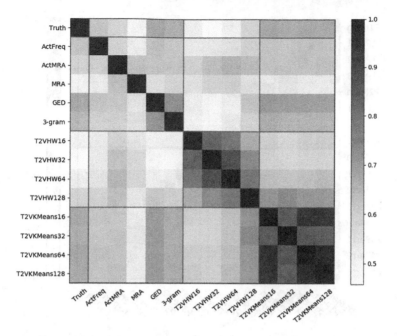

Fig. 6. The pairwise similarity of each of the clustering solutions, as measured by the Rand Index (RI).

4.2 Process Model Comparison with *model2vec*

Next, an empirical illustration is provided of the application of *model2vec* for the calculation of process model similarity. The setup is as follows: four distinct process models were randomly generated using PLG2 [8], using default settings. The characteristics of the main models are provided in Table 1. For each of these four main models, three random "perturbations" were generated by evolving the main model slightly. This is done by a process called "Process Evolution", originally conceived for concept drift detection. In total, this gives us a set of sixteen models, of which four groups of four models are naturally related. All models are made available online (see footnote 1).

Table 1. Characteristics of the four main models.

Model	Activities	Exclusive gateways	Parallel gateways	Loops
P1Main	23	10	0	1
P2Main	25	4	4	0
P3Main	17	4	0	0
P4Main	21	8	0	1

First activity name is always 'Activity A', second 'Activity B', etc.

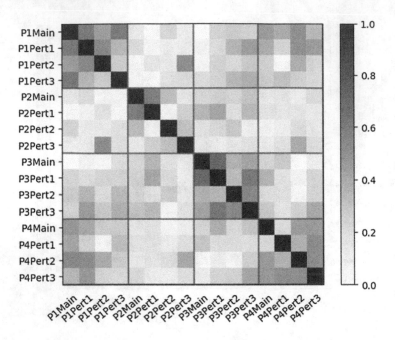

Fig. 7. The pairwise similarity of each of the process model representations, as measured by the cosine distance between the model vectors.

As described in Sect. 2.4, random walks were generated for each of the process models. More specifically, considering a process model as an undirected graph, 100 random walks per element were generated with a fixed length of 50. Observe that this is largely in line with random walk generation in graph representation learning [16] and that more complex and parameterized random walk procedures could be developed, but are out of scope for this work. The result of the random walk procedure is a set of walks, with each walk representing a string of activities (represented by their name) and gateways (represented by their type) covered during the walk. Subsequently, a representation for the activities, gateways and models is then learned using all of the walks tagged with an identifier for the model they were sampled on, as represented in Fig. 4. The feature vector length was set to 32.

The relationship between each of the models is captured by the similarity between the model vectors, i.e. the learned representations for these models. In Fig. 7, the pairwise cosine similarity between each of the model vectors is visualized. *model2vec* correctly identifies the similarity between most pairs of process models originating from models 1, 3 and 4. For model 2, the relationship between the main model and the first perturbation is the only one reflected in the distance between the learned model vectors.

For comparison, Fig. 8 contains the pairwise similarity of an existing process model similarity technique, based on behavioural profiles [32], a frequently applied alternative technique in the setting of process model comparison.

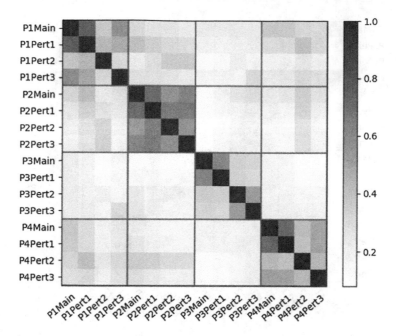

Fig. 8. The pairwise similarity of each of the models, as measured by their behavioural profiles.

Figure 9 presents the pairwise similarity based on the graph matching algorithm proposed by [12]. Both alternative techniques score well in this setting.

To summarize, this illustrative experiment has shown that the representations learned by *model2vec* can help distinguish process models, although the currently obtained results cannot yet improve on dedicated process model similarity measuring alternatives. However, this is also due to the particularities of the small experimental setup in which models with a similar or identical set of activity labels was used, and where the perturbations only relate to control-flow. As such, our technique, making abstraction of the actual control-flow and thus leaning closer to the concept of label similarity, is put at a disadvantage. Furthermore, given the wide range of alternative configurations of *model2vec*, it can be expected that our approach can outperform these methods in the future.

5   Related Work

As related work, we solely focus on the application of neural networks and representation learning in BPM. At the application level, a comparison with the state-of-the-art is performed in Sect. 3. Only a limited number of studies have already applied artificial neural network based techniques in a business process context. Notably, in [15], recurrent neural networks are applied towards a predictive setting, in order to predict the next occurring event in an event sequence, which

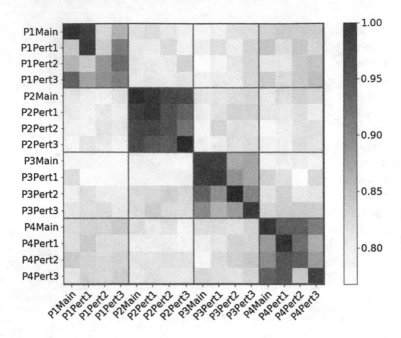

Fig. 9. The pairwise similarity of each of the models, as measured by the graph matching similarity. Due to randomness in the computation, the results are not symmetric.

hence fits with our discussion regarding the use of such networks in Sect. 2.1. The constructed network is here used as-is, however, with no embeddings being extracted from them as low-dimensional representations for subsequent analysis settings, as we propose here.

In [34], a comparable approach is proposed (using LSTM), also with the goal towards predicting the next event in a trace, but extended with the possibility to predict its time stamp and hence the full remaining time of a partial trace. The authors show that their technique outperforms existing monitoring techniques, which supports our claim that neural network based setups, and the inherent representations they learn, indeed offer a powerful approach in a variety of settings. In [23] a similar setup is applied and linked to the setting of process planning.

As mentioned above, although the use of recurrent neural networks and other artificial neural network based techniques have been explored in the setting of business processes, a representation-centric approach has so far not been widely investigated. In [20], word2vec is applied on generated event sequences from an "event connection graph", constructed first, to learn event representations, with potential applications being explored in the domain of health care. In [18], word2vec is applied in the context of learning features from events similarly as described in Sect. 2.1, which are then used as inputs for a recurrent neural network towards solving a classification problem.

6   Discussion and Future Work

From the overview provided in the previous section, we note that although some applications investigating a word2vec-inspired setup on event data have been explored, setups describing representation learning for activities, traces, event logs, or process models as outlined above have not, and hence provide an apt and exciting avenue for future research, as was empirically illustrated by Sect. 4. In this paper, we have proposed a number of architectures towards learning representations for activities, traces, event logs and process models, together with two experiments showing the strong potential of representation learning in process mining.

Importantly, the architectures allow for the unsupervised learning of distributed representation vectors which can be considered as general purpose representations since they are not tailored towards a specific use case, e.g. predicting the next activity. We believe that this initial work has outlined a number of valuable first steps towards representation learning for business processes, with ample opportunities for follow-up work, both in terms of additional use cases such as predictive process monitoring, as well as in terms of technical contributions, especially related to interpretability and the inclusion of other data dimensions.

Interpretability: Common approaches with regards to interpretability of representational learning include applying a second-level dimensionality reduction technique to visualize representational vectors in two-dimensional space and inspect them accordingly. This is a typical use case for word2vec in the area of text mining and hence is also directly applicable on *act2vec*. However, note that a more applicable level of analysis is situated on the trace, rather than activity level in our context. Extracting easy to understand or visual insights at this level is more challenging, as simple dimensionality reduction techniques will fail to satisfactory express the underlying representations. Similar as with doc2vec, analysis involving the similarity between trace vectors can be performed (i.e. to query "most similar traces"), though we foresee ample opportunity to extend upon this by e.g. more adequate visualization approaches.

Other data dimensions: As a second prime opportunity for future work, we emphasize that we have focused herein only on control-flow aspects of an event log (traces and activities). Nevertheless, the proposed architectures also permit incorporating other data dimensions present (such as performer information, time information, and other elements). In addition, as highlighted earlier, the possibility exists to exploring alternative architectural definitions to explicitly take into account the ordering of activities in a sequence or sequences in an event log. Mapping various tasks (process classification, runtime prediction, process monitoring, or modeling support, for instance) to architectural best-practices will form an interesting avenue for future research.

References

1. van der Aalst, W.M.P.: Business process management: a comprehensive survey. ISRN Softw. Eng. **2013** (2013)
2. van der Aalst, W.M.P.: Process Mining - Data Science in Action, 2nd edn. Springer, Heidelberg (2016). https://doi.org/10.1007/978-3-662-49851-4
3. van der Aalst, W.M.P., Schonenberg, M.H., Song, M.: Time prediction based on process mining. Inf. Syst. **36**(2), 450–475 (2011)
4. Bengio, Y., Ducharme, R., Vincent, P., Jauvin, C.: A neural probabilistic language model. J. Mach. Learn. Res. **3**(Feb), 1137–1155 (2003)
5. Bose, R.P.J.C., van der Aalst, W.M.P.: Context aware trace clustering: towards improving process mining results. In: SDM, pp. 401–412 (2009)
6. Bose, R.P.J.C., van der Aalst, W.M.P.: Trace clustering based on conserved patterns: towards achieving better process models. In: Rinderle-Ma, S., Sadiq, S., Leymann, F. (eds.) BPM 2009. LNBIP, vol. 43, pp. 170–181. Springer, Heidelberg (2010). https://doi.org/10.1007/978-3-642-12186-9_16
7. Jagadeesh Chandra Bose, R.P., van der Aalst, W.M.P.: Abstractions in process mining: a taxonomy of patterns. In: Dayal, U., Eder, J., Koehler, J., Reijers, H.A. (eds.) BPM 2009. LNCS, vol. 5701, pp. 159–175. Springer, Heidelberg (2009). https://doi.org/10.1007/978-3-642-03848-8_12
8. Burattin, A.: PLG2: multiperspective process randomization with online and offline simulations. Proceedings of the BPM Demo Track 2016, pp. 1–6 (2016)
9. Cho, K., et al.: Learning phrase representations using RNN encoder-decoder for statistical machine translation. arXiv preprint arXiv:1406.1078 (2014)
10. De Weerdt, J., vanden Broucke, S., Vanthienen, J., Baesens, B.: Active trace clustering for improved process discovery. IEEE Trans. Knowl. Data Eng. **25**(12), 2708–2720 (2013)
11. Di Francescomarino, C., Dumas, M., Maggi, F.M., Teinemaa, I.: Clustering-based predictive process monitoring. IEEE Trans. Serv. Comput. (2016)
12. Dijkman, R., Dumas, M., García-Bañuelos, L.: Graph matching algorithms for business process model similarity search. In: Dayal, U., Eder, J., Koehler, J., Reijers, H.A. (eds.) BPM 2009. LNCS, vol. 5701, pp. 48–63. Springer, Heidelberg (2009). https://doi.org/10.1007/978-3-642-03848-8_5
13. Dijkman, R., Dumas, M., Van Dongen, B., Krik, R., Mendling, J.: Similarity of business process models: metrics and evaluation. Inf. Syst. **36**(2), 498–516 (2011)
14. Doersch, C., Gupta, A., Efros, A.A.: Unsupervised visual representation learning by context prediction. In: 2015 IEEE International Conference on Computer Vision, ICCV 2015, Santiago, Chile, 7–13 December 2015, pp. 1422–1430 (2015)
15. Evermann, J., Rehse, J., Fettke, P.: Predicting process behaviour using deep learning. Decis. Support Syst. **100**, 129–140 (2017)
16. Grover, A., Leskovec, J.: node2vec: scalable feature learning for networks. In: Proceedings of the 22nd ACM SIGKDD International Conference on Knowledge Discovery and Data Mining, pp. 855–864. ACM (2016)
17. Günther, C.W., van der Aalst, W.M.P.: Fuzzy mining – adaptive process simplification based on multi-perspective metrics. In: Alonso, G., Dadam, P., Rosemann, M. (eds.) BPM 2007. LNCS, vol. 4714, pp. 328–343. Springer, Heidelberg (2007). https://doi.org/10.1007/978-3-540-75183-0_24
18. Hake, P., Zapp, M., Fettke, P., Loos, P.: Supporting business process modeling using RNNs for label classification. In: Frasincar, F., Ittoo, A., Nguyen, L.M., Métais, E. (eds.) NLDB 2017. LNCS, vol. 10260, pp. 283–286. Springer, Cham (2017). https://doi.org/10.1007/978-3-319-59569-6_35

19. Hamilton, W., Ying, Z., Leskovec, J.: Inductive representation learning on large graphs. In: Guyon, I., et al. (eds.) Advances in Neural Information Processing Systems 30, pp. 1024–1034. Curran Associates, Inc. (2017)
20. Hong, S., Wu, M., Li, H., Wu, Z.: Event2vec: learning representations of events on temporal sequences. In: Chen, L., Jensen, C.S., Shahabi, C., Yang, X., Lian, X. (eds.) APWeb-WAIM 2017. LNCS, vol. 10367, pp. 33–47. Springer, Cham (2017). https://doi.org/10.1007/978-3-319-63564-4_3
21. Le, Q., Mikolov, T.: Distributed representations of sentences and documents. In: International Conference on Machine Learning, pp. 1188–1196 (2014)
22. Maggi, F.M., Di Francescomarino, C., Dumas, M., Ghidini, C.: Predictive monitoring of business processes. In: Jarke, M., et al. (eds.) CAiSE 2014. LNCS, vol. 8484, pp. 457–472. Springer, Cham (2014). https://doi.org/10.1007/978-3-319-07881-6_31
23. Mehdiyev, N., Lahann, J., Emrich, A., Enke, D., Fettke, P., Loos, P.: Time series classification using deep learning for process planning: a case from the process industry. Procedia Comput. Sci. **114**, 242–249 (2017)
24. Mikolov, T., Chen, K., Corrado, G., Dean, J.: Efficient estimation of word representations in vector space. arXiv preprint arXiv:1301.3781 (2013)
25. Mikolov, T., Sutskever, I., Chen, K., Corrado, G.S., Dean, J.: Distributed representations of words and phrases and their compositionality. In: Advances in Neural Information Processing Systems, pp. 3111–3119 (2013)
26. Mitrovic, S., Singh, G., Baesens, B., Lemahieu, W., De Weerdt, J.: Scalable RFM-enriched representation learning for churn prediction. In: 2017 IEEE International Conference on Data Science and Advanced Analytics (DSAA), pp. 79–88 (2017)
27. Perozzi, B., Al-Rfou, R., Skiena, S.: Deepwalk: online learning of social representations. In: Proceedings of the 20th ACM SIGKDD International Conference on Knowledge Discovery and Data Mining, pp. 701–710. ACM (2014)
28. Polyvyanyy, A., La Rosa, M., Ouyang, C., Ter Hofstede, A.H.: Untanglings: a novel approach to analyzing concurrent systems. Formal Asp. Comput. **27**(5–6), 753–788 (2015)
29. Řehůřek, R., Sojka, P.: Software framework for topic modelling with large corpora. In: Proceedings of the LREC 2010 Workshop on New Challenges for NLP Frameworks, pp. 45–50. ELRA, Valletta, May 2010
30. Rong, X.: word2vec parameter learning explained. arXiv preprint arXiv:1411.2738 abs/1411.2738 (2014)
31. Schoknecht, A., Oberweis, A.: Ls3: Latent semantic analysis-based similarity search for process models. Enterp. Model. Inf. Syst. Archit.-Int. J. Concept. Model. **12**, 1–2 (2017)
32. Smirnov, S., Weidlich, M., Mendling, J.: Business process model abstraction based on behavioral profiles. In: Maglio, P.P., Weske, M., Yang, J., Fantinato, M. (eds.) ICSOC 2010. LNCS, vol. 6470, pp. 1–16. Springer, Heidelberg (2010). https://doi.org/10.1007/978-3-642-17358-5_1
33. Tax, N., Sidorova, N., Haakma, R., van der Aalst, W.M.P.: Event abstraction for process mining using supervised learning techniques. In: Bi, Y., Kapoor, S., Bhatia, R. (eds.) IntelliSys 2016. LNNS, vol. 15, pp. 251–269. Springer, Cham (2018). https://doi.org/10.1007/978-3-319-56994-9_18
34. Tax, N., Verenich, I., La Rosa, M., Dumas, M.: Predictive business process monitoring with LSTM neural networks. In: Dubois, E., Pohl, K. (eds.) CAiSE 2017. LNCS, vol. 10253, pp. 477–492. Springer, Cham (2017). https://doi.org/10.1007/978-3-319-59536-8_30
35. Van Dongen, B.: BPI challenge 2015 (dataset) (2015). https://data.4tu.nl/repository/uuid:31a308ef-c844-48da-948c-305d167a0ec1

Who Is Behind the Model?
Classifying Modelers Based on
Pragmatic Model Features

Andrea Burattin[1]([✉]), Pnina Soffer[2], Dirk Fahland[3], Jan Mendling[4],
Hajo A. Reijers[3,5], Irene Vanderfeesten[3], Matthias Weidlich[6],
and Barbara Weber[1,7]

[1] Technical University of Denmark, Kgs. Lyngby, Denmark
`andbur@dtu.dk`
[2] University of Haifa, Haifa, Israel
[3] Eindhoven University of Technology, Eindhoven, The Netherlands
[4] Vienna University of Economics and Business, Vienna, Austria
[5] Vrije Universiteit Amsterdam, Amsterdam, The Netherlands
[6] Humboldt-University, Berlin, Germany
[7] University of Innsbruck, Innsbruck, Austria

Abstract. Process modeling tools typically aid end users in generic,
non-personalized ways. However, it is well conceivable that different types
of end users may profit from different types of modeling support. In this
paper, we propose an approach based on machine learning that is able
to classify modelers regarding their expertise while they are creating a
process model. To do so, it takes into account pragmatic features of the
model under development. The proposed approach is fully automatic,
unobtrusive, tool independent, and based on objective measures. An
evaluation based on two data sets resulted in a prediction performance of
around 90%. Our results further show that all features can be efficiently
calculated, which makes the approach applicable to online settings like
adaptive modeling environments. In this way, this work contributes to
improving the performance of process modelers.

Keywords: Process modeling · Classification of modelers
Model layout

1 Introduction

Process models play an important role in the analysis, redesign, and implementation of business processes [1,2]. The creation of process models is a design
activity [3], in which a modeler constructs a mental model of a given domain
and externalizes it using a specific modeling tool (including the modeling notation) [4]. This design activity involves deciding which elements to use, which
names to give them, where to position them, and how to connect them. We also
refer to this activity as modeling.

© Springer Nature Switzerland AG 2018
M. Weske et al. (Eds.): BPM 2018, LNCS 11080, pp. 322–338, 2018.
https://doi.org/10.1007/978-3-319-98648-7_19

Modeling is not for free. The activity of creating a model imposes a substantial cognitive load on the limited information processing capacity of the modeler's brain [5]. In particular, cognitive load depends upon various factors including task characteristics, modeler characteristics, and tool characteristics. Modeling research has hardly considered the latter so far [6], and indeed, tools do not anticipate personal differences when they support the modeler [7–10]. However, personalized support could be highly beneficial for novices who require tips and guidance, while experts would perceive this as a distraction. If the profile of the modeler is known, such support can significantly improve performance [11].

In this paper, we lay foundations towards a personalization of modeling tool support. Our key idea is to support an on-the-fly classification of modelers by their expertise level while they interact with the tool. To this end, we identify a set of pragmatic modeling features that presumably reflect the expertise of the modeler in activity-centric, flow-based process models with AND/XOR gateways. We evaluate the relevance of these features using real-world modeling traces of BPMN models in order to classify the modelers as novices or experts. With a classification accuracy of 90%, our results demonstrate the feasibility of personalized support.

The remainder of the paper is structured as follows: Sect. 2 presents background information and related work; Sect. 3 describes our approach to classify modelers. Section 4 evaluates the classification technique on two real datasets and Sect. 5 concludes the paper.

2   Background and Related Work

2.1   Process Modeling as a Design Activity

Creating a process model constitutes a complex cognitive design activity. During this design activity a process modeler solves a problem of how to represent a described process as a process model, using the syntax of a specific modeling language. As a problem solving task, this entails the formation of a mental representation of the problem domain and externalizing this representation as a process model [4,12]. Doing this, the modeler interacts with the modeling environment to create the process model. More precisely, the modeler performs a sequence of modeling interactions (like the creation of an activity or an edge or the movement of an element) resulting into (intermediate) models [13]. The resulting (intermediate) models can be characterized by properties referring to their syntax, semantics, or pragmatics [14]. A graphical representation of these interactions over time, and possible artifacts obtained, is given in Fig. 1.

2.2   Expertise in Modeling

Differences between novices and experts have been intensively studied in the context of various tasks and artifacts. However, throughout the body of research that deals with expertise-related differences, no single and agreed upon criterion

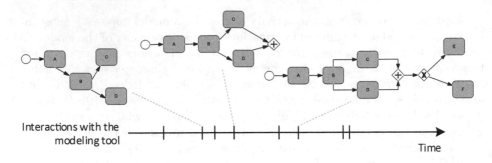

Fig. 1. Interactions with a modeling tool result in different intermediate models.

is used for distinguishing experts from novices. In fact, in [15], the authors define and test the effect of different dimensions of expertise, such as familiarity with a modeling language, intensity of modeling engagement, and knowledge of modeling concepts. In particular, they examined the distinction between students and practitioners, indicating that students exhibit different patterns of interaction with models as compared to practitioners. Bearing all this in mind, we rely on studies that concern differences in task performance between experts and novices for establishing expected differences in our study.

Concerning problem solving in Physics, in [16] authors discuss differences in the strategies employed by experts and novices, related to the differences in the mental problem representation they form and the retrieval of appropriate concepts and solution procedures from long term memory. These differences between novices and experts, in the availability and ease of retrieval of relevant concepts and solution strategies, pertain to many other areas.

In the area of conceptual modeling and process modeling, two main tasks are distinguished: reading (understanding) a model and creating a model. The differences between experts and novices in reading and understanding models have been studied [17], with a general indication that novices and experts have different notational needs [18]. Novices have more difficulties to recognize semantic patterns from graphics and tie them to long term memory concepts. Thus, in general, reading a model entails a higher cognitive load for a novice than for an expert [6,15]. In particular, this cognitive load and the understanding performance can be affected by graphical properties and layout of the model [15].

Novice and expert difference in creating models have also been indicated. For conceptual models, in [19], authors found that experts focus on generating a holistic understanding of the problem, making abstractions of problem characteristics by categorizing problem descriptions before developing the solution. Novices had difficulties in integrating parts of the problem descriptions and mapping them into knowledge structures. In [20], authors present the results of an empirical study aimed at identifying the most typical set of errors frequently committed by novice systems analysts in four commonly used UML artifacts.

In general, these errors can be interpreted as indicating a difficulty in making abstractions, which leads to a focus on specific functional details rather than on solution principles. A good use of graphical cues and layout creation by experts is indicated by [18] in the context of graphical programming. They claim that expert's categorization skills and ability to organize information on the basis of underlying abstractions are reflected in the expert's ability take advantage of secondary notation cues to enable them to recognize sub-term groupings.

Generally speaking, the above discussed differences between experts and novices can lead to two main conclusions. First, experts and novices would benefit from different kinds of support and guidance while creating a model. Second, it should be possible to distinguish and identify whether a certain model is being created by a novice or by an expert, as elaborated next.

3   Identifying the Expertise Level of Process Modelers

Many research efforts over the years have been devoted to personalizing systems based on user properties, organized in a *user model* (e.g., [21]). In this paper, as a first step towards a personalized modeling support, we aim at the most basic, simplest possible user model: a binary classification into *novice* or *expert*.

3.1   Classifying Modelers: Requirements and Design Considerations

Following user modeling literature [21], we identify 4 requirements for an approach for the classification of modelers expertise level:

R1 The approach should be based on objective measures, rather than on modelers' self-assessment of their expertise level;
R2 The approach should be unobtrusive towards the end user, and not involve additional efforts of modelers (e.g., for providing information);
R3 The approach should work online, and be applicable to intermediate (incomplete) models, since modelers are likely to learn and improve their skills over time;
R4 The approach should not depend on a particular modeling tool.

With these four requirements, we turn to assess available approaches for classifying a modeler in terms of expertise level.

One approach is to rely on self-assessment and to ask the modeler to classify himself/herself, e.g., by choosing a predefined profile in the modeling environment. Such an assessment is, however, neither based on objective measures (R1), nor applicable in an online setting since it is not automated (R3).

A second approach is to elicit this information based on a questionnaire regarding relevant modeler-specific features (like modeling experience, domain knowledge, cognitive abilities). For example, [11] used a questionnaire-based modeler's cognitive style for a personalized modeling support. The need to provide such information might, however, raise privacy issues, and seem obtrusive by the modeler (R2). Moreover, due to the manual efforts needed it is not applicable as an online approach (R3).

A third approach is the classification of modelers based on neuro-physiological measures [22]. For example, [23] used the Alpha and Theta signals of an EEG to quantify programmer's expertise. While such an approach can be automated and is based on objective measures of cognitive load, it is intrusive and is not applicable outside of a lab setting (R2).

A fourth approach is classification based on differences in modeling behavior derived from the recorded interactions with the modeling platform. For example, [24] showed that the presence of prior domain knowledge facilitates the creation of an internal representation of the process to be modeled and is associated with shorter initial comprehension phases. Moreover, Martini et al. [25] showed that inexperienced modelers had significantly more comprehension and modeling phases when compared to more experienced modelers. The drawback of relying on detected modeling behavior is, however, its tool dependence (R4).

Finally, a fifth approach is using the (intermediate) modeling artifact as a basis for classification. Considering only model features provides tool independence, since the properties that can serve as features are derived from the model without a need for knowledge about tool interactions. Features related to syntax and semantics (e.g., presence of deadlocks and lack of synchronization) are less suited for online settings. This is since existing metrics assume certain properties of the model (e.g., all elements are fully connected and the model is sound), assumptions that are mainly applicable to complete models rather than to intermediate ones (R3). In contrast, pragmatic properties, that capture the alignment of process model elements or the way gateways are used, fulfill all the above mentioned requirements and will be used as a basis for our classification approach.

3.2 Overview of Our Approach

Classification problems have been extensively studied in the literature [26] and most approaches assume a *feature vector* as input (cf. Sect. 3.3 for the considered features). Figure 2 depicts the general idea proposed by our approach: process modelers interact with the modeling tool to construct a BPMN diagram. Examples of interactions are the creation of an activity or edge or the movement of an element. Hereby, the process model gradually evolves over time resulting into different (intermediate) models. Input to the classification is one such (intermediate) model. In step ① the set of features used for the classification is extracted from the model. Then, in step ②, this features vector is given as input to a trained classification model, which returns the likelihood that the model (i.e., the features vector) has been created by a novice or by an expert.

3.3 Feature Engineering

Feature engineering is considered the "art of creating predictor variables" [27]: this human-driven process requires iterations of brainstorming possible features and studying their impact on the quality of the model. We manually inspected

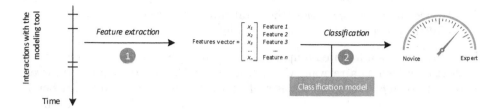

Fig. 2. Our approach to classify the expertise level of process modelers.

several models generated by both experts and novices and we investigated the most relevant differences. Moreover, as a consequence of requirement of being applicable in an online setting (R3) (i.e., support for online measurements) features should be computed efficiently. Thus, we also took into consideration the complexity of computing them.

The first group of features we identified considers the alignment of fragments. The reasoning behind these features is that the alignment of fragments helps model comprehension [28]. Poorly aligned fragments obfuscate the process model and make it difficult to recognize the patterns used. In the context of this paper, we define a *fragment* as a SESE component [29] with at least two tasks. We say that a fragment is *aligned* if the coordinates of its entry and its exit components (i.e., the coordinates of the center of entry and exit components) are within a certain threshold. Formally, two elements a and b located at (x_a, y_a) and (x_b, y_b) are aligned if $\min\{|x_a - x_b|, |y_a - y_b|\} \leq 20\text{px}$. Based on these definitions we can define the following features:

F1 *Alignment of fragments*: ratio of aligned SESE fragments over the total number of fragments in the BPMN model;

F2 *Percentage of activities in aligned fragments*: ratio of activities belonging to aligned SESE fragments over the total number of activities;

F3 *Percentage of activities in not aligned fragments*: ratio of activities belonging to not aligned SESE fragments over the total number of activities.

The next group of features considers the type and usage of the gateways. The usage of implicit gateways as well as the reuse of gateways has been pointed out as bad modeling practice [28,30,31]. Specifically, we call a gateway *explicit* if it is represented as a BPMN gateway element. An *implicit* gateway, in turn, is a BPMN task element with more than 1 entering (or exiting) connections (i.e., sequence flows). Finally, a *reused* gateway is a gateway (either implicit or explicit) which serves as both split and join, i.e., it has more than 1 incoming and more than 1 outgoing edges. Based on these definitions we implemented the following features:

F4 *Number of explicit gateways*: the total number of explicit gateways in the BPMN model;

F5 *Number of implicit gateways*: the total number of implicit gateways in the BPMN model;

F6 *Number of reused gateways*: the total number of reused gateways (either implicit or explicit) in the BPMN model.

Moreover, we have a feature group concerning the style of the edges in a model, which is perceived as a relevant visual feature [31,32]. We distinguish between *edges* and *segments*: an edge consists of at least one segment and each bend-point introduces one additional segment into the corresponding edge. Then we can define the following features:

F7 *Percentage of orthogonal segments*: ratio of the segments that are "orthogonal" (i.e., either vertical or horizontal within a threshold of 10px) over the total number of segments;
F8 *Percentage of crossing edges*: ratio of edges that are crossed by some other edge over the total number of edges.

The last set of features considers the process "as a whole" and therefore concerns more global properties:

F9 *M-BP*: this measure (in the interval $[0,1]$) computes the extent to which the layout of the model is consistent with the temporal logical ordering of corresponding activities [31,33], computed using the algorithm in [34];
F10 *Number of ending points*: the number of nodes with at least one incoming edge, but no outgoing connection [28].

In sum, a *modeling session*, i.e., the modeling exercise of one person, is a series of model snapshots $m_1, m_2, m_3, \ldots$ over time, also called *intermediate models*. Each model, in turn, is characterized by a vector of 10 numerical features (we say that $m_1, m_2, m_3, \ldots$ is *represented* by $F_1, F_2, F_3, \ldots$), extracted in step ①. Those will be used to distinguish the expertise level of modelers.

3.4 Model Classification

For a given intermediate model, the described features were calculated and passed in step ② to a classification model. We used neural networks [35] as classification mechanism. This well studied classification model can be used to approximate any discrete-valued target functions: the "universal approximation theorem" [36] proves that multilayer feedforward networks with as few as a single hidden layer are universal approximators and therefore represent the most general tool available[1]. What is interesting of these models is their capability of dealing with complex decomposition of high dimensional spaces into smaller ones (in our case, we move from a 10-dimensions space to a binary problem). Additionally, by introducing specific topologies (i.e., hidden layers) it is possible to increase the chances of making the problem easier to solve (i.e., transforming the data space into a linear separable one).

We used a feed-forward neural network, with one hidden layer comprising 50 neurons. The input layer contains 10 neurons, one for each feature of the vectorial

---

[1] See external appendix for details: https://doi.org/10.5281/zenodo.1251633.

representation of the model. The output layer contains 2 neurons, whose values distinguish between the two classes (i.e., *novice* or *expert*). The rationale behind the topology of the network comes from an experimental phase, where different configurations have been verified: there is no scientific result defining a generally optimal structure. Still, there are results suggesting that 1 hidden layer can be enough [37]. Concerning the number of units, literature [38] suggests a number of hidden nodes proportional to the number of training samples over a multiple (between 2–10) of the sum of input/output nodes. In our case, we decided to focus on our smallest dataset, with 1000 samples, and 2 as multiplier.

4   Evaluation

In this section, we evaluate the accuracy of the classification for predicting if the modeler is a novice or an expert. As evaluation input, we use data sets that document modeling processes of students (as proxies for novices) and practitioners (as proxies for experts). Note that our approach is independent of the question whether students and practitioners always represent valid proxies. For a recent discussion of student/practitioner differences, refer to [15].

The described approach has been implemented in a Java application[2] and tested on 2 real datasets[3]. For the implementation, we used the libraries of the Weka toolkit[4] with the corresponding multilayer perceptron. The multilayer perceptron was trained with a learning rate of 0.3 and a learning momentum of 0.2. Additionally, we set the number of training epochs to 500.

4.1   Description of Datasets

We performed our tests on datasets collected in several modeling sessions in 2010. Novice data was collected through the participation of students from Eindhoven University of Technology. Expert data, in turn, was collected as part of a Dutch BPM round-table event in Eindhoven, as well as in Berlin with practitioners[5]. In both settings, the experts were recruited from our network of industry practitioners, who where experienced modelers and highly familiar with BPMN. The modeling sessions were ran at the universities in a controlled setting. Participants were aware that the exercises were meant to assess their competence, but did not know that the tool tracked each modeling step. A textual description of the processes to be modeled was provided and subjects were asked to model a BPMN model representing the described process. In this paper we analyze data referring to two process descriptions.

---

[2] The complete source code of the implementation is available at https://github.com/DTU-SPE/ExpertisePredictor4BPMN.

[3] The dataset is available at https://doi.org/10.5281/zenodo.1194780.

[4] See http://www.cs.waikato.ac.nz/ml/weka/.

[5] Please note that the data collected from the practitioners has not been published before. Moreover, the model features used as basis for this paper have not been reported before, neither for students nor for practitioners.

Table 1. Number of modeling sessions and (intermediate) models for each experiment and expertise level included in our datasets.

	Experts		Novices	
	Sessions	Interm. models	Sessions	Interm. models
mortgage-1	31	7299	144	36141
pre-flight	39	4856	118	14147

In the first model (called pre-flight) participants were asked to represent the steps conducted by an airplane's crew before take-off. The reference implementation contains 12 activities and 10 gateways[6]. In the second model (called mortgage-1) participants were asked to represent a mortgage application process with 26 activities and 20 gateways (see footnote 6). While the pre-flight process is fairly simple and only comprises sequences, parallel branches and several optional activities without any nesting, the mortgage-1 example is more complex. It contains a long loop back as well as several levels of nesting depth. In addition, the mortgage-1 process has several outcomes, i.e., rejection because of pre-existing mortgage, rejection through employee, offer not updated, and customer accepts offer. For modeling these processes we used Cheetah [39] which is able to record all interactions with the modeling platform and allows to reconstruct all intermediate models. Thus, for each modeling session (i.e., a single modeling exercise) we have multiple (intermediate) models based on which the features were calculated.

The number of modeling sessions and the number of available models are reported in Table 1. As we can see, the actual number of sessions and models differs between expertise levels and between modeling tasks. For this reason, we used a small fragment of randomly selected models for each experiment.

Before turning to the results of the classification, we compare the differences between the 2 groups of users (i.e., novices and experts) in our 2 datasets (considering intermediate models created during the last 70% of the modeling session). Specifically, we used the Mann-Whitney U Test to understand whether the features described in Sect. 3.3 are proper discriminators of the 2 groups. Results of the test are reported in Table 2 and clearly show that statistically significant differences between novices and experts exist for all 10 considered features. Table 3 reports the descriptive statistics for the two analyzed datasets with mean, standard deviation (SD) and standard error (SE) for each feature and both experts and novices. The descriptive statistics clearly depict, for both datasets (i.e., mortgage-1 and pre-flight), that experts prefer to align elements (cf. values of F1, F2, F3), experts prefer explicit gateways over implicit ones (cf. F4 and F5) and they reuse less gateways when compared to novices (cf. F6). Additionally, novices model less orthogonal segments (cf. F7) and also keep the layout less consistent with the temporal logical ordering of activities (cf. F9).

---

[6] Graphical representations on the appendix: https://doi.org/10.5281/zenodo.1251633.

Table 2. Mann-Whitney U Test on mortgage-1 and pre-flight datasets. The feature codes refer to the descriptions in Sect. 3.3.

(a) mortgage-1

Feature	W	p
F1	1.524e+8	< .001
F2	1.459e+8	< .001
F3	1.199e+8	< .001
F4	1.572e+8	< .001
F5	1.179e+8	< .001
F6	1.359e+8	< .001
F7	1.645e+8	< .001
F8	1.034e+8	< .001
F9	1.688e+8	< .001
F10	1.627e+8	< .001

(b) pre-flight

Feature	W	p
F1	3.650e+7	< .001
F2	3.934e+7	< .001
F3	3.339e+7	< .001
F4	4.128e+7	< .001
F5	3.252e+7	< .001
F6	3.257e+7	< .001
F7	3.850e+7	< .001
F8	3.463e+7	< .001
F9	3.870e+7	< .001
F10	3.331e+7	< .001

Table 3. Descriptive group statistics for experts and novices for the two datasets analyzed. The feature codes refer to the descriptions in Sect. 3.3.

Feature	Group	mortgage-1			pre-flight		
		Mean	SD	SE	Mean	SD	SE
F1	Experts	0.862	0.256	0.003	0.817	0.311	0.004
	Novices	0.807	0.261	0.001	0.764	0.361	0.003
F2	Experts	0.459	0.17	0.002	0.505	0.243	0.003
	Novices	0.434	0.177	0.0009	0.441	0.259	0.002
F3	Experts	0.09	0.164	0.002	0.078	0.151	0.002
	Novices	0.1	0.151	0.0008	0.098	0.188	0.002
F4	Experts	11.901	4.54	0.053	6.84	2.665	0.038
	Novices	10.194	4.468	0.024	5.937	2.515	0.021
F5	Experts	1.309	1.487	0.017	0.371	0.845	0.012
	Novices	1.576	1.614	0.008	0.495	1.079	0.009
F6	Experts	0.344	0.615	0.007	0.501	1.079	0.015
	Novices	0.316	0.627	0.003	0.471	0.773	0.006
F7	Experts	0.712	0.223	0.003	0.572	0.267	0.004
	Novices	0.603	0.179	0.0009	0.494	0.18	0.002
F8	Experts	0.008	0.026	0.0003	0.012	0.041	0.0006
	Novices	0.022	0.044	0.0002	0.008	0.035	0.0003
F9	Experts	0.95	0.067	0.0008	0.95	0.103	0.001
	Novices	0.877	0.126	0.0007	0.906	0.125	0.001
F10	Experts	2.743	1.14	0.013	1.598	0.88	0.013
	Novices	2.267	1.012	0.005	1.64	0.911	0.008

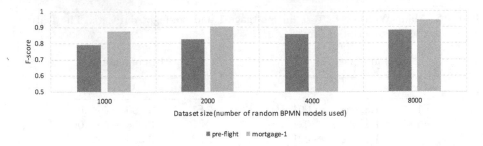

Fig. 3. F-scores on 10-fold cross validation tests, performed on different dataset sizes (models randomly selected), for the two tasks.

4.2   Prediction on Single Intermediate Models

The problem at hand is a binary classification problem (i.e., *novice* vs *expert*). Therefore, to characterize the quality of our predictions, we used the F-score (also known as F_1) measure. This measure is the harmonic mean of *precision* and *recall* and is suitable for capturing the classification quality [40]. We created different datasets with a different number of models (*up to* 1000, 2000, 4000 and 8000 BPMN models) considering BPMN models created during the last 70% of the modeling session (i.e., we discarded the intermediate models created during the initial 30% of the modeling session, to avoid almost-empty models). Then, using a 10-fold cross validation [26] we computed the average performance for each fold. Figure 3 depicts the outcomes of our tests. Clearly, the larger the dataset size, the better the outcome, since the system is able of better approximating the classification function. With the largest datasets, we were able to achieve an F-score of at least 0.88 (for pre-flight) and 0.94 (for mortgage-1). The comparison with 5 other classifiers is reported in the external appendix of this paper together with further implementation details (see footnote 1). To validate the necessity of our feature set, Fig. 4 shows, in turn, the F-scores, if subsets of features were used. The classification performance is notably worse when only subsets are used, as the classification model is not capable of accurately discriminating based just on these features. Additionally, the trend is not monotonically increasing but fluctuating, which suggests that the learning model is probably too complex given the data and it started to overfit the task, thus decreasing the performance. Furthermore, Fig. 5 shows the pairwise Pearson correlation coefficients for all features. Overall, there is little indication for linear correlations between the features, suggesting that they may indeed capture complementary aspects. To demonstrate that our approach (in line with requirement R3) is applicable in an online setting we conducted a performance analysis. The computation of the features was performed by instrumenting Cheetah to compute the features and measure the time. Only the time to compute the actual feature is taken into account. To compute the features, 18341 samples were randomly taken from the mortgage-1 dataset, which is the bigger of the datasets used. The test has been performed on a standard laptop with Windows 10 Enterprise and Java

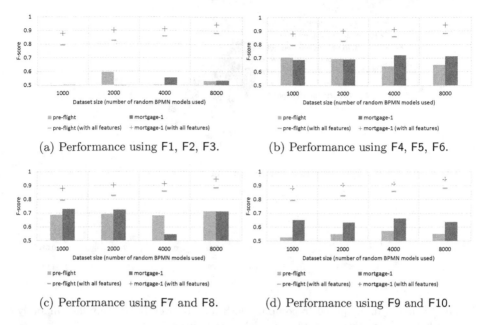

(a) Performance using F1, F2, F3.   (b) Performance using F4, F5, F6.

(c) Performance using F7 and F8.   (d) Performance using F9 and F10.

Fig. 4. F-scores on 10-fold cross validation tests performed on different dataset sizes, considering different subsets of features. Each chart also reports, for comparison purposes, the values obtained using all features.

1.8, Processor Intel Core i7-7500U 2.7GHz and 16 GB RAM. During the test, a typical usage was maintained (i.e., no dedicated computation for the test, just to simulate a modeling environment, with several other software applications running at the same time). Figure 6 shows the average time required to compute each feature. Our results demonstrate that the calculation of most features is very fast. Feature F9, in turn, is more time consuming (93.12 ms). Still, all 10 features can be calculated in just a bit more than 100ms, which is certainly sufficient for application in the intended use cases.

4.3 Prediction of Modeler Expertise in the Entire Session

Our trained classifier predicts if model m_i was created by an expert if $exp(\boldsymbol{F}_i) = 1$ or by a novice if $exp(\boldsymbol{F}_i) = 0$ with an accuracy of 0.88–0.94. In this section we discuss how early $exp(\cdot)$ can predict that the modeler is an expert in a modeling session (from the features $\boldsymbol{F}_i$ of a partial model m_i).

To be robust against temporary changes in the classification over a few model snapshots, we derived a smoothing expert classifier $exp'(\boldsymbol{F}_i)$ that takes the average of $exp(\cdot)$ over the last N feature snapshots in the modeling sessions $\boldsymbol{F}_i, \boldsymbol{F}_{i-1}, \ldots, \boldsymbol{F}_{i-N+1}$. If $exp'(\boldsymbol{F}_i) > k$ for a threshold k, we consider m_i as an "expert model" in the session.

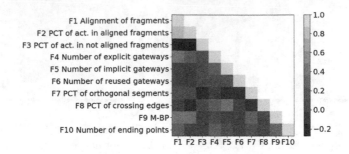

Fig. 5. Correlation coefficient matrix for all features.

Fig. 6. Average time required to compute each feature over 18341 samples randomly selected from the mortgage-1 dataset (which is the biggest).

Using the $exp(\cdot)$ classifier trained on 8000 random samples from the last 70% of a modeling sessions (based on Fig. 4) and choosing $N = 20$ and $k = 0.6$ (based on hyper-parameter optimization) led to the following results: in the first 25% of any modeling session, $exp(\cdot)$ predicts any models to be from an "expert", i.e., exp errs on the "expert side". In the last 70% of a modeling session for mortgage-1 (pre-flight):

- exp' correctly predicted "expert" in 100% (94.5%) of the modeling sessions;
- in 90% (76.9%) of the sessions, this prediction occurred between 0.3-0.4 (0.3-0.55) of the modeling time at an average of 0.32 (0.35); exp' remained stable at "expert" until the end from 0.3-0.45 (0.3-0.75) of the modeling time onwards at an average of 0.33 (0.45);
- exp' falsely predicts "experts" for a short period in 13.8% (25.4%) of the modeling sessions of novices; the false prediction is stable until the end in just <1% (5%) of the novice sessions.

4.4 Discussion and Limitations

First and foremost, the presented approach to classify modelers meets the requirements R1-R4 put forward in Sect. 3.1. The classification is grounded in objective measures (R1), as the classification features capture properties of (intermediate) models only (e.g., the ratio of aligned model fragments or the type and usage of gateways). Feature calculation and classification from the model alone is fast (a bit more than 100 ms) rendering our approach unobtrusive (R2). Modelers do not have to spend any additional effort and, unless it is desired, would not even be aware of the classification. Our experiments further highlighted that the features are well-suited to classify, potentially incomplete, intermediate models. We thus in addition conclude that the approach can be used online (cf. R3), coping with learning and improvement effects of modelers. Moreover, our approach is independent of a specific tool (cf. R4) as the respective features can be computed in any tool for the creation of activity-centric, flow-based process models with AND/XOR gateways.

Turning to the actual classification results, the general trend is encouraging, with an F-score of 0.88 for pre-flight and 0.94 for mortgage-1. Our results suggest that in terms of classification the mortgage-1 task seems easier when compared to the pre-flight one. This is plausible since the pre-flight lacks complex behavioral structures (i.e., no nested blocks or loops). Therefore, models of novices and experts do not differ so much as for more complex models.

Our results have impact both for modeling theory and practice. The ability to distinguish in an automated manner between groups of modelers (i.e., novices or experts) has potential applications in the context of teaching scenarios or as part of an adaptive modeling editor that classifies the user while modeling and adjusts itself based on the classification. The accuracy of the smoothing online-predictor exp' to distinguish novices and experts already early in the modeling session supports this idea; while reliable classification is only available after a third of the modeling session, it may be exactly at the right time to identify novices and offer support. The false positive rate of up to 25% in temporary expert classifications suggests that users need to stay in control of adaptations of the modeling environment.

The assessment of expertise and professional capabilities is needed for different purposes: for recruitment, for deciding on assignment of employees to tasks, for team formation [41], or for forming relatively uniform groups for training. Current approaches for this assessment (e.g., based on success in a modeling task) suffer limitations – specifically biases that stem from differences in domain knowledge, from the specific modeling task selected for the assessment, or from accidental success or failure. The approach we present overcomes these limitations by considering a combination of features which are evidence-based and less subject to conscious and intentional manipulation: it would be very difficult to intentionally introduce bias in such assessment.

Although the feature extraction was conducted in the context of Cheetah platform, it is not dependent on a specific modeling tool, but can be generalized to any other BPMN-based process modeling environment. The general approach could also be applied to modeling notations other than BPMN, however, in this case the feature extraction would need to be adapted to the specific notation.

The features F1-F10 considered in this paper *together* are relevant for accurately classifying expert and novice modelers: sub-groups of features show significantly lower F-scores than all features combined (Fig. 4), and the features discriminate even partial expert and novice models (Sect. 4.3). Other classifiers can be used for the same set of features with the same or even better accuracy (see footnote 1). A threat to generalizability to larger models and applicability in individual cases may be the thresholds used in F1–F3 and F7. Validating the features against more (and larger models) and modeler preferences is subject of future work.

Another limitation of our work is that we used the same tasks for training and prediction. This is a setting that is applicable to teaching or recruitment scenarios, where many modelers work on the same modeling task. To improve the generalizability of the approach and to make it applicable in settings where models are more heterogeneous, inter-model predictions are required. Some of the features considered in our prediction depend on the model (e.g., the number of gateways) and thus have to be adapted for inter-model predictions.

5   Conclusion and Future Work

In this paper we demonstrated that novices and experts can be differentiated to a large extent based on how they lay out their model. By basing our classification approach on a model's layout, we were able to provide an approach that is automated, based on objective-measures, that is unobtrusive, and independent of a particular modeling tool. Our performance analysis further demonstrated that the approach is applicable in online settings. With this paper we focused on inter-model classification.

Future work will generalize the approach by additionally considering intra-model classification. We plan to continue the feature engineering and selection processes by considering features that capture the evolution of model properties over time. In parallel, different prediction techniques can also be investigated.

Acknowledgements. This research was funded by the Austrian Science Fund (FWF): P26140–N15 and P26609N15.

References

1. Burton-Jones, A., Meso, P.: The effects of decomposition quality and multiple forms of information on novices' understanding of a domain from a conceptual model. J. AIS **9**(12), 748–802 (2008)
2. Fettke, P.: How conceptual modeling is used. Commun. AIS (CAIS) **25**, 571–592 (2009)
3. Recker, J., Safrudin, N., Rosemann, M.: How novices design business processes. Inf. Syst. **37**(6), 557–573 (2012)
4. Soffer, P., Kaner, M., Wand, Y.: Towards understanding the process of process modeling: theoretical and empirical considerations. In: Daniel, F., Barkaoui, K., Dustdar, S. (eds.) BPM 2011. LNBIP, vol. 99, pp. 357–369. Springer, Heidelberg (2012). https://doi.org/10.1007/978-3-642-28108-2_35
5. Wickens, C.D., Hollands, J.G.: Engineering Psychology and Human Performance, 3rd edn. Pearson, London (1999)
6. Figl, K.: Comprehension of procedural visual business process models a literature review. Bus. Inf. Syst. Eng. **59**, 41–67 (2017)
7. Koschmider, A., Reijers, H.A.: Improving the process of process modelling by the use of domain process patterns. Enterp. IS **9**(1), 29–57 (2015)

8. Koschmider, A., Hornung, T., Oberweis, A.: Recommendation-based editor for business process modeling. Data Knowl. Eng. **70**(6), 483–503 (2011)
9. Weber, B., Reichert, M., Rinderle-Ma, S.: Change patterns and change support features - enhancing flexibility in process-aware information systems. Data Knowl. Eng. **66**(3), 438–466 (2008)
10. Gschwind, T., Koehler, J., Wong, J.: Applying patterns during business process modeling. In: Dumas, M., Reichert, M., Shan, M.-C. (eds.) BPM 2008. LNCS, vol. 5240, pp. 4–19. Springer, Heidelberg (2008). https://doi.org/10.1007/978-3-540-85758-7_4
11. Claes, J., Vanderfeesten, I.T.P., Gailly, F., Grefen, P., Poels, G.: The structured process modeling method (SPMM) what is the best way for me to construct a process model? Decis. Support Syst. **100**, 57–76 (2017)
12. Claes, J., Vanderfeesten, I., Pinggera, J., Reijers, H.A., Weber, B., Poels, G.: Visualizing the Process of process modeling with PPMCharts. In: La Rosa, M., Soffer, P. (eds.) BPM 2012. LNBIP, vol. 132, pp. 744–755. Springer, Heidelberg (2013). https://doi.org/10.1007/978-3-642-36285-9_75
13. Pinggera, J., et al.: Styles in business process modeling: an exploration and a model. Softw. Syst. Model. **14**, 1055–1080 (2013)
14. Krogstie, J.: Quality of models. In: Krogstie, J. (ed.) Model-Based Development and Evolution of Information Systems, pp. 205–247. Springer, London (2012). https://doi.org/10.1007/978-1-4471-2936-3_4
15. Mendling, J., Recker, J.C., Reijers, H., Leopold, H.: An empirical review of the connection between model viewer characteristics and the comprehension of conceptual process models. Inf. Syst. Front., 1–25 (2018)
16. Larkin, J., McDermott, J., Simon, D.P., Simon, H.A.: Expert and novice performance in solving physics problems. Science **208**(4450), 1335–1342 (1980)
17. Reijers, H.A., Mendling, J.: A study into the factors that influence the understandability of business process models. IEEE Trans. Syst. Man Cybern. - Part A: Syst. Hum. **41**(3), 449–462 (2011)
18. Petre, M.: Why looking isn't always seeing: readership skills and graphical programming. Commun. ACM **38**(6), 33–44 (1995)
19. Batra, D., Davis, J.G.: Conceptual data modelling in database design: similarities and differences between expert and novice designers. Int. J. Man Mach. Stud. **37**(1), 83–101 (1992)
20. Narasimha, B., Leung, F.S.: Assisting novice analysts in developing quality conceptual models with UML. Commun. ACM **49**(7), 108–112 (2006)
21. Jawaheer, G., Weller, P., Kostkova, P.: Modeling user preferences in recommender systems: a classification framework for explicit and implicit user feedback. ACM Trans. Interact. Intell. Syst. **4**(2) (2014). Article no. 8
22. Riedl, R., Léger, P.-M.: Fundamentals of NeuroIS-Information Systems and the Brain. SNPBE. Springer, Heidelberg (2016). https://doi.org/10.1007/978-3-662-45091-8
23. Crk, I., Kluthe, T., Stefik, A.: Understanding programming expertise: an empirical study of phasic brain wave changes. ACM Trans. Comput.-Hum. Interact. **23**(1), 2:1–2:29 (2016)
24. Pinggera, J.: The process of process modeling. Ph.D. thesis, University of Innsbruck (2014)
25. Martini, M., Pinggera, J., Neurauter, M., Sachse, P., Furtner, M.R., Weber, B.: The impact of working memory and the process of process modelling on model quality: investigating experienced versus inexperienced modellers. Sci. Rep. **6** (2016). Article no. 25561

26. Aggarwal, C.C.: Data Mining. Springer, Cham (2015). https://doi.org/10.1007/978-3-319-14142-8
27. Baker, R.: Big Data and Education. Columbia University, New York (2015)
28. Mendling, J., Reijers, H.A., van der Aalst, W.M.P.: Seven process modeling guidelines (7PMG). Inf. Softw. Technol. **52**(2), 127–136 (2010)
29. Polyvyanyy, A.: Structuring process models. Ph.D. thesis, University of Potsdam (2012)
30. Haisjackl, C., Soffer, P., Lim, S.Y., Weber, B.: How do humans inspect BPMN models: an exploratory study. Softw. Syst. Model. **17**, 655–673 (2016)
31. Bernstein, V., Soffer, P.: Identifying and quantifying visual layout features of business process models. In: Gaaloul, K., Schmidt, R., Nurcan, S., Guerreiro, S., Ma, Q. (eds.) CAISE 2015. LNBIP, vol. 214, pp. 200–213. Springer, Cham (2015). https://doi.org/10.1007/978-3-319-19237-6_13
32. Gschwind, T., Pinggera, J., Zugal, S., Reijers, H.A., Weber, B.: A linear time layout algorithm for business process models. JVLC **25**(2), 117–132 (2014)
33. Figl, K., Strembeck, M.: On the importance of flow direction in business process models. In: Proceedings of ICSOFT-EA, pp. 132–136 (2014)
34. Burattin, A., Bernstein, V., Neurauter, M., Soffer, P., Weber, B.: Detection and quantification of flow consistency in business process models. SoSyM **17**(2), 633–654 (2017)
35. Mitchell, T.M.: Machine Learning. McGraw-Hill, New York City (1997)
36. Hornik, K.: Approximation capabilities of multilayer feedforward networks. Neural Netw. **4**(2), 251–257 (1991)
37. Huang, G.B.: Learning capability and storage capacity of two-hidden-layer feedforward networks. IEEE Trans. Neural Netw. **14**(2), 274–281 (2003)
38. Hagan, M., Demuth, H., Beale, M., De Jesús, O.: Neural Network Design (2014). Oklahoma
39. Pinggera, J., Zugal, S., Weber, B.: Investigating the process of process modeling with cheetah experimental platform. In: Proceedings of the ER-POIS, pp. 13–15 (2010)
40. Manning, C.D., Raghavan, P., Schütze, H.: Introduction to Information Retrieval, 1st edn. Cambridge University Press, Cambridge (2008)
41. Niknafs, A., Berry, D.: The impact of domain knowledge on the effectiveness of requirements engineering activities. Empir. Softw. Eng. **22**(1), 80–133 (2017)

Finding the "Liberos": Discover Organizational Models with Overlaps

Jing Yang[1], Chun Ouyang[2], Maolin Pan[1], Yang Yu[1($\boxtimes$)],
and Arthur H. M. ter Hofstede[2]

[1] Sun Yat-sen University, Guangzhou, China
yangj357@mail2.sysu.edu.cn, {panml,yuy}@mail.sysu.edu.cn
[2] Queensland University of Technology, Brisbane, Australia
{c.ouyang,a.terhofstede}@qut.edu.au

Abstract. Organizational mining aims at gaining insights for business process improvement by discovering organizational knowledge relevant to the performance of business processes. A key topic of organizational mining is the discovery of organizational models from event logs. While it is common for modern organizations to have employees sharing roles and responsibilities across different internal groups, most of the existing methods for organizational model discovery are unable to identify such overlaps. The overlapping resources are likely to be generalists in an organization. Existing findings in process redesign best practices have proven that generalists can help increase the flexibility of a business process (similarly to the flexibility of the role of "libero" in certain team sports). In this paper we propose an approach capable of discovering organizational models with overlaps and thus helping identify generalists in an organization. The approach builds on existing cluster analysis techniques to address the underlying technical challenges. Through experiments on real-life event logs the applicability and effectiveness of the proposed method are evaluated.

Keywords: Process mining · Organizational mining
Organizational model mining · Overlapping clustering

1 Introduction

Process mining enables data-driven process analysis using the massive amount of event log data captured by information systems in today's organizations. Various techniques have been developed to help extract insights about the actual business processes with the ultimate goal to improve process performance as well as the organizations' business performance. While the main focus of process mining is on the control-flow perspective, recent years have seen research devoted to mining other aspects such as the organizational context of business processes.

Organizational mining focuses on discovering organizational knowledge, including e.g. organizational structures and human resources relevant to the

© Springer Nature Switzerland AG 2018
M. Weske et al. (Eds.): BPM 2018, LNCS 11080, pp. 339–355, 2018.
https://doi.org/10.1007/978-3-319-98648-7_20

performance of a business process, from event log data [1]. In any organization where humans play a dominant role, organizational mining helps managers gain a better understanding of the *de facto* grouping of human resources and their interactions thus to improve the related business processes. The importance of such organizational knowledge in process improvement is also emphasized by the fact that 10 of the 29 best practices in process redesign proposed in [2] are concerned with the structure and population (i.e. resources) of an organization.

Hence, an interesting research topic concerns the discovery of organizational models from event log data. Given the fact that in many real-life event logs, only limited information about process execution is provided, it is challenging to derive the actual organizational model (e.g. an organizational chart) in an organization. However, it is possible to recognize groups of resources that have similar characteristics relevant to the performance of a business process. For example, in [1] the authors propose a resource grouping mechanism based on how frequently the human resources carry out the same tasks, and suggest that the discovered organizational groups can be relevant to roles and functional units in which employees possess similar skills and knowledge to perform the tasks.

To date there have been a number of research efforts on mining organizational models from event logs (e.g. [1,3,4]), whereas almost all of these existing studies have made an assumption of *disjoint* organizational groups, which means that each resource is a member of *a single* organizational group. In fact, in many real-world organizations it is common to have employees who possess multiple skills to share roles and responsibilities across organizational groups. More generally, modern organizations emphasize the importance of having smooth and active communication among various functional units, and achieve so by setting up cross-department roles to enhance the coordination [5]. From the viewpoint of organizational structures, resources working across different organizational groups form the *overlap* between the groups. From the viewpoint of process improvement, such resources are likely to be the so-called *generalists* – a special category of resources that can help increase the flexibility of a business process [2]. In terms of flexibility, we consider the generalists to carry out a role similar to the role of "libero" in certain team sports.

In this paper we propose an approach for the discovery of organizational models from event logs, which allows the sharing of human resources between different organizational groups. By relaxing the assumption of disjoint organizational groups (applied in most of the existing work), new discovery algorithms are developed to address the challenges arising from dealing with the potential overlaps between organizational groups. Based on the characteristics of the problem of interest, a couple of existing cluster analysis techniques (from the field of data mining) are chosen and applied in our discovery algorithms. Experiments are conducted on an implementation of the discovery algorithms, using real-life event logs, to evaluate the applicability and effectiveness of our approach.

The contribution of our work is twofold. On the one hand, the discovered organizational model with potential overlaps is a better reflection of the actual organizational grouping of resources relevant to process execution, and hence it will enable more insightful resource performance analysis. On the other hand, identifying resources that belong to more than one organizational group from event logs presents a novel data-driven approach to the discovery of generalists in an organization and their organizational positioning (i.e. in which organizational groups they perform in practice). Finding the information about generalists will help improve resource utilization and also serve as an important step for action-able process improvement. For example, one strategy for process improvement is to keep such resources free when possible, which guarantees flexibility in the distribution of work [6].

The rest of the paper is organized as follows. Section 2 provides a review of the related work on the topic. Section 3 introduces basic concepts and pre-liminary notions. In Sect. 4, we present our approach for mining organizational models with overlaps, and in Sect. 5 we discuss the experiments and analyze the evaluation results. Finally, Sect. 6 concludes the paper and outlines future work.

2 Related Work

The research considering the organizational perspective of process mining origi-nates from the work by van der Aalst et al. [7], in which several types of inter-resource relationship metrics are defined for deriving resource social networks from event logs. Based on the analysis of resource social networks, Song and van der Aalst [1] propose the conceptual framework of organizational mining as a sub-field of process mining, within which three research dimensions of organiza-tional mining are proposed: *discovery, conformance checking* and *extension.*

Discovery refers to constructing models that reflect the reality. In the con-text of organizational mining, these models include organizational models, social networks and resource assignment/allocation rules. *Organizational model min-ing* focuses on finding the grouping of resources (employees), e.g. who belongs to which functional unit [1,8], who plays what roles [3,9] or holds what social positions in collaboration [10]. Recently, the work of Appice [8] introduces an approach for mining organizational models using a community detection tech-nique, which makes no assumption about each resource belonging to a single group. To the best of our knowledge, this is so far the only existing approach capable of deriving organizational models with potential overlaps.

The discovery of social networks emphasizes the use of social network analysis to help understand the structure of communication between individual resources as well as between organizational groups [4,7,11]. The research presented in [12, 13] studies the discovery of rules related to staff assignment (who is allowed to do which tasks) and runtime activity distribution (to whom a specific task is allocated) to help with diagnosis and optimization of pre-defined rules.

In addition, there is also existing research concerning the organizational per-spective of business processes at the level of individual resources. For example, in [14] the authors analyze the correlation between the workload of individual resources and their performance, and in [15] the authors propose a framework for analyzing and evaluating different resource behaviors in order to provide insights towards more informed resource-related decisions for performance improvement.

3   Preliminaries

Here we present several preliminary concepts necessary for describing the problem, following the conceptual framework of organizational mining defined by Song and van der Aalst [1]. A typical event log usually consists of a set of uniquely identifiable cases corresponding to the instances of an underlying business process. Each case contains a sequence of events that describe the activities carried out by some resources. Table 1 gives an example fragment of an event log recorded by a process-aware information system. Each row refers to one single event, which is described using attributes such as activity label, timestamp, and identity of the originating resource[1].

Table 1. An example fragment of an event log.

Case ID	Event ID	Activity label	Resource	Timestamp
c_1	e_1	Register request	John	2018/01/03 10:59:06
c_1	e_2	Examine thoroughly	Mike	2018/02/03 11:10:13
c_1	e_3	Decide	Clare	2018/02/21 15:43:32
c_1	e_4	Reject request	John	2018/02/22 10:35:52

Definition 1 (Event Log [7]). *Let T be a set of tasks and R be a set of resources. $E \subseteq T \times R$ is the set of events that denote the execution of tasks by originator resources. For any event $e \in E$, $\pi_t(e) \in T$ is the task being executed (or the activity) in e and $\pi_r(e) \in R$ is the originator resource of e. $C = E^*$ is the set of possible event sequences (traces describing a case). $L = \mathcal{B}(C)$ is an event log, where $\mathcal{B}(C)$ is the set of all bags (multi-sets) over C.*

In Definition 1 we do not take into account the ordering of events in a case. We focus on two standard attributes of an event – task and resource identity. We use them to build a simple "profile" for each resource, which reflects the history of the resource performing activities. Accordingly, a *performer by activity matrix* can be used to represent the profiles of a set of resources given an event log.

Definition 2 (Performer by Activity Matrix, adapted from [7]). *Given an event log L, let $\{e_1, ..., e_n\}$ be the set of all possible events recorded in L. The performer by activity matrix is an integer-valued matrix X of size $|R| \times |T|$, in which each row vector corresponds to the execution history of activities for a specific resource. Each element of X denotes the count of frequencies of a resource $r_i \in R$ conducting a specific task $t_j \in T$, defined as:*

$$X_{ij} = \Sigma_{1 \leqslant k \leqslant n} \begin{cases} 1, & \text{if } \pi_r(e_k) = r_i \text{ and } \pi_t(e_k) = t_j \\ 0, & \text{otherwise} \end{cases}$$

where $1 \leqslant i \leqslant |R|$ and $1 \leqslant j \leqslant |T|$.

---

[1] For illustration purposes, resource name is used in the example in Table 1.

Simply consider the example fragment of an event log shown in Table 1. The performer by activity matrix build from this example based on Definition 2 is shown in Table 2. Below, we propose a generic and simple definition of an organizational group as a non-empty group of human resources (i.e. employees) in an organization. For each organizational group, we define a membership indicator associated with each resource to specify whether or not the resource belongs to the group.

Definition 3 (Organizational Group). *Let R be a set of (human) resources in an organization, an organizational group can be defined as $G \subseteq R$ and $G \neq \varnothing$. Given an organizational group G, for any $r \in R$, we define a membership indicator function $\mathbb{I}_G : R \to \{0, 1\}$ where $\mathbb{I}_G(r) = 1$ if $r \in G$ and 0 otherwise.*

Finally, we define the concept of organization model. It is simply considered as one entire group of several organizational groups defined in the above.

Definition 4 (Organizational Model). *An organizational model O is a set that consists of a finite number of (k) organizational groups $\{G_1, \ldots, G_k\}$. For any resource r that is part of the organizational model O, r belongs to one or more than one organizational group in O. That is, $\forall r \in \bigcup_{G \in O} G$, $\sum_{G \in O} \mathbb{I}_G(r) \geqslant 1$.*

As mentioned before, most of the existing studies in organizational mining apply the assumption of disjoint organizational groups in an organization, and hence they require that each resource should only belong to a single organizational group (i.e. $\forall r \in \bigcup_{G \in O} G$, $\sum_{G \in O} \mathbb{I}_G(r) = 1$). In Definition 4, our focus is to relax such assumption by recognizing that resources may belong to more than one organizational group in reality and thus to allow potential overlaps between different organizational groups.

4   Approach

Organizational model mining aims at recognizing groups of resources having similar characteristics. We concern the connection between this and the purpose of cluster analysis in data mining, which is to group a set of data objects into multiple clusters such that objects within a cluster have high similarity but are dissimilar to those in other clusters [16]. As a relatively mature field, there exist various types of techniques developed to provide solutions for different requirements and contexts. Since our intention is to derive results in which one resource may be member of more than a single organizational group, we select the technique of *overlapping clustering*, which allows flexible assignment of one data object to multiple clusters. In this paper, we design an approach adopting the idea of overlapping clustering to solve the problem of discovering organizational model with overlaps. Figure 1 gives an overview of the three-phased procedure. We start from constructing the performer by activity matrix that characterizes the resources. Then we transfer the problem into cluster analysis and apply the selected model and algorithm to produce the clustering result, from which we derive an organizational model as the end result.

Fig. 1. The designed procedure for discovering organizational model with overlaps.

4.1 Characterizing Resources

Given an event log, we construct the performer by activity matrix by directly following Definition 2 and determine the execution frequencies while iterating over the events. Table 2 shows the result of deriving the matrix using the example event log fragment in Table 1 as input.

Table 2. The performer by activity matrix built from the example event log fragment.

	Activity 1	Activity 2	Activity 3	Activity 4
	Register request	Examine thoroughly	Decide	Reject request
John	1	0	0	1
Mike	0	1	0	0
Clare	0	0	1	0

Once the performer by activity matrix has been built, we need to select a measure for quantifying the similarity between any two resources by comparing the corresponding row vectors, in order to further group similar resources and derive an organizational model. Some variants of distance-based metrics provide meaningful measures in a process mining context. The Hamming distance, for example, accounts for whether or not two resources have executed the same types of tasks. Meanwhile, correlation-based metrics such as Pearson's correlation coefficient provide a view of statistical correlation. The choice of similarity measure should be done depending on the purpose and context of analysis.

For the next step, we apply the clustering techniques in order to obtain the clusters of resources. Two possible solutions are presented then. Since these two vary in terms of the deciding the final clusters, we will describe how to derive the end result, i.e. the output organizational model, respectively.

4.2 Solution 1: Cluster Analysis Using a Mixture Model

We first elaborate on how to correlate the current problem with the concepts of overlapping clustering. The concept of probabilistic cluster and the hypothesis of mixture models are commonly used in cluster analysis to characterize the flexible assignment of one object to multiple clusters simultaneously. The hypothesis states that the latent categories hidden in the data objects could be mathematically represented using a series of distribution functions [16]. Each data object is related to each latent category by a sampling probability, and is viewed as a sample drawn from a mixture of distributions. In the context of our problem, we can regard the execution history of activities (i.e. the row vector in the performer by activity matrix corresponding to a resource) as the result of a resource following the work patterns of the organizational group(s) it belongs to. If the resource is indeed a member of several different groups, then its execution history of activities should be the consequence of multiple work patterns. We may therefore adopt the hypothesis of mixture models as an idea for a solution. First, cluster the resources by leveraging the performer by activity matrix along with the specified similarity measure and find the distribution function for each cluster, then for each row vector we calculate a sampling probability related with each cluster, which could be used to decide the membership of the resource.

Following the idea we could apply a classic Gaussian mixture model (GMM) as the first solution. In GMM we assume a Gaussian distribution for each latent category, and apply the well-founded EM algorithm [16] to fit the mixture model using the performer by activity matrix. EM works in an iterative fitting process, which starts with a random initialization and updates the mixture model greedily towards a higher value of the goal function (the likelihood of sampling all the vectors using the current model). The mixture model converges as the goal function value no longer increases or updates by a very trivial scale.

Using the converged mixture model, we can calculate the posterior probability of a row vector relating with each cluster, and take the result as the sampling probability. However, for actually deciding the membership of a resource, we need to choose a threshold to be applied on the probability value, which determines if the resource belongs to one or several of the groups. For example, if the chosen threshold value is 0.5, then the resource should belong to a group only if its related sampling probability is larger or equal to 0.5.

For the basic solution using GMM, we notice some problems related to its configuration. Before starting the fitting process, it requires us to decide the number of clusters upfront. This should be done based on the control of granularity we desire: with a higher number it enables us to discover more fine-grained groups, which may be the very specific roles or small workgroups, whereas a lower number of clusters would possibly lead to finding departments at a higher level. Another problem concerns the thresholding step applied for the purpose of deciding resource membership. It is hard to determine an effective level of probability value that decides whether a resource *indeed* belongs to an organizational group or not: for instance, for a fitted GMM we could calculate the result that an involved employee Jack has the probability value of 0.49 that he belongs to Group 1, and 0.51 that he belongs to Group 2. The question is: how should we actually decide Jack's membership given these numbers? Selecting an appropriate threshold value may become a challenging task, since the scale of the estimated posterior probabilities lack a solid interpretation in the context of organizational model mining. We therefore present another overlapping clustering algorithm that addresses the challenge.

4.3 Solution 2: Cluster Analysis Using a More Generative Model

Consider the example of deciding Jack's membership illustrated before. The use of mixture models like GMM poses the challenge of configuring proper threshold parameter, which may hinder us from directly applying the method for discovering organizational models. The challenge arises from the underlying hypothesis of mixture models: when we view the row vector corresponding to a resource as a data object being clustered, the posterior probabilities that we use for later deriving membership only indicate the possibilities of having the current data object sampled from each of the distributions *independently* [16]. Hence a mixture model may fail in well characterizing the reality that, for resources with multiple memberships across several groups, their execution history of activities results from the *joint* effect of all the work patterns of the groups.

Without shifting from the general concepts of both organizational model mining and overlapping clustering, we seek to find a more natural and descriptive model that avoids deriving membership from probabilities, and constitutes a better solution for the current problem.

The Model-based Overlapping Clustering (MOC) model [17] bases itself on the same concepts of probabilistic clusters as GMM does, but without employing the hypothesis of having objects sampled from a mixture of distributions related with the latent categories. Instead, a boolean-valued membership vector is defined directly for each of the objects to be clustered, of which the values are inferred after fitting the model with the data. In comparison with mixture models, the MOC model is a more natural generative model for overlapping clustering. In MOC the data objects being clustered are hypothesized to be generated by simultaneously considering multiple components, as each of the components refers to a part of the model that relates to one of the latent categories to be discovered (similar to the distribution functions).

Algorithm 1 depicts the procedure of applying the MOC model to the current problem. We omit some of the mathematical details here for brevity, for which one may refer to [17] for a more in-depth explanation. Given n resources and the related event log, we assume that the performer by activity matrix X and similarity measure have been decided prior to running the algorithm, and the granularity of analysis has been specified already, i.e. k groups to be discovered. The algorithm starts by an initial estimate of the membership matrix M, which is usually initialized in a random manner. Another model parameter to be initialized is a matrix that represents the active status of each component in the MOC model, denoted as A, for which random initialization will be fine.

Algorithm 1: Applying MOC for Discovering an Organizational Model

---

Input:
- $\{r_1, \ldots, r_n\}$: the n resources involved;
- X: the constructed performer by activity matrix, also assuming that the similarity measure has been specified accordingly;
- k: the number of organizational groups expected to be discovered (depending on the desired granularity).

Output: O: the resulting organizational model consisting of k groups.

    `// Step 1: Initialize the membership parameter`

1 **Initialize** an $n \times k$ boolean value matrix M, where each of the n row vectors indicates the membership of a corresponding resource

2 **Initialize** a $k \times d$ real value matrix A that denotes the active status of each component in the model

    `// Step 2: Fit the model to the data through iterative updating`
    `   until convergence`

3 **repeat**

    `// Update A by direct computing the value from X and M`
    `   (cf. [17])`

4     $A \leftarrow update\,(A, X, M)$

    `// Update M by searching a setting that maximizes the selected`
    `   similarity measure`

5   **for** $i = 1$ *to* n **do**

6         $M_i \leftarrow \underset{M_i \in \{0,1\}^k}{\text{argmax}}\ \text{SIMILARITY\_MEASURE}\,(X_i, M_i A)$

7   **end**

    `// Calculate the goal function value using the log-likelihood`
    `   (cf. [17])`

8     $L = \log P\,(X, M, A)$

9   Calculate the increase ΔL by comparing with the last iteration

10 **until** ΔL *is sufficiently small*

    `// Step 3: Derive the resulting organizational model utilizing the`
    `   membership matrix`

11 **Initialize** k empty sets $G_1, G_2, \ldots, G_k$

12 **for** $i = 1$ *to* n **do**

13   **for** $j = 1$ *to* k **do**

14   **if** $M_{ij} = true$ **then**

15             $G_j \leftarrow G_j \cup \{r_i\}$

16   **end**

17   **end**

18 **end**

19 **return** $O = \{G_1, G_2, \ldots, G_k\}$

After the initialization of the model parameters we proceed to the iterative process for fitting the model to the data (Line 3–10). At each iteration we first update the value of A directly using the current M and X [17]. In the next step, for each membership vector M_i we try to find a value that maximizes the metric value. The search may be time-consuming when the desired group number k is large, however certain algorithms could be plugged in here to speed up the search process [17]. When the appropriate setting of M has been obtained, we calculate the value of the goal function defined here as the log-likelihood (Line 7), and compute the increase in comparison to the result of the last iteration. The iterative updating process stops when convergence is reached, i.e. the increase in the goal function value is sufficiently small.

With the fitted model we can now derive the end result in a straightforward way, since the membership of all the n resources has been determined as the value of the $n \times k$ membership matrix M. Therefore, we just need to simply assign the resources to the corresponding ones of the k sets (Line 11–18), and return the resulting sets as the discovered organizational groups.

Comparing to the more naïve solution of GMM, the solution using MOC model avoids introducing probabilities as the degree of resource membership, and therefore addresses the challenge of having to select thresholds. Given the event log and resources to be analyzed, users would only need to focus on the resource profiling phase, and then set up the expected number of groups. The end result will be an organizational model containing the exact number of groups as required, where overlaps are allowed to exist.

5   Evaluation

5.1   Experiment Design

Both solutions (applying either GMM or MOC) have been implemented in a standalone demo[2]. We evaluated their feasibility on real-life event log data. We aim at giving empirical validation on whether the proposed solutions work effectively in discovering organizational models when there indeed exist overlaps among organizational groups.

Event Logs. Different from the evaluation methods in the previous research on the problem (cf. [1,8,11]), the purpose of the validation here requires us to be aware of the "ground truth" information relevant to the internal groups in an organization a priori. For this purpose we picked two sets of real-life event logs, namely "WABO" and "Volvo". The background of these event log datasets are as follows:

- WABO: The event log from the WABO dataset contains the records of the receiving phase of an environmental permit application process in an anonymous municipality within the CoSeLoG project [18].
- Volvo: This dataset includes event logs generated from the problem management system VINST of Volvo Belgium, which was originally released for the BPI Challenge 2013 [19]. It contains the event logs that describe several business processes handling incidents and problems in the IT-services delivered and/or operated by Volvo IT. We choose the event log related with the process managing the open problems for experiment use.

---

[2]   https://github.com/royyjing/bpm-2018-Yang_Find.

The event logs are recorded in the IEEE standard XES format [20], and include an extended event attribute termed `org:group`, which indicates the group identity of the resource that triggered the event. We recognize that the ground truth organizational models can be extracted by utilizing this information of identities, which can then serve as the reference models for our experiments. To do this we first filter out the events with missing values on `org:group` (including both null and invalid ones). Then we extract the ground truth organizational model by putting resources together into groups accordingly, based on the `org:group` values they relate to as event originators. Table 3 gives a brief overview of the preprocessed event logs, along with some basic statistics of the extracted reference models: the average size of groups (Avg. group size), and the average number of groups that a resource belongs to (Avg. membership). One may recognize immediately the existence of overlaps in the reference models after inspecting the basic statistics shown in the table. A further comparison on Avg. membership reveals that the overlapping condition is less obvious in the Volvo case (Avg. membership 1.176 while WABO has a value of 3.886), suggesting considerably fewer employee resources possessing multiple group identities in Volvo IT.

Table 3. Overview of the event logs and the extracted reference models.

Event log	Cases	Events	Activities	Resources	Organizational groups	Avg. group size	Avg. membership
WABO	1,348	6,641	27	44	9	19.0	3.886
Volvo	818	2,331	5	239	11	25.5	1.176

Experiment Setups. We conducted the experiments using the comparison method. Two methods proposed in previous research are selected as baseline: a traditional partitioning method that produces disjoint organizational models [1], namely MJA; and a community detection based method developed by Appice [8] that is capable of deriving organizational models with possible overlaps, namely Commu. We examine if the organizational models discovered from the same source of event logs using GMM and MOC can better capture the reality, i.e. more similar to the reference models.

To start with, we build the performer by activity matrix, and choose the Pearson's correlation coefficient as the metric for similarity measure. Since the setup of the algorithms involved in evaluation may vary, we decided to configure the parameters for each algorithm separately, as long as they produce resulting organizational models with exactly the same number of organizational groups discovered as that in the reference ground truth.

Evaluation Metrics. For the purpose of comparing between the results of discovery and the reference models to assess the effectiveness of different methods, we consider adopting extrinsic evaluation metrics. One example is the entropy measure [1], which can be used for measuring the scale of difference between a generated model and the referenced one. However, as the current research has been extended to the overlapping situation, the entropy measure becomes inappropriate as well as many other commonly used extrinsic measures. We therefore turn to the extended BCubed metrics (including BCubed Precision, Recall and F-measure) [21], as they are applicable for evaluation on the overlapping cases. From an organizational model mining point of view, the meaning of the BCubed metrics can be interpreted as follows:

1. BCubed Precision represents the ratio of how many resources in a same discovered organizational groups belong to the same actual groups. A higher value of BCubed Precision means fewer mistaken assignments in the discovered organizational model.
2. BCubed Recall represents the ratio of how many resources from a same actual groups are assigned to the same discovered organizational groups. A higher value of BCubed Recall means more resources with the same actual group identities are placed together by the mining algorithm.
3. BCubed F-measure is a combination of BCubed Precision and Recall, defined as the harmonic average of the two.

Besides the BCubed metrics, we also want to compare the basic statistics of the discovered organizational model (Avg. group size and Avg. membership), with those of the ground truth model.

5.2   Comparing with the Disjoint Partitioning Method

In the first experiment we wish to compare our solutions with the disjoint partitioning method MJA. The idea behind MJA is to view the resources as vertices in a graph, and connect weighted edges between them based on the measured similarity values. By eliminating certain edges by a threshold value, the original graph is further partitioned into several connected components, which are taken as organizational groups that constitute the final organizational model. The result generated from MJA is obviously disjoint.

Table 4 shows the evaluation results measured by the BCubed metrics. From the table we can see that MJA obtains higher precision rates. However, the disjoint nature of MJA prevents it from recognizing the fact that similar resources may possibly share more than one group identities in an overlapping organizational model. Thus, for MJA, similar resources are clustered into one group only, which lead to the relatively lower recall.

On the other hand, the proposed solutions using either GMM or MOC have comparatively lower precision yet higher recall values. It can be explained that both overlapping clustering based algorithms tend to put more resources into the groups, which is consistent with the larger group sizes shown in Table 5. This leads to the better recall rates, but at the same time makes the discovered organizational groups contain relatively members being mistakenly assigned, which directly cause the lower precision of GMM and MOC.

Moreover, for the Volvo case we notice that even the baseline MJA produces a relatively lower precision, and the situation of recall rates mentioned above becomes even more significant. The reason is due to the large total number of resources compared to the much smaller number of activity types (239 resources compared to 5 activity types). The smaller number of activity types leads to fewer columns in the performer by activity matrix, and may therefore weaken the effect of measuring similarity.

Despite the observation that GMM and MOC may tend to sacrifice some precision rate and bring mistaken assignments, from Table 5 we can draw a conclusion – the overlapping-clustering-based solutions are able to derive an overlapping organizational model that captures the reality, whereas methods like MJA holding the assumption of disjoint organizational model are not.

Nevertheless, we still have the following questions: How effective are our solutions comparing to other solutions that can also produce overlapping organizational models? Will the other solutions also encounter the problem of unsatisfying precision? We will explore the answers to these questions through the following experiment and analysis.

Table 4. Results of comparing with MJA on the BCubed metrics.

Event log	BCubed Precision			BCubed Recall			BCubed F-measure		
	MJA	GMM	MOC	MJA	GMM	MOC	MJA	GMM	MOC
WABO	**0.814**	0.624	0.757	0.213	**0.812**	0.735	0.337	0.706	**0.745**
Volvo	**0.496**	0.186	0.24	0.397	**0.944**	0.94	**0.441**	0.31	0.382

Table 5. Results of comparing with MJA on the grouping statistics.

Event log	Avg. group size				Avg. number of membership			
	Ground truth	MJA	GMM	MOC	Ground truth	MJA	GMM	MOC
WABO	19.0	4.9	28.4	22.8	3.886	1	5.818	4.659
Volvo	25.5	21.7	146.5	110.9	1.176	1	6.745	5.105

5.3   Comparing with the Community-Detection Based Method

In this experiment we choose as baseline a community detection based approach [8] which we refer to as Commu. Our goal is to make a comparison between the effectiveness of Commu and our approach. Commu is based on social network analysis techniques rather than cluster analysis, but shares the same purpose of grouping cohesive resources into communities that represent the internal organizational groups. It applies the linear network model with the Louvain algorithm, and derives organizational models which allow the existence of overlapping communities (organizational groups).

Tables 6 and 7 show the evaluation results of this experiment. By observing the average number of membership we first confirm that the baseline method Commu indeed generates overlapping results. For the BCubed metrics, we notice that GMM performs roughly the same as Commu, whereas MOC performs better than Commu in both cases. And the grouping statistics show that the models produced by using either GMM or MOC are more realistic compared with Commu.

Meanwhile, we learn from the tables that Commu also produced a result of low precision and oversize groups, as in the Volvo case, and even worse while comparing with GMM and MOC (refer to the grouping statistics in Table 7).

In general, we may conclude that our approach is more effective as a solution to discovering organizational models with overlaps, compared to the community detection based method. Nevertheless, as both methods have the shortcoming of introducing mistaken assignment of resources to groups causing low precision and unrealistic group sizes, further work is needed to address this shortcoming.

Table 6. Results of comparing with Commu on the BCubed metrics.

Event log	BCubed Precision			BCubed Recall			BCubed F-measure		
	Commu	GMM	MOC	Commu	GMM	MOC	Commu	GMM	MOC
WABO	0.718	0.624	**0.757**	0.651	**0.812**	0.735	0.683	0.706	**0.745**
Volvo	0.195	0.186	**0.24**	**0.948**	0.944	0.94	0.324	0.31	**0.382**

Table 7. Results of comparing with Commu on the grouping statistics.

Event Log	Avg. group size				Avg. number of membership			
	Ground truth	Commu	GMM	MOC	Ground truth	Commu	GMM	MOC
WABO	19.0	28.8	28.4	**22.8**	3.886	5.886	5.818	**4.659**
Volvo	25.5	152.6	146.5	**110.9**	1.176	7.025	6.745	**5.105**

5.4   Discussion

We can draw some interesting insights considering results from both experiments conducted. The first conclusion concerns the comparison of effectiveness between GMM and MOC. It has been evaluated through the experiments that MOC performs better, indicated by the higher precision and F-measure, along with the grouping characteristics being more similar to the ground truth model. Taking into consideration that it requires no cumbersome decision to set up the extra threshold parameter when applying MOC, we conclude that MOC will serve as a better solution than GMM for discovering organizational models with overlaps.

On the other hand, we also realize that for our solution, there exists a shortcoming which would become significant when the latent organizational model is less overlapped. We infer the possible reasons behind it as twofold. The first one concerns the relatively fewer types of activities compared to the number of resources. The second concerns the lack of constraints on the number of groups allowed for each resource to be assigned to. As the former is limited by the content of the event log, we discuss the remedy for the latter.

Given no constraints, both GMM and MOC may try to relate resources to many organizational groups as long as the goal function value is being optimized. This eventually causes the unrealistic mining result in which one resource is a member of considerably many organizational groups simultaneously, diverging from the reality that some resources may possess few or no shared group identities, as in the Volvo case. To solve this, a natural idea is to set up the constraints to mitigate the problem of involving too many resources. Yet this would require more prior knowledge of the underlying organizational structure to implement. Nevertheless, we argue that such an improvement needs only slight modification on the current solution. For GMM, it requires the proper threshold value. For MOC, heuristics are to be introduced to prune the search space in updating the estimate of membership. Another remedy could be mixing application of the proposed solution with the traditional disjoint method: Given an organizational model mining task with the performer by activity matrix has been built along with the specified similarity measure, one may first mine a disjoint model using the traditional method, and utilize the obtained model statistics for the guided initialization of the parameters. Then, apply GMM or MOC to discover an organizational model with potential overlaps. We plan to leave the exploration for improvement to our future research on the topic.

6   Conclusion

Organizational model mining techniques enable the discovery of organizational models from event logs. In this paper, we relax the assumption of disjoint organizational groups held by existing methods and discover organizational models in which individual resources may share multiple group identities. We refer to overlapping clustering techniques and introduce two solutions, GMM and MOC, for deriving organizational models with overlaps. Results from experiments on real-life event log data demonstrate the applicability and effectiveness of the methods. We also recognize the potential limitation of our solution and conclude the reasons behind it, which lead to identifying the potential heuristics for further amending the current approach.

In future work we will consider the following aspects: (1) to improve our approach by effectively incorporating the identified heuristics; (2) to link the current research with performance analysis on generalist resources; (3) to conduct evaluation on more real-life cases.

Acknowledgements. This work is supported by the National Key Research and Development Program of China (Grant No. 2017YFB0202200); the National Natural Science Foundation of China (Grant No. 61572539); the Research Foundation of Science and Technology Plan Project in Guangdong Province (Grant No. 2016B050502006);

and the Research Foundation of Science and Technology Plan Project in Guangzhou City (Grants No. 2016201604030001, 201704020092).

References

1. Song, M., van der Aalst, W.M.P.: Towards comprehensive support for organizational mining. Decis. Support Syst. **46**(1), 300–317 (2008)
2. Reijers, H., Mansar, S.L.: Best practices in business process redesign: an overview and qualitative evaluation of successful redesign heuristics. Omega **33**(4), 283–306 (2005)
3. Jin, T., Wang, J., Wen, L.: Organizational modeling from event logs. In: International Conference on Grid and Cooperative Computing (GCC), pp. 670–675 (2007)
4. van Zelst, S.J., van Dongen, B.F., van der Aalst, W.M.P.: Online discovery of cooperative structures in business processes. In: Debruyne, V., et al. (eds.) OTM 2016. LNCS, vol. 10033, pp. 210–228. Springer, Cham (2016). https://doi.org/10.1007/978-3-319-48472-3_12
5. Daft, R.L.: Organization Theory and Design, 10th edn. South-Western, Cengage Learning, Mason (2010)
6. van der Aalst, W.M.P., van Hee, K.: Workflow Management: Models, Methods, and Systems. MIT Press, Cambridge (2002)
7. van der Aalst, W.M.P., Reijers, H.A., Song, M.: Discovering social networks from event logs. Comput. Support. Coop. Work (CSCW) **14**(6), 549–593 (2005)
8. Appice, A.: Towards mining the organizational structure of a dynamic event scenario. J. Intell. Inf. Syst. **50**(1), 165–193 (2018)
9. Burattin, A., Sperduti, A., Veluscek, M.: Business models enhancement through discovery of roles. In: IEEE Symposium on Computational Intelligence and Data Mining (CIDM), pp. 103–110 (2013)
10. Liu, R., Agarwal, S., Sindhgatta, R.R., Lee, J.: Accelerating collaboration in task assignment using a socially enhanced resource model. In: Daniel, F., Wang, J., Weber, B. (eds.) BPM 2013. LNCS, vol. 8094, pp. 251–258. Springer, Heidelberg (2013). https://doi.org/10.1007/978-3-642-40176-3_21
11. Ferreira, D.R., Alves, C.: Discovering user communities in large event logs. In: Daniel, F., Barkaoui, K., Dustdar, S. (eds.) BPM 2011. LNBIP, vol. 99, pp. 123–134. Springer, Heidelberg (2012). https://doi.org/10.1007/978-3-642-28108-2_11
12. Rinderle-ma, S., van der Aalst, W.M.P.: Life-cycle support for staff assignment rules in process-aware information systems. Technical report 213, TU Eindhoven (2007)
13. Schönig, S., Cabanillas, C., Jablonski, S., Mendling, J.: A framework for efficiently mining the organisational perspective of business processes. Decis. Support Syst. **89**, 87–97 (2016)
14. Nakatumba, J., van der Aalst, W.M.P.: Analyzing resource behavior using process mining. In: Rinderle-Ma, S., Sadiq, S., Leymann, F. (eds.) BPM 2009. LNBIP, vol. 43, pp. 69–80. Springer, Heidelberg (2010). https://doi.org/10.1007/978-3-642-12186-9_8
15. Pika, A., Leyer, M., Wynn, M.T., Fidge, C.J., ter Hofstede, A.H.M., van der Aalst, W.M.P.: Mining resource profiles from event logs. ACM Trans. Manag. Inf. Syst. **8**(1), 1:1–1:30 (2017)
16. Han, J., Pei, J., Kamber, M.: Data Mining: Concepts and Techniques. Elsevier, Amsterdam (2011)

17. Banerjee, A., Krumpelman, C., Ghosh, J., Basu, S., Mooney, R.J.: Model-based overlapping clustering. In: Proceedings of the Eleventh ACM SIGKDD International Conference on Knowledge Discovery in Data Mining, pp. 532–537 (2005)
18. Buijs, J.: Receipt phase of an environmental permit application process. (WABO), CoSeLoG project (2014)
19. Steeman, W.: BPI challenge 2013 (2013)
20. IEEE: IEEE Standard for eXtensible Event Stream (XES) for Achieving Interoperability in Event Logs and Event Streams. Technical report. IEEE Std 1849-2016, November 2016
21. Amigó, E., Gonzalo, J., Artiles, J., Verdejo, F.: A comparison of extrinsic clustering evaluation metrics based on formal constraints. Inf. Retr. **12**(4), 461–486 (2009)

Track III: Digital Process Innovation

On the Synergies Between Business Process Management and Digital Innovation

Amy Van Looy[(⊠)][iD]

Department of Business Informatics and Operations Management,
Faculty of Economics and Business Administration, Ghent University,
Tweekerkenstraat 2, 9000 Ghent, Belgium
Amy.VanLooy@UGent.be

Abstract. Digital innovation brings about change, both economically and socially. In response, businesses are changing their way of working because of increasing opportunities and the speed of new IT. While previous research statistically identified a positive link between Business Process Management (BPM) and digital innovation (DI), this article focuses on extending the role of BPM to DI. Based on 19 expert interviews, we observed how organizations apply the BPM success factors to realize DI, and which problems are experienced. Common obstacles were revealed, e.g. employee resistance, little top management support, and alignment issues (environmental, strategic and business-IT alignment). The results are presented in a framework that academics and practitioners can apply to create more synergies between BPM and DI. Our findings shed new light on the way BPM can be implemented in a digital economy, and how DI can be positioned in a more stable position by means of BPM.

Keywords: Business process management · Process innovation
Digital innovation · Digital transformation · Critical success factor
Adoption · Expert panel

1 Introduction

Digital innovation can drastically impact on the traditional way of working [1, 2]. E.g., Uber challenges the taxi sector, whereas Airbnb does the same for the hotel sector, Spotify for the music industry and Netflix for the television industry. New technologies also evolve fast [3, 4]. E.g., online payments were innovative about ten years ago but have become mainstream, while some supermarkets (e.g. Amazon Go) are experimenting with automatic payments for check-out free or cashier-less stores. Gartner [5] particularly predicts three IT trends: (1) artificial intelligence (AI) for smart robots, conversational user interfaces and drones, (2) transparently immersive experiences like augmented or virtual reality, and (3) digital platforms like blockchains and IoT.

An interesting question is how the Business Process Management (BPM) field can deal with digital innovation and prepare for the future. While scholars acknowledge a link between BPM and digital innovation [1] or have statistically investigated the degree of this relationship [2], most studies rather provide general explanations without

© Springer Nature Switzerland AG 2018
M. Weske et al. (Eds.): BPM 2018, LNCS 11080, pp. 359–375, 2018.
https://doi.org/10.1007/978-3-319-98648-7_21

elaborating on specific aspects. In response, [6] did a Delphi study to start translating the BPM critical success factors to a digital innovation context, however, without proving which factors contribute more to digital innovation, and why. Since both BPM and digital innovation can help organizations reach their strategic goals in an efficient way [7–9], it is crucial to investigate the role between BPM and digital innovation in more detail, to find synergies, and to think about alternative ways to apply BPM. For this purpose, we address two research questions:

- RQ1. How do organizations apply the critical success factors (CSFs, or capability areas) of BPM to realize digital innovation?
- RQ2. Which practical problems or obstacles are experienced in RQ1?

We conducted exploratory research to better understand organizations' underlying motivations based on interviews with practitioners experienced in combining BPM with digital innovation. Answering these research questions will help gain insight into the changing role of BPM and the practical problems, as well as providing more knowledge about specific aspects that contribute to both BPM and digital innovation. The uncovered ideas will be bundled in a framework that academics and practitioners can apply to create synergies between BPM and digital innovation based on CSFs.

This article continues by providing the research background of BPM and digital innovation in Sect. 2, followed by describing our qualitative research approach in Sect. 3. The findings are then presented (Sect. 4) and discussed (Sect. 5).

2   Research Background

2.1   Innovation

[10: p. 1334] define innovation as *"the multi-stage process whereby organizations transform ideas into new/improved products, services or processes in order to advance, compete and differentiate themselves successfully in their marketplace"*. The 4Ps of innovation are [11]: (1) process innovation (our main focus), (2) product/service innovation (i.e. business process outputs), as well as (3) position innovation and (4) paradigm innovation (3 and 4 are out-of-scope). Per innovation type, two dimensions exist [11]: (1) from incremental innovation (doing what we do better) to radical innovation (new to the world), and (2) from components level to system level.

An innovation typically follows generic stages in an innovation process from idea generation to realization, e.g. discovery, development, diffusion, and impact [3]. Alternatively, the management innovation process framework of [12] has four stages: (1) motivation, (2) invention, (3) implementation, and (4) theorization for legitimation. Each stage requires actions of internal and external stakeholders. If organizations do not follow innovation stages, their innovations will probably be less successful [3].

Nonetheless, innovation success depends on multiple aspects besides the innovation process. Such CSFs may also differ between services and products. [13] state that service organizations should put more emphasis on "launch proficiency", "absorptive capacity" and "organizational design", while product organizations should first focus on "product advantage", "market orientation" and then "launch proficiency".

2.2 Digital Innovation and Digital Transformation

2.2.1 Definitions

[3: p. 330] define digital innovation (DI) as "*a product, process, or business model that is perceived as new, requires some significant changes on the part of adopters, and is embodied in or enabled by IT*". DI is thus a kind of innovation that implies business transformations supported by IT [14]. In fact, DI can use IT in two ways: (1) during the innovation process, and/or (2) to describe (fully or partly) the innovation process outcomes (as process innovation or product/service innovation). Hence, innovation outcomes are not necessarily digital [15]. E.g., 3D printers allow experimenting with product prototypes, resulting in more flexible innovation processes.

2.2.2 Strategies

The literature presents different DI strategies. E.g., [7] define two digital strategy types based on IT investments (i.e. general IT investments and IT outsourcing), both seen as the budget relative to competitors and throughout the years. Alternatively, [16] present two other digital strategies for traditional organizations (i.e. by customer engagement and by data-driven solutions to anticipate customer needs). Both are to be executed by two technology-enabled assets (i.e. an operational system for efficiency and a digital services platform for agility). Next, the digital transformation framework [14] shows four common dimensions in a digital transformation strategy: (1) use of technologies, (2) value creation, (3) structural changes, and (4) financial aspects.

Two user-driven innovation strategies are gaining importance [17, 18]: (1) a Lean start-up (in the solution space) and (2) design thinking (in the problem space). A Lean start-up is a new (sub) organization that starts from a business model canvas with testable hypotheses on multiple dimensions (e.g. customers, partners, value, costs, revenues). It focuses on "*testing hypotheses, gathering early and frequent customer feedback, and showing minimal viable products to prospects*" [17: p. 67]. Only a proven model will be executed, building a formal organization. On the other hand, design thinking focuses on general innovations and starts from a challenge or problem with yet unknown customers (instead of a business model). It focuses more on ideation to inductively generate solutions during the innovation process, e.g. by means of user journeys or causal maps [18]. Both approaches cope with testing prototypes rapidly and iteratively, in line with agile software development and SCRUM thinking.

2.2.3 Critical Success Factors

Although the innovation management literature is broad, the CSFs for DI are still under investigation. To some extent, we can derive CSFs from the strategies of Sect. 2.2.2, e.g. customer engagement and data-driven solutions [16] or the dimensions of [14]. Based on 11 IT adoption models, [19] uncovered 42 CSFs across five groups for IT innovation adoption in the public sector: (1) perceived technology, (2) support, (3) external forces, (4) collaboration, and (5) organizational factors. Alternatively, [8] offer a generic diagnostic tool (i.e. as a DI index) with three main groups of CSFs and five sub groups: (1) a product group (i.e. user experience and value proposition), (2) an environment group (i.e. digital evolution scanning), and (3) an organization group (i.e.

skills and improvisation). Nonetheless, DI implies transformations in business processes and work environments.

2.3 Business Process Management and Digital Process Innovation

2.3.1 Definitions

Introducing IT and redesigning business processes were two approaches that shaped organizations during the information revolution in the 1990s [20, 21], and they still remain. [22: p. x] explain the link with IT by positioning BPM as "*a cross-disciplinary field, striking a balance between business management and IT aspects*" in order to apply this knowledge to business processes (as organizational work) and achieve strategic objectives (e.g. competitive advantage or long-term success) [4]. While business processes are present in every organization [22], the level of BPM adoption and process innovation depends on an organization's environment [23].

2.3.2 Strategies

[4] sees merit in advancing towards more value-driven BPM, namely starting from an organization's strategy to realize people- and technology-based process executions. Besides value-driven BPM, [24] also sees a future for ambidextrous BPM (i.e. extending exploitative BPM with explorative BPM to become open to innovation) and customer process management (i.e. broadening an internal process view to an external view to better consider stakeholder experience). Alternatively, [25] propose three directions for BPM to evolve: (1) technology-driven, data-driven or intelligent BPM (e.g. process mining), (2) human-driven, social-driven or collaboration BPM (e.g. social software), and (3) case-driven BPM (i.e. case management for knowledge-intensive and unstructured processes). These different BPM directions give evidence that the innovative capacity of BPM has become more relevant, given the fast evolutions in business environments and new IT [1]. Particularly, while BPM used to primarily focus on (semi)-structured business processes, DI can be stimulated by focusing more on data and less structured business processes [1, 6].

2.3.3 Critical Success Factors

The literature presents different approaches for addressing the BPM CSFs: (1) based on theories, (2) from the perspective of problems or issues, (3) using a mathematical formula, and (4) based on maturity model assessments and roadmaps. We subsequently give one example per approach. First, [23] situates the BPM discipline within the theories of: (1) contingency, (2) dynamic capabilities and (3) task-technology fit, and claims that a continued fit between an organization's business processes and its business environment is required. Secondly, [26] identified strategic, tactical and operational issues for BPM adoption. Thirdly, [27] offer a mathematical approach to evaluate BPM implementation. According to their research, the four most important CSFs are: (1) strategic alignment, (2) top management support, (3) project management, and (4) a collaborative environment. Fourthly, maturity models such as CMMI see CSFs as capability areas to be measured and/or improved. Given the huge amount of maturity models, [9] summarized the capability areas of 69 process-centric maturity models to obtain a comprehensive list of 17 sub capability areas across six main areas: (1) process modeling, (2) process deployment, (3) process optimization, and a process-oriented (5) culture and (6) structure. Their list can be seen as a generic BPM index with CSFs. Thus, a large number of different CSFs exist for BPM implementation. This research elaborates on these CSFs to interpret their relevance for DI.

3 Methodology

3.1 Research Method Selection

"Qualitative research uses a naturalistic approach to understand complex phenomena in a context-specific setting" [28: p. 600]. It is preferred over quantitative research to explore and obtain detailed information about complex phenomena, like a digitalizing business context. We relied on an expert panel with semi-structured, in-depth interviews. Expert panels are an efficient and concentrated method to collect data, during which the interviewed experts are the source of information [29]. While subject matter experts can add new or specialized knowledge from practice, the inclusion of a broad panel also reveals different perspectives (e.g. about BPM and DI) [30].

In contrast to a case study approach, the respondents could draw on their experience acquired throughout their career (instead of one organization) to enrich the data. Next, a Grounded Theory approach was not chosen because we started from the BPM body of knowledge. Since the BPM CSFs have been identified in previous works, we deliberately decided to conduct a single research round which contrasts to a Delphi approach. Delphi studies have multiple rounds and usually relate to another type of research question, e.g. to find consensus about yet an unknown list of topics, CSFs or criteria instead of collecting a wide variety of experiences. Finally, we did not opt for focus groups to better understand individual opinions without group pressure.

3.2 Selection of Respondents

We conducted one-hour, face-to-face interviews with 19 West-European experts during November 2017. Potential respondents were identified within our professional networks and via LinkedIN, and were invited in their role as: (1) BPM manager with knowledge about DI, (2) DI manager with knowledge about BPM, or (3) IT consultant combining BPM with DI. Given such role specification, the experts had a minimum experience of five years in either BPM or DI. Appendix A shows more details about the experts' profiles, indicating different sectors, roles and years of experience. In total, 85 candidates were contacted, resulting in a response rate of 22.35%. Our panel size is adequate to surpass the minimum size (i.e. a point on which data saturation generally occurs) [31, 32] and the panel size in other BPM studies [26, 33].

3.3 Variables

Our main variable was a BPM index with CSFs or capability areas that we intend to translate into a DI context. We deliberately opted for a similar framework as [2] to be able to relate our qualitative explanations to their quantitative findings. This BPM capability framework is based on multiple process-centric maturity models (instead of a single maturity model) [9]. Moreover, this decision can complement the work of [6] by focusing more on similarities and problems experienced by practitioners.

The semi-structured interview scheme had 11 main questions (one per CSF): (1) PDCA, (2) strategic alignment, (3) external relationships, (4) role of the process owner, (5) skills of the process owner, (6) skills of the process participants, (7) process-oriented values, (8) process-oriented HR appraisals and rewards, (9) top management support, (10) process-oriented organization chart, and (11) specific BPM governance body. For each BPM CSF, we asked the following sub questions:

- *"How can this aspect play a role for digital innovation, and why?"* (RQ1)
- *"Which practical problems or obstacles are experienced?"* (RQ2)
- *"Can you give examples?"* (RQ1, RQ2).

3.4 Coding

The interview transcripts were analyzed in Nvivo. Since each interview (sub) question can be traced back to a specific research question, we started by assigning one node (i.e. tag or label) per interview question (i.e. construct). About 1,200 sub nodes were inductively created to differentiate between the answers (i.e. ideas), which were then aggregated into higher-level nodes (i.e. categories or themes). Finally, the experts per aggregated node were counted to describe the findings [34]. It was important to detect patterns, e.g. to identify which BPM CSF can play a role for DI. We also looked at major differences in expert opinions, and considered initial background questions to better understand an expert's reasoning [34].

3.5 Evaluation Criteria

Regarding dependability (aka reliability), [35] propose investigator triangulation to let multiple observers read and interpret the same set of transcripts. Our study relied on multiple observers, namely one research coordinator and 59 Master students in IT management who followed a mandatory BPM course, and who followed the same interview protocol. Each interview was taken by a different group of circa five Master students, who also typed out the recorded interview together. Since multiple students were present per interview, they could help each other in asking additional questions and obtaining correct interpretations. Afterwards, the transcripts were also read and analyzed by all other groups individually, and then peer reviewed. Also the research coordinator did the coding in parallel. Credibility (aka internal validity) was enhanced by deriving the variables from the literature (Sect. 3.3), training the interviewers, carefully selecting respondents, keeping close notes on decisions, and repositioning the findings with respect to the body of knowledge (Sect. 5). Although our data rely on a single interview round, we obtained multiple perspectives from BPM and DI experts referring to some degree of data triangulation [36]. Also the minimum number of 12 interviews was surpassed [31, 32]. Regarding confirmability (aka measurement validity), the interviewers were asked to regularly summarize the respondents' answers to let the participants review our interpretations and to strengthen face validity with additional sub questions. The use of Nvivo also facilitates outsiders to verify our coding efforts, if necessary. Finally, transferability (aka external validity) is typically more difficult for qualitative than quantitative studies. Although our respondents have experience in different sectors, roles and years of experience (Appendix A), generalization will be limited to the covered business settings in West-Europe.

4 Results

4.1 PDCA Lifecycle

The traditional process lifecycle (inspired by Deming's PDCA cycle) represents iterative stages through which any business process evolves: (1) PLAN, (2) DO, (3) CHECK, and (4) ACT. PDCA can be applied to the DI process and the new products/services resulting from DI (2 experts). Some similarities are present, e.g., BPM and DI both strive for customer orientation, quality and operational excellence (5 experts), and a synergy between People-Process-System (3 experts). One expert referred to a stronger focus on the PLAN phase (design), while another expert stated that the CHECK phase (monitoring) will gain in importance. In the PLAN phase, the business case will take a more prominent role (4 experts). The PLAN and ACT phase can profit from existing methods and techniques like root cause analyses to consider potential solutions (2 experts), but also new ones such as: out-of-the-box thinking (2 experts) and co-creation with employees, customers and other stakeholders (1 expert). For the DO and CHECK phase, three experts referred to Robotic Process Automation (RPA) and advanced data analytics (AI). In general, however, the PDCA cycles will become faster in a DI context to obtain more iterative, agile and shorter cycles (11 experts). A more pragmatic view is needed with experiments and pilots (6 experts).

Regarding the experienced PDCA problems, four experts referred to resistance of inflexible employees, customers and stakeholders, while three other experts stated that resistance is people-dependent (e.g. with generation issues between older and younger people). Three experts mentioned that organizations generally give too little attention to process/project measurement. Organizations also seem to cope with identifying and involving the right people (2 experts), or they continue working on unsuccessful initiatives (1 expert). Also the lack of out-of-the-box training was mentioned (1 expert). Furthermore, problems may arise since new IT evolves fast and is still immature, resulting in uncertain outcomes (1 expert). Another expert agreed that it remains difficult to keep process designs up-to-date, while reuse across business processes is sometimes difficult. Other issues related to time constraints (4 experts), budget constraints (2 experts), privacy (1 expert) and security (1 expert).

4.2 Strategic Alignment

BPM and DI seem to share strategic reasons, such as aiming at: (1) service delivery to customers and stakeholders (6 experts), operational excellence (2 experts), or expansion (2 experts). Three experts also discussed the need for finding a balance between value-driven and cost-driven strategies. Since strategic decisions are sector-dependent,

BPM and DI should consider an organization's environment (3 experts). Another aspect in common is that IT enables the realization of (process and business) strategies (2 experts). Also complementary methods and techniques exist, like: (1) strategic maps with different levels (3 experts), (2) a business cases including different business model canvas perspectives (1 expert), and (3) KPI dashboards used for brainstorming (1 expert). Finally, two experts summarized the general need for a strategic vision with clear leadership and employee coaching.

Also for this CSF, many experts mentioned problems regarding cultural resistance or change management (6 experts), and issues regarding time (4 experts), budget (2 experts), and resources (2 experts). Organizations still seem to struggle with finding a balance: (1) between short- and long-term thinking (2 experts), and (2) between new and old (1 experts). In additional, two other experts warned for the risks of surfing from hype to hype, resulting in unstructured results. Other common problems relate to strategy realization, because organizations might create a gap between strategy and practice (2 experts), or miscommunicate about the direction an organization wants to go (2 experts). While another expert referred to changing customer expectations, two experts explained that (sector-related) legislation may impose certain work procedures or platforms. Finally, one expert added that organizations should strategically think more about online versus offline work alternatives, given today's availability of IT.

4.3   External Relationships

Since external relationships are crucial for both BPM and DI, many experts mentioned opportunities like co-creation, win-win collaboration and ecosystems (8 experts). New ways for customer differentiation were seen as assets because thinking in terms of end customers is key for both BPM and DI (5 experts). Another expert referred to the increasing impact of (online) customer reviews. Regarding suppliers, also DI can involve outsourcing challenges (2 experts) or needs well-defined Service Level Agreements or SLAs (1 expert). In general, the experts acknowledged that new IT facilitates communication and document exchange (2 experts), and emphasized the importance of systems integration, authentic sources, standards (3 experts). Nonetheless, also for this CSF, experts recognized the role of an organization's environment by stating that the degree of external collaboration is sector-dependent (4 experts).

The most important problems are external stakeholders having a slower pace or needing more acceptance time (5 experts), and the unavailability of IT-minded or creative partners (2 experts). Other experts added that such problems are, however, usually people-dependent (e.g. generation issues) (2 experts) or sector-dependent (e.g. legislation imperatives) (1 expert). Two experts perceived external time pressures since online communication can speed up decision-making and doing business. Other problems were miscommunication (1 expert), lack of trust (1 expert) or little attention to risk management (e.g. regarding the impact of customer reviews) (1 expert).

4.4   Role of the Process Owner

A process owner is responsible for the PDCA cycle and performance of a particular business process. The experts agreed that this role is especially useful to help create ideas and analyze the pros and cons of (IT) opportunities (5 experts). From this innovation perspective, other experts added the need for working in varied or inter-disciplinary teams to think more in terms of end customers (5 experts), or they stated that the process owner serves as a single point of contact (SPOC), also for IT (2 experts). Because of this importance to actually sit together and having an overarching view (1 expert), the process owner often needs to function as a change and people manager (2 experts). Additionally, two other experts referred to the faster pace for managing changes and the increasing flexibility in problem-solving. Three experts argued that the process owner should be a business person (3 experts) whose role increases because IT offers new ways to monitor-measure-document a process (1 expert).

The most important problem was the silo mentality or pyramidal hierarchy in many organizations that causes sub optimization (9 experts). Also for this CSF, resistance to change was considered as a serious bottleneck, both resistance of functional managers (2 experts) and of the process owner (2 experts). While most experts saw the process owner's cross-department view as an advantage, one expert mentioned that this view is still limited to one business process. Other problems relate to work overload (3 experts), communication problems (e.g. mail overload) (1 expert), and the lack of a competence overview or finding competent people to create a team (1 expert). Three other experts asserted that many process owners have insufficient DI knowledge or skills. Similarly, one expert stated the risk for over-digitalization when process owners do non-strategic things or with only a low ROI. Finally, business-IT alignment problems were discussed by two experts.

4.5 Skills of the Process Owner

A process owner needs: (1) skills related to DI and new IT (8 experts), (2) BPM skills (5 experts), (3) communication and people skills (3 experts), (4) project management skills (2 experts), and (5) IT skills. Five experts also emphasized on-the-job learning, while three experts mentioned learning from best practices, success stories and benchmarking. This varied set of skills helps prevent boredom (1 expert).

Regarding skill-related problems, three experts acknowledged that a process owner usually lacks experience with new IT or that new IT is still too abstract. In fact, a process owner is used to document rather than to create (1 expert). Additionally, a process owner seems to experience difficulties in balancing new ideas with existing business experience. Two experts also referred to a general lack of people skills, e.g. by having difficulties in communicating evident things or by missing empathy to ask the right questions to stimulate out-of-the-box thinking. His/her role also covers borderline tasks with other managers (e.g. innovation manager, strategic manager and/or project manager) (1 expert). A recurrent problem, also for this CSF, seems business-IT alignment when a process owner thinks too much from an IT solution instead of a business solution (1 expert). Other problems relate to time constraints (1 expert) and the fact that knowledge and information sharing becomes international (1 expert).

4.6   Skills of the Process Participants (or Process Workers)

For regular process participants, the experts saw benefit in: (1) a combination of general BPM and IT/DI skills to acquire basic knowledge and awareness (5 experts), (2) training about one's role in the entire processes involved to obtain an end-to-end view (5 experts), and (3) training on how to use operating systems (2 experts). Other important aspects were: (1) self-steering skills and employee involvement via brainstorming (4 experts), (2) communication initiatives (e.g. newsletter or fora) to translate BPM issues for all employees (1 expert), and (3) problem-solving skills (1 expert). Finally, one expert acknowledged the increasing need for lifelong learning.

Most skill-related problems were considered people-dependent, such as generation issues or resistance from employees and managers (5 experts). Two experts also mentioned stress when personals benefits are not communicated or when employees fear for job losses. Furthermore, employees can become demotivated when their bottom-up suggestions are not followed (1 expert). While it can be difficult to be good at both BPM and IT/DI skills (1 expert), some experts also referred to over processing problems (1 expert) and/or a tension between creativity and efficiency (1 expert). One expert admitted that people tend to overestimate one's own skills and knowledge.

4.7   Process-Oriented Values

BPM and DI have common values, such as: (1) an external focus prevailing over an internal focus (10 experts), (2) trust and collaboration (9 experts), and (3) empowerment to increase motivation, possibly with self-steering teams (4 experts). DI also challenges the empowerment value towards more entrepreneurship (2 experts), and a climate that allows failures and creativity (1 expert). For both BPM and DI, the organization needs to be open-minded with people open to change (2 experts). Above all, those values need to be linked to a corporate strategy (3 experts) and being well-communicated (2 experts). As such, employees will gain a more positive feeling of belonging/contributing to something, and thus reaching more engagement (3 experts).

Some problems exist with process-oriented values for a DI context when (1) they are too internally focused (1 expert), or when (2) they are killed by a hierarchical or silo mentality (1 expert). Also some general problems were mentioned, such as: (1) resistance (2 experts), (2) too abstract values (2 experts), (3) a lack of communication and/or training (1 expert), or (4) managers who allow bypassing new ways of working (1 expert). Three experts mentioned time and budget constraints for IT platforms that stimulate values. Also experiments cost money and can fail. Nevertheless, two experts stated that the level of experimentation and the level of resistance is sector-dependent.

4.8   Process-Oriented HR Appraisals and Rewards

For both BPM and DI, HR appraisals and rewards should be closely linked to the corporate strategy and values (5 experts) to stimulate cross-departmental thinking (2 experts). BPM and DI should be part of formal job descriptions (4 experts), while seniority becomes less important (1 expert). Instead, it is important that employees and managers also have competence-related objectives (i.e. how to do their job) (1 expert), and receive feedback linked to training (as in a learning organization) (4 experts). Thus, to stimulate BPM and DI, an organization should also focus on human capital and charity (1 expert). It can choose between financial compensations (7 experts) and/or non-financial incentives (e.g. team events, success stories, best practices) (4 experts). DI may add interesting incentives to BPM by gamification initiatives (e.g. to collect-discuss-like ideas, badges) (1 expert), by co-creation initiatives (e.g. an innovation box with profit-sharing) (2 experts), and by self-steering teams (e.g. an innovation book with a corporate credit card) (2 experts). Those incentives illustrate that intrinsic motivation and involvement are key for both BPM and DI (4 experts), and that organizations should rather use carrots than sticks (1 expert).

When appraisals and rewards are linked to BPM and DI thinking, organizations can experience problems with being objective, and so risk stimulating discrimination or jealously among employees (5 experts). In fact, it remains difficult to translate financial and non-financial rewards systems to individuals and teams (1 expert). Moreover, organizations being too results-oriented cannot cope with failures or trial-and-error (1 expert). Regular problems arise when organizations: (1) focus too much on financial compensations (2 experts) or (2) have too many incentives (while culture and training count) (2 experts). Two other experts explained that organizations can turn inflexible due to too many rules. Another problem occurs when managers have different priorities (1 expert) or when employees need more control (2 experts). The latter is people-dependent, and seems to be mostly observed for older people who are less used to be challenged. HR decisions can also be sector-dependent, e.g. financial salaries in the public sector are regulated (1 expert). While one expert referred to budget and resource constraints, non-financial incentives can be an option for many organizations. Nonetheless, process-oriented appraisals and rewards may not be required for all employees, but organizations can start with the BPM/DI managers, optimization/innovation teams or a specialized digitalization department (2 experts).

4.9 Top Management Support

While all experts agreed that top management support is crucial, two experts explicitly pleaded for a combined interest in or commitment to BPM and DI at the top. Eight experts even stated that BPM/DI initiatives without top management support are doomed to fail or take too long. Nonetheless, also middle managers play an important role, e.g. for empowerment and self-steering teams (5 experts). Thus, it is fruitful to have at least one top manager being the believer, sponsor or strategic coordinator to walk the talk given the importance of soft aspects (6 experts). Such a BPM/DI believer is also important for communication and to derive investment decisions and strategic objectives across departments and business processes (5 experts).

The biggest problem occurs when top managers agree on BPM and DI, but without corresponding budget, resources or infrastructure (6 experts). Employees do not always get the required time and trust aside from their regular work. Another prominent problem is a lack of vision or focus, with top managers only looking at trends or short-term results (5 experts). Three experts mentioned problems when a CxO role is missing for BPM and DI, or when too many related CxO roles exist (e.g. CFO, CIO, COO, chief digital officer, DI officer). Furthermore, problems arise when organizations focus too much on external consultants who stay temporarily and are less familiar with the corporate culture (1 expert). On the other hand, in many organizations, top managers do not give explicit sponsorship or sufficient commitment (3 experts), or even resist when they are no digital natives (i.e. generation issues also exist among top managers) (4 experts). Moreover, two experts explained that business-IT alignment problems also occur within the Board, e.g. when business and IT people do not understand each other, when innovation ideas rather come from the IT side, or when Board members are not diverse enough to be open for BPM and DI. Such problems are also sector-dependent, e.g. when BPM is rather used for quality assurance or just as a way to generate results and end products (1 expert).

4.10   Process-Oriented Organization Chart (Organogram)

A process-oriented organization chart provides organizations with more transparency about responsibilities, SPOCs (i.e. process owners) and degrees of freedom, which is also important for DI (5 experts). It facilitates an easier detection of common interests between process owners, functional managers and top managers (1 expert), and allows to clearly position temporary projects, bodies, start-ups (1 expert). Although three experts observed faster decision-making and empowerment in flatter organizations, four other experts elaborated on different chart options (e.g. virtual, global, departmental, matrix). One expert described a project-based "Squads & Tribes" model to facilitate business-IT alignment, namely a model in which process owners invite people with particular skills in business and IT. Process sponsorship (board level) and process ownership can then participate in operational excellence and transformation projects. In any case, both BPM and DI seem to profit from an organogram oriented towards customers and results (rather than tasks) (2 experts).

Regarding chart-related problems, four experts referred to tensed relationships between process owners and functional managers, e.g. when process owners only have an informal or uninfluential role and/or when discussions cope with power over resources and knowledge. Given an organogram's impact, three experts also referred to resistance of employees and/or (top) managers for this CSF. Nonetheless, an organogram strongly depends on other CSFs like culture and job descriptions (HR) (2 experts). Other observed problems are: (1) risks or incidents when being too focused on new products or immature IT (1 expert), (2) different time zones to collaborate (1 expert), or (3) when only temporary projects or bodies are supported (1 expert). Finally, two experts concluded that this CSF is size-dependent, and possibly only required for large-sized organization. They added that, especially in pyramidal organograms, people sometimes have different roles to capture the notion of process ownership.

4.11   Specific BPM Governance Body

The final BPM CSF includes a program manager for coordinating all process owners as well a Center of Excellence (CoE) or competence center with internal experts in BPM/DI-related methods and techniques. The program manager and CoE help obtain a helicopter view, coordinate and share best practices and knowledge (6 experts). Two experts also referred to standardization and quality assurance reasons. Eight experts argued that the CoE should not merely provide training but especially support in BPM and DI (8 experts). The program managers and CoE should thus closely collaborate with process owners and employees (2 experts). Eleven experts explicitly emphasized that the CoE should be close to business and end customers, with IT serving a supportive role. The CoE can also take different forms, e.g. with a BPM sub group and IT sub group, or as virtual coordination teams and even as an online toolbox (6 experts). One expert suggested that some CoE functions can be done by external consultants, e.g. to give initial advice about BPM/DI possibilities. Nonetheless, one expert asserted that the choice for a centralized CoE or decentralized teams is sector-dependent (e.g. standardization and quality assurance versus starting from scratch to meet market expectations). Three experts referred to size-dependence. E.g., smaller organizations may not have enough volume (i.e. employees, processes, resources) for many owners, who can coordinate themselves. Similarly, IT solutions are not always the cheapest solutions, and are thus dependent on a business case that considers an organization's context (e.g. size). In sum, all experts agreed that coordination is vital.

Not all organizations can afford a CoE due to cost constraints (3 experts) or experience a lack of expert skills (1 expert). If a CoE is present, it also risks to be unknown by other employees when acting as an ivory tower (2 experts). Hence, the CoE members also need people skills to translate BPM/DI methods and techniques to practice (1 expert). Otherwise, organizations usually face resistance from departments, process owners, employees or even customers (4 experts), although resistance is people-dependent (e.g. generation issues) (1 expert). It also appears to be difficult to keep business processes up-to-date (2 experts). Furthermore, one expert mentioned that a generic approach is not always possible due to process-dependent issues. E.g., DI opportunities differ from business process to business process. Another problem arises when the CoE is too IT-driven or located in the IT department (3 experts). Organizations also experience IT architecture problems to link activities and processes with systems. Finally, two experts warned for knowledge losses when the CoE is outsourced or if a decentralized approach is taken.

5 Discussion and Conclusion

Table 1 synthesizes the most important synergies between BPM and DI along generic CSFs and common strategic dimensions (Sect. 2). This framework helps structure the BPM-DI debate and can be used by practitioners when incorporating DI into their BPM practices to overcome frequently occurring problems or obstacles.

Table 1. A summary of the most important synergies between BPM and digital innovation.

DI aspects		BPM aspects	BPM-DI aspects
[8]	[14]	[2]	Interviews
User experience	Use of technologies	1/PDCA 3/External relationships 7/Process-oriented values	• End customer is key • Business case (PLAN) • Monitoring data (CHECK) • Customer differentiation
Value proposition	Value creation	2/Strategic alignment	• Value-driven versus cost-driven
Digital evolution scanning		3/External relationships 4/Role of the process owner 11/BPM governance body	• People-Process-System • Involvement, co-creation
Skills		5/Skills of the process owner 6/Skills of process participants 8/Process-oriented appraisals and rewards	• BPM/DI training for process owners • BPM/DI awareness for all employees • Project-based skills • Self-steering skills • Incentives (also non-financial)
Improvisation		1/PDCA	• More flexible, agile cycles • More experimentation, out-of-the-box thinking
	Structural changes	4/Role of the process owner 9/Top management support 10/Process-oriented chart 11/BPM governance body	• Process owners are SPOCs • Formal coordination and support • Dedicated CxO role
	Financial aspects	9/Top management support	• Sufficient budget, time, resources • Soft aspects

Three recurrent problems were observed across multiple CSFs: (1) organization-dependence (i.e. sector, size), (2) person-dependence (i.e. resistance of employees, managers, stakeholders), and (3) business-IT alignment problems. The first problem calls for more contingency research or context-aware BPM to further investigate the impact of an organization's environment like size and sector [23]. Solutions for the second problem can be found in change management models or information systems acceptance theories (e.g. by emphasizing personal net benefits) [37, 38], as well as in the literature on HR recruitment to find candidates in line with the corporate mission and vision. Also the management innovation process framework of [12] may help reduce resistance by emphasizing process-oriented values during the "motivation" stage, and by collecting bottom-up ideas from employees and stakeholder instead of only involving a small group in the "invention" stage. Thirdly, business-IT alignment has been previously recognized in the BPM literature as the need for a task-technology fit [23], for which solutions can also be found in corresponding business-IT alignment theories [39].

The present article is intended to strengthen the debate about the changing role of BPM in a digitalized world. It is worthwhile to further investigate the individual CSFs (e.g. by case studies among different sectors, sizes and BPM adoption levels). Also future research on BPM strategies (Sect. 2.3.2) seems fruitful. For instance, the interviews mentioned aspects relevant for value-driven BPM [4, 24], ambidextrous BPM [24], customer process management [24], intelligent BPM [25], collaboration BPM [25], and case-driven BPM [25]. Moreover, the BPM discipline would benefit from stipulating synergies with other management disciplines to better position itself and become more future-proof.

Appendix A: The Profile of Respondents

See Tables A1, A2 and A3.

Table A1. The experts' experience in terms of sectors throughout their career (N = 19).

Sector	Frequency	Sector	Frequency
Agriculture, forestry, fishing	1	Scientific, technical activities	1
Manufacturing of products	11	Administrative/support service	2
Construction	1	Public, defense, social security	3
Electricity, gas, air conditioning	3	Education	1
Wholesale, retail, vehicle repair	3	Human health, social work	2
Transportation, storage	3	Arts, entertainment, recreation	1
ICT	13	Other services	2
Financial, insurance	3		

Table A2. The experts' experience in terms of functional roles throughout their career (N = 19).

Role	Frequency	Role	Frequency
BPM manager	5	BPM and digital innovation manager	2
Digital innovation manager	2	BPM manager and IT consultant	2
IT consultant	6	Top manager (e.g. CEO, founder)	2

Table A3. The experts' experience in terms of seniority in BPM and digital innovation (N = 19).

Years	Involvement in BPM	Involvement in digital innovation
0–5	3	8
>5–10	7	6
>10–15	4	3
>15–20	5	1
>20	0	1

References

1. Schmiedel, T., vom Brocke, J.: Business process management: potentials and challenges of driving innovation. In: vom Brocke, J., Schmiedel, T. (eds.) BPM - Driving Innovation in a Digital World. MP, pp. 3–15. Springer, Cham (2015). https://doi.org/10.1007/978-3-319-14430-6_1
2. Van Looy, A.: A quantitative study of the link between business process management and digital innovation. In: Carmona, J., Engels, G., Kumar, A. (eds.) BPM 2017. LNBIP, vol. 297, pp. 177–192. Springer, Cham (2017). https://doi.org/10.1007/978-3-319-65015-9_11
3. Fichman, R.G., Dos Santos, B.L., Zheng, Z.: Digital innovation as a fundamental and powerful concept in the information systems curriculum. MIS Q. **38**(2), 329–343 (2014)
4. Kirchmer, M.: High Performance Through BPM. Springer, Cham (2017). https://doi.org/10.1007/978-3-319-51259-4
5. Gartner: Top trends in the Gartner hype cycle for emerging technologies (2017). https://www.gartner.com/
6. Kerpedzhiev, G., König, U., Röglinger, M., Rosemann, M.: Business process management in the digital age (2017). http://www.bptrends.com/
7. Mithas, S., Tafti, A., Mitchell, W.: How a firm's competitive environment and digital strategic posture influence digital business strategy. MIS Q. **37**(2), 511–536 (2013)
8. Nylén, D., Holmström, J.: Digital innovation strategy. Bus. Horiz. **58**, 57–67 (2015)
9. Van Looy, A., De Backer, M., Poels, G.: A conceptual framework and classification of capability areas for business process maturity. Enterpr. Inf. Syst. **8**(2), 199–224 (2014)
10. Baregheh, A., Rowley, J., Sambrook, S.: Towards a multidisciplinary definition of innovation. Manag. Decis. **47**(8), 1323–1339 (2009)
11. Tidd, J., Bessant, J., Pavitt, K.: Managing Innovation. Wiley, West Sussex (2005)
12. Birkinshaw, J., Hamel, G., Mol, M.: Management innovation. Acad. Manag. Rev. **33**(4), 825–845 (2008)
13. Storey, C., Cankurtaran, P., Papastathopoulou, P., Hultinck, E.J.: Success factors for service innovation: a meta-analysis. J. Prod. Innov. Manag. **33**(5), 527–548 (2016)
14. Matt, C., Hess, T., Benlian, A.: Digital transformation strategies. Bus. Inf. Syst. Eng. **57**(5), 339–343 (2015)
15. Nambisan, S., Lyytinen, K., Majchrzak, A., Song, M.: Digital innovation management. MIS Q. **41**(1), 223–238 (2017)
16. Sebastian, I.M., Ross, J.W., Beath, C., Mocker, M., Moloney, K.G., Fonstad, N.O.: How big old companies navigate digital transformation. MIS Q. Exec. **16**(3), 197–213 (2017)
17. Blank, S.: Why the lean start-up changes everything. Harvard Bus. Rev. **91**(5), 65–72 (2013)
18. Müller, R.M., Thoring, K.: Design thinking vs. lean startup. In: IDMRC Proceedings, pp. 151–164 (2012)
19. Kamal, M.M.: IT innovation adoption in the government sector: identifying critical success factors. J. Enterpr. Inf. Manag. **19**(2), 192–222 (2006)
20. Davenport, T.H.: Process innovation. Harvard Business School, Boston (1993)
21. Hammer, M., Champy, J.: Reengineering the Corporation. HarperCollins Publishers, New York (2003)
22. Dumas, M., La Rosa, M., Mendling, J., Reijers, H.A.: Fundamentals of BPM. Springer, Berlin (2013). https://doi.org/10.1007/978-3-642-33143-5
23. Trkman, P.: The critical success factors of BPM. Int. J. Inf. Manag. **30**, 125–134 (2010)
24. Rosemann, M.: Proposals for future BPM research directions. In: Ouyang, C., Jung, J.-Y. (eds.) AP-BPM 2014. LNBIP, vol. 181, pp. 1–15. Springer, Cham (2014). https://doi.org/10.1007/978-3-319-08222-6_1

25. Lederer, M., Knapp, J., Schott, P.: The digital future has many names. In: ICITM Proceedings, pp. 22–26 (2017)

26. Bandara, W., Indulska, M., Chong, S., Sadiq, S.: Major issues in business process management: an expert perspective. In: ECIS Proceedings, pp. 1240–1251 (2007)

27. Bai, C., Sarkis, J.: A grey-based DEMATEL model for evaluating BPM critical success factors. Int. J. Prod. Econ. **146**(1), 281–292 (2013)

28. Golafshani, N.: Understanding reliability and validity in qualitative research. Qual. Rep. **8** (4), 597–606 (2003)

29. DiCicco-Bloom, B., Crabtree, B.: The qualitative research interview. Med. Educ. **40**(4), 314–321 (2006)

30. Boyce, C., Neale, P.: Conducting in-depth interviews (2006). http://www2.pathfinder.org/

31. Guest, G., Bunce, A., Johnson, L.: How many interviews are enough? Field Methods **18**(1), 59–82 (2006)

32. Galvin, R.: How many interviews are enough? J. Build. Eng. **1**(3), 2–12 (2015)

33. Doebeli, G., Fisher, R., Gapp, R., Sanzogni, L.: Using BPM governance to align systems and practice. BPM J. **17**(2), 184–202 (2011)

34. Saldaña, J.: The Coding Manual for Qualitative Researchers. SAGE Publications, London (2016)

35. Guion, L., Diehl, D., McDonald, D.: Conducting an In-depth Interview, University of Florida (2011). http://edis.ifas.ufl.edu/

36. Flick, U.: Triangulation in qualitative research. In: Flick, U., von Kardoff, E., Steinke, I. (eds.) A Companion to Qualitative Research, pp. 178–183. SAGE Publications, London (2004)

37. DeLone, W.H., McLean, E.R.: The DeLone and McLean model of information systems success: a ten-year update. J. Manag. Inf. Syst. **19**(14), 9–30 (2003)

38. Venkatesh, V., Morris, M.G., Davis, G.B., Davis, F.D.: User acceptance of information technology: toward a unified view. MIS Q. **27**(3), 425–478 (2003)

39. Luftman, J., Kempaiah, R.: An update on business-IT alignment: "A line" has been drawn. MIS Q. Exec. **6**(3), 165–177 (2007)

Effective Leadership in BPM Implementations: A Case Study of BPM in a Developing Country, Public Sector Context

Rehan Syed[✉], Wasana Bandara, and Erica French

Queensland University of Technology, 2 George Street, Brisbane, Australia
{syedrehan.abbaszaidi, w.bandara, e.french}@qut.edu.au

Abstract. Public sector organizations across the globe have shown a keen interest in adopting BPM, yet research studies have identified many obstacles impeding successful BPM outcomes. While leadership has been emphasized as critical for BPM to succeed, it is still an under-researched area in BPM. The limited discourse on BPM leadership is a-theoretical and provides few guidelines on what effective BPM leadership is. This paper views BPM leadership from a Complexity Leadership Theory (CLT) perspective and applies the Actor Network Theory (ANT) to assist in understanding the complex social networks in leading continuous process improvement. Employing an in-depth single case, this study explores a successful BPM initiative in a Sri Lankan public-sector organization. The study results provide a rich understanding of leadership actions that support BPM success, which can be applied by practitioners to support BPM-leadership practice, and for future research investigating the role of leadership within BPM contexts.

Keywords: Business process management · Leadership
Public-sector organizations · Developing countries · Actor-network theory
Complexity leadership theory · Case study

1 Introduction

Public sector organizations across the globe have shown a keen interest in adopting BPM principles and practices [1, 2] as the key solution [3] to effectively handle citizens' demand for better government services—particularly in developing countries [4]. However, the successful implementation of BPM initiatives has been an ongoing challenge in the public sector [5] of developing countries with an approximate failure rate of 85% [6, 7]. Critical success factors for BPM in general [8], and in the public-sector developing country contexts in particular [6], have pointed to top management support and leadership as one of the key elements of success [9]. However, there is a dearth of research to explain the phenomenon of leadership in BPM initiatives [10]. While leadership is a well-defined discipline, a clear, theoretically grounded definition of leadership in BPM is to date absent [9, 10]. Leadership in BPM is defined broadly to include all those who are capable of exerting influence in the organization [11] or positively influence the project goal [12]—specifically, it's a "complex phenomenon, with many internal and external aspects that interact and influence leadership behaviors

© Springer Nature Switzerland AG 2018
M. Weske et al. (Eds.): BPM 2018, LNCS 11080, pp. 376–391, 2018.
https://doi.org/10.1007/978-3-319-98648-7_22

in BPM environments (p. 9)" [10]. Leadership styles and strategies need to fit both with the environment and the intended BPM strategies [13]. In summary, the role of leadership in complex business process change in the public sector, particularly in developing nations, is crucial for success; however, it is under-researched. We ask the question: "What are the effective leadership actions that contribute to the success of BPM initiatives in the public sector of developing countries?" The paper first considers an appropriate perspective on leadership to ensure consistency in identifying and analyzing leadership actions within a BPM context. It uses Actor Network Theory to explore leadership actions in a complex environment and proposes a conceptual model for analysis of leadership in the case organization. After an introduction to the case design and background, the case is analyzed according to the conceptual model.

2   Theoretical Underpinnings and the Conceptual Framework

2.1   Leadership Actions Within Business Process Improvement Initiatives

Public sector BPM initiatives involve interaction with a diverse network of interrelated stakeholders with a variety of conflicting interests. Studies [9, 14] describe the need for leaders to actively manage these networks to avoid delays and resistance to change. In this study, we used Actor Network Theory (ANT) to explore the leadership actions within the complex social networks in a public sector process improvement effort. ANT allows for the study of 'the focal actor' (in this context, a leader) within a social network; it also recognizes the power structures (a key property of public sector organizations [15, 16]) and the global and local networks that exist between the Government and various other organizations, such as international donor organizations, the beneficiary departments, systems implementers, and the external consultants, who play a crucial role in these initiatives, enabling an exploration of these interactions.

ANT has been used to explain the social processes associated with technology implementation, business process change, and information systems in developing countries in varied contexts [14, 17]. The theory aims to identify and explain the process by which "successful networks of aligned interests are created through the enrolment of a sufficient body of allies and the translation of their interests so that they are willing to participate in particular ways of thinking and acting that maintain the network" (p. 42) [18], and to scrutinize the reasons for the failure of networks to establish themselves [19]. With its emphasis on empirical enquiry, ANT allows an analyst to observe the key actors and the relationships between different actors through the phases of the translation process. Latour [20] explained how "actors know what they do and we have to learn from them not only what they do, but how and why they do it" (p. 19).

We focus on the ANT process of translation, which creates the ordering effects in a network. Translation is described [21] as the process of "creating temporary social order or moving between orders through changes in the alignment of interests in a network" (p. 54). The translation process includes four chronological steps: problematisation, interessement, enrolment, and mobilization (see Table 1). It explains how a successful network aligns the actors' interests in the network. Multiple actors interact in a BPM initiative with their individual goals and interests; thus, the translation process can suitably explain the actors' interaction and alignment of interest. A key proposition of this study is that leadership behaviors of the focal actors will determine the success of the translation process in a BPM initiative.

Table 1. Core theoretical concepts applied in this case study (adopted from, [19, 21])

Concept	Definition
Focal actor	Attempts to translate the interests of other actors to their own interests in a network. A focal actor is an actor who initiates the translation process
Actor-network	Heterogeneous network of aligned interests, including people, organizations and standards [19]
Translation	The process of the alignment of the interests of a diverse set of actors with the interests of the focal actor [18]
Problematisation	The first moment of translation, during which a focal actor defines identities and interests of other actors that are consistent with its own interests [18]. In the case of BPM initiatives, the problematisation will be related to control of resistance to change
Interessement	The second moment of translation, which involves negotiating with actors to accept definition of the focal actor [18]. In the case of BPM initiatives, the interessement will involve use of monetary and non-monetary motivational strategies to gain acceptance of the initiative
Enrolment	The third moment of translation, wherein other actors in the network accept (or get aligned to) interests defined for them by the focal actor [18]
Mobilization	Mobilization constitutes methods employed by the focal actor to the legitimacy of spokespersons. Achievement of complete elimination of resistance in a network

2.2 The Role of Different Leadership Actions

Considering the dynamic nature of the environment in which modern organizations exist, it is argued that the traditional models of leadership are ineffective to explain the dynamic, nonlinear, and contextual nature of leadership in organizations [22, 23]. Public-sector settings (particularly in developing-country contexts) are complex and leaders need to manage the challenges created by these complexities. Complexity Leadership Theory (CLT) predicts leadership behaviors in such contexts and recognizes the behavioral and situational influences and dynamic interactions between different elements to form self-organizing systems, change, and adaptation [24]. CLT is defined as "a framework for leadership that enables the learning, creative, and adaptive capacity of complex adaptive systems (CAS) in knowledge-producing organizations or organizational units" (p. 304) [24]. It views an organization as a collection of complex adaptive systems (CAS) and presents leadership as emergent, interactive and dynamic [24]. CLT asserts that effective leadership will create an adaptive space and defined this space as "context and conditions that enable networked interactions to foster the generation and linking of novel ideas, innovation, and learning in a system (p. 12)" [25]. CLT describes three types of leadership behaviors: 'operational leadership' behavior aims to uphold the traditional, bureaucratic hierarchies, alignment, and controls within an organization; the 'enabling leadership' behavior fosters creating and maintaining conditions that enable CAS to actively engage in creative problem solving, adaptability, and learning activities; and, the 'entrepreneurial leadership' behavior that is emergent, interactive, and dynamic, and produces adaptive outcomes (i.e. alliance of people, ideas, technologies, and cooperative efforts) in a social system [25, 26]. Entanglement is defined as a dynamic relationship between operational, enabling, and entrepreneurial leadership [27] and it explains the need for enabling leadership as an interface between operational and entrepreneurial leadership. A key strength of CLT compared to other leadership theories is that it provides a holistic view of leadership implemented at different levels of the organizational hierarchy, as well as the relational interactions that are dynamic in nature and emergent in different situations [27].

2.3   The Conceptual Model

Figure 1 depicts the research model, which incorporates all the elements of the theoretical underpinnings. It provides the basis for investigating how leadership actions (pertaining to the three different leadership types delineated in CLT) across the four phases of the ANT translation process and amongst diverse contextual influences, contribute to successful BPM initiatives in the public sector in developing countries.

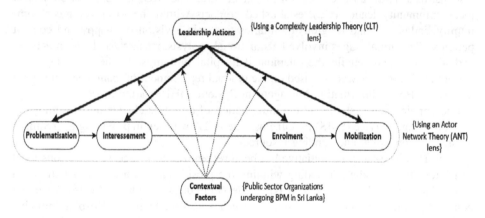

Fig. 1. Conceptual model

3   Study Design and Context

3.1   Case Study Design

In this paper a single in-depth case study in a Sri Lankan government hospital was used as a rich source of evidence to identify leaders and their actions for successful BPM implementation. Data was collected through nine interviews from key stakeholders from February 2017 to July 2017. A range of official documents were also collected and analyzed to augment and triangulate observations from the interviews. The data was analyzed iteratively across three phases using the systematic combining approach [28]. NVivo Version 11 was the primary data management tool, and a comprehensive "coding rule book" [29] was used as a guide to perform the data summarization and grouping; firstly according to the ANT translation sub-processes and secondly, according to the three leadership functions of CLT. Details pertaining to contextual factors were captured separately.

3.2 Introducing the Case Organization

The case study is of an IT-enabled BPM initiative in a regional public hospital in Sri Lanka—Dompe District Hospital. The project took place under the eHealth program initiated by Sri Lanka's apex national agency Information Communication Technology Agency (ICTA) responsible for public sector IT modernization, to improve the efficiency of healthcare services in the country, and has been recognized (nationally and internationally) as a success story that thrived through many challenges [30].

Dompe's transformation from a poor to an exemplary health care service delivery organization was initiated in 2011 by a young Medical Officer (MO) in charge of the hospital; the hospital was notorious for its lack of organization, poor quality of services, and chaos. Most of the staff were close to their retirement and had absolutely no concern for patient care and quality service delivery, leading to a highly unproductive, disorganized, and laid-back organizational culture. The MO, in consultation with a local community leader (CL), established contacts within the local volunteer community, industries, and religious dignitaries to create a vision of 'happy and content patients'. The initial stages involved a situational analysis, research on better practices, and to design and rebuild the dilapidated hospital buildings. In the next stage, the patient care processes were focused on the gradual replacement of manual systems with a Hospital Health Information Management System (HHIMS), the incorporation of an online appointment system, and the introduction of a mobile channeling system using Short Message Service (SMS) for citizens. The ICTA was key in providing the required technical knowledge, and funding resources.

The Dompe e-Hospital initiative has been recognized as a success by various local and international bodies, including winning the presidential award for productivity in 2015. The Hospital was (and still is) in the continuous monitoring & improvement phase [31] of its BPM endeavor. Today, the ICTA uses the Dompe e-Hospital initiative as an exemplar for its national eHealth program. After many challenges in introducing HHIMS to government hospitals, ICTA has now embarked on an ambitious drive to implement the system in 300 hospitals with a centralized digital healthcare management strategy to change the healthcare landscape of the country. Overall, this case demonstrates the feasibility and success of a new service model that is now replicated in the Sri Lankan public healthcare sector.

4   Study Findings

This section describes how the ANT translation process is applied to explain the actions of the focal actors (who demonstrated leadership within the eDompe case study) to design and implement the transformation activities, and how a stable network of heterogeneous actors was formed. There were three focal actors, namely the Medical Officer (MO) [provided the overall leadership, and the main interface between various external par-ties involved.], the eHealth Program Manager of ICTA (ICTA-PgM) [appointed by the ICTA to devise the strategy, design, and execution of the eHealth Program], and the local community leader (CL) [liaison between the local community] as individuals, and at times, other hospital staff also demonstrated leadership. Leadership actions from these four focal actors across the four phases of the translation process are discussed in the following sections.

4.1   Leadership Actions Observed Within the ANT-Problematisation Phase

The problematisation process was initiated by the MO through a situational analysis that identified staff concerns. A general meeting was called to discuss the poor situation of the hospital, at which the MO explained to the staff his vision, his strategy to overcome the issues, and his plan to improve the reputation of the hospital. A total of thirteen leadership actions were observed in the problematisation phase and are grouped around the three CLT leadership behaviors: Operational, Enabling, and Entrepreneurial leadership. Figure 2 depicts how each focal-actor contributed to the identified leadership actions and the coding references (i.e. the number of supporting codes). All leadership actions are explained below.

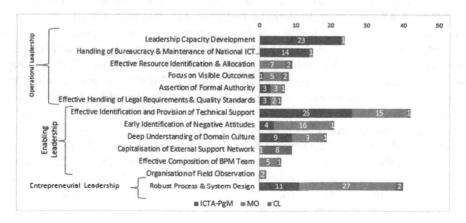

Fig. 2. Leadership actions in the problematisation phase

Six (6) operational leadership actions were observed in the problematisation phase. (i) **Leadership Capacity Development** related to the ICTA-PgM identifying leaders, giving them ownership of the BPM initiative and providing them with support, enabling them to develop leadership capacity and to drive the eHealth implementation. (ii) **Handling of Bureaucracy & Maintaining National ICT Standards** captured the provision of the required technical and logistical support while ensuring that the required rules, regulations and government policies were adhered to for procurement and development. (iii) **Effective Resource Identification & Allocation** relates to the provisioning of funding. The MO interacted with ICTA for software development support and the CL for physical infrastructure development support, funded by local industries and concerned members of the community. (iv) **Focus on Visible Outcomes** involved the MO designing and communicating targets in consultation with the core staff team, and the support of the CL and ICTA-PgM. (v) **Assertion of Formal Authority** relates to the positive and tactful use of formal authority and bureaucracy to support innovation and change. The MO used his personal connections with senior authorities in the organization on multiple occasions to establish his power and influence. His tact avoided internal confrontations. (vi) **Effective Handling of Legal Requirements and Quality Standards**. The 5S concept [35] initially introduced by the CL was re-introduced to the hospital staff by the MO with new emphasis and training.

Six (6) Enabling leadership actions were observed in the problematisation phase. (i) **Effective Identification and Provision of Technical Support** relates not only to the build/implementation of the system but also to provide the right systems training and support. The ICTA-PgM took a prominent role in assisting the MO to identify and devise the necessary technical support. (ii) **Early Identification of Negative Attitudes** were identified and managed. The MO used his experience and knowledge of working in the health sector to identify and analyze the diverse set of issues, rank them, and address each one. The CL supported the MO as a mediator. (iii) **Deep Understanding of Domain Culture**. The MO and ICTA-PgM created strategies for clear and effective communications to fit various subcultures. For example, the ICTA-PgM recognized that it was more effective when ideas were presented to the medical fraternity by somebody who belongs to the group. The CL's deep understanding of the conflicts between different groups of staff in the healthcare system helped him to act as a mediator and mentor. (iv) **Capitalization of External Support Networks**. The MO worked closely with the CL to identify and capitalize on these relationships and connections in planning, strategizing, and executing the technology-enabled transformation of the hospital. (v) **Effective Composition of the BPM Team.** The MO established two sets of cross-sectional teams: the core team, directly involved in transformation activities whose key focus was to identify staff concerns and support the MO in handling those issues; and the support team, to provide back-up. (vi) **Organization of Field observation** visits to other district hospitals that had attempted similar system implementations helped in motivating the staff to understand the potential of re-engineering and ICT to identify the weaknesses in the implementation strategies adopted by those hospitals. The MO closely coordinated with the ICTA-PgM and the software developer to customize the HHIMS to suit the diverse users of the Dompe's healthcare processes (i.e. Doctors, Nurses, Paramedics, Pharmacy staff).

One (1) Entrepreneurial leadership action was observed in the problematisation phase. (i) The MO engaged in **Robust Process & System Design** with the support of his staff and the CL. The MO initiated a number of innovative process improvements, which included changes to the physical layout of the hospital to improve the workflow, re-designing the patient appointment process, and removing manual writing tasks by the doctors.

4.2 Leadership Actions Observed Within the ANT-Interessement Phase

The focal actors used a variety of interessement strategies in this phase to motivate the actors leading to the alignment of interests. Figure 3. Illustrates how each focal-actor contributed to the identified leadership actions and the coding references (i.e. the number of supporting codes). Each leadership action is explained below.

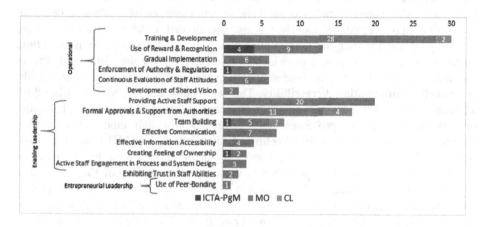

Fig. 3. Leadership actions in the interessement phase

Six (6) Operational leadership actions were observed in this phase. (i) **Training & Development**. Both inbound and outbound training and development programs were organized by the MO and CL to assist staff in developing their technical abilities for the continued use of the new processes and systems, build team attitudes, bridge gaps between different social and cultural groups, and bind them as a single unit. IT training was designed to suit the requirements of the users. (ii) **Use of Rewards & Recognition**. An inclusive environment for staff was created in order to align their interests with those of the organization by using methods like competitions, social and cultural events, awards and certificates, allowing for overtime payment, and by creating champions. The ICTA-PgM encouraged the BPM champions (from within the hospital staff) by getting them formal contracts and promoting their efforts in the state health-care sector. (iii) **Gradual Implementation**. Considering the potential resistance to change, the MO planning the BPM implementation followed an incremental approach to avoid overwhelming staff with new processes and technology. For example, not all patients were registered immediately. (iv) **Enforcement of Authority & Regulations.** This was applied to ensure adherence to rules regarding resource constraints, such as internet access. (v) **Continuous Evaluation of Staff Attitude**. This was used to identify staff issues, potential resistance to change, and to gauge their attitudes towards the BPM initiative. (vi) **Development of Shared Vision.** This was ensured by the MO providing a "very clear road map".

Eight (8) **Enabling leadership** actions were identified in the interessement phase. (i) **Providing Active Staff Support**. The MO organized highly valued around-the-clock support to staff by actively engaging with them to help overcome their technical fears and attitudes. (ii) **Formal Approvals & Support from Authorities**. A good rapport was brokered with senior authorities to secure their support and approval for speedy implementation of planned activities and to bypass rigid bureaucratic requirements. The strategy has also helped control resistance. The CL gained the much-needed financial support from local industry for the infrastructure development. (iii)**Team Building**. The MO realized the importance of a coherent team-based culture and utilized staff with positive attitudes to build and develop team culture, allowing decision-making through team consensus. The CL also contributed to the positive team culture by engaging with the staff via team building exercises. (iv) **Effective Communication.** Open communication channels were created to clearly convey the vision and aims of the initiative and to identify issues and their solutions, such as the end goals. (v) **Effective Information Accessibility.** This was enabled by the MO for all staff for all relevant information through methods like meetings and discussions to help continuous improvement of processes and consideration of staff input when creating process guidelines. (vi) **Creating Feeling of Ownership**. The MO created an environment of shared ownership and responsibility of the new system, by, for example enabling staff to describe their processes to visitors. (vii) **Active Staff Engagement in Process and System Design.** This captured how staff suggestions and involvement were encouraged in every step of developing the processes, ensuring the cooperation of the diverse stakeholders. (viii) **Exhibiting Trust in Staff Abilities.** The MO exhibited his trust in staff's abilities by delegating the responsibilities and work according to their knowledge and skills.

One (1) **Entrepreneurial leadership action** observed in the interessement phase was the **Use of Peer-Bonding.** To develop hospital-wide positive attitudes, the MO opted to use staff with positive attitudes to influence staff who were negative and resistant to change. Peer groups were formed where competent members helped those who were less competent.

4.3   Leadership Actions Observed Within the ANT-Enrolment Phase

An effective interessement strategy leads to enrolment from the actors to form an irreversible network. The focal actors used the following strategies in this phase to stabilize the network. Figure 4 illustrates the leadership actions and the coding references (i.e. the number of supporting codes). Each leadership action is explained below.

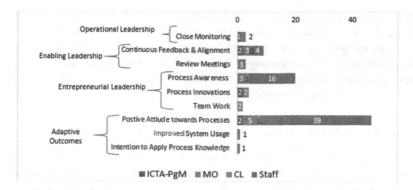

Fig. 4. Leadership actions in the enrolment phase

Only one **operational leadership** action was observed in the Enrolment phase. **Close monitoring** was used to ensure that staff were not worried about using the newly implemented processes and the HHIMS system and were supported if mistakes were made.

Enabling leadership actions observed in this phase include the following. (i) **Continuous Feedback & Alignment strategies** were used to gauge staff feelings and attitudes regarding their assigned tasks, feedback was collected by independent trainers and analyzed by the MO to identify any negative patterns and to develop corrective actions. (ii) **Review meetings** were conducted in the form of brainstorming sessions with the staff and shortages of equipment were overcome by staff suggestions.

A higher focus was on the **entrepreneurial leadership** actions in the enrolment phase. (i) **Process Awareness** comprised of interessement strategies to get staff actively engaged in BPM activities. Open channels of communication were effective in accomplishing a high degree of awareness of the processes and how changes lead towards improvement in efficiency and performance; both staff and patients were happy with the new system. (ii) **Process Innovations.** As use of the processes became popular and improved, there were further improvements in the process design and system features through staff feedback and suggestions. As well, support from the ICTA PgM was utilized, including the use of bar codes, autofill forms, international medical classifications like the ICPC2[1] standard, and the village database. (iii) **Team Work** emerged as a result of continuous training, development, and bonding programs, and led to the emergence of complex adaptive systems in which staff made independent decisions leading to further improved performance.

The effective operational, enabling, and entrepreneurial leadership actions taken by the focal actors produced the following **Adaptive outcomes.** (i) **Positive Attitude towards Processes and ICT System**. The value of re-engineering was appreciated by the staff, and it reversed their negative attitudes and created a positive impact on daily job routines. Staff started to apply process thinking in diverse areas such as the injection rooms and stores. (ii) **Improved System Usage**. The training and peer

---

[1] http://www.who.int/classifications/icd/adaptations/icpc2/en/.

bonding strategies were effective in developing staff confidence in their technical skills leading to positive impact on their use of the new system. (iii) **Intention to Apply Process Knowledge**. The trust and confidence gained by the staff through continuous capacity-building activities encouraged them to get actively involved in suggesting new and innovative ideas to improve processes. This showed the irreversible nature of staff attitudes towards the improvement process.

4.4 Leadership Actions Observed Within the ANT-Mobilization Phase

A network reaches the mobilizing phase when the actors reflect the extensive acceptance of a devised strategy or solution, exhibit the spokesperson behavior in representing the network, and the focal actor cements the alliance by using appropriate methods to ensure irreversibility [17]. The MO's and ICTA-PgM's main objectives were to improve the patient care services and implementation of HHIMS for the eHealth records program using both the human (staff) as well as the non-human (processes and HHIMS) actors.

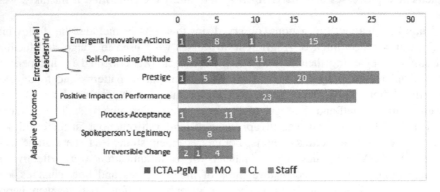

Fig. 5. Leadership actions in the mobilization phase

As shown Fig. 5, there was no operational or enabling leadership actions observed in the mobilization phase of this case study. The **Entrepreneurial leadership** observed in the **Mobilization phase** was twofold: (i) **Emergent Innovative Actions** and (ii) **Self-Organizing Attitude.** The emergence of innovative and self-organizing behavior by staff for the continuous improvement of the system was a result of the positive culture of empowerment created by the focal actors (by moving away from a 'command & control' approach) and the increasing reputation of the hospital.

Five (5) different **Adaptive Outcomes** were observed in this phase. (i) **Prestige**. The national recognition received by the hospital and the MO's recognition of staff achievement helped in creating pride. ICTA ensured that Dompe was used as the role model for other hospitals. The visible results in process efficiency and improved production in all areas of operations had a (ii) **Positive Impact on Performance** of daily operations that has changed patients' attitudes, helped staff move away from old, unproductive practices, and brought order to the earlier chaos in the hospital.

(iii) **Process Acceptance** was also a critical outcome. Because the initiative used the staff domain knowledge, it was sustainable and widely accepted with pride by the staff. The MO made use of his domain knowledge to drive the improvements and encouraged staff to do the same. (iv) **Spokesperson's Legitimacy** explains how an effective alignment of actors' interests will result in the acknowledgment of the focal actor's status as the legitimate spokespersons. The MO achieved this by his approach to re-engineering and maintaining the processes and system. (v) **Irreversible Change** pertained to ensuring that the re-engineered processes were embedded in daily operations and that the entire system was productive, efficient, stable and sustainable. It was critical to make sure that there was no going back and that the change was permanent.

4.5 Contextual Factors Influencing Leadership Action Across the Translation Process

The analysis identified a variety of contextual factors influencing leadership actions across each stage of the translation process (see Table 2 for a summary overview).

Table 2. Contextual factors and stages of translation process

	National Vision & Program of Work	Local Influences	Performance Pressure	Political Influences	External Recognition	Exploitation & Corruption	Resistance to Change	Dilapidated Conditions	Healthcare Culture
Problematisation	9	5	5	6					
Interessement					3	1	10		
Enrolement								12	10
Mobilisation					4				

In the Problematisation phase, four (4) contextual factors were identified. (i) The fact that the initiative was **related** to a **National Vision and program of work** helped to kick start the work and also influenced the leadership actions. For example, the Enterprise System Approach adopted by the focal actors was directly related to achieving the national strategy. (ii) **Local influences**. Staff approaching their retirement stage and some local politicians were quite resistant to the proposed changes, but the involvement of the local community and religious entities led to positive outcomes. The leadership actions—Robust Process & System Design and Early identification of Negative Attitudes—were influenced by the local factors, and lack of understanding of the cultural context would have resulted in major problems. (iii) The government's vision for the healthcare sector and the history (with initial systems implementations by ICTA) of failure put **Performance Pressure** on the leaders. ICTA's strategy to provide the required technical resources and funding for Dompe's BPM initiative was a direct outcome of the performance pressures in ICTA. (iv) There were also a number of negative **Political Influences.**

Three (3) contextual factors were influential in the Interessement phase: (i) **External Recognition** of the first ever successful eHealth system within the national healthcare system by national authorities and the public sector healthcare community. This had a major positive influence on staff changing their opinion about the ICT-enabled BPM endeavour at Dompe. (ii) **Exploitation and Corruption** were mentioned by the participants in relation to a parallel process invented by the local trishaw drivers, who charged patients for arranging the appointments to reduce the long hours spent waiting in queues, and provided financial incentives to hospital staff. **(iii) Resistance to change** persisted amongst some staff members and the national medical association also tried to influence the internal staff to not accept the proposed ICT-enabled BPM. The Government Medical Officers' Association (GMOA) argued that the proposed process changes would increase the doctors' workloads and adversely affect the noble image of the profession. However, the provisioning of latest laptops to the doctors by ICTA refuted GMOA's influence, doctors felt rewarded and ignored the workload concerns. Two (2) contextual factors influenced the leadership actions in the Enrolment phase. (i) **The Dilapidated Conditions**. The contrast between the ad hoc operations and the dilapidated infrastructure prior to the transformation and the new processes and IT system helped demonstrate the positive value of the improvements to the staff. (ii) A key influencing factor was the matching of the system with the **Healthcare Culture** in the country. As mentioned previously, the healthcare workers have a distinct way of thinking and perception towards the ICT systems [32, 33], therefore, the design of the system had to incorporate the cultural issues, such as providing support, incorporating staff feedback, and enabling self-organisation.

External recognition was the main contextual factor identified in the mobilisation phase. The acknowledgment of service excellence by the President of the country and increased recognition from colleagues from other hospitals boosted the staff confidence and helped stabilise the network.

5   Discussion and Conclusions

As discussed earlier, previous studies have identified many internal and external obstacles that could restrict organizations in achieving the intended potential of BPM and highlighted leadership as one of the critical success factors for BPM [6, 34]. However, the current research on BPM (in general) provides very limited understanding of the nature, definition, and properties of leadership in Business Process Management initiatives, especially in the public sector, and developing countries. This study present a series of leadership actions by different leaders (focal actors) according to the core phases of ANT translation process, and how contextual factors may influence them. A total of 36 leadership actions across the phases of Problematisation (13), Interessement (15), Enrolment (6), and Mobilization (2) were identified with (9) contextual factors. The case study findings explained the nature of leadership actions that led to the successful implementation, and sustainability of BPM in a public hospital as well as creating an organization-wide process-centric culture.

The selection and analysis of a real-life situation where BPM is used to achieve a national agenda of citizen well-being contributes to the understanding of the public-sector BPM domain by defining the nature and role of leadership in effective handling of the complexity dynamics associated with BPM initiatives. Despite the specific and narrow scope of the study, we argue that the nature of the leadership actions can provide the basis for the study of leadership-change in any socio-technical phenomenon. According to Wacker [35], correct conceptualization and definitions of constructs is the first step towards building a good theory before statistical verification. The leadership constructs and contextual influences presented in this paper can be operationalized and tested in future research. The conceptual framework and findings presented in this study can be quantitatively validated by measuring the statistical significance and correlations between leadership actions as independent variable and adaptive outcomes as the dependent variables with moderating effect of contextual factors. This paper is part of a larger program of study, and as the next step, a comparative analysis of failed and successful public sector BPM initiatives and the role of leadership is planned for the future.

This study therefore contributes to ANT research by confirming the suitability and usefulness of the Translation Process to explain leadership and network interactions in socio-technical process improvement approaches. This study was also able to show the use of CLT to explain the leadership actions in a socio-technical phenomenon and how a balanced use of operational, enabling, and entrepreneurial leadership created the adaptive space leading to irreversible change and self-organization. The results also confirmed that a high degree of entanglement [27] exists between operational and enabling leadership and shows it can have a positive impact on adaptive outcomes.

The findings of this paper also have considerable practical significance. The detailed identification of leadership actions in a highly successful transformation will assist national IT agencies in developing countries to address the critical need to develop robust leadership capabilities to overcome chronic BPM failure rates. The outcomes of the study can act as a guide for senior management of BPM initiatives in the public sector to lead successful BPM implementations and as a useful framework for designing leadership capacity building for BPM initiatives. The net results can assist developing countries to achieve the socio-technical and financial benefits expected of the ICT-enabled BPM initiatives to provide government services to the citizens in an efficient and effective manner.

References

1. Alves, C., Valença, G., Santana, A.F.: Understanding the factors that influence the adoption of BPM in two Brazilian public organizations. In: Bider, I., et al. (eds.) BPMDS/EMMSAD - 2014. LNBIP, vol. 175, pp. 272–286. Springer, Heidelberg (2014). https://doi.org/10.1007/978-3-662-43745-2_19
2. Kassahun, A.E., Molla, A.: BPR complementary competence for developing economy public sector: a construct and measurement instrument. In: PACIS 2011. Citeseer (2011)
3. Dubey, S.K., Bansal, S.: Critical success factors in implementing BPR in a government manufacturing unit - an empirical study. Int. J. Bus. Manag. 8(2), 107–124 (2013)
4. Rajapakse, J.: e-government adoptions in developing countries: a Sri Lankan case study. Int. J. Electron. Gov. Res. 9(4), 38 (2013)
5. Weerakkody, V., Baire, S., Choudrie, J.: E-government: the need for effective process management in the public sector. In: Proceedings of the 39th Annual Hawaii International Conference on System Sciences, Hawaii, p. 74b. IEEE (2006)
6. Syed, R., et al.: Getting it right! Critical success factors of BPM in the public sector: a systematic literature review. Australas. J. Inf. Syst. 22 (2018)

7. Heeks, R.: Most eGovernment-for-development projects fail: how can risks be reduced? iGovernment-Information Systems, Technology and Government: Working Papers (2003)
8. Buh, B., Kovačič, A., Indihar Štemberger, M.: Critical success factors for different stages of business process management adoption–a case study. Ekonomska istraživanja **28**(1), 243–257 (2015)
9. Syed, R., et al.: What does leadership entail in public sector BPM initiatives of developing nations: insights from an interpretative case study from Sri Lanka. In: 2017 International Conference on Research and Innovation in Information Systems (ICRIIS) (2017)
10. Syed, R., et al.: The status of research on leadership in business process management: a call for action. In: ANZAM 2016: 30th Australian and New Zealand Academy of Management Conference: Under New Management: Innovating for Sustainable and Just Futures, Brisbane (2016)
11. Holloway, M.: Process leadership. In: Jeston, J., Nelis, J. (eds.) Management by Process, pp. 111–131. Butterworth-Heinemann, Oxford (2008). Chap. 4
12. Cha, K.J., Hwang, T., Gregor, S.: An integrative model of IT-enabled organizational transformation: a multiple case study. Manag. Decis. **53**(8), 1755–1770 (2015)
13. Hyötyläinen, T., vom Brocke, J.: Steps to Improved Firm Performance with Business Process Management: Adding Business Value with Business Process Management and Its Systems. Springer, Wiesbaden (2015). https://doi.org/10.1007/978-3-658-07470-8
14. Eka, P.J., Abidin, M.Z.: Opening the black box of leadership in the successful development of local e-government initiative in a developing country. Int. J. Actor-Netw. Theory Technol. Innov. (IJANTTI) **3**(3), 1–20 (2011)
15. Fettke, P., Zwicker, J., Loos, P.: Business process maturity in public administrations. In: vom Brocke, J., Rosemann, M. (eds.) Handbook on Business Process Management 2: Strategic Alignment, Governance, People and Culture. Springer, Heidelberg (2014)
16. Tregear, R., Jenkins, T.: Government Process Management: A review of key differences between the public and private sectors and their influence on the achievement of public sector process management. BPTrends (2007)
17. Díaz Andrade, A., Urquhart, C.: The affordances of actor network theory in ICT for development research. Inf. Technol. People **23**(4), 352–374 (2010)
18. Callon, M.: Some elements of a sociology of translation: domestication of the scallops and the fishermen of St Brieuc Bay. Sociol. Rev. **32**(S1), 196–233 (1984)
19. Walsham, G., Sahay, S.: GIS for district-level administration in India: problems and opportunities. MIS Q. **23**(1), 39–65 (1999)
20. Latour, B.: On recalling ANT. Sociol. Rev. **47**(S1), 15–25 (1999)
21. Sarker, S., Sarker, S., Sidorova, A.: Understanding business process change failure: an actor-network perspective. J. Manag. Inf. Syst. **23**(1), 51–86 (2006)
22. Hunt, J.G., Dodge, G.E.: Leadership déjà vu all over again. Leadersh. Q. **11**(4), 435–458 (2001)
23. Uhl-Bien, M., Marion, R.: "Breaking the frame" even farther: complexity science and lampe theory. In: Research in Multi-Level Issues, pp. 429–442 (2006)
24. Uhl-Bien, M., Marion, R., McKelvey, B.: Complexity leadership theory: shifting leadership from the industrial age to the knowledge era. Leadersh. Q. **18**(4), 298–318 (2007)
25. Uhl-Bien, M., Arena, M.: Complexity leadership: enabling people and organizations for adaptability. Org. Dyn. **46**(1), 9–20 (2017)
26. Uhl-Bien, M., Marion, R.: Complexity Leadership. Part I, Conceptual Foundations. Leadership Horizons, xxiv, 422 p. IAP Information Age Publishing, Charlotte (2008)
27. Uhl-Bien, M., Marion, R.: Complexity leadership in bureaucratic forms of organizing: a meso model. Leadersh. Quart. **20**(4), 631–650 (2009)

28. Dubois, A., Gadde, L.-E.: "Systematic combining" – a decade later. J. Bus. Res. **67**(6), 1277 (2014)
29. DeCuir-Gunby, J.T., Marshall, P.L., McCulloch, A.W.: Developing and using a codebook for the analysis of interview data: an example from a professional development research project. Field Methods **23**(2), 136–155 (2011)
30. Bandara, W., Syed, R., Ranathunga, B., Sampath Kulathilaka, K.B.: People-centric, ICT-enabled process innovations via community, public and private sector partnership, and e-leadership: the case of the Dompe eHospital in Sri Lanka. In: vom Brocke, J., Mendling, J. (eds.) Business Process Management Cases. MP, pp. 125–148. Springer, Cham (2018). https://doi.org/10.1007/978-3-319-58307-5_8
31. Dumas, M., et al.: Fundamentals of Business Process Management. Springer, Heidelberg (2013). https://doi.org/10.1007/978-3-642-33143-5
32. Kulathilaka, S.: "eHospital-Dompe" project–the story of the transformation of a district hospital in Sri Lanka. Sri Lanka J. Bio-Med. Inf. **4**(2) (2013)
33. Maged, A., Weerakkody, V., El-Haddadeh, R.: The impact of national culture on e-government implementation: a comparison case study (2009)
34. Grover, V., et al.: The implementation of business process reengineering. J. Manag. Inf. Syst. **12**, 109–144 (1995)
35. Wacker, J.G.: A theory of formal conceptual definitions: developing theory-building measurement instruments. J. Oper. Manag. **22**(6), 629–650 (2004)

Conceptualizing a Framework to Manage the Short Head and Long Tail of Business Processes

Florian Imgrund$^{(\boxtimes)}$, Marcus Fischer, Christian Janiesch, and Axel Winkelmann

University of Würzburg, Würzburg, Germany
{florian.imgrund,marcus.fischer,christian.janiesch, axel.winkelmann}@uni-wuerzburg.de

Abstract. Naturally limited by resource constraints, most business process management (BPM) initiatives can only improve a few processes at a time. Prioritized based on their importance, feasibility, and dysfunctionality, these processes form the short head of an enterprise's process distribution. Beyond that, a long tail of processes contains a large amount of unmanaged yet imperfect processes. Due to a lack of scope and complexity, established BPM approaches can hardly support enterprises in realizing improvement potentials for these processes. In this research, we draw upon the theory of the long tail of business processes and upon insights from multiple case studies to conceptualize a management framework for hybrid BPM initiatives. The framework addresses organizational and technological requirements for the management of the BPM initiative as well as for the management of individual processes. In summary, we suggest that enterprises must manage their most important processes centrally, while improving others at their place of execution. Technology can provide the means for communication and collaboration and aligns both initiatives.

Keywords: Business process management · Long tail of business processes Management framework · Decentralization

1 Introduction

Business Process Management (BPM) has become an important management discipline to leverage organizational competitiveness by streamlining operations and centering them on customer needs [1]. With a focus on end-to-end processes, BPM breaks up functional silos and departmental boundaries [2] and enables improvements at company level. To support enterprises in identifying, analyzing, improving, and monitoring their processes, research has introduced various concepts, methods, and tools [1, 3].

In practice, we can observe two types of BPM. On the one hand, central BPM entails that a BPM department collects and consolidates information and implements change in a top-down manner [2]. Due to resource constraints and complexity issues, centralized initiatives typically focus on managing a few processes at a time. As a result, they can only improve a process organization partially [1]. On the other hand,

© Springer Nature Switzerland AG 2018
M. Weske et al. (Eds.): BPM 2018, LNCS 11080, pp. 392–408, 2018.
https://doi.org/10.1007/978-3-319-98648-7_23

decentralized BPM suggests transferring tasks from centralized BPM to multiple distributed initiatives that together form a dynamic social system [4]. Enabled by technology, stakeholders can optimize processes at their place of execution and facilitate improvements in large parts of the organization [5]. However, as decentralized BPM is costly and lacks effectiveness, BPM outcomes frequently do not yield expected benefits.

Today, digital transformation facilitates the development of new technologies that provide enterprises with various opportunities to address the growing demands for agility, flexibility, and responsiveness [6]. Although more and more enterprises start to grasp the benefits of technology for communication and collaboration, they struggle to utilize them for leveraging operational performance [7]. Against this backdrop, BPM is frequently considered as a key enabler to prepare and adapt organizational structures toward the demands of digital transformation [8]. However, this requires enterprises to not only improve a few highly important core processes but to establish an operational backbone that fosters speed, quality, and consistency [9]. Due to an inherent tradeoff between quality and quantity, neither central nor decentral BPM can adequately support enterprises in establishing the capabilities to manage digital transformation.

Hence, we draw upon the theory of the long tail of business processes [5] and extend it by proposing a hybrid BPM approach. This entails that enterprises manage some processes centrally, while improving others in multiple distributed initiatives. For the first time, we combine implications from theory with insights from multiple case studies to formulate propositions and construct a hybrid framework for the management of both the short head and long tail of business processes. The framework addresses organizational and technological requirements from the perspective of the BPM initiative as well as from the management of individual processes.

We summarize our research questions as follows:

(1) *What are the shortcomings of established BPM concepts in practice when facing a long tail distribution of business processes?*
(2) *What are propositions to address these limitations and how can they be consolidated into a hybrid BPM framework?*

We structure our paper as follows. In Sect. 2, we introduce fundamental concepts of BPM and introduce the theory of the long tail of business processes. We summarize our research method in Sect. 3 and present findings from multiple case studies in Sect. 4. In Sect. 5, we derive eleven propositions for implementing hybrid BPM and consolidate them into a framework. Section 6 concludes this research by summarizing findings, limitations, and future research potentials.

2   Theoretical Background and Foundations

2.1   Business Process Management

BPM describes a body of methods, techniques, and tools to identify, analyze, improve, monitor, and implement business processes [1]. Typically yielding cost reductions and improvements in various dimensions, such as quality, efficiency, and effectiveness,

BPM has become increasingly relevant for enterprises seeking to achieve or ensure long-term competitiveness [3]. Research has established multiple concepts to guide and support BPM in practice. For example, lifecycle models [1, 10] specify a set of sequential activities to systematically optimize enterprise structures. Maturity models [11, 12] further provide the means to assess operational quality and to identify areas for improvement. Furthermore, generic frameworks such as the Ten Principles of Good BPM [13] or the Six Core Elements of BPM framework [3] specify requirements for successful BPM initiatives. The latter distinguishes the perspectives of *strategic alignment, governance, methods, information technology, people,* and *culture* [3].

While the number of enterprises adopting BPM is growing continuously, several studies report of initiatives that do not deliver expected benefits or that result in project failure. One reason is that most BPM approaches are developed to fit a specific type of business context, focusing primarily on situations with structured processes and clear goals and requirements [14]. However, driven by the emergence of digital technology [7], today's business contexts increasingly demand BPM to address situational requirements [14]. Consequently, it is no longer sufficient to only rely on a top-down designed BPM initiative based on management commitment and the downward integration of functional managers, process owners, and operational staff [1, 5]. Based on the theory of the long tail of business processes, we argue that BPM must facilitate holistic improvements, while remaining easy to manage, maintain, and adjust.

2.2 The Theory of the Long Tail of Business Processes

By performing BPM as a central initiative, most enterprises can only improve a few processes at a time [1, 5]. More specifically, they actively manage processes that have a significant impact on business success and yield benefits that exceed corresponding costs. Many processes in any enterprise are not recognized as this valuable and therefore not considered for improvement. Based on Anderson's concept of long tail economics [15], this phenomenon has been conceptualized as the *long tail of business processes* [5]. Enterprises usually focus on a few processes that they prioritize due to different criteria [16], including their importance, dysfunctionality, or feasibility [1]. These processes form the short head of their process distribution. However, as the amount of processes with a lower value typically exceeds the number of highly valuable processes, they shape the distribution's long tail. To explain decision making during process prioritization, the theory further conceptualizes BPM as a neoclassic utility maximization problem. Enterprises naturally seek to maximize their utility U as a function of expected surplus $E[BPM(x)]$ and costs $C(x)$. Expected surpluses depend primarily on an enterprise's BPM capabilities, including its degree of process orientation and standardization as well as on the availability of technology. The complexity of a company's operations reduces potential surpluses respectively. For economic reasons, companies prioritize processes that are located in the short head of their process distribution. More specifically, they manage all processes with an advantageous proportion of expected surplus to cost of management. The *line of manageability* (x^{LoM}) is a theoretical construct that indicates where both determinants break even. Moving it to the right of the distribution is only possible if the surplus-to-cost ratio improves. This is hardly possible for centralized approaches due to complexity and resource cost. To incorporate additional processes, enterprises must widen the scope of their BPM initiative by drawing upon principles of decentralization. Thereby, independent employees collaborate within and across distributed initiatives to form a self-regulating system that optimizes low-value processes at their place of execution. Processes whose central management was unfeasible before, are now improved continuously in small iterations by way of communication and collaboration. Figure 1 summarizes the theory's main implications.

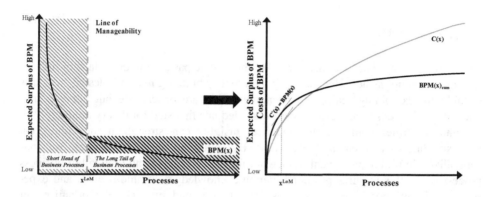

Fig. 1. The long tail of business processes

This study expands the theoretical foundation provided in [5] by combining principles of collaboration with insights from multiple case studies and formulates 11 propositions on the design of hybrid BPM initiatives in practice.

2.3 Approaches to Decentralized Organization of Process Work

As argued above, our work environment and also BPM face fundamental changes due to technological advancements and new opportunities for collaboration and communication. While companies relied on systematic approaches in the past, a more dynamic environment requires them to contextualize BPM [14]. Thus, we conducted a systematic literature search based on established guidelines and concepts [17]. Additionally, we integrated recommendations for conducting literature analysis in interdisciplinary research domains [17].

In general, companies can choose from multiple strategies to create and collect relevant process data. This data is typically contained by multiple entities, including work routines, systems, documents, and stakeholders [18]. To establish a knowledge management system, enterprises must access this data efficiently [19]. For data collection in collaborative environments, research suggests socializing BPM, which augments established approaches with features of social networks [4, 20]. Brambilla et al. [20] further propose that social features increase the scope of BPM. Hence, companies can uncover weak ties, explicate decision making, and improve processes by integrating social feedback [20].

Collaborative production models provide enterprises with the means to coordinate distributed activities within a network of independent stakeholders. In general, *peer-to-peer networks* describe a distributed infrastructure for stakeholders to share resources, services, and contents [21]. *Peer production groups* further use complementary skills, knowledge, and experience for collaborative problem solving [18]. *Crowdsourcing* refers to a sourcing model in which stakeholders obtain goods and services from a large and often rapidly evolving network. Crowdsourcing builds upon peer production to divide tasks between stakeholders and achieve a cumulative result [22].

3   Research Design

Although BPM has been extensively researched in the past, we notice a lack of holistic management approaches. Hence, we apply a theory building research design, which is suitable for examining real-world phenomena that are observable but not yet fully understood. Yin describes case studies as an adequate first step for theory building [23]. Eisenhardt further confirms that practical insights can strengthen the conceptual understanding of a research domain [24]. By following these suggestions, we conceptualize a hybrid management framework for the short head and long tail of business processes. As this requires profound literature knowledge, common sense, and experience [24], we combine implications from theory and practice. We organize our research along three phases:

- **Phase 1:** *We conduct five case studies, in which we analyze the companies' BPM characteristics and corresponding capabilities to manage the long tail of business processes.*
- **Phase 2:** *We consolidate our findings to build a hybrid management framework for the short head and long tail of processes.*
- **Phase 3:** *We derive propositions for enterprises seeking to implement a hybrid BPM approach.*

To address the limitations of case study research, we seek to ensure a high degree of rigor during data collection and analysis procedures. Thus, we conduct interviews using a multiple informants design based on semi-structured questions. All interviews are performed by two research assistants on the phone. The interviews are transcribed and we use a two-step coding and analysis procedure. To mitigate bias, one author codes the data into themes, while the second step of analysis involves summarizing the data for each theme across all cases. This enables us to build a comprehensive collection of case study protocols, which we complement with supplementary data, such as internal presentations and process documentations. Furthermore, we ensure reliability by organizing our findings based on established frameworks and by performing data analysis in multiple iterations. For external validity, we select case companies that differ regarding their size, business focus, and industry. None of the companies or interviewees have a formal relationship with the researchers. Ultimately, we address internal validity by incorporating implications from literature.

4   The Long Tail of Business Processes in Practice

4.1   Data Collection

To identify shortcomings of current BPM approaches and to answer RQ1, we conduct case studies at five different companies. This ensures an adequate variability in our results, which is necessary to derive generalizable implications.

To build theories from case study results, Eisenhardt proposes the use of theoretical sampling to *"(...) choose cases, such as extreme situations and polar types, in which the process of interest is 'transparently observable'"* [24]. *Polar types*, in this context, refer to cases that are likely to replicate or extent the emergent theory [24]. In this study, the sampling strategy is based on the theory of the long tail of business processes [8]. Hence, we derive the two polar types *short head* and *long tail*. First, the polar type short head refers to enterprises that perform BPM as a central initiative. Second, we introduce the polar type long tail to incorporate companies that improve large parts of their organization with decentralized, collaborative BPM. We conduct multiple case studies for each polar type. Due to distinct similarities, this study focuses on those companies that provide the most significant implications for each polar type. Table 1 summarizes the case companies' main characteristics. Each company is unique regarding industry, size, and business focus. In summary, they represent a variety of voices.

Table 1. Polar types for the long tail of business processes theory

Polar type	Industry	Size	Management approach	Selected for in-depth analysis	Number of interviews
Short head	Financial services (C1)	~5,000	Central	Yes	2
	Food industry (C2)	~230	Central	No	2
	Plant construction (C3)	~150	Central	No	3
Long tail	Telecommunications carrier (C4)	~8,000	Decentral	Yes	3
	Manufacturing (C5)	~19,000	Decentral	No	3

Polar Type *Short Head*. We describe the main characteristics of the short head polar type based on implications from a globally operating financial service company that is specialized on offering corporate and specialty insurances. The company employs about 5,000 people and operates as a subsidiary of a German financial conglomerate with more than 140,000 employees. About 10 years ago, it established BPM to facilitate operational responsiveness, quality, and customer satisfaction. It follows a standardized BPM lifecycle approach, which draws upon different methodologies, such as Six Sigma and Lean Management. The company has further established the *Center of BPM Excellence*, which employs a team of process analysts that support, conduct, and coordinate company-wide BPM activities. The BPM team frequently takes on the role of in-house consultants to address organizational issues that were previously discovered by employees. Before processes are added to the architecture, the BPM team assigns them to a quality assurance cycle and, thus, controls their conformance with conventions and guidelines. BPM activities are either initiated by stakeholders or by the schedule of a centrally maintained process improvement roadmap. To deploy BPM projects, the team collects information from stakeholders by questioning them about the as-is condition of their processes. Subsequently, the team constructs a to-be concept, implements necessary changes, and offers training courses to facilitate stakeholder adoption. All employees can access the company's process architecture and use available BPM tools to model, discuss, and share processes with other stakeholders. However, modifications only apply to local process versions and are not part of the official BPM initiative.

Polar Type *Long Tail*. To analyze the long tail polar type, we present the case of a telecommunication carrier that employs more than 8,000 people and is headquartered in Germany. Initially introduced in 2010, the company performs BPM as part of a larger initiative for continuous business improvement, which was originally designed based on TOGAF. The initiative draws upon the assumption that every employee holds valuable knowledge for process improvement. Hence, it is organized as a dynamic and self-regulating system that builds upon collaboration and communication. Further, the company mostly waives hierarchical structures and central regulation. Its BPM department mainly monitors adoption but does not actively conduct or intervene in corresponding activities. Instead, BPM is performed by multiple distributed initiatives that manage and improve processes at their place of execution. To coordinate and support these distributed efforts, the company relies on a BPM tool that provides several features of social networks and enables all employees to view, create, and edit process models. To ensure the availability of necessary skills and expertise, the company offers optional training courses on BPM, tool functionalities, and rules for communication and collaboration. As each employee can participate in process modeling, it yields large repositories of process models as well as their versions and variants. Today, the company's repository contains more than 18,000 models constructed by 1,800 modelers. While these models cover large parts of the company's organizational structure, they frequently exhibit quality issues, which range from redundancies and inconsistencies to varying levels of abstraction. Partially, this is due to lose guidelines and conventions, by which the company seeks to support rather than to overregulate BPM activities.

4.2   Consolidation of Case Study Findings

Following the recommendations of [24], we combine within-case analysis and cross-case search to extract patterns and to construct a a-priori model. To avoid premature or false conclusions, we draw upon the *Six Core Elements of BPM framework* [3] to organize our findings. Table 2 summarizes the main characteristics of BPM initiatives of each polar type.

The case studies confirm that companies can essentially choose from two BPM approaches: *short head* or *long tail*. Due to their different requirements, both approaches are typically viewed as mutual exclusive. While short head BPM manages a few highly important processes with high quality and effectiveness, long tail approaches incorporate large parts of a company's process organization but face quality constraints. Thus, the practical insights from the conducted case studies confirm our theoretical foundation provided in [5]. We can further utilize the identified shortcomings and limitations of current BPM approaches to significantly widen the scope of our framework and to align its implications with practical requirements. Thereby, we set the groundwork to transform the concept into a complete BPM framework in future studies.

Table 2. Characteristics of polar types according to the six core elements of BPM

	Polar type *short head*	Polar type *long tail*
Strategic alignment	BPM focuses on a few important processes, which are managed by a central department	The scope of BPM depends on stakeholder participation instead of organizational resources
	Supported by top-management commitment, the scope and strategy of BPM is continuously aligned to the overall goals of the company	Top management approves BPM but is not actively involved. Initially, BPM is aligned to the company's goals
	A process architecture structures the company's process landscape.	A process architecture supports and connects BPM activities. Adaptations require fundamental changes in the company's business environment
	The BPM department may adapt its scope and coverage	
Governance	Organized in a top-down manner, a central governance regulates BPM activities by defining conventions, guidelines, and best practices	Bottom-up governance provides specifications to guide distributed activities. These include conventions, guidelines, best practices, and indicators
	A central BPM department controls the conformance with BPM specifications	Conformance checking is self-organized and builds upon trust, credibility, and reputation
Methods	BPM follows customized methodologies derived from common frameworks and tools	BPM is based on a lightweight implementation that promotes collaboration and information sharing
	Methods for process design, modeling, implementation, and execution rely on proven practices and are conducted by experts	The tasks of process discovery, analysis, and redesign are transferred to stakeholders that share work by means of collaborative production models
	Knowledge management is organized in a top-down manner. This entails that a central BPM team requests information from employees	Knowledge management is supported by a platform that facilitates knowledge sharing. Employees can reuse the information for individual purposes
	The BPM department centrally controls, maintains, and manages its process repositories. Employees can access the repository (read-rights)	Processes and supplementary data are stored in a central repository, which is organized by high-level guidelines. All employees can access and edit data

(continued)

Table 2. (*continued*)

	Polar type *short head*	Polar type *long tail*
Information technology	Tool support is provided by on-premise software that integrates with the company's IT infrastructure Access to the BPM software is restricted to experts and process analysts	Tool support is based on easy-to-use software tools. The infrastructure builds upon service orientation and thus facilitates responsiveness All employees are granted full access to BPM tools, which support their BPM activities
People	Companies offer differentiated training courses to selected employees that are actively involved in the BPM initiative The BPM department consists of high-skilled process analysts with a formal education for conducting BPM activities	Training offers are available to all interested employees. These offers consist of self-organizing courses with employees as instructors All employees form a network of different backgrounds and skills and collaborate across departments to solve organizational issues
Culture	The BPM team commits to the initiative and shares knowledge to stakeholders if necessary Process awareness is mostly limited to participants of the central BPM initiative	Companies facilitate an open-minded culture that enables teamwork and collaboration Process awareness is fostered actively and perceived as an organizational asset

In addition to analyzing their current BPM configurations, we further prompt the companies for information about perceived limitations. In *short head* initiatives, most companies focus their BPM resources on a limited number of structured processes that involve controlled interactions among participants and generate an immediate impact on an enterprise's value creation. These processes are typically summarized by a process architecture, which connects different parts of their organization and breaks down processes to sub-structures and tasks. At all companies, establishing a process architecture required an initial discovery and prioritization of processes. In such static environments, central BPM initiatives can produce high-quality results, which ensure a continuous improvement of core processes. However, they typically lack the mechanisms to identify, model, and analyze neglected or unknown processes, which can only be improved by uncovering tacit knowledge from process stakeholders. Due to resource constraints, most companies encounter bottlenecks when processing pending projects, even if they involve processes that are part of their process architecture. To increase BPM capacities, some companies rely on hiring external consultants to address the most urgent and important issues. The companies further report that relevant process knowledge remains primarily within the BPM department, as other employees are hardly involved or lack the opportunities for direct participation.

By deploying *long tail* initiatives, companies can overcome most resource constraints and improve large parts of their organizational structure. However, our case companies report that BPM outcomes frequently lack consistency and quality.

Consequently, they cannot realize the full improvement potential in many processes, which can negatively influence productivity, competitiveness, and customer satisfaction in the long term. As collaborative BPM entails transferring tasks and responsibilities to a self-regulating social system, our case companies face several challenges navigating and controlling their initiatives. One company initially waived central regulations to facilitate adoption and participation. While this yielded a large number of process models, results lacked quality, clarity, and comparability. This caused a decreasing participation rate, as stakeholders could not realize the expected benefits. Hence, the company introduced specifications to not only restrict but also guide the distributed BPM activities.

4.3 Discussion of Shortcomings of Analyzed BPM Approaches

Results from our case studies indicate that both types of BPM suffer from various shortcomings. Subsequently, we draw upon the dimensions of the Six Core Elements of BPM framework to discuss these shortcomings and their consequences.

Regarding the dimension of *strategic alignment*, several limitations emerge from the inherent tradeoff between quality and quantity. As central initiatives are limited by resource constraints, widening their scope requires additional staff to be hired, trained or sourced as consultants. However, this does not only increase the complexity of planning, control, and coordination, but can also result in a disadvantageous ratio of costs to additional surpluses generated by managing previously neglected, low-value processes. By implementing a decentral BPM initiative, companies can overcome these resource limitations. However, increasing the quality of BPM outcomes would require more rigid specifications and investments into IT and education.

With respect to *governance*, central BPM requires rigid specifications and top-down control mechanisms. Although this enables companies to improve their most important processes effectively, such approaches lack the flexibility to identify and optimize unstructured or unknown processes. As decentral BPM initiatives transfer most responsibilities to a self-regulating social system, they foster participation and collaboration. However, BPM outcomes frequently lack quality, consistency, and comparability, which can reduce the benefits of BPM in the long term.

In the area of *methods*, central initiatives demand controlled techniques and indicators to identify organizational issues and to prioritize and conduct improvement projects that conform to a company's strategies. However, as collaborative BPM draws upon self-regulation and flexibility, its activities and priorities cannot be linked to a company's objectives directly. Regarding knowledge management, controlled read-only repositories provide a single source of truth and support a basic knowledge diffusion in some areas of an organization. However, they do not operationalize knowledge from distributed stakeholders. Open repositories provide users with the means to view process models and to modify them to their individual needs or perceptions. Although this allows companies to uncover tacit process knowledge, resulting repositories can lack clarity and contain inconsistencies and redundancies.

Regarding the area of *information technology*, central initiatives require tools with advanced functionalities to support process analysts. However, these tools can exceed the capabilities of stakeholders in distributed initiatives. Decentral BPM is conducted at the place of process execution and requires lightweight tools for collaboration, coordination, and communication. Mostly waiving advanced functionalities, these tools are designed to ensure usability and participation based on features of social networks, including commenting, sharing, and tagging. While collaborative tools foster participation and allow companies to manage more processes actively, they only support a limited range of application scenarios.

In the area of *people*, central initiatives involve a few skilled and specialized process analysts, which conduct BPM activities in a top-down manner. Hence, companies can focus training offers on advanced topics. While this facilitates efficiency and enhances the capabilities of the central BPM team, it cannot address the requirements for managing long tail processes. Education offers in decentral approaches focus primarily on conciliating basic BPM concepts for the discovery, analysis, and redesign of processes. However, these formats do not provide the knowhow and expertise to accomplish major improvements in the companies' core processes.

Regarding the dimension of *culture*, companies must establish process orientation, regardless of their selected BPM approach. However, in companies that operate a central initiative, process awareness is limited to a restricted number of process analysts. Incorporating process knowledge from distributed stakeholders requires higher degrees of process orientation, simultaneously placing new demands on participation and collaboration. However, the central BPM team typically lacks the resources to equally address all uncovered improvement potentials, which can yield stakeholder resistance and dissatisfaction. Decentral BPM can address most of these challenges by drawing upon process orientation and awareness as essential components of its overall strategy.

In summary, our case study findings indicate that neither central nor decentral BPM approaches can fully realize the improvement potential in all processes of an organization. Besides presenting our findings, we ask the interviewees to elaborate on opportunities to address these shortcomings and, thereby, derive a set of design principles for holistic BPM. By drawing upon these insights, we formulate a total of eleven propositions to guide the development and implementation of such an approach in research and practice.

5 Conceptualizing a Framework to Manage the Short Head and Long Tail of Business Processes

5.1 Propositions for Managing Hybrid BPM Initiatives

Findings from our case studies indicate that companies struggle to utilize BPM as a means to cope with the requirements of digital transformation and globalization. Thus, we use these implications to formulate propositions (P) on how to improve both the short head and long tail of business processes.

P1: Enterprises must establish a hierarchical system of objectives that guides actions of the central BPM team. Participation of stakeholders in distributed initiatives is ensured by mechanisms that intrinsically stimulate participation.

Hybrid BPM entails that a central initiative manages an enterprise's most important processes, while others are improved by means of collaboration and communication. Aligning the central initiative to an enterprise's overall strategy requires clear goals and complementary extrinsic incentives. By contrast, collaborative BPM builds upon distributed efforts by independent stakeholders within a self-organizing social system. While key processes that are important for the company's success remain centrally managed, all interviewees agree that the central initiative should hand over responsibilities for the management of long tail processes to the respective departments. Yet, to avoid inefficiencies due to missing responsibilities, cross-functional processes require a web-based collaborative platform that foster communication. As indicated by the interviewees of C4 and C5, stakeholders in distributed BPM initiatives should be motivated primarily based on intrinsic incentives. By simply defining objectives and target agreements, enterprises can hamper their willingness to participate and hamper the initiative's success. Instead, they rely on incentives that influence stakeholders' actions indirectly. Hence, enterprises must establish a hierarchical system of objectives during BPM's initiation. While initially determining objectives at company and department level, enterprises must analyze stakeholder opinions and preferences to reduce the gap between individual and company goals subsequently.

P2: Enterprises must determine the scope of their BPM initiatives.

Both initiatives manage different process types. While central BPM focuses on processes that are prioritized by their importance, feasibility, and dysfunctionality, distributed initiatives improve low-value processes that are typically neglected due to a disadvantageous ratio of costs to benefits. For hybrid BPM, the interviewees from C1, C2, and C3 confirmed that enterprises must establish an indicator system to assess the improvement potential of each process and dynamically determine the scope of their central initiative. Thus, they can either use performance-based or non-performance-based methods [16]. As they provide high accuracy at low costs, enterprises should preferably implement performance-based approaches. However, they rely on non-performance-based methods if they lack necessary IT support.

P3: Enterprises must implement a BPM platform that connects both initiatives.

At the core of hybrid BPM, enterprises must implement a BPM platform that connects both initiatives. As central BPM can accomplish multiple purposes, it requires tools with advanced functionalities. To avoid exceeding other stakeholders' capabilities, these functionalities should not be part of the BPM platform, but be implemented in a separate software environment. All interviewees agree that enterprises must ensure a fully integrated IT infrastructure. According to C4 and C5, companies must equip their BPM platform with functionalities for coordination, collaboration, and communication [4, 20]. Among others, these should include a process repository, social networking features, a modeling environment, and a process architecture. There is general agreement that employees must be able to access the platform with user accounts.

P4: Enterprises must establish different specifications for both initiatives. Central BPM demands an active governance control. The BPM platform performs basic conformance checking for outcomes of the decentral initiative.

To accomplish different BPM objectives, such as automation or standardization, the central BPM team must comply with conventions, guidelines, and technical requirements. However, outcomes of decentral BPM are used primarily to facilitate communication among stakeholders. Based on the implications of case studies C4 and C5, strict specifications are less relevant and can hamper BPM adoption and participation. As a result, hybrid BPM demands two sets of specification, with conventions and guidelines on departmental level derived as a subset of those for the central BPM team. This ensures comparability and interoperability of corresponding outcomes. To ensure high-quality results in the central initiative, enterprises rely on top-down mechanisms for quality assurance. By contrast, the BPM platform must provide basic conformance checking features to support activities in distributed initiatives. This should be complemented by a bottom-up governance with stakeholders constantly checking and improving BPM outcomes. Ultimately, regular meetings and events provide a platform for communication and provide the means to align the specifications of both initiatives.

P5: Enterprises must select a single notation and adapt its scope to the needs of both initiatives. Collaborative production models facilitate collaboration and coordination.

To ensure processes' interoperability, enterprises must select a single modeling language for both initiatives. However, adaptations are necessary to fit their specific needs. Furthermore, the modeling environment must support notational specifications and enable stakeholders to model in a drag-and-drop manner. Enterprises must further establish a universally accessible repository, which organizes processes based on the process architecture. Collaborative production models, such as crowdsourcing, can coordinate activities in and between both initiatives. The central BPM team must analyze the outcomes of decentral BPM for challenges and issues. Stakeholders must also be able to flag problems that cannot be solved collaboratively but require the central BPM team.

P6: Enterprises must provide training offers for basic and advanced BPM topics.

To ensure the availability of BPM capabilities, enterprises must deploy a comprehensive education program. While the central BPM team typically consists of high-skilled process analysts, training offers can be limited to advanced topics. However, offers for the decentral initiative must focus on teaching basic competencies, such as process modeling. In addition to classes and learning circles, the BPM platform should provide complementary video tutorials as well as discussion forums. The central BPM team must screen these forums to identify current issues and to adapt training offers.

P7: Enterprises must communicate benefits and ensure transparency.

Similar to collaborative BPM, hybrid approaches rely strongly on stakeholder participation and process awareness. Thus, enterprises must design and implement a communication plan that emphasizes the benefits of BPM and ensures transparency. They must further analyze the platform's user data to identify key users that foster stakeholders' participation and can serve as contact persons for the central BPM team.

Besides the requirements induced by running a holistic approach to BPM, enterprises must also guide BPM activities at operational level.

P8: Enterprises must establish a process architecture to navigate hybrid BPM.

For coordination purposes, enterprises must implement an integrated process architecture. Upper levels of the architecture comprise their most important core and support processes, which are managed by the central BPM team. All processes that are discovered by distributed initiatives are assigned to the architecture's bottom level. Enterprises can thereby establish a hierarchy of abstraction, in which each process of the bottom level connects to a process on higher levels. Thus, the BPM team can operationalize process data from its place of execution. If processes cannot be connected to higher levels, enterprises can identify yet disregarded areas of their organization.

P9: The BPM platform must gather data on roles and responsibilities in a process.

While the central BPM team manages a controlled set of processes located in the upper levels of the process architecture, distributed stakeholders independently identify, analyze, and improve processes at their place of execution. As distributed BPM efforts build upon collaboration, each process must provide information on involved stakeholders and responsibilities. Hence, the BPM platform must gather this data as BPM activities are initiated and notify all involved stakeholders. This increases process quality and consistency and avoids redundancies. The BPM platform must remind users periodically to enter missing data and to complete the processes' information.

P10: The platform must provide approval mechanisms for process adaptations.

When discovering processes, stakeholders are asked to assign them to an enterprise's process architecture. To change these processes, the BPM platform must provide mechanisms for version and variant management. In fact, modifications should only apply to stored process variants. According to the interviewee of C4, the BPM platform must further notify all involved stakeholders about changes and demand their active control and approval. If the approved changes are only relevant for the architecture's bottom level, the new version can be stored subsequently. However, if they affect higher architectural levels, the BPM platform should transfer the approved version to the central BPM team, which then decides whether to implement the recommended changes.

P11: The BPM platform must periodically ask stakeholders about their perceptions of the importance, dysfunctionality, and feasibility of processes they are involved in.

To manage the scope of the central BPM initiative, enterprises rely on consistent information regarding the importance, feasibility, and dysfunctionality of processes within their organizational structure. Regardless of the approach for data collection and analysis, both interviewees from C4 and C5 point out that the BPM platform should provide mechanisms to periodically ask stakeholders about their perceptions about a process. Aggregating this data, enterprises can assign processes to either initiative.

5.2   Consolidated Framework

In the following, we consolidate our findings and propositions into a framework for hybrid BPM (cf. Fig. 2). Our propositions provide initial insights in how companies can leverage BPM to build an integrated strategy for the holistic management of short head and long tail processes. To this end, propositions 1 and 2 summarize organizational requirements to continuously align and adjust the strategy and scope of distributed initiatives to the company's overall objectives. Propositions 3 and 4 specify technological requirements to the BPM platform and the need for a fully integrated system and software environment. Proposition 5 emphasizes the differences during the modeling of short head and long tail processes from both an organizational and technological point of view. Proposition 6 and 7 highlight the need to facilitate and incentivize stakeholder collaboration and participation. At operational level, propositions 8 to 10 define mechanisms to organize processes and to extract implicit process knowledge. Finally, proposition 11 aims at facilitating and incentivizing stakeholder collaboration.

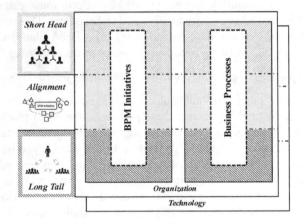

Fig. 2. Framework to manage of the short head and long tail of business processes

Consequently, we distinguish the layers of *organization* and *technology*. While the *organizational* layer provides propositions for introducing hybrid BPM to an enterprise's current organizational structure (cf. P4, P7, P8), the layer of *technology* specifies complementary tools and systems (cf. P3, P9–11). In general, the framework distinguishes between requirements for the implementation of *BPM initiatives* (P1–P7) and the management of operational *business processes* (P8–P11). The framework further provides recommendations to *align* initiatives at company level (*short head*) and department level (*long tail*) (P4, P5) as well as the connection thereof (P3). In line with our case study findings, literature suggests that the contemporary body of knowledge sufficiently addresses the management of clear-cut and structured processes [14] in the short head. To this end, the framework emphasizes the importance of expanding the scope of central BPM with distributed and self-organized initiatives.

The framework stems from observations made at case companies that differ significantly regarding their industry, size, business focus, and BPM configuration. This conforms to the general notion of combining observations from highly different cases to produce generalizable results. Consequently, our framework fits multiple application scenarios and is adaptable for different contexts and situational requirements.

6   Conclusion

Although BPM can drive organizational agility, established concepts face a tradeoff between quality and quantity. We drew upon the theory of the long tail of business processes to propose a hybrid BPM framework, which advises enterprises to continue managing their most important processes centrally, while improving others at their place of execution. We presented results from five case studies to identify good practices as well as shortcomings of BPM concepts. We augmented our findings with implications from literature, formulated eleven propositions and constructed a framework for the management of the short head and long tail of business processes.

This research is not without limitations. First, case studies are suitable for explorative research, but provide only limited capabilities for testing and validation. Additional empirical studies are necessary to provide more definite evidence. Second, our framework builds upon findings from the literature, which we derived by analyzing relevant contributions from related research domains. Although the applied literature search procedure drew upon established guidelines and concepts, we cannot eliminate the possibility that we missed contributions that might have offered additional insights for our study. Third, our framework is of preliminary nature and requires more analyses as well as empirical studies to evaluate the feasibility and impact of the suggested approach in practice. Future research must refine the propositions and the framework's dimensions. We further identified very individual and varying requirements for the built-time and run-time management of processes in an initial evaluation with our case companies. That is, it remains a case-by-case challenge whether to fully implement our propositions. As the framework involves sociological adjustments to the corporate culture, companies should gauge a step by step integration considering contextual factors.

References

1. Dumas, M., Rosa, M.L., Mendling, J., Reijers, H.A.: Fundamentals of Business Process Management. Springer, Berlin (2018). https://doi.org/10.1007/978-3-662-56509-4
2. Becker, J., Kugeler, M., Rosemann, M.: Process Management: A Guide for the Design of Business Processes. Springer, Berlin (2013)
3. Rosemann, M., vom Brocke, J.: The six core elements of business process management. In: vom Brocke, J., Rosemann, M. (eds.) Handbook on Business Process Management 1. IHIS, pp. 105–122. Springer, Heidelberg (2015). https://doi.org/10.1007/978-3-642-45100-3_5
4. Niehaves, B., Plattfaut, R.: Collaborative business process management: status quo and quo vadis. Bus. Process Manag. J. **17**, 384–402 (2011)

5. Imgrund, F., Fischer, M., Janiesch, C., Winkelmann, A.: Managing the long tail of business processes. In: 25th European Conference on Information Systems, Guimarães, pp. 595–610 (2017)
6. Legner, C., et al.: Digitalization: opportunity and challenge for the business and information systems engineering community. BISE **59**, 301–308 (2017)
7. Hess, T., Matt, C., Benlian, A., Wiesböck, F.: Options for formulating a digital transformation strategy. MIS Q. Exec. **15**, 123–139 (2016)
8. Vom Brocke, J., Mendling, J.: Business Process Management Cases: Digital Innovation and Business Transformation in Practice. Springer, Heidelberg (2017). https://doi.org/10.1007/978-3-319-58307-5
9. Sebastian, I.M., Ross, J.W., Beath, C., Mocker, M., Moloney, K.G., Fonstad, N.O.: How big old companies navigate digital transformation. MIS Q. Exec. **16**, 197–213 (2017)
10. Weske, M.: Business Process Management: Concepts, Languages, Architectures. Springer, Berlin (2012). https://doi.org/10.1007/978-3-642-28616-2
11. Rosemann, M., de Bruin, T.: Towards a business process management maturity model. In: 13th European Conference on Information Systems (ECIS), Regensburg (2005)
12. Fisher, D.M.: The business process maturity model: a practical approach for identifying opportunities for optimization. Bus. Process Trends **9**, 11–15 (2004)
13. Kohlborn, T., et al.: Ten principles of good business process management. Bus. Process Manag. J. **20**, 530–548 (2014)
14. vom Brocke, J., Zelt, S., Schmiedel, T.: On the role of context in business process management. Int. J. Inf. Manag. **36**, 486–495 (2016)
15. Anderson, C.: The Long Tail: Why the Future of Business is Selling Less of More. Hyperion (2006)
16. Lehnert, M., Roeglinger, M., Seyfried, J.: Prioritization of interconnected processes. Bus. Inf. Syst. Eng. **60**, 95–114 (2017)
17. Webster, J., Watson, R.T.: Analyzing the past to prepare for the future: writing a literature review. MIS Q. **26**, xiii–xxiii (2002)
18. Benkler, Y.: The Wealth of Networks: How Social Production Transforms Markets and Freedom. Yale University Press, London (2006)
19. Alavi, M., Leidner, D.E.: Review: knowledge management and knowledge management systems: conceptual foundations and research issues. MIS Q. **25**, 107–136 (2001)
20. Brambilla, M., Fraternali, P., Ruiz, C.K.V.: Combining social web and BPM for improving enterprise performances: the BPM4People approach to social BPM. In: 21st International Conference on World Wide Web, Lyon, pp. 223–226. ACM (2012)
21. Schollmeier, R.: A definition of peer-to-peer networking for the classification of peer-to-peer architectures and applications. In: First International Conference on Peer-to-Peer Computing, pp. 101–102 (2001)
22. Howe, J.: The Rise of Crowdsourcing. Wired Mag. **14**, 1–4 (2006)
23. Yin, R.K.: Case Study Research: Design and Methods. Sage, Newbury Park (1989)
24. Eisenhardt, K.M.: Building theories from case study research. Acad. Manag. Rev. **14**, 532–550 (1989)

Using Business Process Compliance Approaches for Compliance Management with Regard to Digitization: Evidence from a Systematic Literature Review

Stefan Sackmann[(✉)], Stephan Kuehnel[(✉)], and Tobias Seyffarth[(✉)]

Martin Luther University Halle-Wittenberg, 06108 Halle (Saale), Germany
{stefan.sackmann, stephan.kuehnel,
tobias.seyffarth}@wiwi.uni-halle.de

Abstract. Business Process Compliance (BPC) means ensuring that business processes are in accordance with relevant compliance requirements. Thus, BPC is an essential part of both business process management (BPM) and compliance management (CM). Digitization has also been referred to as a "digital revolution" that describes a technological change that has extended to many organizational areas and tasks, including compliance. Current efforts to digitize, e.g., by realizing cyber-physical systems, rely on the automation and interoperability of systems. In order for CM not to hamper these efforts, it becomes an increasingly relevant issue to digitize compliance as well.

The managerial perspective of compliance comprises several phases, which together represent a CM life-cycle. Efforts to digitize compliance require bundling interoperable BPC technologies, methods, and tools supporting this life-cycle in a consolidated manner. Several approaches addressing the field of BPC have already been developed and explored. Based on a systematic literature review, we examined these approaches in terms of their suitability for supporting the CM life-cycle phases in support of the digitization of compliance.

The results of our literature review show which CM life-cycle phases are supported by BPC approaches and which phases are the focus of research. Moreover, the results show that a purely sequential clustering, as specified in a CM life-cycle, is not always suitable for the bundling of BPC approaches in support of the digitization of compliance. Consequently, we propose a novel, task-oriented clustering of BPC approaches that is particularly oriented toward interoperability.

Keywords: Compliance management life-cycle · Business process compliance
Digitization

The original version of this chapter was revised: The total values of table 2 and the text on page 417 were initially published with errors. The correction to this chapter is available at https://doi.org/10.1007/978-3-319-98648-7_30

Electronic supplementary material The online version of this chapter (https://doi.org/10.1007/978-3-319-98648-7_24) contains supplementary material, which is available to authorized users.

© Springer Nature Switzerland AG 2018
M. Weske et al. (Eds.): BPM 2018, LNCS 11080, pp. 409–425, 2018.
https://doi.org/10.1007/978-3-319-98648-7_24

1 Introduction

Business Process Compliance (BPC) denotes that business processes adhere to applicable compliance requirements [1]. Therefore, BPC can be seen as a link between business process management (BPM) and compliance management (CM) [2]. When aiming at the potentials of digitization in the field of BPM, it also becomes necessary to digitize compliance. Otherwise, compliance can be expected to hamper digitization efforts and lead to competitive disadvantages [3–5]. In general, digitization can be distinguished as two essential perspectives. First, it describes the conversion of analogue into discrete values with the objective of electronic processing and storage [6]. Second, it is also known as a "digital revolution" that describes a technological change that has extended to many organizational areas and tasks including compliance [6].

Nowadays, CM in general and BPC in particular are challenged by current digitization strategies. On the one hand, business processes become both increasingly supported and controlled by process-aware information systems and managed by process-execution environments [7]. On the other hand, the pressure on business processes to become more agile or flexible increases since digitization allows a distributed process execution and control via the Internet, even across different organizations [8–10]. This might still be a high level of maturity that organizations have not yet realized; however, in the context of digitization, they are moving exactly in this direction [11].

One of the main challenges of digitizing compliance is the use of suitable BPC technologies, methods, and tools to support CM along each of its individual life-cycle phases. A substantial body of research has investigated the state of the art of BPC either from an author-centric perspective (e.g., [12–14]) or from a concept-centric perspective of the specific tasks of BPC (e.g., [15]). Although existing research proposes many technologies, methods, and tools for BPC, it has not yet been clarified whether a comprehensive allocation of BPC approaches to the CM life-cycle phases is possible and whether a "bundling" of interoperable BPC approaches in support of the digitization of compliance can be achieved. In order to investigate this in greater detail, we raise the following research questions:

- RQ1: *Which CM life-cycle phases are supported by which BPC approaches?*
- RQ2: *Which CM life-cycle phases are the focus of BPC research, and which are not?*
- RQ3: *To what extent does a phase-oriented representation of a CM life-cycle allow for the bundling of BPC approaches in support of the digitization of compliance?*

To achieve our research goals, we conducted a literature review according to vom Brocke et al. [16] that resulted in a twofold contribution. First, a concept matrix is presented from which we can derive which CM life-cycle phases are supported by BPC approaches and which phases are the focus of research. Since our literature review shows that BPC approaches are mainly task-oriented and therefore address only specific CM life-cycle phases, we introduce, secondly, a novel task-oriented clustering of the identified BPC approaches in support of the digitization of compliance.

The remainder of the paper is structured as follows. Section 2 describes the procedure for conducting our literature review and presents a concept matrix for the CM life-cycle phases as a first result. Based on this, Sect. 3 provides both a discussion of the suitability of a CM life-cycle for the bundling of BPC approaches in support of the digitization of compliance, and a novel, task-oriented clustering of BPC technologies, methods, and tools. The paper concludes with a summary and a disclosure of the study's limitations in Sect. 4.

2 Literature Review

In order to ensure a rigorous documentation of the literature search, we conducted our study based on the method described by vom Brocke et al. [16]. Consequently, we will discuss our procedure and key findings in relation to each of the following five phases: first, a definition of the review scope (Sect. 2.1); second, a conceptualization of the topic (Sect. 2.2); third, the literature search (Sect. 2.3); fourth, a literature analysis and synthesis (Sect. 2.4); and fifth, research desiderata (Sect. 3).

2.1 Definition of Review Scope

As advised in the first phase of the framework for reviewing literature [16], we defined our review scope based on the taxonomy proposed by Cooper [17], which is shown in Fig. 1. Research outcomes of the examined contributions are the focus of our literature review (1). Since we want to identify BPC approaches that support CM life-cycle phases, the goal of our literature review is the integration of central issues (2). The organization of the review results is conceptual, as the literature synthesis aims to bundle BPC approaches in support of the digitization of compliance (3). Furthermore, we aim to present our results neutrally (4). The audience being addressed represents specialized researchers in the field of BPC and practitioners who are confronted with compliance, its management, and digitization (5). Finally, the coverage of the literature review is exhaustive and selective since we use generic search terms in a variety of common databases (6).

Characteristics	Categories			
(1) focus	research outcomes	research methods	theories	applications
(2) goal	integration	criticism		central issues
(3) organization	historical	conceptual		methodological
(4) perspective	neutral representation		espousal of position	
(5) audience	specialised scholars	general scholars	practitioners	general public
(6) coverage	exhaustive	exhaustive & selective	representative	central/pivotal
			selected category	unselected category

Fig. 1. Defined scope of the literature review

2.2 Conceptualization of Topic

As previously mentioned, BPC is typically described as ensuring that a company's business processes are in accordance with relevant compliance requirements that can stem from internal and external compliance sources including legal regulations, standards, guidelines, policies, contracts, best practices, and so on [18–20]. In the research domain, BPC is also frequently described as mainly addressing the verification of business process models against compliance rules. However, this corresponds to a rather narrow view of BPC since there are also approaches whose central research subject is not focused solely on verification, for example, approaches that concentrate on how to react flexibly to changes in legislation [21] or the impact of legal interpretation on BPC [22].

As part of this literature review, we will take a broader view of BPC and consider approaches that go beyond mere verification. Iterative models or so-called life-cycles have been used and continue to be used in management to structure more comprehensive managerial responsibilities (such as compliance) through phases as well as to control and improve processes, projects, products, tasks, or systems [23]. On the one hand, there are life-cycles and iterative processes that include CM as an important element but were designed for other domains, such as management accounting [24], safety management [25], or risk management [26]. On the other hand, there are research efforts that deal explicitly with the design of CM life-cycles and its phases [2, 27–29]. Although these life-cycles originate from different domains and vary in terms of target and granularity, their core structure is similar to Deming's Plan-Do-Check-Act (PDCA) cycle [30], which is a management method in the form of a multistage iterative process that can be specified for various application areas [31]. Nowadays, various norms and standards propose the application of PDCA-like cycles for different domains, such as ISO/IEC 27001 for information-security management or ISO 31000 for risk management.

The focus of our literature review is to investigate the suitability of BPC approaches for the implementation of a PDCA-like cycle for a comprehensive CM. In order to achieve this, the topic must be conceptualized according to the second phase of the framework for reviewing literature as described by vom Brocke et al. [16]. Since CM life-cycles have already been developed and our goal is not to develop a new one, we analyzed existing literature as a starting point for the conceptualization. For this purpose, vom Brocke et al. [16] recommend following the procedure suggested by Baker [32], who states that "to begin with one should consult those sources most likely to contain a summary or overview of the key issues relevant to a subject." Therefore, we analyzed publications dealing explicitly with life-cycles and phases for CM [2, 27–29]. The publication by Ramezani et al. [27], which builds on the study of [29], has been found to contain the most detailed description of the CM life-cycle and its phases. Therefore, our paper builds on the key concepts of [27], who introduced a multistage sequential CM life-cycle comprising five phases: elicitation, formalization, implementation, checking/analysis, and optimization.

1. *Elicitation.* The elicitation phase is dedicated to the identification of relevant compliance requirements, taking into account organization-specific characteristics such as company size, industry affiliation, or product portfolio.
2. *Formalization.* Compliance requirements identified during the elicitation phase are often presented in abstract and informal text form. In order to enable a tool-supported verification of BPC, compliance requirements have to be converted to a formal, structured, and machine-readable form, i.e., so-called compliance rules.

3. *Implementation.* The requirements formalized in the previous phase must be implemented in a way that allows for a subsequent detection of compliance-rule violations.

4. *Checking/Analysis.* Once the implementation has been completed, the compliance rules can be checked against business processes or respectively the compliance of a business process can be verified.

5. *Optimization.* The results of compliance checking reveal whether a business process model or instance has violated a compliance rule. The detection of a violation triggers an improvement phase, in which it must be determined whether the violation results from a faulty compliance rule (false positive) or is actually based on non-compliance. In the case of actual non-compliance, the business process must be optimized in order to ensure future compliance.

The subsequent literature analysis serves to identify BPC approaches that support the CM life-cycle phases as well as approaches that could be bundled in support of the digitization of compliance. In order to achieve this, the CM life-cycle phases are used as a starting point for constructing a concept matrix according to Webster and Watson [33], based on which the later synthesis of the search results is carried out.

2.3   Literature Search

The following chapter is dedicated to the third phase of the framework for literature reviews according to [16]. In order to identify a broad range of BPC approaches in the search for literature, we used the following generic search term: <<compliance AND "business process">>. The literature search, whose procedure is summarized in Fig. 2, was conducted in pertinent databases including the ACM Digital Library, the AIS Electronic Library, EBSCOHost Academic Search Premier, EBSCOHost Business Source Premier, EBSCOHost Information Science and Technology, and SpringerLink. Initially, the search was performed without restriction in all available search fields, i.e., in title, abstract, keywords, and full text.

After testing the search term, we adapted our search strategy within several databases. Due to the high number of hits (313) in the database of EBSCOHost Business Source Premier, we limited the search in this database to the title. We also received a very high number of hits (9,313) at SpringerLink, which is why we initially limited the search in this database to the title as well. Since a high number of hits (379) still resulted, we adapted the search term at SpringerLink to <<business process compliance>>.

After removing duplicates, we retrieved a total of 232 unique hits; these were evaluated in two review phases. First, each paper was evaluated by two independent researchers according to its title, keywords, and abstract. In the case of matching evaluations, the paper was marked either as relevant or irrelevant. In the case of different evaluation results, the relevance of each paper was discussed. Finally, we excluded 130 publications due to irrelevance in the first review phase. Next, all remaining papers were read in full, and those that did not meet the review scope were omitted. In total, we received 74 relevant publications for further analysis following the second review phase.

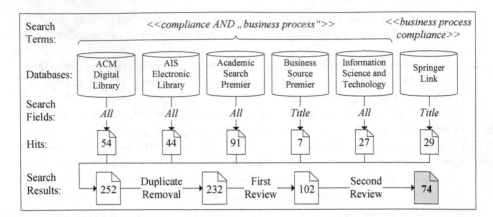

Fig. 2. Literature search process and search results

2.4   Literature Analysis and Synthesis

According to the fourth phase of the framework for reviewing literature [16], the literature found has to be analyzed and synthesized, whereby RQ1 and RQ2 are addressed. To implement this phase, we relied on the concept-centered approach of Webster and Watson [33], who propose compiling a concept matrix for the purpose of synthesis. Table 1 shows the concept matrix of our study and presents the identified literature and its classification according to the concepts introduced in Sect. 2.2. Each relevant paper is evaluated against these concepts in a separate line of Table 1, whereby an "x" identifies the assignment to a concept.

Table 1. Concept matrix

	E	F	I	C	O		E	F	I	C	O		E	F	I	C	O
[34]	x			x		[35]		x		x		[36]				x	x
[37]				x		[38]		x		x		[39]				x	
[40]				x		[41]	x	x		x		[42]				x	
[43]		x		x		[44]		x		x	x	[45]				x	
[46]				x		[47]		x		x	x	[48]				x	
[49]				x		[50]		x		x	x	[51]	x				
[52]	x			x	x	[18]	x	x		x		[53]	x		x		
[54]	x			x	x	[55]		x		x		[56]	x			x	x
[57]				x	x	[58]		x		x	x	[59]				x	
[60]	x			x		[61]		x		x		[62]	x			x	
[63]	x			x		[64]	x	x		x	x	[65]	x			x	
[66]	x			x		[67]	x	x		x		[68]	x				
[69]	x			x		[19]		x		x	x	[70]	x				

(continued)

Table 1. (*continued*)

	E	F	I	C	O		E	F	I	C	O		E	F	I	C	O
[71]				x		[72]	x		x	x		[73]				x	
[74]	x	x			x	[75]			x			[76]		x			
[77]		x		x		[78]	x		x	x		[79]		x			
[80]				x		[81]	x		x			[82]		x			
[83]		x		x		[84]	x		x	x		[85]				x	
[86]				x		[87]	x					[88]		x		x	
[89]				x		[90]	x	x		x		[1]		x			
[91]				x		[92]	x					[93]		x		x	
[94]				x		[95]	x		x			[21]	x				x
[96]		x		x		[97]	x		x			[98]		x		x	
[99]		x				[100]	x					[101]				x	x
[22]	x	x				[102]	x										

E: Elicitation | F: Formalization | I: Implementation | C: Checking/
Analysis | O: Optimization

The evaluation of the concept matrix shows that the majority of the publications are concerned with compliance checking (78.4%), the formalization of compliance requirements (67.6%), or both (44.6%). In contrast, relatively few publications (24.3%) deal with improving and adapting business processes in the case of actual non-compliance, i.e., with the optimization phase. An even smaller proportion of publications (13.5%) deal with the elicitation of compliance requirements. Furthermore, none of the approaches explicitly addresses the implementation phase, which is focused on the integration of already formalized compliance rules and takes place before the actual compliance checking.

3   Discussion

This chapter addresses RQ3 and therefore analyzes to what extent the sequential phase-oriented CM life-cycle allows for the representation of "bundled" BPC approaches in support of the digitization of compliance. The classification conducted in the previous section led to several difficulties, as not all research work could be clearly assigned to the CM life-cycle phases.

- First, the formalization phase does not distinguish between purely formal approaches for the representation of compliance requirements in logical languages, such as approaches based on Linear Temporal Logic and semiformal approaches, such as those based on graphical modeling languages.
- Second, it has been shown that a multitude of research approaches has been assigned to the checking and analyzing phase due to its generic characterization. In particular, this phase does not explicitly distinguish between approaches with a more preventive character, i.e., so-called forward compliance checking approaches,

and approaches with a more detective or reactive character, i.e., so-called backward compliance checking approaches. Ramezani et al. [27] briefly mention a similar distinction in their article, but refrain from including it in the life-cycle, describing it in greater detail or discussing its effects on the other phases. However, the distinction between forward and backward compliance checking seems to be an important issue, as it determines the necessity of a purely retrospective optimization phase. Since backward compliance checking approaches check business process instances after they have been executed [103], the business process must be adapted retrospectively as part of the subsequent optimization phase in order to prevent future non-compliance. By contrast, forward compliance checking approaches attempt to prevent the occurrence of non-compliance [103] so that either no business process adaptation is necessary or it is done at the design or runtime phase of the business process and not retrospectively. This raises the question of whether a purely sequential arrangement of the CM life-cycle phases is always appropriate.

- Third, our classification has shown that there are approaches that focus on a different meaning of the term optimization and are concerned with improving the cost-efficiency and effectiveness of ensuring compliance [46, 71, 78, 94]. Due to the different understanding of terms, these approaches could not be assigned to the optimization phase of the CM life-cycle.
- Fourth, the implementation phase is originally described as follows [27]: "To ensure that an IS complies with a given requirement, its formalization (the formalized compliance rules) has to be implemented in a way that allows detecting if an execution violates some compliance rule". Since the implementation is located after formalization and even before checking and analyzing, this description can best be interpreted as a guideline. The phase neither corresponds to implementation in terms of software development nor instantiation in terms of design science research. All in all, there is no technical equivalent to this phase, and therefore, there are no suitable BPC approaches.

In summary, the CM life-cycle provides a useful foundation for our work but reveals two major shortcomings in the context of this study. On the one hand, the abstraction levels of the CM life-cycle phases differ, such as the rather broad focus of the checking and analysis phase and the rather narrow focus of the optimization phase. On the other hand, ensuring BPC is not always tied to a purely sequential flow through all CM life-cycle phases. The necessity of the implementation phase can be questioned in general, while the optimization phase can be done in different ways. For example, if forward compliance checking approaches are used, the optimization phase cannot be executed retrospectively or can even be skipped completely. Due to the shortcomings, we revised our initial concepts and synthesized the relevant papers again.

The formalization phase was renamed as Specification and divided into the sub-concepts *Formal Specification* and *Semiformal Specification*. The checking and analysis phase was renamed as Compliance Checking and divided into the sub-concepts *Forward Compliance Checking* and *Backward Compliance Checking*. Two new concepts were derived from the optimization phase. On the one hand, this involves the concept of *Metrics*, which is focused on quantitative measurement and optimization approaches for the cost-efficiency and effectiveness of compliance. On the other hand, this involves the concept of Business Process Adaptation, which essentially aligns with Ramezani et al.'s [27] idea of optimization and is dedicated to the revision and remodeling of business processes. Since Business Process Adaptation already includes approaches dedicated to automation, which is particularly important with regard to digitization, we divided it into the sub-concepts *Manual Business Process Adaptation* and *Automatic Business Process Adaptation*. The elicitation phase was renamed as Internalization and equally divided into the sub-concepts *Manual Internalization* and *Automatic Internalization*.

Table 2 shows an excerpt of the revised concept matrix, whereby an "x" identifies the assignment of a research work to a concept. The last line in Table 2 contains the total number of all assigned approaches per column. The complete revised concept matrix can be accessed in the supplement to this paper.[1]

Table 2. Excerpt of the revised concept matrix

Reference	Manual internalization	Automatic internalization	Formal specification	Semiformal specification	Forward compliance checking	Backward compliance checking	Manual business process adaptation	Automatic business process adaptation	Metrics
[18]		x	x		x				
[34]	x				x				
[37]						x			
[46]						x			x
[50]		x			x		x		
[54]		x			x		x		
[56]		x			x			x	
[58]		x			x			x	
[70]				x					
[78]		x			x		x		x
[86]					x				
...	...	...	...	...	...	...	...	...	...
Sum	10	1	44	6	51	7	10	7	7

The evaluation of the revised concept matrix shows that the most intensive research is conducted in the area of forward compliance checking (68.9%) and the formal specification of compliance requirements (59.5%). While at least 13.5% of the approaches still deal with the manual identification of compliance requirements and manual business process adaptation, the concepts semiformal specification, backward compliance checking, automatic business process adaptation, and metrics are under-represented, with proportions between 8.1% and 9.5%. It is evident that there is a need for further research in these areas. A negative outlier and thus clearly underrepresented is the automatic identification of compliance requirements, to which only one approach could be assigned within the scope of our analysis.

In the course of the discussion, it has already been pointed out that a purely sequential order of CM life-cycle phases is not suitable for the bundling of BPC approaches in support of the digitization of compliance. In particular, attention must be

---

[1] The supplement is accessible via goo.gl/HPpoCK.

paid to the interoperability of BPC approaches since the digitization of compliance can only be achieved by bundling different technologies, tools, and methods so that all tasks of CM can be addressed comprehensively. However, such a bundling of approaches requires a task-oriented view of CM, which is not provided by the phase-oriented representation in the form of a sequential CM life-cycle.

In Fig. 3, we propose a novel, task-oriented clustering of BPC approaches. It is particularly oriented toward the representation of connections between concepts and thus toward the bundling of BPC approaches in support of the digitization of compliance. In Fig. 3, concepts are marked as ellipses, and sub-concepts are represented by dashed circles. For approaches addressing several concepts, a connecting line between the dashed circles is drawn. The numbers in brackets represent the number of approaches assigned to one or more concepts. For example, there are 51 approaches addressing forward compliance checking, 7 approaches addressing metrics, and 3 approaches addressing both.

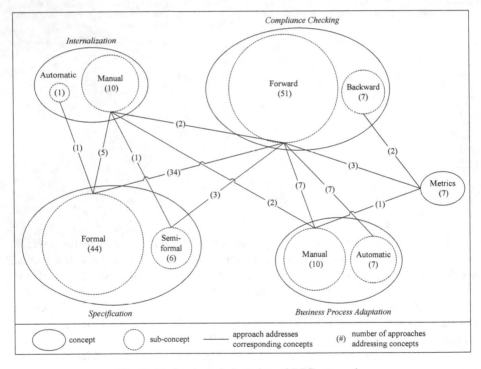

Fig. 3. Task-oriented clustering of BPC approaches

Our task-oriented clustering shows that there are already several approaches addressing various aspects of CM from a task-oriented perspective. The focus is particularly on the formal specification of compliance requirements that are subsequently used for forward compliance checking (46.0%). In addition, 19.0% of the approaches deal with forward compliance checking and business process adaptation, of which

9.5% deal with automatic and the same number with manual adaptations. All other connections can be interpreted in the same way but are addressed only by specific approaches in a few cases (between 1.4% and 6.8%). Altogether, it turns out that there is already a multitude of BPC approaches addressing several task-oriented concepts of CM. Nevertheless, future research will need to develop further approaches that address several task-oriented concepts of CM simultaneously or establish interoperability between existing BPC approaches in support of the digitization of compliance.

4 Conclusion

Digitization denotes a technological change that has extended to many organizational areas and tasks including compliance. Current efforts to digitize rely on the automation and interoperability of systems. In order for CM not to hamper these efforts, it becomes an increasingly relevant issue to digitize compliance as well. One of the greatest challenges in connection with the digitization of compliance is the bundling of suitable technologies, methods, and tools in support of CM along each of its individual life-cycle phases. Based on a structured literature review, we examined BPC approaches in relation to their suitability to support the CM life-cycle proposed by Ramezani et al. [27]. A well-known shortcoming of any literature review is that whether all relevant work has actually been found cannot be verified. However, by rigorously documenting the literature search following the methods described by vom Brocke et al. [16] and Webster and Watson [33], comprehensibility is provided in a scientific manner.

Our results show that BPC approaches supporting the life-cycle phases check and analyze as well as formalize are dominant in research whereas no approach addresses the implementation phase. Moreover, our results demonstrate that several CM life-cycle phases should be specified more precisely. For example, the phase checking and analysis should explicitly distinguish between forward and backward compliance checking and the phase optimization between metrics and business process adaptation.

Special attention should be paid to interoperability since the digitization of compliance can only be achieved by bundling different BPC approaches so that all tasks of a CM life-cycle can be addressed comprehensively. It turns out that such a bundling of approaches requires a task-oriented view of CM, which is not always provided by the phase-oriented representation in the form of a sequential CM life-cycle. Consequently, we proposed a novel, task-oriented clustering of BPC approaches that is particularly oriented toward interoperability.

Potential for further research is seen in two major fields. On the one hand, there is a need for the development of further approaches that address several task-oriented concepts of CM simultaneously or establish further interoperability between existing BPC approaches in support of the digitization of compliance. On the other hand, the proposed task-oriented clustering of BPC approaches requires a procedure model that addresses business- and technology-related management decisions and proposes necessary and/or possible steps fostering the digitization of BPC. The development of such a procedural model is the consequent next step of our research.

References

1. Schaefer, T., Fettke, P., Loos, P.: Control patterns. Bridging the gap between is controls and BPM. In: ECIS (2013)
2. El Kharbili, M., Stein, S., Markovic, I., Pulvermüller, E.: Towards a framework for semantic business process compliance management. In: Proceedings of GRCIS 2008 (2008)
3. Bamberger, K.A.: Technologies of compliance. Risk and regulation in a digital age. Texas Law Rev. **88**, 669 (2010)
4. Legner, C., et al.: Digitalization. Opportunity and challenge for the business and information systems engineering community. BISE **59**, 301–308 (2017)
5. Imgrund, F., Fischer, M., Janiesch, C., Winkelmann, A.: Approaching digitalization with business process management. In: Proceedings of the MKWI, pp. 1725–1736 (2018)
6. BarNir, A., Gallaugher, J.M., Auger, P.: Business process digitization, strategy, and the impact of firm age and size. The case of the magazine publishing industry. J. Bus. Ventur. **18**, 789–814 (2003)
7. Reichert, M., Weber, B.: Enabling Flexibility in Process-Aware Information Systems. Springer, Heidelberg (2012). https://doi.org/10.1007/978-3-642-30409-5
8. Weber, I., Xu, X., Riveret, R., Governatori, G., Ponomarev, A., Mendling, J.: Untrusted business process monitoring and execution using blockchain. In: La Rosa, M., Loos, P., Pastor, O. (eds.) BPM 2016. LNCS, vol. 9850, pp. 329–347. Springer, Cham (2016). https://doi.org/10.1007/978-3-319-45348-4_19
9. Fridgen, G., Radszuwill, S., Urbach, N., Utz, L.: Cross-organizational workflow management using blockchain technology. Towards applicability, auditability, and automation. In: 51st Annual Hawaii International Conference on System Sciences (HICSS-51) (2018)
10. Fdhila, W., Rinderle-Ma, S., Knuplesch, D., Reichert, M.: Change and compliance in collaborative processes. In: 12th IEEE International Conference on Services Computing (SCC 2015), pp. 162–169 (2015)
11. Zaplata, S., Haman, K., Kottke, K., Lamersdorf, W.: Flexible execution of distributed business processes based on process instance migration. J. Syst. Integr. **1**(3), 3–16 (2010)
12. Hashmi, M., Governatori, G., Lam, H.-P., Wynn, M.T.: Are we done with business process compliance. State-of-the-art and challenges ahead. Knowl. Inf. Syst. 1–55 (2018). https://doi.org/10.1007/s10115-017-1142-1
13. Fellmann, M., Zasada, A.: State-of-the-art of business process compliance approaches. In: ECIS (2014)
14. El Kharbili, M.: Business process regulatory compliance management solution frameworks. A comparative evaluation. In: Proceedings of the Eighth Asia-Pacific Conference on Conceptual Modelling, vol. 130, pp. 23–32 (2012)
15. Becker, J., Delfmann, P., Eggert, M., Schwittay, S.: Generalizability and applicability of model-based business process compliance-checking approaches. A state-of-the-art analysis and research roadmap. Bus. Res. **5**, 221–247 (2012)
16. vom Brocke, J., Simons, A., Niehaves, B., Riemer, K., Plattfaut, R., Cleven, A.: Reconstructing the giant. On the importance of rigour in documenting the literature search process. In: ECIS, pp. 2206–2217 (2009)
17. Cooper, H.M.: Organizing knowledge syntheses: a taxonomy of literature reviews. Knowl. Soc. **1**, 104–126 (1988)

18. Knuplesch, D., Ly, L.T., Rinderle-Ma, S., Pfeifer, H., Dadam, P.: On enabling data-aware compliance checking of business process models. In: Parsons, J., Saeki, M., Shoval, P., Woo, C., Wand, Y. (eds.) ER 2010. LNCS, vol. 6412, pp. 332–346. Springer, Heidelberg (2010). https://doi.org/10.1007/978-3-642-16373-9_24
19. Liu, Y., Müller, S., Xu, K.: A static compliance-checking framework for business process models. IBM Syst. J. **46**, 335–361 (2007)
20. Governatori, G., Sadiq, S.: The Journey to Business Process Compliance (2009)
21. Gong, Y., Janssen, M.: From policy implementation to business process management. Principles for creating flexibility and agility. Gov. Inf. Q. **29**, S61–S71 (2012)
22. Ghanavati, S., Hulstijn, J.: Impact of legal interpretation on business process compliance. In: Proceedings of the First International Workshop on TEchnical and LEgal Aspects of Data pRIvacy, pp. 26–31 (2015)
23. King, W.R., Cleland, D.I.: Life-cycle management. In: Cleland, D.I., King, W.R. (eds.) Project Management Handbook, pp. 191–205. Wiley, New York (1988)
24. Hermanson, R.H., Edwards, J.D., Maher, M.: Accounting Principles. A Business Perspective, Financial Accounting (2015, 2011). (Chaps. 1–8)
25. Roughton, J., Crutchfield, N.: Safety Culture. An Innovative Leadership Approach. Elsevier Science, New York (2013)
26. Heldman, K.: Project Manager's Spotlight on Risk Management. Wiley, New York (2010)
27. Ramezani, E., Fahland, D., van der Werf, J.M., Mattheis, P.: Separating compliance management and business process management. In: Daniel, F., Barkaoui, K., Dustdar, S. (eds.) BPM 2011. LNBIP, vol. 100, pp. 459–464. Springer, Heidelberg (2012). https://doi.org/10.1007/978-3-642-28115-0_43
28. Elgammal, A., Turetken, O.: Lifecycle Business Process Compliance Management: A Semantically-Enabled Framework (2015)
29. Giblin, C., Liu, A.Y., Müller, S., Pfitzmann, B., Zhou, X.: Regulations expressed as logical models (REALM). In: Proceedings of the 2005 Conference on Legal Knowledge and Information Systems (JURIX 2005), pp. 37–48 (2005)
30. Deming, W.E.: Out of the Crisis. Massachusetts Institute of Technology Center for Advanced Engineering Study, Cambridge (1986)
31. Moen, R., Norman, C.: Evolution of the PDCA Cycle (2006)
32. Baker, M.J.: Writing a literature review. Mark. Rev. **1**, 219–247 (2000)
33. Webster, J., Watson, R.T.: Analyzing the past to prepare for the future. Writing a literature review. MIS Q. **26**, xiii–xxiii (2002)
34. Accorsi, R., Lowis, L., Sato, Y.: Automated certification for compliant cloud-based business processes. Bus. Inf. Syst. Eng. **3**, 145 (2011)
35. Governatori, G., Rotolo, A.: How do agents comply with norms? In: Web Intelligence and Intelligent Agent Technologies, pp. 488–491 (2009)
36. Schultz, M.: Enriching process models for business process compliance checking in ERP environments. In: vom Brocke, J., Hekkala, R., Ram, S., Rossi, M. (eds.) DESRIST 2013. LNCS, vol. 7939, pp. 120–135. Springer, Heidelberg (2013). https://doi.org/10.1007/978-3-642-38827-9_9
37. Accorsi, R., Stocker, T., Müller, G.: On the exploitation of process mining for security audits. The process discovery case. In: Proceedings of the 27th Annual ACM Symposium on Applied Computing, pp. 1462–1468 (2012)
38. He, Q.: Detecting runtime business process compliance with artifact lifecycles. In: Ghose, A., et al. (eds.) ICSOC 2012. LNCS, vol. 7759, pp. 426–432. Springer, Heidelberg (2013). https://doi.org/10.1007/978-3-642-37804-1_45

39. Seeliger, A., Nolle, T., Schmidt, B., Mühlhäuser, M.: Process compliance checking using taint flow analysis. In: Proceedings of the International Conference on Information Systems (2016)

40. Accorsi, R., Wonnemann, C.: Strong non-leak guarantees for workflow models. In: Proceedings of the 2011 ACM Symposium on Applied Computing, pp. 308–314 (2011)

41. Höhenberger, S., Riehle, D., Delfmann, P.: From legislation to potential compliance violations in business processes. Simplicity matters. In: ECIS (2016)

42. Seeliger, A., Nolle, T., Mühlhäuser, M.: Detecting concept drift in processes using graph metrics on process graphs. In: Proceedings of the 9th Conference on Subject-Oriented Business Process Management, p. 6:1 (2017)

43. Kumar, A., Liu, R.: A rule-based framework using role patterns for business process compliance. In: Bassiliades, N., Governatori, G., Paschke, A. (eds.) RuleML 2008. LNCS, vol. 5321, pp. 58–72. Springer, Heidelberg (2008). https://doi.org/10.1007/978-3-540-88808-6_9

44. Höhn, S.: Model-based reasoning on the achievement of business goals. In: Proceedings of the 2009 ACM Symposium on Applied Computing, pp. 1589–1593 (2009)

45. Song, L., Wang, J., Wen, L., Kong, H.: Efficient semantics-based compliance checking using LTL formulae and unfolding. J. Appl. Math. 2013(1), 1–24 (2013)

46. Shamsaei, A., Pourshahid, A., Amyot, D.: Business process compliance tracking using key performance indicators. In: zur Muehlen, M., Su, J. (eds.) BPM 2010. LNBIP, vol. 66, pp. 73–84. Springer, Heidelberg (2011). https://doi.org/10.1007/978-3-642-20511-8_7

47. Hummer, W., Gaubatz, P., Strembeck, M., Zdun, U., Dustdar, S.: An integrated approach for identity and access management in a SOA context. In: Proceedings of the 16th ACM Symposium on Access Control Models and Technologies, pp. 21–30 (2011)

48. Thi, T.T.P., Helfert, M., Hossain, F., Le Dinh, T.: Discovering business rules from business process models. In: Proceedings of the 12th International Conference on Computer Systems and Technologies, pp. 259–265 (2011)

49. Awad, A., Barnawi, A., Elgammal, A., Elshawi, R., Almalaise, A., Sakr, S.: Runtime detection of business process compliance violations. An approach based on anti patterns. In: Proceedings of the 30th Annual ACM Symposium on Applied Computing, pp. 1203–1210 (2015)

50. Namiri, K., Stojanovic, N.: Pattern-based design and validation of business process compliance. In: Meersman, R., Tari, Z. (eds.) OTM 2007. LNCS, vol. 4803, pp. 59–76. Springer, Heidelberg (2007). https://doi.org/10.1007/978-3-540-76848-7_6

51. Turetken, O., Elgammal, A., van den Heuvel, W.-J., Papazoglou, M.: Enforcing compliance on business processes through the use of patterns. In: ECIS (2011)

52. Awad, A., Goré, R., Hou, Z., Thomson, J., Weidlich, M.: An iterative approach to synthesize business process templates from compliance rules. Inf. Syst. 37, 714–736 (2012)

53. Turetken, O., Elgammal, A., van den Heuvel, W.-J., Papazoglou, M.P.: Capturing compliance requirements. A pattern-based approach. IEEE Softw. 29, 28–36 (2012)

54. Awad, A., Weidlich, M., Weske, M.: Visually specifying compliance rules and explaining their violations for business processes. J. Vis. Lang. Comput. 22, 30–55 (2011)

55. Knuplesch, D., Reichert, M., Kumar, A.: A framework for visually monitoring business process compliance. Inf. Syst. 64, 381–409 (2017)

56. Schumm, D., Turetken, O., Kokash, N., Elgammal, A., Leymann, F., van den Heuvel, W.-J.: Business process compliance through reusable units of compliant processes. In: Daniel, F., Facca, F.M. (eds.) ICWE 2010. LNCS, vol. 6385, pp. 325–337. Springer, Heidelberg (2010). https://doi.org/10.1007/978-3-642-16985-4_29

57. Barnawi, A., Awad, A., Elgammal, A., El Shawi, R., Almalaise, A., Sakr, S.: Runtime self-monitoring approach of business process compliance in cloud environments. Cluster Comput. **18**, 1503–1526 (2015)
58. Ghose, A., Koliadis, G.: Auditing business process compliance. In: Krämer, B.J., Lin, K.-J., Narasimhan, P. (eds.) ICSOC 2007. LNCS, vol. 4749, pp. 169–180. Springer, Heidelberg (2007). https://doi.org/10.1007/978-3-540-74974-5_14
59. Wang, Y., Kelly, T., Lafortune, S.: Discrete control for safe execution of IT automation workflows. In: Proceedings of the 2nd ACM SIGOPS/EuroSys European Conference on Computer Systems 2007, vol. 41, pp. 305–314 (2007)
60. Basin, D., Klaedtke, F., Müller, S., Zalinescu, E.: Monitoring metric first-order temporal properties. J. ACM (JACM) **62**, 15:1 (2015)
61. Knuplesch, D., Reichert, M., Kumar, A.: Visually monitoring multiple perspectives of business process compliance. In: Motahari-Nezhad, H.R., Recker, J., Weidlich, M. (eds.) BPM 2015. LNCS, vol. 9253, pp. 263–279. Springer, Cham (2015). https://doi.org/10.1007/978-3-319-23063-4_19
62. Witt, S., Feja, S., Speck, A., Prietz, C.: Integrated privacy modeling and validation for business process models. In: Proceedings of the 2012 Joint EDBT/ICDT Workshops, pp. 196–205 (2012)
63. Becker, J., Bergener, P., Delfmann, P., Eggert, M., Weiss, B.: Supporting business process compliance in financial institutions. A model-driven approach. In: WI Proceedings (2011)
64. Letia, I.A., Goron, A.: Model checking as support for inspecting compliance to rules in flexible processes. J. Vis. Lang. Comput. **28**, 100–121 (2015)
65. Hashmi, M., Governatori, G., Wynn, M.T.: Normative requirements for business process compliance. In: Davis, J.G., Demirkan, H., Motahari-Nezhad, H.R. (eds.) ASSRI 2013. LNBIP, vol. 177, pp. 100–116. Springer, Cham (2014). https://doi.org/10.1007/978-3-319-07950-9_8
66. Becker, J., Bergener, P., Delfmann, P., Weiss, B.: Modeling and checking business process compliance rules in the financial sector. In: ICIS Proceedings (2011)
67. Letia, I.A., Groza, A.: Compliance checking of integrated business processes. Data Knowl. Eng. **87**, 1–18 (2013)
68. Zoet, M., Versendaal, J.: Business rules management solutions problem space: situational factors. In: PACIS 2013 Proceedings (2013)
69. Becker, J., Delfmann, P., Dietrich, H.-A., Steinhorst, M., Eggert, M.: Business process compliance checking. Applying and evaluating a generic pattern matching approach for conceptual models in the financial sector. Inf. Syst. Front. **18**, 359–405 (2016)
70. Zur Muehlen, M., Indulska, M., Kamp, G.: Business process and business rule modeling languages for compliance management. A representational analysis. In: Proceeding ER 2007 Tutorials, Posters, Panels and Industrial Contributions, pp. 127–132 (2007)
71. Bhamidipaty, A., Narendra, N.C., Nagar, S., Varshneya, V.K., Vasa, M., Deshwal, C.: Indra. An integrated quantitative system for compliance management for IT service delivery. IBM J. Res. Dev. **53**, 6:1–6:12 (2009)
72. Lohmann, N.: Compliance by design for artifact-centric business processes. Inf. Syst. **38**, 606–618 (2013)
73. Becker, J., Bergener, P., Breuker, D., Delfmann, P., Eggert, M.: An efficient business process compliance checking approach. In: Nüttgens, M., Gadatsch, A., Kautz, K., Schirmer, I., Blinn, N. (eds.) TDIT 2011. IAICT, vol. 366, pp. 282–287. Springer, Heidelberg (2011). https://doi.org/10.1007/978-3-642-24148-2_19

74. Boella, G., Janssen, M., Hulstijn, J., Humphreys, L., van der Torre, L.: Managing legal interpretation in regulatory compliance. In: Proceedings of the Fourteenth International Conference on Artificial Intelligence and Law, pp. 23–32 (2013)
75. Loreti, D., Chesani, F., Ciampolini, A., Mello, P.: Distributed compliance monitoring of business processes over MapReduce architectures. In: Proceedings of the 8th ACM/SPEC on International Conference on Performance Engineering Companion, pp. 79–84 (2017)
76. Riesner, M., Pernul, G.: Supporting compliance through enhancing internal control systems by conceptual business process security modeling. In: ACIS 2010 Proceedings (2010)
77. Bräuer, S., Delfmann, P., Dietrich, H.-A., Steinhorst, M.: Using a generic model query approach to allow for process model compliance checking. An algorithmic perspective. Wirtschaftsinformatik Proceedings 2013 (2013)
78. Lu, R., Sadiq, S., Governatori, G.: Measurement of compliance distance in business processes. Inf. Syst. Manag. **25**, 344–355 (2008)
79. Knuplesch, D., Reichert, M., Ly, L.T., Kumar, A., Rinderle-Ma, S.: Visual modeling of business process compliance rules with the support of multiple perspectives. In: Ng, W., Storey, V.C., Trujillo, J.C. (eds.) ER 2013. LNCS, vol. 8217, pp. 106–120. Springer, Heidelberg (2013). https://doi.org/10.1007/978-3-642-41924-9_10
80. Brucker, A.D., Hang, I., Lückemeyer, G., Ruparel, R.: SecureBPMN. Modeling and enforcing access control requirements in business processes. In: ACM Symposium on Access Control Models and Technologies (SACMAT), pp. 123–126 (2012)
81. de Masellis, R., Maggi, F.M., Montali, M.: Monitoring data-aware business constraints with finite state automata. In: Proceedings of the 2014 International Conference on Software and System Process, pp. 134–143 (2014)
82. Rosemann, M., Zur Muehlen, M.: Integrating risks in business process models. In: ACIS 2005 Proceedings (2005)
83. Corea, C., Delfmann, P.: Detecting compliance with business rules in ontology-based process modeling. Wirtschaftsinformatik Proceedings 2017 (2017)
84. Alaküla, M.-L., Matulevičius, R.: An experience report of improving business process compliance using security risk-oriented patterns. In: Ralyté, J., España, S., Pastor, Ó. (eds.) PoEM 2015. LNBIP, vol. 235, pp. 271–285. Springer, Cham (2015). https://doi.org/10.1007/978-3-319-25897-3_18
85. Rozsnyai, S., Slominski, A., Lakshmanan, G.T.: Discovering event correlation rules for semi-structured business processes. In: Proceedings of the 5th ACM International Conference on Distributed Event-Based System, pp. 75–86 (2011)
86. Ly, L.T., Rinderle-Ma, S., Knuplesch, D., Dadam, P.: Monitoring business process compliance using compliance rule graphs. In: Meersman, R., et al. (eds.) OTM 2011. LNCS, vol. 7044, pp. 82–99. Springer, Heidelberg (2011). https://doi.org/10.1007/978-3-642-25109-2_7
87. Mishra, S., Weistroffer, H.R.: A framework for integrating sarbanes-oxley compliance into the systems development process. CAIS **20**, 44 (2007)
88. Sandner, T., Kehlenbeck, M., Breitner, M.H.: An implementation of a process-oriented cross-system compliance monitoring approach in a SAP ERP and BI environment. In: ECIS (2010)
89. D'Aprile, D., Giordano, L., Martelli, A., Pozzato, G.L., Rognone, D., Dupré, D.T.: Business process compliance verification: an annotation based approach with commitments. In: De Marco, M., Te'eni, D., Albano, V., Za, S. (eds.) Information Systems, pp. 563–570. Physica, Heidelberg (2012). https://doi.org/10.1007/978-3-7908-2789-7_61

90. de Moura Araujo, B., Schmitz, E.A., Correa, A.L., Alencar, A.J.: A method for validating the compliance of business processes to business rules. In: Proceedings of the 2010 ACM Symposium on Applied Computing, pp. 145–149 (2010)

91. D'Aprile, D., Giordano, L., Gliozzi, V., Martelli, A., Pozzato, G.L., Theseider Dupré, D.: Verifying business process compliance by reasoning about actions. In: Dix, J., Leite, J., Governatori, G., Jamroga, W. (eds.) CLIMA 2010. LNCS (LNAI), vol. 6245, pp. 99–116. Springer, Heidelberg (2010). https://doi.org/10.1007/978-3-642-14977-1_10

92. Sadiq, S., Governatori, G., Namiri, K.: Modeling control objectives for business process compliance. In: Alonso, G., Dadam, P., Rosemann, M. (eds.) BPM 2007. LNCS, vol. 4714, pp. 149–164. Springer, Heidelberg (2007). https://doi.org/10.1007/978-3-540-75183-0_12

93. Gómez-López, M.T., Gasca, R.M., Pérez-Álvarez, J.M.: Compliance validation and diagnosis of business data constraints in business processes at runtime. Inf. Syst. **48**, 26–43 (2015)

94. Doganata, Y.N., Curbera, F.: A method of calculating the cost of reducing the risk exposure of non-compliant process instances. In: Proceedings of the First ACM Workshop on Information Security Governance, pp. 7–12 (2009)

95. de Nicola, A., Missikoff, M., Smith, F.: Towards a method for business process and informal business rules compliance. J. Softw.: Evol. Process **24**, 341–360 (2012)

96. Elgammal, A., Turetken, O., Heuvel, W.-J., Papazoglou, M.: Formalizing and appling compliance patterns for business process compliance. Softw. Syst. Model. **15**, 119–146 (2016)

97. Pham, T.A., Le Thanh, N.: An ontology-based approach for business process compliance checking. In: Proceedings of the 10th International Conference on Ubiquitous Information Management and Communication, p. 56:1 (2016)

98. Gong, P., Knuplesch, D., Feng, Z., Jiang, J.: A rule-based monitoring framework for business processes compliance. Int. J. Web Serv. Res. **14**, 81–103 (2017)

99. Semmelrodt, F., Knuplesch, D., Reichert, M.: Modeling the resource perspective of business process compliance rules with the extended compliance rule graph. In: Bider, I., et al. (eds.) BPMDS/EMMSAD -2014. LNBIP, vol. 175, pp. 48–63. Springer, Heidelberg (2014). https://doi.org/10.1007/978-3-662-43745-2_4

100. Knuplesch, D., Reichert, M.: A visual language for modeling multiple perspectives of business process compliance rules. Softw. Syst. Model. **16**, 715–736 (2016)

101. Gheorghe, G., Massacci, F., Neuhaus, S., Pretschner, A.: GoCoMM. A governance and compliance maturity model. In: WISG 2009, pp. 33–38 (2009)

102. Giordano, L., Martelli, A., Dupré, D.T.: Temporal deontic action logic for the verification of compliance to norms in ASP. In: Proceedings of the 14th International Conference on Artificial Intelligence and Law, pp. 53–62 (2013)

103. Cabanillas, C., Resinas, M., Ruiz-Cortés, A.: Hints on how to face business process compliance. Actas de los Talleres de las Jornadas de Ingeniería del Software y Bases de Datos (JISBD) **4**, 26–32 (2010)

Big Data Analytics as an Enabler of Process Innovation Capabilities: A Configurational Approach

Patrick Mikalef[✉] and John Krogstie

Norwegian University of Science and Technology,
Sem Saelandsvei 9, 7491 Trondheim, Norway
{patrick.mikalef, john.krogstie}@ntnu.no

Abstract. A central question for information systems (IS) researchers and practitioners is if, and how, big data can help attain a competitive advantage. Anecdotal claims suggest that big data can enhance a firm's incremental and radical process innovation capabilities; yet, there is a lack of theoretically grounded empirical research to support such assertions. To address this question, this study builds on the Resource-Based View and examines the fit between big data analytics resources and organizational contextual factors in driving a firm's process innovation capabilities. Survey data from 202 chief information officers and IT managers working in Norwegian firms is analyzed by means of fuzzy set qualitative comparative analysis (fsQCA). Results demonstrate that under different patterns of contextual factors the significance of big data analytics resources varies, with specific combinations leading to high levels of incremental and radical process innovation capabilities. These findings suggest that IS researchers and practitioners should look beyond direct effects, and rather, identify key combinations of factors that lead to enhanced process innovation capabilities.

Keywords: Big data analytics · Process innovation capabilities
fsQCA · Resource-based view · Contingency theory

1 Introduction

The domain of big data analytics has received increasing attention from academics and business practitioners over the past few years. By analyzing large volumes of unstructured data from multiple sources, actionable insights can be generated that help firms transform their business and gain an edge over their competition [1]. Such insights are particularly relevant, especially in dynamic and high-paced business environments, in which the need to continuously innovate is enhanced [2]. Quickly recognizing the potential of big data analytics, organizations have targeted their initiatives towards improving the efficiency and quality of their processes. Nevertheless, effective use of big data analytics within organizations, is argued not only to help create incremental improvements to existing processes, but also to lead to help develop exploration or radical process innovation capabilities [3]. As a result, the successful application of big data analytics towards both incremental and radical process

© Springer Nature Switzerland AG 2018
M. Weske et al. (Eds.): BPM 2018, LNCS 11080, pp. 426–441, 2018.
https://doi.org/10.1007/978-3-319-98648-7_25

innovation capabilities can help organizations redefine their business and outperform their competitors [4]. The competence to successfully pursue process innovation represents an important capability, particularly for organizations that are exposed to a dynamic business environment. Despite this, organizations today are still facing challenges concerning the firm-wide deployment of big data analytics initiatives, and a difficulty to align their new investments towards the attainment of strategic goals [5].

Recognizing this issues that many companies face, several research commentaries have been written emphasizing the importance of delving into the whole spectrum of aspects that surround big data analytics [6, 7]. Nevertheless, empirical studies on the topic is are still quite limited, particularly in explaining how specific organizational goals should be achieved, and what factors influence their attainment [8]. Past literature reviews on the broader IS domain have demonstrated that there are multiple factors that should be taken into account when examining the value of IT investments [9], and especially in relation to process management and innovation [10–12]. Literature in the area of IT business value has predominantly used the notion of IT capabilities to refer to the broader context of technology within firms, and the overall proficiency in leveraging and mobilizing the different resources and capabilities [13]. The main idea underlying the concept is that when firms manage to obtain valuable bundles of resources, they will develop the capacity to effectively utilize them towards strategic objectives [14]. We therefore deem it necessary to identify and explore the domain specific aspects that are relevant to big data analytics within the business context and examine the ways in which they add value [15].

Building on these gaps, we seek to explore the importance that different combinations of big data analytics capability resources have on enhancing a firm's process innovation capabilities. While external requirements have always prompted change and innovation, new technologies of the digital age represent a key source of numerous affordances for process innovations today [16]. In fact, fundamental business transformations are often a result of integrating IT into business processes. The big data age offers manifold opportunities to promote process innovation, but to do so first requires identifying the value-creating resources [16]. In doing so, we differentiate between incremental and radical process innovation capabilities [4, 17], since the types of goals they are targeted towards will likely influence the importance of different combinations of resources [12]. In addition, we include contextual factors in our examination pertinent to the internal and external environment of the firm. Building on a sample of 202 survey responses from IT managers in Norwegian firms, we employ a configurational theory approach and examine the patterns of elements that lead to high levels incremental and radical process innovation capabilities. We do so through the novel methodological tool fsQCA, which allows the examination of such complex phenomena and the reduction of solutions to a core set of elements.

In Sect. 2 we provide an overview of literature on big data analytics capabilities, process innovation capabilities, and some of the most important contextual factors when examining process innovation practices. In Sect. 3, we introduce the logic of configurational theories, and develop a set of propositions that guide this research. Section 4 defines the methodology of the study, including the data, measurements, and reliability and validity tests, while in Sect. 5 we present the results of the fsQCA analyses. In closing, we draw on the theoretical and practical implications of this study.

2 Background

2.1 Big Data Analytics Capabilities

A lot has been written on the relationship between different IT investments and business process management (BPM) projects [18]. The general consensus in the academic and research community is that IT acts as the enabler and the facilitator of changes identified in BPM projects [10]. Specifically, Sambamurthy et al. [19] suggest that IT could be driving the modularization and atomization of business processes and enabling their combination and recombination to create new business processes. Recent articles have begun to develop this idea on a more theoretically grounded basis, both in relation to how the value of IT investments should be measured [10], as well as on how the context shapes this relationship [12]. It is generally accepted that while a strong IT capability may be more appropriate in assessing effects on business process management, the value of some resources that comprise it may be of a greater or lesser importance depending on the context of examination [12]. Despite the extended work the impact of IT capabilities on business process management and innovation [19, 20], empirical studies examining the enabling role of big data analytics are still scarce.

Past research has shown that when assessing the business value of IT investments, it is critical to take a broader view and capture all the underlying factors that enable effective and efficient use of IT as a differentiator of firm success [13]. Studies that examine the effects of a firms IT capability, typically base their theoretical assumptions and operationalization on the Resource-Based View (RBV) of the firm [21], which argues that a competitive advantage emerges from unique combinations of resources that are economically valuable, scarce, and difficult to imitate. Likewise, the main premise on which the notion of IT capability is built, is that while resources can be easily replicated, distinctive firm-specific capabilities cannot be readily assembled through markets, and can thus, constitute a source of a sustained competitive advantage [22]. Since the aim of this study is to define the main resources have an impact on process innovation capabilities, the choice of the RBV as the underlying theoretical framework is deemed as suitable. Consequently, we define big data analytics capability as the ability of the firm to capture and analyze data towards the generation of actionable insights, by effectively deploying its data, technology, and talent through firm-wide processes, roles and structures.

Building on prior studies of the RBV and on IT capabilities literature, we identify between three broad types of resources, tangible (e.g. physical and financial resources), human skills (e.g. employees skills and knowledge), and intangible (e.g. organizational culture and organizational learning) [23]. Regarding tangible resources, data, technology and other basic resources are considered critical for success. Despite the defining characteristics of big data being its volume, variety, and velocity, a common concern amongst IT strategists and data analysts are the quality and availability of the data they analyze [24]. It is also critical for firms to possess the necessary infrastructure for storing, sharing, and analyzing data, as well as analytics methods to turn data into insight [7]. Finally, basic resources such as financial support are necessary at all stages of big data projects, particularly when considering the long lag effects they have in producing measurable business value [15]. When it comes to human skills, literature recognizes that both technical and managerial-oriented skills are necessary to derive value from big data investments [25]. Specifically, regarding technical skills, Davenport and Patil [26] emphasize on the importance that the emerging job of the data scientist will have in the next few years throughout a number of industries. Yet, while technical skills are important, one of the most critical aspects of data science is the ability of data-analytic thinking and strategic planning based on data-driven insight [2]. In relation to intangible resources, a data-driven culture and organizational learning are widely regarded as important components of effective deployment of big data initiatives [8]. For firms that have deployed big data projects, a data-driven culture has been suggested to be a key factor in determining overall success and alignment with organizational strategy [27]. Yet, a complementary facet of governance is organizational learning, primarily due to the constantly changing landscape in terms of technologies and business practices, which require firms to infuse the idea of continuous learning into their fabric [28].

2.2 Process Innovation Capabilities

While BPM has traditionally emphasized on promoting incremental improvements through efficiency and effectiveness on business processes through standardization, automation, and optimization, there is also a stream of research that highlights the potential for radical process innovations [29, 30]. In today's dynamic globalized business arena, process innovation is important for at least two reasons. First, process innovation is closely associated with product innovation [31]. Developing new products often requires changes in either existing processes, or even forming new ones when they involve techniques that are novel to the firm. In their empirical investigation, Fritsch and Meschede [32] show that process innovation has a positive effect on product innovation, and that by fostering process innovation, a firm will be able to improve its product quality or even to produce entirely new products. Second, the value of process innovation is proportional to the level of output produced by a given firm. Hence, as industries mature and increase their numbers and frequencies of use of their business processes, they have increased incentives to pursue process innovation [31].

In this study we examine a firm's process innovation capability, which is defined as a firm's ability, relative to its competitors, to apply the collective knowledge, skills, and resources to innovation activities relating to new processes, in order to create added value for the firm [33]. We identify two main types of process innovation capabilities, incremental and radical [34]. An incremental process innovation capability is defined as an organizations ability to reinforce and extend its existing expertise in processes, by significantly enhancing or upgrading them [35]. On the other hand, a radical process innovation capability is focused around the ability of the firm to make current/existing processes obsolete through the introduction of novel ones [36]. Literature on BPM has focused quite heavily on methods for product business process innovation, yet, there is considerable skepticism on whether maturity models such the Capability Maturity Model (CMM) are able to capture the need for business process innovation [37]. Rather, the elements underpinning BPM and process innovation, emphasize on the importance of taking into account a holistic view, including the culture, IT, and people [38].

2.3 Contextual Factors

While the role of context has been researched extensively in the fields of information systems and organizational studies, it is still at a very early stage in the field of BPM [12]. While not explicitly the focus of many studies, contradicting findings pinpoint the contextuality of results [39], placing the context of examination as an important aspect that should be considered when looking at process management and particularly process innovation outcomes. Building on this, the principle of context awareness has been identified as a key perspective for successful BPM implementations [39]. This perspective is rooted in contingency theory [40], which assumes that there is not one universal best way to manage business processes, but rather, that management practices and resources should fit the organization and the external environment [12]. Similar views on process innovation have been found to be true on studies that adopt an strategic management and organizational research perspective [41].

A first contextual factor that is examined in this study is the goal of the organization, since goals directly influence the business process management practices and resources that are most suitable [12]. Several authors make the distinctions between exploitation and exploration, or else incremental and radical process innovation capabilities [42, 43]. Since the process of developing either incremental or radical innovations differs fundamentally, managers need to select and adapt their approach depending on the goal, thereby constituting the focus as an important contextual factor. Another important group of contextual factors have to do with the external environment of the organization. Particularly, the uncertainty of the environment is critical to consider since under such conditions organizations need to reconfigure the way the operate and emphasize more on analytical and research capabilities. Finally, an important group of contextual factors relate to the organization itself. Based on contingency theory, the size of the organization plays an important role since, typically, larger organizations require more formalized processes that cross vertical and horizontal functions than smaller firms [12]. Finally, type of industry is considered to be an important contextual factor, since practices and resources that may be effective in one industry may not be the most suitable in another [11].

3 Research Approach

Following the studies described above, research has begun to examine how these contextual factors coalesce in order to produce both types of process innovation outcomes for firms, incremental and radical [44]. Particularly in relation to the emerging area of big data analytics, little is known about what are the core resources that help drive a firm's process innovation capabilities, and even less regarding the role of internal and external factors in shaping these requirements [25, 45]. While it may be useful to consider separate elements of context and examine their influence on outcomes of business process innovation, it is also important to research their combinations to derive context patterns that are more meaningful than any single dimension would be in isolation.

Configurational theories are a newly applied approach in the field of IS which are best suited for examining holistic interplays between elements of a messy, and nonlinear nature [46]. The aim of configuration theory is to identify patterns and combinations of variables and reveal how their synergistic effects lead to specific outcomes. Configurations occur by different combinations of causal variables that affect an outcome of interest. The main difference of configuration theory is that it views elements through a holistic lens so that they must be examined simultaneously, and is therefore particularly attractive for context-related studies in which there is a complex causality. Contrarily to variance and process theories, configuration theory supports the concept of equifinality, meaning that the same outcome can be a result of one or more sets of configuration patterns. Additionally, configuration theory includes the notion of causal asymmetry, meaning that the combination of elements leading to the presence of an outcome may be different than those leading to an absence of the outcome [46].

4   Methodology

4.1   Data

To explore the combination of factors that lead to strong process innovation capabilities, a survey instrument was developed and administered to key informants within firms. We conducted a pre-test in a small-cycle study with 23 firms to examine the statistical properties of the measures. Through the pre-test procedure we were able to assess the face and content validity of items and to make sure that key respondents would be in place to comprehend the survey as intended. For the main study, we used a population of 500 firms from a list of Norway's largest companies, measured in terms of revenue (Kapital 500). Each of these firms was contacted by phone in order to get contact details of the most appropriate key respondent (e.g. chief information officer, chief technology officer) and inform them about the purpose of this research. To ensure a collective response, the respondents were instructed to consult other employees within their firms for information that they were not knowledgeable about. The data collection process lasted for approximately four months (February 2017–July 2017), and on average completion time of the survey was 14 min. A total of 213 firms started to complete the survey, with 202 providing complete responses (Table 1).

Table 1. Sample characteristics

Factors	Sample (N = 202)	Proportion (%)
Industry		
Bank & financials	28	13.8%
Consumer goods	22	10.8%
Oil & gas	21	10.4%
Industrials (construction & industrial goods)	19	9.4%
ICT and telecommunications	11	5.4%
Technology	9	4.4%
Media	9	4.4%
Transport	8	3.9%

(continued)

Table 1. (*continued*)

Factors	Sample (N = 202)	Proportion (%)
Other (shipping, consumer services etc.)	75	37.1%
Firm size (number of employees)		
1–9	1	0.5%
10–49	34	16.8%
50–249	36	17.8%
250+	131	64.8%

With non-response bias being a common problem in large-scale questionnaire studies, we took measures both during the collection of the data to ensure we had a representative response rate, as well as after the concluding of the data gathering. All participants were given an incentive to partake in the study, and were promised a personalized report benchmarking their firms' performance in a number of areas to industry means [47]. After the initial invitation to take part in the survey, respondents were re-contacted on three occasions with two-week interval between each reminder. After the data collection was finished, and to ensure that no bias existed within data, we compared between early and late responses at the construct level to verify that no significant differences existed. To do so, we constructed two groups of responses, those who replied within the first three weeks and those that replied in the final three weeks. Through t-test comparisons between group means, no significant differences were detected. In addition, no significant differences were found between responding and non-responding firms in terms of size and industry. Considering that all data were collected from a single source at one point in time, and that all data consisted of perceptions of key respondents, we controlled for common method bias following the guidelines of Chang et al. [48]. *Ex-ante*, respondents were assured that all information they provided would remain completely anonymous and confidential, and that any analysis would be done on an aggregate level for research purposes solely. *Ex-post*, we used Harman's one factor test, which indicated that a single construct could not account for the majority of variance.

4.2   Construct Definition and Measurement

We build on the notion of big data analytics capability from the study of Gupta and George [8] to determine all relevant resources [49]. The concept distinguishes between the three underlying pillars which are big data-related tangible, human skills, and intangible resources. Each of these groups of factors is very distinct and comprises of a unique set of variables. Specifically, within the tangle big data resources, we distinguish between data, technology, and basic resources. With regards to human skills, we identify two mains categories, technical and managerial skills. Finally, in relation to intangible resources, we include a data-driven culture and the intensity of organizational learning as two core resources. Each of the previously mentioned concepts is measured on a 7-point Likert scale, in accordance to the study of Gupta and George [8].

The degree of environmental uncertainty was assessed through three measures; dynamism (DYN), heterogeneity (HET), and hostility (HOST) [50]. Dynamism is defined as the rate and unpredictability of environmental change. Heterogeneity reflects the complexity and diversity of external factors, such as the variety of customer buying habits and the nature of competition. Hostility is defined as the availability of key resources and the level of competition in the external environment. All constructs were measured as latent variables on a 7-point Likert scale.

A process innovation capability is defined in the context of the skills and knowledge needed to effectively absorb, master and improve existing processes and to create new ones. We measured process innovation capability through two first-order latent construct; incremental process innovation capability (INC) and radical process innovation capability (RAD). Incremental process innovation capability was measured with three indicators assessing an organizations capability to reinforce and extend its existing expertise in processes. Likewise, radical process innovation capability was assessed through three indicators that asked respondents to evaluate their organization's ability to make current processes [36].

Firm size was measured as a binary variable in accordance with recommendations of the European Commission (2003/361/EC) with SME's including micro (0–9 employees), small (10–49 employees), and medium (50–249 employees) enterprises, and large being those with more than 250 employees. Large firms were assigned the value 1, while SME's were represented with 0. The industry was further grouped into product and service industries, in which 1 connotes a product industry, and 0 a service industry.

4.3 Reliability and Validity

Since the research design contains both reflective and formative constructs, we used different assessment criteria to evaluate each. For first-order reflective latent constructs we conducted reliability, convergent validity, and discriminant validity tests. Reliability was gauged at the construct and item level. At the construct level we examined Composite Reliability (CR), and Cronbach Alpha (CA) values, and confirmed that their values were above the threshold of 0.70. Indicator reliability was assessed by examining if construct-to-item loadings were above the threshold of 0.70. To establish convergent validity, we examined if AVE values were above the lower limit of 0.50, with the smallest observed value being 0.59 which greatly exceeds this threshold. We examined for the presence of discriminant validity through three ways. The first looked at each constructs AVE square root to verify that it is greater than its highest correlation with any other construct (Fornell-Larcker criterion). The second tested if each indicator's outer loading was greater that its cross-loadings with other constructs [51]. Recently, Henseler et al. [52] argued that a new criterion called the heterotrait-monotrait ratio (HTMT) is a better assessment indicator of discriminant validity. Values below 0.85 are an indication of sufficient discriminant validity, hence, the obtained results confirm discriminant validity. The abovementioned outcomes suggest that first-order reflective measures are valid to work with and support the appropriateness of all items as good indicators for their respective constructs.

For formative indicators, we first examined the weights and significance of their association with their respective construct. While all of the indicators weights for data and basic resources were statistically significant, one of the three indicators weights (BR2) of the technology construct was found to be non-significant. According to Cenfetelli and Bassellier [53], formative constructs are likely to have some indicators with non-significant weights. Their suggestion is that a non-significant indicator should be kept providing that the researchers can justify its importance. Since the technology construct is proposed as an aggregate of three items, where each captures a different big data-related technology, we believe that it is critical to include the indicator in the model as it makes a distinct contribution. A similar approach is followed by Gupta and George [8] in their operationalization of big data analytics capability. Next, to evaluate the validity of the items of formative constructs, we followed MacKenzie et al. [54] and Schmiedel et al. [55] guidelines using Edwards [56] adequacy coefficient (R_a^2). To do so we summed the squared correlations between formative items and their respective formative construct and then divided the sum by the number of indicators. All R_a^2 value exceeded the threshold of 0.50, suggesting that the majority of variance is shared with the overarching construct, and that the indicators are valid representations of the construct. Next, we examined the level to which the indicators of formative constructs presented multicollinearity. Variance Inflation Factor (VIF) values below 10 suggest low multicollinearity, however, a more restrictive cutoff of 3.3 is used for formative constructs Petter et al. [57]. All values were below the threshold of 3.3 indicating an absence of mutlicollinearity.

5   Analysis

5.1   Methodology and Calibration

To determine what big data analytics resources are most important in the formation of process innovation capabilities among different environmental and contextual conditions, this study employs a fuzzy-set Qualitative Comparative Analysis (fsQCA). FsQCA follows the principles of configurational theories which allow for the examination of interplays that develop between elements of a messy and non-linear nature [58]. As such, it is important to isolate what combination of factors and conditions contribute towards firms developing strong incremental and radical process innovation capabilities. The first step of the fsQCA analysis is to calibrate dependent and independent variables into fuzzy or crisp sets. These fuzzy sets may range anywhere on the continuous scale from 0, which denotes an absence of set membership, to 1, which indicates full set membership. Crisp sets are more appropriate in categorical variables that have two, and only two options. The procedure followed of transforming continuous variables into fuzzy sets is grounded on the method proposed by Ragin [59]. According to the procedure, the degree of set membership is based on three anchor values. These represent a full set membership threshold value (fuzzy score = 0.95), a full non-membership value (fuzzy score = 0.05), and the crossover point (fuzzy score = 0.50) [60]. Since this study uses a 7-point Likert scale to measure constructs, the guidelines put forth by Ordanini et al. [61] are followed to calibrate them into fuzzy sets. Therefore, full membership thresholds are set for values over 5.5, the cross over point is set at 4, and full non-membership values at 2.5. The size-class of firms is coded as 1 for large enterprises and 0 for SME's, while product-based companies are marked with 1, and service-oriented ones with 0.

5.2 Results

We performed two separate fsQCA analyses, one for each dependent variables of interest, that is high incremental and radical process innovation capabilities. Each analysis produces a truth table of 2^k rows, where k represents the number of predictor elements, and each row stands for a possible combination (solution). Solutions that have a consistency level lower than 0.80 are disregarded [62]. In addition, a minimum of three cases for each solution is set [60]. Having established these parameters, the fsQCA analyses are then performed using high incremental and radical process innovation capabilities as the dependent variables. The outcomes of the fuzzy set analysis are presented in Table 2. The solutions are presented in the columns with the black circles (●) denoting the presence of a condition, the crossed-out circles (⊗) indicating an absence of it, while the blank spaces represent a "don't care" situation in which the causal condition may be either present or absent [63].

Table 2. Configurations for high incremental and radical process innovation capabilities

Configuration	Solution					
	Incremental process innovation capability			Radical process innovation capability		
	1	2	3	1	2	3
Big data analytics resources						
Data	●	●	●	●	●	●
Technology	●					
Basic resources		●	●			
Technical skills	●	●	●	●		●
Managerial skills				●	●	●
Organizational learning					●	
Data-driven culture					●	●
Environment						
Dynamism		●	●		●	●
Heterogeneity	●		●	●		
Hostility				●	●	
Context						
Size	●		●	●	●	⊗
Industry	●	⊗	⊗	●	⊗	⊗
Consistency	0.872	0.868	0.927	0.906	0.823	0.815
Raw coverage	0.237	0.202	0.161	0.241	0.197	0.152

(continued)

Table 2. (*continued*)

Configuration	Solution					
	Incremental process innovation capability			Radical process innovation capability		
	1	2	3	1	2	3
Unique coverage	0.207	0.139	0.135	0.187	0.122	0.124
Overall solution consistency	0.825			0.841		
Overall solution coverage	0.479			0.427		

With regards to a firm's incremental innovation capabilities, solution 1 corresponds to large firms of a large size-class that operate in product-based industries, and with a business environment characterized by high heterogeneity. For these companies the presence of strong data, technology, and technical skills is found to be a solution for achieving high incremental process innovation capabilities. The next two solutions, 2 and 3, represent firms that are service-oriented. Specifically, in solution two, for companies that operate in dynamic environments, data and technical skills are found to be important, including the presence of solid basic resources. This solution is size-class independent, meaning that it could apply to either large firms or SME's. Solution 3 on the other hand concerns large firms in the service industry that operate in highly dynamic and heterogeneous environments. For these firms, the presence of strong data and basic resources, technology and technical skills is shown to lead to a high incremental process innovation capability. It is important to note that an appropriate lens to consider the ways these resources are leveraged in such environments is the dynamic capabilities view of the firm [64].

Concerning configurations that lead to high radical process innovation capabilities, there are also three solutions. Solution applies to firms that are in product-based industries and belong to the large size-class. The environment that they operate in is characterized by high heterogeneity and hostility. For these firms the presence of strong data resources, coupled with solid technical and managerial skills yields high radical process innovation capabilities. Solutions 2 and 3 correspond to service industry firms. Specifically, solution 2 is about large firms operating in dynamic and hostile conditions. In such settings, the presence of strong data resources, along with solid managerial skills, organizational learnings and a mature data-drive culture are the cornerstones of achieving high radical process innovation capabilities. Solution 3 on the other hand highlights the conditions for firms of the SME size-class, that conduct business in dynamic markets. These firms rely on the presence of strong data resources along with mature technical and managerial skills, and a solid data-drive culture.

6 Discussion

While much has been written about the strategic value of big data analytics, studies dedicated to empirical evidence on the real business value of such investments at the firm level remain scarce. Our understanding about, if, and how big data analytics can help support process innovation is still at a rudimentary state. To explore this topic, the present study built on a theory-driven conceptualization of big data analytics, and identified the main resources that underpin the notion. In addition, following literature on context-aware process design success factors, we explore how the context and big data analytics resources coalesce to drive firm process innovation capabilities. Grounded on a configurational theory approach that enables us to examine such interactions, we analyze responses from 202 IT managers of Norwegian firms and derive different solutions through which high levels of incremental and radical process innovation capabilities are attained.

From a theoretical perspective, the findings of this study add to existing literature in several ways. First, it demonstrates how a contingency approach can be empirically explored in the context of business process management. Such contextual factors are seldom investigated in quantitative studies with regards to business process innovation [10]. Second, despite much anecdotal claims concerning the enabling effect that big data analytics have on strengthening existing or developing new business processes, there is still limited empirical research to consolidate them [3]. Our findings show that different combinations of big data-related resources have a greater or lesser significance depending on the context and the type of process innovation capability they are targeted towards. More precisely, we find that more technological and technical resources contribute towards delivering incremental process innovation capabilities, whereas a firm-wide data-driven culture and strong managerial data analytic skills are critical when it comes to radical process innovation capabilities.

From a practical point of view these results suggest that managers should develop different strategies in relation to their big data analytics initiatives, depending on the types of business process innovation they aim to achieve, while also taking into account the contingencies of the environment and the organization. Specifically, the results suggest that when it comes to radical process innovation capabilities, data governance practices should encourage the breakdown of organizational silos and promote the notion of data-driven decision-making at all levels of the organization [65]. In addition, managerial knowledge on data-driven initiatives and the potential application of big data to organizational problems should be encouraged through targeted seminars and training. Contrarily, for incremental process innovations to emerge, managers should focus on technical excellence in terms of human skills and tangible resources. For these types of process innovations, strong technical skills are critical, since gaining insight to produce incremental improvements likely requires expertise in skills that are domain specific.

While the results of this study shed some light on the relationship of big data analytics resources and process innovation capabilities, they must be considered under their limitations. First, our sample comprises of companies operating in Norway and belonging to the 500 largest in terms of revenue. It is highly likely that firms that operate on a smaller scale will have different configurations of factors that drive process innovation capabilities. Second, while we differentiate between incremental and radical process innovations, we do not control for the different types of processes in terms of their domain area. The different functional areas in which big data analytics are applied are likely to yield different results and require varying configurations of resources to enhance or create innovative business processes. Third, although fsQCA allows us to examine the configurations of resources and the contextual factors under which they produce process innovation capabilities, the significance of each resource, as well as the process through which it produces this outcome is not well explained. A complementary study suing a qualitative approach would likely reveal more insight on how value is produced from such investments.

Acknowledgements.

 This project has received funding from the European Union's Horizon 2020 research and innovation programme, under the Marie Sklodowska-Curie grant agreement No. 704110.

Appendix A. Survey Instrument

The survey instrument can be found here: https://goo.gl/4y4QVr.

References

1. McAfee, A., Brynjolfsson, E., Davenport, T.H.: Big data: the management revolution. Harvard Bus. Rev. **90**, 60–68 (2012)
2. Prescott, M.: Big data and competitive advantage at Nielsen. Manag. Decis. **52**, 573–601 (2014)
3. Fosso Wamba, S., Mishra, D.: Big data integration with business processes: a literature review. Bus. Process Manag. J. **23**, 477–492 (2017)
4. Ortbach, K., Plattfaut, R., Poppelbuß, J., Niehaves, B.: A dynamic capability-based framework for business process management: theorizing and empirical application. In: 2012 45th Hawaii International Conference on System Science (HICSS), pp. 4287–4296. IEEE (2012)
5. Kiron, D.: Lessons from becoming a data-driven organization. MIT Sloan Manag. Rev. **58** (2017)
6. Constantiou, I.D., Kallinikos, J.: New games, new rules: big data and the changing context of strategy. J. Inf. Technol. **30**, 44–57 (2015)
7. Mikalef, P., Pappas, I.O., Krogstie, J., Giannakos, M.: Big data analytics capabilities: a systematic literature review and research agenda. Inf. Syst. e-Bus. Manag. 1–32 (2017)
8. Gupta, M., George, J.F.: Toward the development of a big data analytics capability. Inf. Manag. **53**, 1049–1064 (2016)
9. Schryen, G.: Revisiting IS business value research: what we already know, what we still need to know, and how we can get there. Eur. J. Inf. Syst. **22**, 139–169 (2013)
10. Trkman, P.: The critical success factors of business process management. Int. J. Inf. Manag. **30**, 125–134 (2010)
11. Trkman, P.: Increasing process orientation with business process management: critical practices'. Int. J. Inf. Manag. **33**, 48–60 (2013)
12. vom Brocke, J., Zelt, S., Schmiedel, T.: On the role of context in business process management. Int. J. Inf. Manag. **36**, 486–495 (2016)
13. Bharadwaj, A.: A resource-based perspective on information technology capability and firm performance: an empirical investigation. MIS Q. **24**, 169–196 (2000)

14. Ravichandran, T., Lertwongsatien, C.: Effect of information systems resources and capabilities on firm performance: a resource-based perspective. J. Manag. Inf. Syst. **21**, 237–276 (2005)
15. Mikalef, P., Framnes, V.A., Danielsen, F., Krogstie, J., Olsen, D.H.: Big data analytics capability: antecedents and business value. In: Pacific Asia Conference on Information Systems (2017)
16. Schmiedel, T., vom Brocke, J.: Business process management: potentials and challenges of driving innovation. In: vom Brocke, J., Schmiedel, T. (eds.) BPM - Driving Innovation in a Digital World. MP, pp. 3–15. Springer, Cham (2015). https://doi.org/10.1007/978-3-319-14430-6_1
17. Mikalef, P., Boura, M., Lekakos, G., Krogstie, J.: Complementarities between information governance and big data analytics capabilities on innovation. In: European Conference on Information Systems (ECIS), Portsmouth. AIS (2018)
18. Vom Brocke, J., Rosemann, M.: Handbook on Business Process Management. Springer, Heidelberg (2010). https://doi.org/10.1007/978-3-642-00416-2
19. Sambamurthy, V., Bharadwaj, A., Grover, V.: Shaping agility through digital options: reconceptualizing the role of information technology in contemporary firms. MIS Q. **27**, 237–263 (2003)
20. Wang, N., Liang, H., Zhong, W., Xue, Y., Xiao, J.: Resource structuring or capability building? An empirical study of the business value of information technology. J. Manag. Inf. Syst. **29**, 325–367 (2012)
21. Bhatt, G.D., Grover, V.: Types of information technology capabilities and their role in competitive advantage: an empirical study. J. Manag. Inf. Syst. **22**, 253–277 (2005)
22. Lu, Y., Ramamurthy, K.: Understanding the link between information technology capability and organizational agility: an empirical examination. MIS Q. **35**, 931–954 (2011)
23. Grant, R.M.: The resource-based theory of competitive advantage: implications for strategy formulation. Calif. Manag. Rev. **33**, 114–135 (1991)
24. Janssen, M., van der Voort, H., Wahyudi, A.: Factors influencing big data decision-making quality. J. Bus. Res. **70**, 338–345 (2017)
25. Wamba, S.F., Gunasekaran, A., Akter, S., Ren, S.J.F., Dubey, R., Childe, S.J.: Big data analytics and firm performance: effects of dynamic capabilities. J. Bus. Res. **70**, 356–365 (2017)
26. Davenport, T.H., Patil, D.: Data scientist: the sexiest job of the 21st century. Harvard Bus. Rev. **90**, 70–76 (2012)
27. LaValle, S., Lesser, E., Shockley, R., Hopkins, M.S., Kruschwitz, N.: Big data, analytics and the path from insights to value. MIT Sloan Manag. Rev. **52**, 21 (2011)
28. Vidgen, R., Shaw, S., Grant, D.B.: Management challenges in creating value from business analytics. Eur. J. Oper. Res. **261**, 626–639 (2017)
29. Vom Brocke, J., Schmiedel, T.: BPM-Driving Innovation in a Digital World. Springer, Heidelberg (2015). https://doi.org/10.1007/978-3-319-14430-6
30. Recker, J.C., Rosemann, M.: Systemic ideation: a playbook for creating innovative ideas more consciously. 360° Bus. Transform. J. **13**, 34–45 (2015)
31. Adner, R., Levinthal, D.: Demand heterogeneity and technology evolution: implications for product and process innovation. Manag. Sci. **47**, 611–628 (2001)
32. Fritsch, M., Meschede, M.: Product innovation, process innovation, and size. Rev. Ind. Organ. **19**, 335–350 (2001)
33. Hogan, S.J., Soutar, G.N., McColl-Kennedy, J.R., Sweeney, J.C.: Reconceptualizing professional service firm innovation capability: scale development. Ind. Mark. Manag. **40**, 1264–1273 (2011)

34. Ettlie, J.E., Bridges, W.P., O'keefe, R.D.: Organization strategy and structural differences for radical versus incremental innovation. Manag. Sci. **30**, 682–695 (1984)
35. Gallouj, F., Savona, M.: Innovation in services: a review of the debate and a research agenda. J. Evol. Econ. **19**, 149 (2009)
36. Subramaniam, M., Youndt, M.A.: The influence of intellectual capital on the types of innovative capabilities. Acad. Manag. J. **48**, 450–463 (2005)
37. Smith, H., Fingar, P.: Process management maturity models. Business Process Trends (2004)
38. Rosemann, M., vom Brocke, J.: The six core elements of business process management. In: vom Brocke, J., Rosemann, M. (eds.) Handbook on Business Process Management 1. IHIS, pp. 105–122. Springer, Heidelberg (2015). https://doi.org/10.1007/978-3-642-45100-3_5
39. Vom Brocke, J., Schmiedel, T., Recker, J., Trkman, P., Mertens, W., Viaene, S.: Ten principles of good business process management. Bus. Process Manag. J. **20**, 530–548 (2014)
40. Donaldson, L.: The Contingency Theory of Organizations. Sage, Thousand Oaks (2001)
41. Ortt, J.R., van der Duin, P.A.: The evolution of innovation management towards contextual innovation. Eur. J. Innov. Manag. **11**, 522–538 (2008)
42. vom Brocke, J., Seidel, S., Tumbas, S.: The BPM curriculum revisited. BPTrends, April 2015
43. Rosemann, M.: Proposals for future BPM research directions. In: Ouyang, C., Jung, J.-Y. (eds.) AP-BPM 2014. LNBIP, vol. 181, pp. 1–15. Springer, Cham (2014). https://doi.org/10.1007/978-3-319-08222-6_1
44. Pöppelbuß, J., Plattfaut, R., Niehaves, B.: How do we progress? An exploration of alternate explanations for BPM capability development. CAIS **36**, 1 (2015)
45. Torres, R., Sidorova, A., Jones, M.C.: Enabling firm performance through business intelligence and analytics: a dynamic capabilities perspective. Inf. Manag. (2018)
46. Fiss, P.C.: A set-theoretic approach to organizational configurations. Acad. Manag. Rev. **32**, 1180–1198 (2007)
47. Sax, L.J., Gilmartin, S.K., Bryant, A.N.: Assessing response rates and nonresponse bias in web and paper surveys. Res. High. Educ. **44**, 409–432 (2003)
48. Chang, S.-J., Van Witteloostuijn, A., Eden, L.: From the editors: common method variance in international business research. J. Int. Bus. Stud. **41**, 178–184 (2010)
49. Mikalef, P., Krogstie, J.: Big data governance and dynamic capabilities: the moderating effect of environmental uncertainty. In: Pacific Asia Conference on Information Systems (PACIS). AIS, Yokohama (2018)
50. Newkirk, H.E., Lederer, A.L.: The effectiveness of strategic information systems planning under environmental uncertainty. Inf. Manag. **43**, 481–501 (2006)
51. Farrell, A.M.: Insufficient discriminant validity: a comment on Bove, Pervan, Beatty, and Shiu (2009). J. Bus. Res. **63**, 324–327 (2010)
52. Henseler, J., Ringle, C.M., Sarstedt, M.: A new criterion for assessing discriminant validity in variance-based structural equation modeling. J. Acad. Mark. Sci. **43**, 115–135 (2015)
53. Cenfetelli, R.T., Bassellier, G.: Interpretation of formative measurement in information systems research. MIS Q. **33**(4), 689–707 (2009)
54. MacKenzie, S.B., Podsakoff, P.M., Podsakoff, N.P.: Construct measurement and validation procedures in MIS and behavioral research: integrating new and existing techniques. MIS Q. **35**, 293–334 (2011)
55. Schmiedel, T., Vom Brocke, J., Recker, J.: Development and validation of an instrument to measure organizational cultures' support of business process management. Inf. Manag. **51**, 43–56 (2014)
56. Edwards, J.R.: Multidimensional constructs in organizational behavior research: An integrative analytical framework. Organ. Res. Methods **4**, 144–192 (2001)

57. Petter, S., Straub, D., Rai, A.: Specifying formative constructs in information systems research. MIS Q. **31**(4), 623–656 (2007)
58. Fiss, P.C.: Building better causal theories: a fuzzy set approach to typologies in organization research. Acad. Manag. J. **54**, 393–420 (2011)
59. Ragin, C.C.: Qualitative comparative analysis using fuzzy sets (fsQCA). In: Configurational Comparative Methods, vol. 51 (2009)
60. Woodside, A.G.: Moving beyond multiple regression analysis to algorithms: calling for adoption of a paradigm shift from symmetric to asymmetric thinking in data analysis and crafting theory. Elsevier (2013)
61. Ordanini, A., Parasuraman, A., Rubera, G.: When the recipe is more important than the ingredients: a qualitative comparative analysis (QCA) of service innovation configurations. J. Serv. Res. **17**, 134–149 (2014)
62. Ragin, C.C.: Fuzzy-Set Social Science. University of Chicago Press, Chicago (2000)
63. Mikalef, P., Pateli, A., Batenburg, R.S., Wetering, R.V.D.: Purchasing alignment under multiple contingencies: a configuration theory approach. Ind. Manag. Data Syst. **115**, 625–645 (2015)
64. Mikalef, P., Pateli, A.: Information technology-enabled dynamic capabilities and their indirect effect on competitive performance: Findings from PLS-SEM and fsQCA. J. Bus. Res. **70**, 1–16 (2017)
65. Mikalef, P., Van de Wetering, R., Krogstie, J.: Big Data enabled organizational transformation: the effect of inertia in adoption and diffusion. In: Business Information Systems (BIS) (2018)

Track III: Method Analysis and Selection

Assessing the Quality of Search Process Models

Marian Lux[1,2]($\boxtimes$), Stefanie Rinderle-Ma[1], and Andrei Preda[2]

[1] ds:UniVie, Faculty of Computer Science, University of Vienna, Vienna, Austria
{marian.lux,stefanie.rinderle-ma}@univie.ac.at
[2] LuxActive KG, Vienna, Austria
office@myoha.at

Abstract. Search processes are highly individual business processes reflecting the search behavior of users in search systems. The analysis of search processes is a promising instrument in order to improve customer journeys and experience. The quality of the analysis results depends on the underlying data, i.e., logs and the search process models. However, it is unclear what quality means with respect to search process logs and models. This paper defines search process models and revisits existing process model and log quality metrics. A metric for search process models is proposed that assesses their complexity and degree of common behavior. In order to compare metrics for search process models different logs and search processes are generated by using ontologies for user guidance during search process execution and for post processing of the logs. Based on an experiment with users in the tourism setting different logs and models are created and compared.

1 Introduction

It is a big asset for companies to know and understand their business processes. Process mining offers a bundle of promising techniques for discovering and analyzing business processes [1]. Business processes are ubiquitous and vary in their nature ranging from short-running and rather rigid administrative processes to highly individual processes such as patient treatment processes [12] and customer journeys describing the user interactions with the company [26]. Recent case studies show that process mining techniques can be successfully applied in order to derive customer journey processes from system logs, e.g., in the entertainment domain [25], in banking [3], and in the tourism domain [16]. Customer journey processes often imply search activities by the users, such as searching for activities when planning a trip. The search behavior can be captured as a *search process* [14] where each of the activities represents a search term a user has looked for through the search system provided by the company. Analyzing such search processes can provide valuable insight for companies [16] and answering the following **analysis questions (AQ)** through search processes:

© Springer Nature Switzerland AG 2018
M. Weske et al. (Eds.): BPM 2018, LNCS 11080, pp. 445–461, 2018.
https://doi.org/10.1007/978-3-319-98648-7_26

- What is the typical customer search behavior?
- What are critical customer touch points?

These **AQ** can influence the customer satisfaction which helps companies to win in the market to increase their revenue [18].

The high variety in search terms, however, might lead to discovered search process models of paramount complexity (also referred to as spaghetti models [1]). Thus the promise of gaining valuable insight might be repealed by non-interpretable process models. Hence, assessing and improving the quality of discovered search process models is the prerequisite to reach analysis goals in a meaningful way. Further on, measures to reduce the complexity of the discovered search process models through pre- and post-processing of the analyzed search logs can be taken (cf. general data quality issues in process mining [1]). For search processes, operational ambiguities might be one major source of data quality issues as caused by description of process activities at different abstraction levels [11], but also by the usage of different languages or due to homonyms and synonyms [29]. Hence, this paper aims at assessing the quality of discovered search process models with respect to the **AQ**, considering how ontologies can be exploited for improving the quality of discovered search processes. In detail:

Q1. How to measure the quality of search process models discovered by process mining techniques with respect to the **AQ**?
Q2. Does the quality of search process models increase when users are supported by an ontology during the search process?
Q3. Does the quality of search process models increase when using an ontology for log post processing?

This paper has an empirical focus with a concrete application setting, but also necessitates the creation of artifacts. As such the developed concepts are application-independent.

Research Method and Contribution: The paper follows design science research (cf. [34]). The relevance of the research problem is underpinned by practical applications from tourism [16], entertainment [25], and banking [3] as well as by literature, specifically on process model quality, e.g., [31] and process mining, e.g., [1]. The following artifacts are created to answer RQ 1–3. At first, a notion of search process models as well as a quality metric specifically tailored towards search process models with respect to **AQ** are proposed in Sect. 2 balancing complexity and clustering in search process models. Section 3 introduces the concept of using ontologies during process execution and for post processing of logs. These artifacts are then evaluated based on an experiment with users in Sect. 4: it creates 4 types of logs for different modi operandi, i.e., for executing search processes in the tourism domain with and without using an ontology and combining these logs with or without post processing. The logs are compared statistically and different metrics are applied to assess the effectiveness of using ontologies on the quality of the resulting logs as well as the feasibility of the newly proposed metric. The paper continues with a discussion in Sect. 5 and a related work discussion in Sect. 6. It concludes in Sect. 7.

2   Search Process Models and Quality Metrics

Search Process Models: One type of human-driven and highly individual processes are search processes. The manifestation of real-world search processes are process logs as, e.g., stored by a tourism platform. Basically, process logs store events that refer to the execution of process activities together with the time stamp of execution and possibly further information such as the originator [1]. The events are grouped for the different process instances based on a case id. One can analyze the logs directly, but as we are particularly interested in typical customer behavior and touch points ($\rightarrow$ **AQ**) also the models behind these logs are of high interest. To the best of our knowledge no formal definition for search process models exists. In information science, informally, an (information) search process is defined as *"the user's constructive activity of finding meaning from information in order to extend his or her state of knowledge on a particular problem or topic. It incorporates a series of encounters with information within a space of time rather than a single reference incident."* [14]. In web (usage) mining, a log-based view is taken: a search or *"query trail qt comprises a user's query q (consisting of a sequence of terms* $\{t_1,\ t2,\ \ldots,\ t_{|q|}\}$*"* [2]. We converge and elaborate both definitions into a (graph-based) search process model:

Definition 1 (Search process model). *Let S be set of all search terms. A search process model is defined as directed graph $SP := (N, E, l)$ where*

- *N is a set of nodes*
- *$E \subseteq N \times N$ denotes the set of control edges*
- *$l : N \mapsto S$ denotes a function that maps each node to its label, i.e., $\forall n \in N$ n is a search activity, i.e., the node n represents the search for a certain search term and is labelled with this search term respectively.*

Figure 1a depicts a small example for a search process model in the tourism domain. The search terms label the process activities, e.g., *bicycling* or *sports shop*, meaning that – after searching for *active* – a user has searched for *bicycling* followed by searching for *sports shop*.

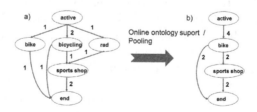

Fig. 1. Example search process model from tourism.

One major difference to typical business processes is that a search process is not set out upfront to manage the behavior, but develops individually for each search during runtime. As we aim at ex post analysis of process data, we abstain from defining process instance states for search processes, but rather count the number of executions and annotate the control edges with this information as done for, for example, dependency graphs [33]. For the example shown in Fig. 1a, 2 users searched for *bicycling* where one user followed up by searching for *sports shop*. Under the precondition that *active* is the starting point for all searches, overall 4 search processes (and instances) were conducted in this example. This perception of the number of executions is suitable for, e.g., process analysis regarding question such as "what are the main search paths taken by the users within the platform".

The example depicted in Fig. 1 illustrates the tackled research problem. Even in this small example, 3 different search terms were used in order to describe the same concept, i.e., *bicycling*, *bike*, and *rad* (the latter being the German word for bike). If the goal is to analyze user behavior it could be more interesting to consolidate these terms into one term as depicted in Fig. 1b where the 3 aforementioned terms have been pooled into search term *bike*. Here it can be seen more easily that 4 users were searching for a term related to concept *bike* and 2 users followed up looking for *sports shop*.

Quality Metrics: How can a metric assessing the quality of search process models with respect to the **AQ** be defined? We argue that one metric cannot assess all quality aspects of a process model at the same time as they might even be contradicting (e.g., showing all details vs. abstraction). The aim of the metric proposed in the following is to emphasize those properties of the model that relate to the **AQ**. Here, specifically, quality aspects refer to the comprehensibility of and the degree of common behavior in the search process models as well as the semantic enrichment of the process logs. Comprehensibility is tied to the complexity of the search process model by reducing the "spaghetti degree". The degree of common behavior is reflected by clustering of activities in the model and the log. The quality metric does not refer to other quality aspects such as how well the discovered search process models reflect the underlying process logs (cf. fitness for process conformance [27]). In the following, the metric is constructed by considering existing business process quality metrics for assessing the complexity and metrics for process log quality for assessing the clustering. In Sect. 4 the new metric is then evaluated against selected existing metrics.

Graph metrics can be applied to assess the complexity of a process model [20]. Transferring this to a search process model $SP := (N, E, l)$ one can consider size ($|N|$), diameter (length of the longest path in SP), structuredness (share of nodes in structured blocks), separability (share of cut vertices in SP), and cyclicity (number of nodes in cycles in SP). For measuring the relation or connection between activities in process models coupling and cohesion have been proposed by [30]. Coupling measures *"how strongly the activities in a workflow process are related, or connected, to each other"*. The connection is measured based on the information elements shared by the activities. The cross-connectivity metric assigns weights to nodes and edges to reflect their connectivity [31]: nodes are weighed based on the number of outgoing edges, edges by the product of the weights of source and target nodes. Cross-connectivity seems promising for the envisioned quality metrics in terms of expressing the role of a node in a network and indicating clusters in the process models. Contrary, for search processes, coupling and cohesion are not meaningful in the context of this paper as no information objects are currently considered for search processes. However, such metrics are promising for future analysis.

Process log and model quality is a major concern in process mining [1]. Different techniques have been proposed to deal with "spaghetti degree", including pre-processing of logs, process mining techniques, and post processing of logs. An example for pre-processing of logs is trace clustering [15] where logs can be clustered along certain criteria, e.g., for a certain process duration or where certain activities were executed. A process mining technique that aims at reducing the complexity of the mined models is the Fuzzy Miner [10]. It employs the principles of aggregation, abstraction, emphasis, and customization. Post processing as suggested by [6] also works with filtering, i.e., abstraction from details, in order to simplify the discovered models. From the principles of Fuzzy Miner and post processing aggregation and abstraction will be chosen for the assessment of search process models with respect to **AQ**. Moreover, the size of the logs will be considered in the proposed metric.

As an outcome of the above discussion, a quality metric for search processes shall incorporate ingredients of process model quality assessing the complexity and connection as well as the existence of clusters as used for process mining, i.e., the frequency of activity execution and the degree of the associate node as well as the number of overall activity executions in order to rate the frequency the activity of interest has been executed. Further on the overall number of events is incorporated to consider the overall diversity of the search process, formally:

Definition 2 (Search Process Quality Metric). *Let $SP = (N, E, l)$ be a search process and let L be a log created by executing instances on SP. Let further $|L|$ be the number of all events contained in L and A be the set of distinct activities having been executed in L. Then the search process quality metric spm(n) for a node $n \in N$ is defined as*

$$spm(n) := 1 - \frac{degree(n) * |A|}{freq * |L|}$$

where freq denotes the number of executions of n.
Search process quality metric spm(SP) for SP turns out as:

$$spm(SP) := \frac{\sum_{n \in N} spm(n)}{|N|}$$

Search process quality metric spm(SP) of a path $p = < n_1, \ldots, n_k > n_i \in N (k \geq 2)$ can be determined similarly, i.e., by

$$spm(SP) := \frac{\sum_{n_i, i=1, \ldots, k} spm(n_i)}{k}$$

Note that the metric avoids isolated nodes being considered as paths. By construction, $spm(n) \in [-1; 1]$ holds as $|A| \leq |L|$ and $degree \leq 2 * freq$.

3   Using Ontologies for User Support and Pooling

We consider the usage of ontologies to improve the process model quality of mined search processes. For that, we distinguish between two approaches where the same given ontology could be applied, (a) during the search functionality, when users are entering search terms and (b) when post processing event logs from search terms which were entered by users through the search process.

For (a) – without the support by an ontology – the user has no guidance and in a broader sense no recommendations for entering search terms. When mining resulting search processes in the sequel, the discovered process models might get complex because of the possibly infinite options for search terms. For (b), the search system has to possibly process synonyms and different languages which are known to pose challenges on later process mining [13], for example, resulting in different activities which have the same meaning and hence unnecessarily pump up the complexity of the search process models. In order to foster (a) and (b), we develop a meta concept which is responsible for recommendations when a user is searching and for post processing of event logs on search processes (cf. Fig. 2). Both approaches, (a) and (b) are implemented for a commercial tourism platform within the CustPro [35] project which aims to analyze the customer journey process of tourists where (a) is already used in the live system by tourists and (b) is implemented for evaluation purpose of this work but is also on the agenda to be implemented in the live system. The backend is written in *Java* as *RESTful Web-Service* and the ontology support with reasoning was conducted with the *Java* framework *Apache Jena*[1]. The frontend was developed in *HTML5* and *JavaScript* as single-page application[2] and talks to the backend with *AJAX*. Therefore the frontend was accessible in the web browser and the UI was optimized for mobile devices.

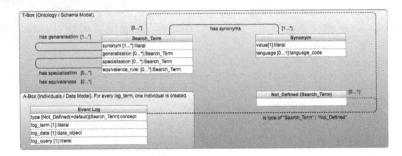

Fig. 2. Ontology with the capability of reasoning and possibilities for post processing.

As set out in Fig. 2, a T-Box model [24] is defined which contains the knowledge base for possible search terms in a specific domain. There are only two

---

[1] https://jena.apache.org/index.html.

[2] http://itsnat.sourceforge.net/php/spim/spi_manifesto_en.php.

concepts defined, which makes the ontology easy to implement and maintain. The first concept is called *Search_Terms*. It consists of the elements *synonyms*, *generalisations*, *specialisations*, and *equivalence rules*. Synonyms are literals in different languages which are referring to the possible search terms in the logs for the concept. Every *Search_Terms* concept has at least one *synonyms* element and every element describes the concept equally in contrast to e.g., SKOS [32] with property *skos:prefLabel*. Generalizations and specializations refer to the concept itself for defining relations between search terms and are optional. Also equivalence rules for defining relations, e.g., for combinations of search terms or relations which cannot be defined in the outlined elements before, or for generating suggestions for search terms, e.g., based on combinations of search terms entered by a user before, are provided and also optional. The other concept, *Not_Defined* does not contain any knowledge. It acts as helper for post processing the event logs. Non-matching terms between the ontology and the logs are flagged with this concept.

Online Usage With Ontology: The T-Box model in Fig. 2 contains the ontology and thus a knowledge base for search terms in a specific domain. For every performed user search query, which can include multiple search terms, online suggestions are created during the search process for user support. Through rules in the ontology, search term specific combinations in a search query determine suggestions by using logical rule reasoning (e.g., a query containing *"mountain"* and *"sports"* results in a suggestion for *"hiking"*). Further, search term combinations like *"x, y"* in the same search query are separated into two search terms and for every search term based on its matching synonyms, specializations and generalizations are determined by using hierarchical reasoning (e.g., a specialization of *"bike"* is *"mountainbike"*) for suggestions. As result, the user gets online suggestions based on the last performed search query with the given ontology.

Post Processing With Ontology: The A-Box [24] model in Fig. 2 contains the event logs which are individuals and defined with the concept *Event Log*. For every log query, which can include multiple search terms, a separate individual per search term is created. Hence, an event log with the search query *"x, y"* will be separated into two individuals, one for the term x and one for y. Each individual consists of the elements *log_term, log_data, log_query* and in the initial phase belongs to the concept *Not_Defined*. *log_term* refers to the search term, which will be processed by using logical rules with the T-Box model (ontology). *log_query* contains the original search query from the user and *log_data* acts as symbolic placeholder for further data from the origin log entry. We applied a logical rule which searches for matching synonyms between a given *log_term* and all *Search_Terms* from the ontology. If there is a match, the type *"Not_Defined"* from the individual is removed, if not already happened before, and the found *Search_Term* is added as type to the individual. After the rule was applied to all individuals, there is a lookup for individuals which do not contain the type *"Not_Defined"*. For these individuals, the initial search strings of their corresponding logs are replaced with the class names from their *Search_Term* types.

Therefore the logs contain pooled search queries for reducing the complexity of mined process models.

In this paper, we employ a controlled ontology (as defined by the analyst). Hence employing post processing does not lead to loss of information in the resulting models when compared to online usage. Contrary to online usage, in case of using an ontology that has not been defined by the analyst, post processing might not reflect the user's intention during search. In general, a user interface for the online ontology support through the search process could positively impact the frequency of its usage as well the quantity of observed logs.

4   Experiment

This section presents the design and execution of the experiment to evaluate the proposed artifacts, i.e., the metrics and the ontology support algorithms. The experiment bases on the implementation described in Sect. 3 and is conducted with subjects in a real world scenario.

4.1   Experimental Setup

The experiment was conducted with students of one course of the Bachelor Computer Science at the University of Vienna. This leads to a relatively homogeneous group of participants and can be regarded as sufficient with respect to knowledge on working with a tourism app. Overall, 93 students participated in the experiment. From an experimental point of view, 2 independent groups are required, i.e., one group working with ontology support and one without. Due to organizational reasons (the course is held in 4 groups), 4 independent groups were built where 2 worked with and 2 without ontology support.

Each of the 4 groups has the same scenario and task to accomplish: Every participant plans touristic activities on an imaginary three day stay from Friday to Monday in a hotel as tourist in the tourism region Mondsee in Austria. The subject writes down titles of activities on an empty schedule which was handed out [35]. The titles of the activities are searched in a search application which contains touristic activities. The experiments started with an introduction of 5 min explaining the task of the experiment. Then the subjects had a 10 min time frame to use their own mobile device (smartphone, tablet, notebook) to search for activities in the provided search application. Two of the four groups received ontology support in their search function. The search application and the ontology were provided in German. The search logs were recorded in the respective time frames and for every group. Different cases were recorded in the event log entries for distinguishing them. In Table 1, the groups of the subjects are depicted.

Note that there is a difference between subjects and devices. The reason for that is, that two subjects can share the same device and one subject can use multiple devices. In the following, only the used devices are further addressed

Table 1. Groups of subjects in experiment.

Group number	Online usage with ontology	Number of different devices (case device)	Number of subjects
1	No	19	20
2	Yes	24	24
3	No	27	24
4	Yes	24	25

because the number corresponds to the recorded event logs during the experiment. As explained before, the groups with and without ontology support are to be compared. For this purpose the 4 groups from Table 1 are merged into 2 logs, i.e., group number 1 and 3 and group number 2 and 4 (cf. Table 2).

Table 2. Merged groups of subjects in experiment.

Log name	Group numbers	Online usage with ontology	Number of different devices (case device)	Number of activities	Number of events
Log 1	1+3	No	46	117	246
Log 3	2+4	Yes	48	116	331

We also implemented a web service, according to Sect. 3, for post processing the obtained event logs, which uses the same ontology as in the experiment for online supporting the subjects on their search functionality in groups 2 and 4. Therefore *log 2* contains post processed event logs from *log 1* which means that both logs originate from the same log recording. The scheme is analogous between *log 4* and *log 3*. Overall, this results in the 4 logs shown in Fig. 3. These logs with their designated log names (*log 1–log 4*) build the basis for further analysis.

Post processing ⇨ Online usage ⇩	Without ontology	With ontology
Without ontology	log 1	log 2
With ontology	log 3	log 4

Fig. 3. Experimental log creation.

4.2 Statistical Comparison of Logs

t-tests [21] are applied to compare the logs from Fig. 3 with respect to the mean of occurred events per *case device*. *Hypthesis I (HI): A higher mean in the logs results when providing online usage with ontology support.* As the subjects tend to perform more search queries because based on the recommendation for further search terms, the user does not have to think about formulating queries. Formulating *H1* is justified by the number of total events as in Table 2. *Hypthesis II (HII): A lower mean results when post processing the logs with ontology support because of pooling search queries.*

First we compared *log 1* with *log 3* to prove a statistical effect on online usage with ontology support. With a one tailed t-test and a 90% confidence interval, we obtained significance for *HI*. There was no evidence on a 95% confidence interval and as well on a two tailed test. Second, we compared *log 1* with *log 2* to prove a statistical effect on post processing logs with ontology support. HII cannot be accepted on a 90% confidence interval. Further log comparisons were not suitable because, as mentioned before, online usage with ontology support tends to increase and post processing with ontology support tends to decrease the mean. The complete t-test is shown in the supplemental material [35] (cf. folder "*T-Test*").

4.3   Process Model Quality Metrics

For assessing the process model quality of each log from Fig. 3, we first applied selected process model quality metrics, i.e., *size* and *diameter* (cf. Sect. 2) as they provide an overview on the complexity of the models. Here, we also applied a filter, which counts only the 20% most frequent activities from each log (abstraction). We chose that percentage because of the *pareto principle*, which shows that in many cases, ranging from the economy to the nature behavior, 80% of causes are produced by 20% of activities [22]. Thus, depending on the given log with its distribution of activities and number of events, only activities with a specific frequency are included in the calculations. Table 3 shows the results which are explained in the following. For *log 1* and *log 2* the filtering has no effect, i.e., the least occurrence of activities remains 1. *log 1* had a *size* of 117 and the *size* of *log 2* was 110. The *diameter* was 19 for both logs. This shows that applying post processing has only a slight effect on the number of activities and no effect on the diameter. For *logs 3* and 4, filtering had an effect, i.e., *log 3* filtered on an activity occurrence ≥ 2 and *log 4* an activity occurrence ≥ 3. With respect to the metrics, this results in a notably reduced size, i.e., for *log 3* a reduction of the size from 116 to 41 and of the diameter from 19 to 15 and for *log 4* a reduction of the size from 109 to 26 and of the diameter from 19 to 13. Hence, it can be interpreted that online ontology support has a considerable effect on reducing the process model metrics size and diameter when using filtering. There is also to mention, that each of *log 1* and *log 2* contained 33 variants of paths where only 3 variants contained more than 1 *case device*. Nearly the same picture was discovered on *log 3* and as well *log 4*. Each of them had 39 variants and only 2 variants contained more than 1 *case device*. This is a good indicator, which shows how highly individual a user performed search process can be.

As next step for assessing the process model quality of each log, we applied our defined *Search Process Quality Metric* from Sect. 2. Table 4 summarizes the

Table 3. Results of applying regular process model quality metrics.

Log name	Size unfiltered	Diameter unfiltered	Size filtered	Diameter filtered
Log 1	117	19	117	19
Log 2	110	19	110	19
Log 3	116	19	41	15
Log 4	109	19	26	13

results. In regard to the values of the path it is to be noted that there was as well a filter applied which includes only the 20% most frequent activities. But we also excluded paths which contain solely the search terms "*" or *FIRST_RUN_-_*" in any combination. We did the latter because: the search term *FIRST_RUN_-_*" signals, that after loading the application the first time on a device, an automatic "*"-search is performed. A "*"-search in the logs signals that the user just hit the search button without considering a search term as query input. Thus, such a path, where no search term was entered by a subject has no meaning in our case and was filtered out. As we can see, from *log 1* to *log 4* there was an increase in the quality of the logs. We can also see that the online usage of an ontology had a bigger impact than post processing with ontology. The difference between *log 1* and *log 2* is 0, 058 on *spm(SP) unfiltered* where the difference between *log 1* and *log 3* is 0, 228. The values of *spm(SP) filtered* have a greater impact through the online usage with ontology. In comparison to the regular process model quality metrics (size and diameter), we can see an improvement in terms of meaningfulness by using our *Search Process Quality Metric* for identifying clusters and quality comparisons of logs. All values in Table 4 increased when using ontology support. We can conclude, that the ontology usage improved the process model quality in general and on specific paths.

Table 4. Results of applying *Search Process Quality Metrics.*

Log name	spm(SP) unfiltered entire search process	spm(SP) filtered path highest value	spm(SP) filtered path lowest value
Log 1	0,145	0,714	0,107
Log 2	0,203	0,731	0,166
Log 3	0,373	0,781	0,412
Log 4	0,415	0,794	0,481

For visual inspection, process models were discovered for each of the 4 logs from Fig. 3 using *Disco* (cf. Fig. 4) which uses an adapted version of the Fuzzy Miner, called *Disco miner*, which is geared towards discovering clusters in process

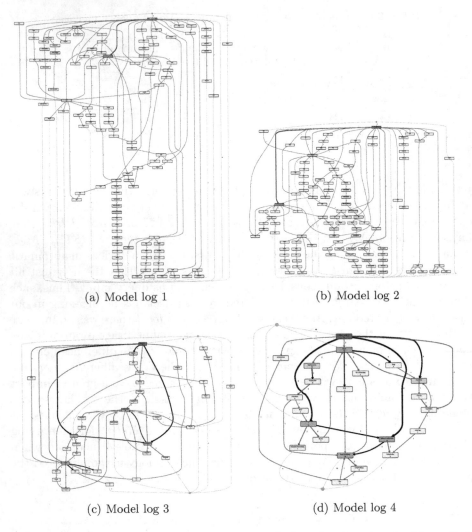

(a) Model log 1   (b) Model log 2

(c) Model log 3   (d) Model log 4

Fig. 4. Visual inspection of process discovery results based on logs 1 to 4 with enabled filter for showing 20% of most frequent activities and only most dominant paths.

model[3]. The process models show only the most frequent activities and the most dominant paths in their process map. For this the *paths slider* was set to 0% to show only dominant connections between activities that have occurred and the *activities slider* was set to 20% to show only the 20% most frequent activities in the mined process map. Furthermore the absolute frequency of an activity is visualized using color strength. Only from visual inspection the resulting model for *log 4* seems to be less complex and to contain more clusters when compared to logs 1–3.

---

[3] https://fluxicon.com/blog/2012/05/say-hello-to-disco/.

Finally, activities are selected through visual inspection, i.e., those showing high clustering based on color strength, and analyzed using the *Search Process Quality Metric*. We started with the models for logs before and after post processing. Search term *"Kino"* (cinema), for example, is compared for *log 1* with value 0,049 and *log 2* with value 0,285 as well search term *"essen"* (eat) with value 0,239 from *log 1* which was pooled to the term *"Gastronomie"* (gastronomy) with the value 0,292. We also compared the model from *log 3* with the search terms *"Restaurant"* (restaurant) with value 0,299 and *"Berg"* (mountain) with value 0,509 with *log 4* and their scorings for *"Restaurant"* with value 0,506 and *"Berg"* with value 0,539. These results confirm that post processing with ontology usage has a positive impact on search process quality. Then selected activities are compared for the same process model. For the model of *log 4*, we chose two frequent search terms. The first is *"Restaurant"* with term frequency of 19 and a *spm(n)* of 0,567. The second one is *"Freizeitaktivität"* (leisure activity) with term frequency of 24 and a *spm(n)* of 0,575. Both terms had a lower value than the term *"Ball"* (ball) with value of 0,78 despite the lower frequency of only 3 but with a clearer path (lower degree). For the terms *"Familie"* (family) with a term frequency of 4 and *"Restaurant"* with a term frequency of 8 and with the same value of *spm(n)*, which was 0,506, nearly the same value was indicated than for the two very frequent terms *"Restaurant"* and *"Freizeitaktivität"*. The reason for that is, that on *"Familie"* and *"Restaurant"* the incoming and outgoing nodes were better clustered. We can conclude that the *Search Process Quality Metric* supports to discover and rank important paths and fragments in terms of clustering in search processes. The results, figures and calculations from the experiment can be found in the supplemental material [35] (cf. folder *"Experiment"*).

5   Limitations and Threats to Validity

The results could be improved if the ontology increases in terms of its size and relevance to its domain where the search process is executed for. Moreover, the sample size of the experiment could have been too small for filtering and multiple languages. We will investigate the relation between filtering and sample size in future work. Because of the small sample size and therefore the small number of available subjects, we decided to run the experiment in one language, i.e., German, because the ontology contains a different number of classes and labels for describing them per language. Also rules for suggestions of combined search terms are not the same per language. This could be an explanation why post processing does not show a comparable impact to another study in the tourism domain on a live system that was conducted using German and English at the same time. Find the discovered models in Fig. 5. Though there is not enough space to discuss this study in detail, the models give an impression that post processing using an ontology resolves ambiguities with respect to language. Another limitation of the proposed approach is the missing handling of compound nouns in search queries with mixed search terms that contain both, compound nouns and single nouns. In the tourism domain, we have had to deal with this problem and added hyphens between the search terms. The corresponding tourism ontology contains also hyphens for compound nouns, defined as synonyms. For example, if user enters search term *"nature sights"* it is modified to *"nature-sights"* and the label in the ontology is exact the same. But in the particular case of the tourism platform, we have to mention, that search queries with multiple nouns are very seldom. Nevertheless we are planning future research activities, to deal with compound nouns in search queries. A starting point would be the work presented in [28].

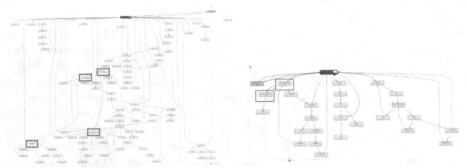

(a) Model without post processed log. 4 red squares show different search terms

(b) Model after post processing the log. 4 squares from (a) are pooled to 2 squares

Fig. 5. Mined search processes from logs which contain different languages. Without post processing the log (*left*) and after post processing the log (*right*) which reduces the complexity of the model by pooling the search terms from different languages (*English and German*). The models can be found in the supplemental material [35] (cf. folder *"Figure5Detail"*.)

6   Related Work

Process mining algorithms have become mature and efficient [1]. There are various ways for simplifying discovered processes [6]. For improving the quality of process mining results, apart from implementing constantly improving algorithms [4], it seems obvious to improve also the event log quality. This can be performed in different kind of approaches [11,17]. One of them is the promising research field of semantic technologies [8], which is also quite proven very well. Process mining, combined with semantic technologies can improve the meaningfulness and therefore the quality of mined processes reasonably well [1,19]. Most approaches, for enhancing process mining with semantic technologies, are using ontologies behind the scene for log preparation, e.g., by mapping process labels from event logs to hierarchical links in ontologies [5]. Ontologies are also used, to reduce the complexity of discovered process models by dealing with synonyms, hierarchies, reasoning or constraints [13]. But most ontologies, which are used for creating or enhancing semantic logs are either complex and thus burdensome [23] or too domain specific [7] for using them in different domains in industrial software solutions [29]. As opposed to all the aforementioned approaches this work addresses search processes where the search terms are defined by the users in an arbitrary manner, and not by the application in form of predefined labels as addressed mostly in literature and research [11]. Semantics, together with keyword search is also already covered in literature [9], but without log preparation and meta ontologies, that are easy to adapt with a minimum of effort in the widest possible range of domains and industries. Quality metrics from literature have been discussed in Sect. 2.

7   Conclusion

Case studies from different domains emphasize the potential of process mining for customer journey understanding and improvement. This work assesses and improves the quality of mined search processes as important brick in customer journeys. Regarding *RQ 1*, a newly proposed quality metric for search processes rates the complexity of the output combined with an assessment of the existence of clusters. The experiment evaluates the feasibility of the metric in comparison with results from visual inspection and existing metrics. Moreover, the experiment evaluates quality improvement when using ontologies for online user support (*RQ 2*) as well as for post processing the resulting logs (*RQ 3*). The experiment is demonstrated by a case study in the tourism domain. In summary, *RQ 2* has been positively demonstrated in Sect. 4.2 by analyzing the mean of occurred events per *case device* and in Sect. 4.3 – where the results have been significantly reinforced by using filtering – by showing through existing metrics, visual inspection of mined process models, and the proposed quality metric, that the complexity of process models decreases and clusters improve. The same improvements could also be shown in Sect. 4.3 for *RQ 3*, but with slightly less evidence for the experiment because of the limitations as pointed out in Sect. 5. Overall, by answering *RQ 1–3*, it can be seen that complexity of and clustering in search processes improve, in particular, in combination with filtering. In future work, we will also consider time with respect to new metrics and experiment designs and measure the impact of an ontology on how accurate search results are for the users *"to get their jobs done"*. Moreover, the mined search processes will be further analyzed with respect to their differences in relation to context variables such as location or weather.

Acknowledgment. This work has been partly conducted within the CustPro project funded by the Vienna Business Agency.

References

1. van der Aalst, W.M.P.: Process Mining - Data Science in Action, 2nd edn. Springer, Heidelberg (2016). https://doi.org/10.1007/978-3-662-49851-4
2. Bailey, P., White, R.W., Liu, H., Kumaran, G.: Mining historic query trails to label long and rare search engine queries. TWEB 4(4), 15:1–15:27 (2010)
3. Celonis: Process mining story postfinance: Optimizing the customer journey in banking (2018). https://youtu.be/qJ2NcdZSxA4
4. Dixit, P., Buijs, J.C., van der Aalst, W.M., Hompes, B., Buurman, H.: Enhancing process mining results using domain knowledge. In: SIMPDA, pp. 79–94 (2015)
5. Dunkl, R.: Data improvement to enable process mining on integrated non-log data sources. In: Moreno-Díaz, R., Pichler, F., Quesada-Arencibia, A. (eds.) EURO-CAST 2013. LNCS, vol. 8111, pp. 491–498. Springer, Heidelberg (2013). https://doi.org/10.1007/978-3-642-53856-8_62
6. Fahland, D., van der Aalst, W.M.P.: Simplifying discovered process models in a controlled manner. Inf. Syst. 38(4), 585–605 (2013)
7. Fernandez, F.M.H., Ponnusamy, R.: Data preprocessing and cleansing in web log on ontology for enhanced decision making. Indian J. Sci. Technol. 9(10) (2016). http://www.indjst.org/index.php/indjst/article/view/88899
8. Fürber, C.: Data Quality Management with Semantic Technologies. Springer, Wiesbaden (2015). https://doi.org/10.1007/978-3-658-12225-6
9. Gulla, J.A.: Applied Semantic Web Technologies. Auerbach Publications, Boca Raton (2011)
10. Günther, C.W., van der Aalst, W.M.P.: Fuzzy mining – adaptive process simplification based on multi-perspective metrics. In: Alonso, G., Dadam, P., Rosemann, M. (eds.) BPM 2007. LNCS, vol. 4714, pp. 328–343. Springer, Heidelberg (2007). https://doi.org/10.1007/978-3-540-75183-0_24
11. Ingvaldsen, J.E., Gulla, J.A.: Industrial application of semantic process mining. Enterp. Inf. Syst. 6(2), 139–163 (2012)
12. Kaes, G., Rinderle-Ma, S.: Generating data from highly flexible and individual process settings through a game-based experimentation service. In: Datenbanksysteme für Business, Technologie und Web, pp. 331–350 (2017)
13. Koschmider, A., Oberweis, A.: Ontology based business process description. In: EMOI-INTEROP, pp. 321–333 (2005)
14. Kuhlthau, C.C.: Inside the search process: information seeking from the user's perspective. Am. Soc. Inf. Sci. 42(5), 361–371 (1991)
15. de Leoni, M., van der Aalst, W.M.P., Dees, M.: A general process mining framework for correlating, predicting and clustering dynamic behavior based on event logs. Inf. Syst. 56, 235–257 (2016)
16. Lux, M., Rinderle-Ma, S.: Problems and challenges when implementing a best practice approach for process mining in a tourist information system. In: BPM 2017 Industry Track. CEUR, vol. 1985, pp. 1–12 (2017)
17. Ly, L.T., Indiono, C., Mangler, J., Rinderle-Ma, S.: Data transformation and semantic log purging for process mining. In: Ralyté, J., Franch, X., Brinkkem-per, S., Wrycza, S. (eds.) CAiSE 2012. LNCS, vol. 7328, pp. 238–253. Springer, Heidelberg (2012). https://doi.org/10.1007/978-3-642-31095-9_16
18. Maechler, N., Neher, K., Park, R.: From touchpoints to journeys: seeing the world as customers do, March 2016. http://bit.ly/2AAzjcJ
19. Medeiros, A.K.A., et al.: An outlook on semantic business process mining and monitoring. In: Meersman, R., Tari, Z., Herrero, P. (eds.) OTM 2007. LNCS, vol. 4806, pp. 1244–1255. Springer, Heidelberg (2007). https://doi.org/10.1007/978-3-540-76890-6_52

20. Mendling, J., Strembeck, M.: Influence factors of understanding business process models. In: Abramowicz, W., Fensel, D. (eds.) BIS 2008. LNBIP, vol. 7, pp. 142–153. Springer, Heidelberg (2008). https://doi.org/10.1007/978-3-540-79396-0_13
21. Mertens, W., Pugliese, A., Recker, J.: Quantitative Data Analysis: A Companion for Accounting and Information Systems Research. Springer, Cham (2016). https://doi.org/10.1007/978-3-319-42700-3
22. Moore, H.: Cours d'économie politique. Ann. Am. Acad. Polit. Soc. Sci. **9**(3), 128–131 (1897). http://bit.ly/2FGyANa
23. Pedrinaci, C., Domingue, J., Alves de Medeiros, A.K.: A core ontology for business process analysis. In: Bechhofer, S., Hauswirth, M., Hoffmann, J., Koubarakis, M. (eds.) ESWC 2008. LNCS, vol. 5021, pp. 49–64. Springer, Heidelberg (2008). https://doi.org/10.1007/978-3-540-68234-9_7
24. Petnga, L., Austin, M.: An ontological framework for knowledge modeling and decision support in cyber-physical systems. Adv. Eng. Inform. **30**(1), 77–94 (2016)
25. Pmig, Y., Yongil, L.: Customer Journey Mining (2018)
26. Richardson, A.: Using customer journey maps to improve customer experience. Harvard Bus. Rev. **15**(1), 2–5 (2010)
27. Rozinat, A., van der Aalst, W.M.P.: Conformance checking of processes based on monitoring real behavior. Inf. Syst. **33**(1), 64–95 (2008)
28. Silverstein, C., Marais, H., Henzinger, M., Moricz, M.: Analysis of a very large web search engine query log. In: ACm SIGIR Forum, vol. 33, pp. 6–12. ACM (1999)
29. Thomas, O., Fellmann, M.: Semantic process modeling - design and implementation of an ontology-based representation of business processes. Bus. Inf. Syst. Eng. **1**(6), 438–451 (2009)
30. Vanderfeesten, I.T.P., Reijers, H.A., van der Aalst, W.M.P.: Evaluating workflow process designs using cohesion and coupling metrics. Comput. Ind. **59**(5), 420–437 (2008)
31. Vanderfeesten, I., Reijers, H.A., Mendling, J., van der Aalst, W.M.P., Cardoso, J.: On a quest for good process models: the cross-connectivity metric. In: Bellahsène, Z., Léonard, M. (eds.) CAiSE 2008. LNCS, vol. 5074, pp. 480–494. Springer, Heidelberg (2008). https://doi.org/10.1007/978-3-540-69534-9_36
32. W3C: SKOS simple knowledge organization system reference, August 2009. https://www.w3.org/TR/2009/REC-skos-reference-20090818/
33. Weijters, A.J.M.M., Ribeiro, J.T.S.: Flexible heuristics miner (FHM). In: Computational Intelligence and Data Mining, pp. 310–317 (2011)
34. Wieringa, R.: Design Science Methodology for Information Systems and Software Engineering. Springer, Heidelberg (2014). https://doi.org/10.1007/978-3-662-43839-8
35. WST: Supplementary material (2018). http://gruppe.wst.univie.ac.at/projects/CustPro/

Predictive Process Monitoring Methods: Which One Suits Me Best?

Chiara Di Francescomarino[1], Chiara Ghidini[1], Fabrizio Maria Maggi[2(✉)], and Fredrik Milani[2]

[1] FBK-IRST, Via Sommarive 18, 38050 Trento, Italy
{dfmchiara,ghidini}@fbk.eu
[2] University of Tartu, Liivi 2, 50409 Tartu, Estonia
{f.m.maggi,milani}@ut.ee

Abstract. Predictive process monitoring has recently gained traction in academia and is maturing also in companies. However, with the growing body of research, it might be daunting for data analysts to navigate through this domain in order to find, provided certain data, what can be predicted and what methods to use. The main objective of this paper is developing a value-driven framework for classifying predictive process monitoring methods. This objective is achieved by systematically reviewing existing work in this area. Starting from about 780 papers retrieved through a keyword-based search from electronic libraries and filtering them according to some exclusion criteria, 55 papers have been finally thoroughly analyzed and classified. Then, the review has been used to develop the value-driven framework that can support researchers and practitioners to navigate through the predictive process monitoring field and help them to find value and exploit the opportunities enabled by these analysis techniques.

Keywords: Predictive process monitoring · Process mining
Value-driven framework

1 Introduction

Process mining is a family of methods to analyze business processes based on their observed behavior recorded in *event logs*. In this setting, an event log is a collection of *traces*, each representing one execution of the process (a.k.a. a *case*). A trace consists of a sequence of timestamped events, each capturing the execution of an activity. Each event may carry a payload consisting of attribute-value pairs such as the resource(s) involved in the execution of the activity, or other data recorded with the event. Since process mining is a relatively young discipline, the open challenges in this field are still many [1,33]. In particular, one of these challenges defined in [33] is about "providing operational support"

F. M. Maggi and F. Milani—This research is supported by the Estonian Research Council Grant IUT20-55.

M. Weske et al. (Eds.): BPM 2018, LNCS 11080, pp. 462–479, 2018.
https://doi.org/10.1007/978-3-319-98648-7_27

and, in particular, about the definition of techniques for supporting the three main operational support activities, i.e., *detect*, *predict* and *recommend*. Very recently, researchers have started focusing on the development of techniques supporting the operational support activity *predict* a.k.a. predictive process monitoring techniques.

Predictive process monitoring [41] is a branch of process mining that aims at predicting at runtime and as early as possible the future development of ongoing cases of a process given their uncompleted traces. As demonstrated in [47], recently, a wide literature about predictive process monitoring techniques has become available. This large number of studies has triggered the need to order and classify what has already been done, as well as to identify gaps in the current literature, thus supporting and guiding researchers towards the future advances in this research field. In addition, although selected companies have integrated predictive methods in their business processes [32], the potential for greater impact is still very real. However, due to the large availability of techniques, it might be daunting for companies to navigate through this domain. In the quest to find, provided certain data, what can be predicted and what methods to use, companies can easily get lost.

As such, this paper has the objective to develop, based on the review of existing research, a value-driven framework for classifying existing predictive process monitoring methods, thus supporting, on the one hand, researchers who need to crystallize the existing work to identify the next challenges of this research area, and, on the other hand, companies that need to be guided to find the best solutions within the wide literature available.

In particular, the main research question we want to answer is (**RQ**): *"How can the body of relevant academic publications within the field of predictive process monitoring be classified as a framework?"*. This question is answered through a systematic literature review that aims at identifying the state of the art of predictive process monitoring. We conducted the literature review based on a standard protocol defined in [35]. Starting from a keyword-based search from electronic libraries, we identified around 780 papers and, going through multiple filtering rounds, we selected 55 papers that were analyzed and categorized. The review was then used to develop the value-driven framework that supports researchers and practitioners to identify the methods that fit their needs. For example, one of the discriminative characteristics of the classified methods is the type of prediction. Once defined the type of prediction, the input data required is taken into consideration. The next step is understanding the family of algorithms applied, whether tools are available, the domain in which they have been applied and so on. We started from 4 main dimensions that were, in our opinion, the most relevant in the categorization of the identified methods (type of prediction, type of input, family of algorithms and tool availability) and we expanded them while progressing with the literature review.

The remainder of the paper is organized as follows. Section 2 describes the background about predictive process monitoring, while Sect. 3 presents the literature review protocol and its results. Sect. 4 provides a classification of the identified methods. In Sect. 5, the framework is presented and discussed. Finally, Sect. 6 concludes the paper.

2   Predictive Process Monitoring

The execution of business processes is generally subject to internal policies, norms, best practices, regulations, and laws. For this reason, compliance monitoring is an everyday imperative in many organizations. Accordingly, a range of research proposals have addressed the problem of assessing whether a process execution complies with a set of requirements [40]. However, these monitoring approaches are *reactive*, in that they allow users to identify a violation only *after it has occurred* rather than supporting them in *preventing* such violations in the first place.

Based on an analysis of historical execution traces, *predictive process monitoring* methods [41] continuously provide the user with predictions about the future of a given ongoing process execution. The forward-looking nature of predictive monitoring provides organizations with new frontiers in the context of compliance monitoring. Indeed, knowing in advance about violations, deviances and delays in a process execution would allow them to take preventive measures (e.g., reallocating resources) in order to avoid the occurrence of those situations and, as often happens, to avoid money loss.

3   Systematic Literature Review Protocol

The systematic review protocol of our literature review specifies the research questions, the search protocol, and the selection criteria predominantly following the guidelines provided by Kitchenham [35]. The work was divided into two phases. In the first phase, two researchers designed the review protocol. Here, the research questions were formulated, the electronic databases identified, the inclusion and exclusion criteria defined, and the data extraction strategy formulated. The second phase was conducted by two other researchers who reviewed the protocol, ran the searches, filtered the list of papers, produced the final list of papers and extracted the data. The data extraction was initially independently conducted by the two researchers. The results were then compared and discussed. If differences were noted, they were reconciled first by discussions, then with the participation of the researchers involved in the first phase and, if needed, by contacting the author team of the paper in question.

The main research question (**RQ**): "*How can the body of relevant academic publications within the field of predictive process monitoring be classified as a framework?*" is decomposed in the following four sub-questions. The first research question seeks to identify the different aspects of business processes that can be predicted by means of predictive process monitoring techniques. As such, the first research question is formulated as **RQ1**: "*what aspects of business processes do predictive process monitoring techniques predict?*". The next two research questions concern the algorithms employed in predictive process monitoring techniques. The second is **RQ2**: *"what input data do predictive process monitoring algorithms require?"*. The third research question is **RQ3**: *"what are the main families of algorithms used in predictive process monitoring techniques?"*. Finally, the last research question focuses on tools for predictive process monitoring i.e., **RQ4**: *"what are the tools that support predictive process monitoring?"*.

The first research question is motivated by the necessity of knowing what can be predicted when guiding companies or researchers in the selection of a predictive process monitoring technique. The second research question is motivated by the fact that, to apply a certain technique, different types of inputs can be required. It is therefore crucial to know what information is needed to run each technique. Most of the techniques require an event log as input, but the information contained in it is often requested to be provided at different levels of granularity. The third research question is mainly relevant for researchers for classifying existing methods in different families, which could help to understand their strengths and limitations, and to identify the next challenges in this field. Finally, tool support, investigated in the fourth research question, is relevant for running the proposed methods in academy and industry contexts. Starting from these dimensions that were, in our opinion, the most relevant for categorizing the existing predictive process monitoring methods, while analyzing the literature, some additional relevant aspects were identified to discriminate among the methods under examination such as the type of output, the metrics used for evaluating the algorithms and the domains where the methods were evaluated.

To answer these research questions, we defined a search string to query some electronic libraries. We followed the guidance given by [35], which resulted in using the keywords "predictive", "prediction", "business process", and "process mining". The keywords "predictive" and "prediction" were derived from the research questions. However, the literature on predictive analysis is vast and encompasses areas outside of the domain of business processes. As such, we added the keyword "business process". Finally, predictive process monitoring concerns "process mining" and, therefore, this keyword was added. To retrieve as many results as possible, we intentionally left out additional keywords such as "monitoring", "technique", and "algorithm" so not to limit the search. The keywords were used to formulate the following boolean search string: ("predictive" OR "prediction") AND ("business process" OR "process mining").

Following the definition of the search string, the electronic libraries were chosen. The databases selected were Scopus, SpringerLink, IEEE Xplore, Science Direct, ACM Digital Library, and Web of Science. These were selected as they cover scientific publications within the field of computer science.

The results[1] were exported into an excel sheet for processing. The first filtering was for duplicates. Duplicate studies are those that appeared in more than one electronic database with identical title by the same author(s) [36]. Having performed this filtering, 779 papers were identified. Next, the list was filtered

---

[1] The queries were run on October 20, 2017.

based on the title of the study. Studies clearly out of scope were removed. In detail, we filtered (i) all documents that are not proper research papers, but rather editorial or book introductions (e.g., "25 Years of Applications of Logic Programming in Italy", or "Introduction to Process Mining"); (ii) all studies related to completely different research areas (e.g., "A RFID-based recursive process mining system for quality assurance in the garment industry", or "Process modeling and optimization of complex systems using Scattered Context Grammars"). After this filter, 186 papers remained. Position papers and papers published in workshops were excluded as their contribution is less mature as compared to full papers published in conferences and journals. At this stage, 162 papers remained. We then filtered the papers by looking at the abstracts to further assess their relevance. Following this step, 77 papers remained. After this, we used the inclusion criterion *"does the study propose a novel algorithm or technique for predictive process monitoring?"*. We found 50 papers that complied with this criterion. Finally, we added 5 relevant papers to these 50 papers, via a backward reference search.

From the final list of 55 papers, we extracted standard meta-data (title, authors, year of publication, number of citations, and type of publication). For each paper, the type of prediction considered was extracted in accordance with **RQ1**. Some methods can predict several aspects of a business process. In such cases, all aspects were recorded. Secondly, information about the input data was extracted. Process mining requires that logs contain at least a unique case id, activity names and timestamps. However, if a method requires additional data for analysis, this was extracted and noted (**RQ2**). Thirdly, we examined the type of algorithm. The underlying family of each identified predictive process monitoring method was extracted to address **RQ3**. We also extracted information about the validation of each method. The quality of a method depends indeed on its validation. As such, data about validation was extracted. In addition, if validated, data about if the log was synthetic or from real-life (including the industry domain) was extracted as well. Data about tool support was also extracted. This information considers if there is a plug-in or a stand-alone application for the proposed predictive process monitoring technique. This data relates to **RQ4**. In addition to the above data, the type of output, the metrics used for evaluating the algorithms and if the datasets used were public or not was annotated.[2]

4   Systematic Literature Review Results

The literature review reveals that the number of publications on predictive process monitoring has significantly increased in the last few years. By analyzing the meta-data extracted, we found that out of 55 papers analyzed 14 were published between 2006 and 2013, whereas 41 in the last 4 years (17 journals and 38 conferences in total).

---

[2] The data extracted was entered into an excel sheet and is available for download at https://docs.google.com/spreadsheets/d/1l1enKhKWx_3KqtnUgggr Pl1aoJMhvmy9TF9jAM3snas/edit#gid=959800788 .

The main dimension that is typically used to classify predictive process monitoring techniques is the type of prediction [21]. In the following sections, some of the research studies identified through our systematic literature review and characterizing three prediction type macro-categories, i.e., numeric, categorical and next activities predictions, are presented.

4.1   Numeric Predictions

We can roughly classify the studies dealing with numeric predictions in two groups, based on the specific type of predictions returned:

- time predictions;
- cost predictions.

Time Predictions. The group of studies focusing on the time perspective is a rich group. Several studies, in this context, rely on explicit models. In [2], the authors present a set of approaches in which transition systems, which are built based on a given abstraction of the events in the event log, are annotated with time information extracted from the logs. In particular, information about elapsed, sojourn, and remaining time is reported for each state of the transition system. The information is then used for making predictions on the completion time of an ongoing trace. Further extensions of this approach are proposed in [53,54], where the authors apply machine learning techniques to annotate the transition systems. In detail, in [53], the transition systems are annotated with machine learning models such as Naïve Bayes and Support Vector Regression models. In [54], instead, the authors present, besides two completely new approaches based on Support Vector Regression, a refinement of the study in [53] that also takes into account data. Moreover, the authors evaluate the three proposed approaches both on stationary and non-stationary (i.e., characterized by evolving conditions) processes. Other extensions of the approach presented in [2], which also aim at predicting the remaining time of an ongoing trace, are the studies presented in [28,30]. In these studies, the annotated transition system is combined with a context-driven predictive clustering approach. The idea behind predictive clustering is that different scenarios can be characterized by different predictors. Moreover, contextual information is exploited in order to make predictions, together with control-flow [28] or control-flow and resources [30].

Another approach based on the extraction of (explicit) models (*sequence trees*) is presented in [10] predicting the completion time and the next activity of a current ongoing case. Similarly to the predictive clustering approach, the sequence tree model allows for clustering traces with similar sequences of activities (control-flow) and building a predictor model for each node of the sequence tree by leveraging data payload information. In [56], the authors use generally distributed transitions stochastic Petri nets (GDT-SPN) to predict the remaining time of a case. In detail, the approach takes as inputs a stochastic process model, which can be known in advance or inferred from historical data, an ongoing trace, and some other information as the current time in order to make predictions on the remaining time. In [57], the authors also exploit the elapsed time since the last event in order to make more accurate predictions on the remaining time and to estimate the probability of missing a deadline.

Differently from the previous approaches, in [20], the authors only rely on the event log in order to make predictions. In detail, they develop an approach for predicting the remaining cycle time of a case by using non-parametric regression and leveraging activity duration and occurrences as well as other case-related data. In [4,11], the contextual clustering-based approach presented in [28,30] is updated in order to address the limitation of transition system-based approaches requiring the analyst to choose the log abstraction functions, by replacing the transition system predictor with standard regression algorithms. Moreover, in [11], the clustering component of the approach is further improved in order to address scalability and accuracy issues. In [49], Hidden Markov Models (HMM) are used for making predictions on the remaining time. A comparative evaluation shows that HMM provides more accurate results than annotated transition systems and regression models. In [59], *inter-case feature predictions* are introduced for predicting the completion time of an ongoing trace. The proposed approaches leverage not only the information related to the ongoing case, but also the status of other (concurrent) cases (e.g., the number of concurrent cases) in order to make predictions. The proposed encodings demonstrated an improvement of the results when applied to two real-life event logs.

A prediction type that is very close to remaining time is the prediction of the delay of an ongoing case. In [62], queuing theory is used to predict possible delays in business process executions. The authors propose approaches that either enhance traditional approaches based on transition systems, as the one in [2], to take queueing effects into account, or leverage properties of queue models.

Cost Predictions. A second group of studies focuses on cost predictions. Also in this group, we can find studies explicitly relying on models as the study in [66]. In such a study, cost predictions are provided by leveraging a process model enhanced with costs (i.e., a frequent-sequence graph enhanced with costs) taking into account information about production, volume and time.

4.2   Categorical Predictions

The second family of prediction approaches predicts categorical values. In this settings, two main specific types of predictions can be identified:

- risk predictions;
- categorical outcome predictions.

Risk Predictions. A first large group of studies falling under the umbrella of outcome-oriented predictions, deals with the prediction of risks.

Also in this case, an important difference among state-of-the-art approaches is the existence of an explicit model guiding the prediction. For example, in [15], the authors present a technique for reducing process risks that requires a process model as input. The idea is supporting process participants in making risk-informed decisions, by providing them with predictions related to process executions. Decision trees are generated from logs of past process executions, by taking into account information related to data, resources and execution frequencies provided as input with the process model. The decision trees are then traversed and predictions about risks returned to the users. In [12,13], two extensions of the study in [15] are presented. In detail, in [12], the framework for risk-informed decisions is extended to scenarios in which multiple cases run concurrently. In particular, in order to deal with the risks related to different cases of a process, a technique that uses integer linear programming is exploited to compute the optimal assignment of resources to tasks to be performed. In [13], the study in [15] is extended so that the process executions are not considered in isolation anymore, but, rather, the information about risks is automatically propagated to similar running cases of the same process in real-time in order to provide early runtime predictions.

In [44], three different approaches for the prediction of constraint violations in a case are investigated: machine learning, constraint satisfaction and QoS aggregation. The authors, beyond demonstrating that all the three approaches achieve good results, identify some differences and propose to combine them. Results on a real case study show that combining these techniques actually allows for improving the prediction accuracy.

Other studies, devoted to risk prediction, do not take into account explicit models. For instance, in [50], the authors make predictions about time-related process risks by identifying and leveraging process risk indicators (e.g., abnormal activity execution time or multiple activity repetition) by applying statistical methods to event logs. The indicators are then combined by means of a prediction function, which allows for highlighting the possibility of transgressing deadlines. In [51], the authors extend their previous study by introducing a method for configuring the process risk indicators. The method learns from the outcomes of completed cases the most suitable thresholds for the process risk indicators, thus taking into account the characteristics of the specific process and, therefore, improving the accuracy.

Categorical Outcome Predictions. A second group of predictions relates to the fulfillment of predicates. Almost all studies falling under this category do not rely on any explicit model. For example, in [41], a framework for predicting the fulfillment (or the violation) of a predicate in an ongoing execution is introduced. Such a framework makes predictions by leveraging: (i) the sequence of events already performed in the case; and (ii) the data payload of the last activity of the ongoing case. The framework is able to provide accurate results, although it demands for a high runtime overhead. In order to overcome such a limitation, the framework has been enhanced in [18] by introducing a clustering preprocessing step in which cases sharing a similar behaviour are clustered together. A predictive model - a classifier - for each cluster is then trained with the data payload of the traces in the cluster. In [17], the framework is enhanced in order to support users by providing them with a tool for the selection of the techniques and the hyperparameters that best suit their datasets and goals.

In [39], the authors consider traces as complex symbolic sequences, i.e., sequences of activities each carrying a data payload consisting of attribute-value pairs. By starting from this assumption, the authors focus on the comparison of different feature encoding approaches, ranging from traditional ones, such as counting the occurrences of activities and categorical attributes in each trace, up to more complex ones, relying on HMM. In [67], the approach in [39] is enhanced with clustering, by proposing a two-phase approach. In the first phase, prefixes of historical cases are encoded as complex symbolic sequences and clustered. In the second phase, a classifier is built for each cluster. At runtime, (i) the cluster closest to the current ongoing process execution is identified; and (ii) the corresponding classifier is used for predicting the case outcome (e.g., whether the case is normal or deviant).

In [65], in order to improve prediction accuracy, unstructured (textual) information, contained in text messages exchanged during process executions, is also leveraged, together with control and data flow information. In particular, different combinations of text mining (bag-of-n-grams, Latent Dirichlet Allocation and Paragraph Vector) and classification (Random Forest and logistic regression) techniques are proposed and exercised. In [48], an approach based on evolutionary algorithms is presented. The approach is based on the definition of process indicators (e.g., whether a case is completed on time or whether it is reopened). At training time, the process indicators are computed, the training event log encoded and the evolutionary algorithms applied for the generation of a predictive model composed of a set of decision rules. At runtime, the current trace prefix is matched against the decision rules in order to predict the correct class for the ongoing running case.

4.3   Next Activities Predictions

A third more recent family of studies deals with predicting the sequence of the future activities and their payload given the activities observed so far, as in [19,22,23,54,64]. In [54], the authors propose an approach for predicting the sequence of future activities of a running case by relying on an annotated data-aware transition system, obtained as a refinement of the annotated transition system proposed in [2].

Other approaches, e.g., [22,23,64], make use of RNNs with LSTM (Recurrent Neural Networks with Long Short-Term Memory) cells. In particular, in [22,23], an RNN with two hidden layers trained with back propagation is presented, while, in [64], an LSTM and an encoding based on activities and timestamps is leveraged to provide predictions on the next activities and their timestamps. Finally, the study in [19] investigates how to take advantage of possibly existing a-priori knowledge for making predictions on the sequence of future activities. To this aim, an LSTM approach is equipped with the capability of taking into account also some given knowledge about the future development of an ongoing case.

5 Value-Driven Framework for Selecting Predictive Process Monitoring Methods

Tables 1 and 2 report the devised framework.[3] By reading it from left to right, in the first column, we find the **prediction type**. Our review shows that the algorithms can be categorized according to six main types of high level categories of prediction types. The first category, `time prediction`, encompasses all the different aspects of process execution time such as `remaining time` or `delay`. The second main category of identified prediction types is related to `categorical outcome`(s). Such methods predict the probability of a certain predefined outcome, such as if a case will lead to a disruption, to the violation of a constraint,

Table 1. Predictive process monitoring framework: `time` and `categorical outcome` predictions

Pred. type	Det. Pred. type	Input 1	Input 2	Input 3	Tool	Domain	Family of algorithms 1	Family of algorithms 2	Family of algorithms 3	Refer.
time	maint. time activity delays	event log (with timestamps)	process model		N	automotive	time series	probabilistic model		[58]
					N	financial telecomm.	queueing theory	transition system		[61,62]
						telecomm.	stat. analysis			[6]
						public admin.	transition system			[3,2]
					ProM plugin	customer supp.	stochastic Petri net			[55]
						financial customer supp.	regression	classification		[69]
	rem. time	event log (with timestamps) with data			N	unspecified	pattern mining			[10]
					Y but unavail.	unspecified	classification	time series		[9]
					Y	financial public admin.	regression	classification		[68]
						healthcare	stochastic Petri net			[60]
						public admin.	regression			[20]
						public admin.	transition system	regression	classification	[53]
					ProM plugin	customer supp. public admin. financial	transition system	regression		[54]
						financial logistics	stochastic Petri net			[56,57]
		event log (with timestamps) with data	inter-case metrics		Y	healthcare manufacturing	regression			[59]
		event log (with timestamps) with data and contextual information			Y but unavail.	logistics	clustering	regression		[31]
					Y	logistics	clustering	pattern mining transition system		[5]
					ProM plugin	logistics	clustering	transition system		[30]
			labeling funct.	proc. model	ProM plugin	no validation	classification			[29]
			labeling funct.	proc. model	ProM plugin		classification			[38]
categorical outcome	outcome	act. durations and routing probab.	threshold(s)	proc. model	N	synthetic	simulation	stat. analysis		[63]
		event log	labeling funct.		Y implem.	financial automotive	prob. automata			[7]
		event log (with timestamps) with data			N	logistics	neural network	constraint-sat.	QoS aggregation	[44]
						healthcare	probab. automata	classification		[39]
						logistics	classification			[8]
						synthetic	classification			[34]
			labeling funct.		Y but unavail.	synthetic	prob. automata			[37]
						synthetic	classification	neural network		[42]
						healthcare	clustering	classification		[18]
					Y	no valid.	stat. analysis			[24]
						logistics	stat. analysis			[46]
						financial public admin.	classification			[68]
					ProM pl.	healthcare	classification			[41]
						healthcare	clustering	classification		[17]
					ProM and Camunda pl.	automotive healthcare	evol. algorithm			[48]
			threshold(s)		Y but unavail.	unspecified	classification			[9]
					Y	domotic	stat. analysis			[25]
		event log (with timestamps) with data and contextual information	labeling funct.	proc. model	Y but unavail. ProM plugin	healthcare	clustering	classification		[16]
			threshold(s)		ProM plugin	logistics	no validation	classification		[38]
					ProM plugin	logistics	clustering	transition system		[28]
			clusters of behav.		N	logistics manufacturing	clustering	classification		[27]
		event log (with timestamps) with data and unstructured text	labeling funct.		N	financial	classification	text mining		[65]
	next activity	event log (with timestamps) with data			N	unspecified	pattern mining			[10]
					Y	domotic	stat. analysis			[26]
		event log (with timestamps) with data and contextual information	labeling funct.	proc. model	Y ProM plugin	domotic	stat. analysis			[25]
						no valid.	classification			[38]
	last value of an attribute	event log (with timestamps) with data and contextual information	labeling funct.	proc. model	ProM plugin	no valid.	classification			[38]

---

[3] For space limitations, in this article, an abridged version of the framework is presented. The complete version of the framework includes additional data and is available for download at https://docs.google.com/spreadsheets/d/1l1enKhKWx_3KqtnUgggrPl1aoJMhvmy9TF9jAM3snas/edit#gid=959800788.

Table 2. Predictive process monitoring framework: `sequence of outcomes/values`, `risk`, `inter-case metrics`, `cost`

Pred. type	Det. Pred. type	Input 1	Input 2	Input 3	Tool	Domain	Family of algorithms 1	Family of algorithms 2	Family of algorithms 3	Refer.
sequence of outcomes/values	sequence of future activities	event log (with timestamps)			Y implem.	customer supp. financial public admin.	neural network			[64]
					Y but unavail.	financial automotive customer supp.	neural network			[43]
					Y	financial automotive	neural network			[22]
			backgr. knowledge		Y impl.	healthcare automotive financial public admin. customer supp.	neural network			[19]
					Y	financial automotive	neural network			[23]
		event log (with timestamps) with data			ProM plugin	customer supp. public admin. financial	neural network			[54]
		event log (with timestamps) with data and contextual information			Y but unavail.	logistics	clustering	regression		[31]
	sequence of future activity timestamps	event log (with timestamps)	backgr. knowledge		N	customer supp. financial public admin.	neural network			[64]
					Y impl.	healthcare automotive financial public admin. customer supp.	neural network			[19]
					Y	financial automotive	neural network			[23]
risk	risk	event log (with timestamps)	labeling funct.		Y but unavail.	logistics	clustering	classification		[11]
					Camunda pl.	financial	similarity-weight. graph	stat. analysis		[13]
			threshold(s)		N	transport logistics	neural network			[45]
		event log (with timestamps) with data			ProM plugin	financial logistics	stochastic Petri net			[57]
			labeling funct.			no valid.	classification			[15]
				proc. model Yawl pugin		logistics unspecified	evol. algorithm			[14]
inter-case metr.	inter-case metrics	event log (with timestamps)			Y but unavail.	logistics	clustering	regression		[11]
			labeling funct.		ProM plugin	unspec.	regression			[52]
		event log (with timestamps) with data	threshold(s)		N	transport logistics	neural net.			[45]
						no valid.	classification	regression	time series	[71]
					Y but unavail.	unspec.	classification	time series		[71]
	workload	event log (with timestamps) with data and contextual information	labeling funct.		ProM plugin	no valid.	classification			[9]
cost	cost	event log (with timestamps) with data	threshold(s)		N	transport logistics	neural net.			[45]
		event log (with timestamps) with resources	cost schema		ProM plugin	no valid.	trans. system	stat.analysis		[70]

or whether it will be delayed. The third type of prediction type is related to `sequence of next outcomes/values`. These predictions focus on the probability that future sets of events will occur in the execution of a case. The fourth prediction type is `risk`. When elimination of risks is not feasible, this type of predictions allows for reducing and managing risks. The fifth prediction type pertains to `inter-case metrics`. The final category is related to `cost` predictions.

The next step (second column) in the framework concerns the **input data**. Event logs containing different types of information should be provided as input to the different methods (e.g., `event logs (with timestamps)`, `event logs (with timestamps) with data`). In some cases, together with the event log, other inputs are required. For instance, in case of the outcome-based predictions, the `labeling function`, e.g., the specific predicate or category to be predicted, is usually required.

The framework also considers the existence of **tool support**. If a tool has been developed, using, evaluating, and understanding the applicability, usefulness and potential benefits of a predictive process monitoring technique becomes easier. Given that tool support is provided, the framework captures the type of support provided such as whether the tool is a stand-alone application or a plug-in of a research framework, e.g., a `ProM plug-in`.

The validation of the algorithms on logs can take different forms. For instance, it can be achieved by using `synthetic logs`. Such validations can be considered as "weaker" as they do not necessarily mirror the complexity and variability of `real-life logs`. Algorithms tested on real-life logs reflect industry logs the best and are, therefore, considered as "stronger". Furthermore, the suitability of an algorithm is better to be validated on logs from the same industry domain as the one of the company seeking to use it. As such, the framework makes note of the **domain** from which the logs originate. When a domain is specified, it indicates that the algorithm has been tested on a log from that domain. If no domain is specified, the algorithm has not been validated on a real-life log.

At the heart of each predictive process monitoring method lies the specific algorithm used to implement it. The **family of algorithms** might matter when assessing advantages and limitations of an approach and as such, it is incorporated in the framework. The specific algorithm is not listed in the framework, but rather the foundational technique it is based on, such as `regression`, `neural networks`, or `queuing theory`.

The proposed framework has two main benefits. First, it can be used by companies to identify, along the above outlined parameters, the most suitable predictive process monitoring method(s) to be used in different scenarios. Secondly, it can be used by researchers to have a clear structuration and assessment of the existing techniques in the predictive process monitoring field. This assessment is crucial to identify gaps in the literature and relevant research directions to be further investigated in the near future. For example, the framework shows that only one paper discusses techniques that take advantage of possibly existing a-priori knowledge for making predictions [19]. Further investigation is also needed for what concerns the use of incremental learning algorithms as a way to incrementally construct predictive models by updating them whenever new cases become available, which is a crucial topic, but only discussed in [42]. Another direction for future research is to further investigate the use of inter-case features for constructing a predictive model. This means that scenarios should be taken into consideration where not only the information related to the ongoing case, but also the status of other (concurrent) cases (e.g., the number of concurrent cases) are considered in order to make predictions (this type of techniques is only discussed in [59]).

6   Conclusion

Predictive process monitoring approaches have been growing quite fast in the last few years. If, on the one hand, such a spread of techniques has provided researchers and practitioners with powerful means for analyzing their business processes and making predictions on their future, on the other hand, it could be difficult for them to navigate through such a complex and unknown domain.

By means of a systematic literature review in the predictive process monitoring field, we provide data analysts with a framework to guide them in the selection of the technique that best fit their needs.

The main threat to validity of our work refers to the potential selection bias and inaccuracies in data extraction and analysis typical of literature reviews. In order to minimize such issues, our systematic literature review carefully adheres to the guidelines outlined in [35]. Concretely, we used well-known literature sources and libraries in information technology to extract relevant works on the topic of predictive process monitoring. Further, we performed a backward reference search to avoid the exclusion of potentially relevant papers. Finally, to avoid that our review was threatened by insufficient reliability, we ensured that the search process could be replicated by other researchers. However, the search may produce different results as the algorithm used by source libraries to rank results based on relevance may be updated.

In the future, we plan to empirically evaluate the proposed framework with users in order to assess its usefulness in real contexts. Furthermore, we would like to extend the existing framework with other dimensions of interest for the academic and the industrial world. The presented literature review can indeed be considered as a basis where to incorporate broader theoretical perspectives and concepts across the predictive process monitoring domain that might help future research endeavors to be well-directed.

References

1. van der Aalst, W.M.P.: Process Mining - Data Science in Action, 2nd edn. Springer, Heidelberg (2016). https://doi.org/10.1007/978-3-662-49851-4
2. van der Aalst, W.M.P., Schonenberg, M.H., Song, M.: Time prediction based on process mining. Inf. Syst. **36**(2), 450–475 (2011)
3. van der Aalst, W.M.P., Pesic, M., Song, M.: Beyond process mining: from the past to present and future. In: Pernici, B. (ed.) CAiSE 2010. LNCS, vol. 6051, pp. 38–52. Springer, Heidelberg (2010). https://doi.org/10.1007/978-3-642-13094-6_5
4. Bevacqua, A., Carnuccio, M., Folino, F., Guarascio, M., Pontieri, L.: A data-adaptive trace abstraction approach to the prediction of business process performances. In: ICEIS, vol. 1. SciTePress (2013)
5. Bevacqua, A., Carnuccio, M., Folino, F., Guarascio, M., Pontieri, L.: A data-driven prediction framework for analyzing and monitoring business process performances. In: Hammoudi, S., Cordeiro, J., Maciaszek, L.A., Filipe, J. (eds.) ICEIS 2013. LNBIP, vol. 190, pp. 100–117. Springer, Cham (2014). https://doi.org/10.1007/978-3-319-09492-2_7
6. Bolt, A., Sepúlveda, M.: Process remaining time prediction using query catalogs. In: Lohmann, N., Song, M., Wohed, P. (eds.) BPM 2013. LNBIP, vol. 171, pp. 54–65. Springer, Cham (2014). https://doi.org/10.1007/978-3-319-06257-0_5
7. Breuker, D., Matzner, M., Delfmann, P., Becker, J.: Comprehensible predictive models for business processes. MIS Q. **40**(4), 1009–1034 (2016)
8. Cabanillas, C., Di Ciccio, C., Mendling, J., Baumgrass, A.: Predictive task monitoring for business processes. In: Sadiq, S., Soffer, P., Völzer, H. (eds.) BPM 2014. LNCS, vol. 8659, pp. 424–432. Springer, Cham (2014). https://doi.org/10.1007/978-3-319-10172-9_31

9. Castellanos, M., Salazar, N., Casati, F., Dayal, U., Shan, M.-C.: Predictive business operations management. In: Bhalla, S. (ed.) DNIS 2005. LNCS, vol. 3433, pp. 1–14. Springer, Heidelberg (2005). https://doi.org/10.1007/978-3-540-31970-2_1

10. Ceci, M., Lanotte, P.F., Fumarola, F., Cavallo, D.P., Malerba, D.: Completion time and next activity prediction of processes using sequential pattern mining. In: Džeroski, S., Panov, P., Kocev, D., Todorovski, L. (eds.) DS 2014. LNCS (LNAI), vol. 8777, pp. 49–61. Springer, Cham (2014). https://doi.org/10.1007/978-3-319-11812-3_5

11. Cesario, E., Folino, F., Guarascio, M., Pontieri, L.: A cloud-based prediction framework for analyzing business process performances. In: Buccafurri, F., Holzinger, A., Kieseberg, P., Tjoa, A.M., Weippl, E. (eds.) CD-ARES 2016. LNCS, vol. 9817, pp. 63–80. Springer, Cham (2016). https://doi.org/10.1007/978-3-319-45507-5_5

12. Conforti, R., de Leoni, M., La Rosa, M., van der Aalst, W.M.P., ter Hofstede, A.H.M.: A recommendation system for predicting risks across multiple business process instances. Decis. Support Syst. **69**, 1–19 (2015)

13. Conforti, R., Fink, S., Manderscheid, J., Röglinger, M.: PRISM – a predictive risk monitoring approach for business processes. In: La Rosa, M., Loos, P., Pastor, O. (eds.) BPM 2016. LNCS, vol. 9850, pp. 383–400. Springer, Cham (2016). https://doi.org/10.1007/978-3-319-45348-4_22

14. Conforti, R., ter Hofstede, A.H.M., La Rosa, M., Adams, M.: Automated risk mitigation in business processes. In: Meersman, R., et al. (eds.) OTM 2012. LNCS, vol. 7565, pp. 212–231. Springer, Heidelberg (2012). https://doi.org/10.1007/978-3-642-33606-5_14

15. Conforti, R., de Leoni, M., La Rosa, M., van der Aalst, W.M.P.: Supporting risk-informed decisions during business process execution. In: Salinesi, C., Norrie, M.C., Pastor, Ó. (eds.) CAiSE 2013. LNCS, vol. 7908, pp. 116–132. Springer, Heidelberg (2013). https://doi.org/10.1007/978-3-642-38709-8_8

16. Cuzzocrea, A., Folino, F., Guarascio, M., Pontieri, L.: A multi-v2016a multi-view multi-dimensional ensemble learning approach to mining business process deviances. In: IJCNN (2016)

17. Di Francescomarino, C., Dumas, M., Federici, M., Ghidini, C., Maggi, F.M., Rizzi, W.: Predictive business process monitoring framework with hyperparameter optimization. In: Nurcan, S., Soffer, P., Bajec, M., Eder, J. (eds.) CAiSE 2016. LNCS, vol. 9694, pp. 361–376. Springer, Cham (2016). https://doi.org/10.1007/978-3-319-39696-5_22

18. Di Francescomarino, C., Dumas, M., Maggi, F.M., Teinemaa, I.: Clustering-based predictive process monitoring. IEEE Trans. Serv. Comput. **PP**(99) (2016)

19. Di Francescomarino, C., Ghidini, C., Maggi, F.M., Petrucci, G., Yeshchenko, A.: An eye into the future: leveraging a-priori knowledge in predictive business process monitoring. In: Carmona, J., Engels, G., Kumar, A. (eds.) BPM 2017. LNCS, vol. 10445, pp. 252–268. Springer, Cham (2017). https://doi.org/10.1007/978-3-319-65000-5_15

20. van Dongen, B.F., Crooy, R.A., van der Aalst, W.M.P.: Cycle time prediction: when will this case finally be finished? In: Meersman, R., Tari, Z. (eds.) OTM 2008. LNCS, vol. 5331, pp. 319–336. Springer, Heidelberg (2008). https://doi.org/10.1007/978-3-540-88871-0_22

21. Dumas, M., Maggi, F.M.: Enabling process innovation via deviance mining and predictive monitoring. In: vom Brocke, J., Schmiedel, T. (eds.) BPM - Driving Innovation in a Digital World. MP, pp. 145–154. Springer, Cham (2015). https://doi.org/10.1007/978-3-319-14430-6_10

22. Evermann, J., Rehse, J.-R., Fettke, P.: A deep learning approach for predicting process behaviour at runtime. In: Dumas, M., Fantinato, M. (eds.) BPM 2016. LNBIP, vol. 281, pp. 327–338. Springer, Cham (2017). https://doi.org/10.1007/978-3-319-58457-7_24

23. Evermann, J., Rehse, J.R., Fettke, P.: Predicting process behaviour using deep learning. Decis. Support Syst. **100**, 129–140 (2017)

24. Feldman, Z., Fournier, F., Franklin, R., Metzger, A.: Proactive event processing in action: a case study on the proactive management of transport processes (industry article). In: ACM DEBS (2013)

25. Ferilli, S., Esposito, F., Redavid, D., Angelastro, S.: Predicting process behavior in WoMan. In: Adorni, G., Cagnoni, S., Gori, M., Maratea, M. (eds.) AI*IA 2016. LNCS (LNAI), vol. 10037, pp. 308–320. Springer, Cham (2016). https://doi.org/10.1007/978-3-319-49130-1_23

26. Ferilli, S., Esposito, F., Redavid, D., Angelastro, S.: Extended process models for activity prediction. In: Kryszkiewicz, M., Appice, A., Ślęzak, D., Rybinski, H., Skowron, A., Raś, Z.W. (eds.) ISMIS 2017. LNCS (LNAI), vol. 10352, pp. 368–377. Springer, Cham (2017). https://doi.org/10.1007/978-3-319-60438-1_36

27. Folino, F., Greco, G., Guzzo, A., Pontieri, L.: Mining usage scenarios in business processes: outlier-aware discovery and run-time prediction. Data Knowl. Eng. **70**(12), 1005–1029 (2011)

28. Folino, F., Guarascio, M., Pontieri, L.: Discovering context-aware models for predicting business process performances. In: Meersman, R., et al. (eds.) OTM 2012. LNCS, vol. 7565, pp. 287–304. Springer, Heidelberg (2012). https://doi.org/10.1007/978-3-642-33606-5_18

29. Folino, F., Guarascio, M., Pontieri, L.: Context-aware predictions on business processes: an ensemble-based solution. In: Appice, A., Ceci, M., Loglisci, C., Manco, G., Masciari, E., Ras, Z.W. (eds.) NFMCP 2012. LNCS (LNAI), vol. 7765, pp. 215–229. Springer, Heidelberg (2013). https://doi.org/10.1007/978-3-642-37382-4_15

30. Folino, F., Guarascio, M., Pontieri, L.: Discovering high-level performance models for ticket resolution processes. In: Meersman, R., et al. (eds.) OTM 2013. LNCS, vol. 8185, pp. 275–282. Springer, Heidelberg (2013). https://doi.org/10.1007/978-3-642-41030-7_18

31. Folino, F., Guarascio, M., Pontieri, L.: Mining predictive process models out of low-level multidimensional logs. In: Jarke, M., et al. (eds.) CAiSE 2014. LNCS, vol. 8484, pp. 533–547. Springer, Cham (2014). https://doi.org/10.1007/978-3-319-07881-6_36

32. Halper, F.: Predictive analytics for business advantage. TDWI Research (2014)

33. van der Aalst, W.M.P., et al.: Process mining manifesto. In: Daniel, F., Barkaoui, K., Dustdar, S. (eds.) BPM 2011. LNBIP, vol. 99, pp. 169–194. Springer, Heidelberg (2012). https://doi.org/10.1007/978-3-642-28108-2_19

34. Kang, B., Kim, D., Kang, S.H.: Real-time business process monitoring method for prediction of abnormal termination using KNNI-based LOF prediction. Expert Syst. Appl. **39**(5), 6061–6068 (2012)

35. Kitchenham, B.: Procedures for performing systematic reviews. Keele UK Keele Univ. **33**(2004), 1–26 (2004)

36. Kofod-Petersen, A.: How to do a structured literature review in computer science. Ver. 0.1, 1 October 2012

37. Lakshmanan, G.T., Shamsi, D., Doganata, Y.N., Unuvar, M., Khalaf, R.: A Markov prediction model for data-driven semi-structured business processes. Knowl. Inf. Syst. **42**(1), 97–126 (2015)

38. de Leoni, M., van der Aalst, W.M.P., Dees, M.: A general framework for correlating business process characteristics. In: Sadiq, S., Soffer, P., Völzer, H. (eds.) BPM 2014. LNCS, vol. 8659, pp. 250–266. Springer, Cham (2014). https://doi.org/10.1007/978-3-319-10172-9_16

39. Leontjeva, A., Conforti, R., Di Francescomarino, C., Dumas, M., Maggi, F.M.: Complex symbolic sequence encodings for predictive monitoring of business processes. In: Motahari-Nezhad, H.R., Recker, J., Weidlich, M. (eds.) BPM 2015. LNCS, vol. 9253, pp. 297–313. Springer, Cham (2015). https://doi.org/10.1007/978-3-319-23063-4_21

40. Ly, L.T., Maggi, F.M., Montali, M., Rinderle-Ma, S., van der Aalst, W.M.P.: Compliance monitoring in business processes: functionalities, application, and tool-support. Inf. Syst. **54**, 209–234 (2015)

41. Maggi, F.M., Di Francescomarino, C., Dumas, M., Ghidini, C.: Predictive monitoring of business processes. In: Jarke, M., et al. (eds.) CAiSE 2014. LNCS, vol. 8484, pp. 457–472. Springer, Cham (2014). https://doi.org/10.1007/978-3-319-07881-6_31

42. Maisenbacher, M., Weidlich, M.: Handling concept drift in predictive process monitoring. In: IEEE SCC, pp. 1–8. IEEE Computer Society (2017)

43. Mehdiyev, N., Evermann, J., Fettke, P.: A multi-stage deep learning approach for business process event prediction. In: CBI, vol. 01, July 2017

44. Metzger, A., et al.: Comparing and combining predictive business process monitoring techniques. IEEE Trans. Syst. Man Cybern.: Syst. **45**(2), 276–290 (2015)

45. Metzger, A., Föcker, F.: Predictive business process monitoring considering reliability estimates. In: Dubois, E., Pohl, K. (eds.) CAiSE 2017. LNCS, vol. 10253, pp. 445–460. Springer, Cham (2017). https://doi.org/10.1007/978-3-319-59536-8_28

46. Metzger, A., Franklin, R., Engel, Y.: Predictive monitoring of heterogeneous service-oriented business networks: the transport and logistics case. In: Proceedings of SRII, SRII 2012 (2012)

47. Márquez-Chamorro, A.E., Resinas, M., Ruiz-Cortés, A.: Predictive monitoring of business processes: a survey. IEEE Trans. Serv. Comput. 1 (2017). https://doi.org/10.1109/TSC.2017.2772256

48. Márquez-Chamorro, A.E., Resinas, M., Ruiz-Cortés, A., Toro, M.: Run-time prediction of business process indicators using evolutionary decision rules. Expert Syst. Appl. **87**, 1–14 (2017)

49. Pandey, S., Nepal, S., Chen, S.: A test-bed for the evaluation of business process prediction techniques. In: 7th International Conference on Collaborative Computing: Networking, Applications and Worksharing (CollaborateCom), October 2011

50. Pika, A., van der Aalst, W.M.P., Fidge, C.J., ter Hofstede, A.H.M., Wynn, M.T.: Predicting deadline transgressions using event logs. In: La Rosa, M., Soffer, P. (eds.) BPM 2012. LNBIP, vol. 132, pp. 211–216. Springer, Heidelberg (2013). https://doi.org/10.1007/978-3-642-36285-9_22

51. Pika, A., van der Aalst, W.M.P., Fidge, C.J., ter Hofstede, A.H.M., Wynn, M.T.: Profiling event logs to configure risk indicators for process delays. In: Salinesi, C., Norrie, M.C., Pastor, Ó. (eds.) CAiSE 2013. LNCS, vol. 7908, pp. 465–481. Springer, Heidelberg (2013). https://doi.org/10.1007/978-3-642-38709-8_30

52. Pika, A., van der Aalst, W.M.P., Wynn, M.T., Fidge, C.J., ter Hofstede, A.H.M.: Evaluating and predicting overall process risk using event logs. Inf. Sci. **352–353**, 98–120 (2016)

53. Polato, M., Sperduti, A., Burattin, A., de Leoni, M.: Data-aware remaining time prediction of business process instances. In: 2014 International Joint Conference on Neural Networks (IJCNN), July 2014

54. Polato, M., Sperduti, A., Burattin, A., de Leoni, M.: Time and activity sequence prediction of business process instances. Computing (2018)
55. Rogge-Solti, A., Vana, L., Mendling, J.: Time series Petri net models. In: Ceravolo, P., Rinderle-Ma, S. (eds.) SIMPDA 2015. LNBIP, vol. 244, pp. 124–141. Springer, Cham (2017). https://doi.org/10.1007/978-3-319-53435-0_6
56. Rogge-Solti, A., Weske, M.: Prediction of remaining service execution time using stochastic Petri nets with arbitrary firing delays. In: Basu, S., Pautasso, C., Zhang, L., Fu, X. (eds.) ICSOC 2013. LNCS, vol. 8274, pp. 389–403. Springer, Heidelberg (2013). https://doi.org/10.1007/978-3-642-45005-1_27
57. Rogge-Solti, A., Weske, M.: Prediction of business process durations using non-Markovian stochastic Petri nets. Inf. Syst. **54**, 1–14 (2015)
58. Ruschel, E., Santos, E.A.P., de Freitas Rocha Loures, E.: Mining shop-floor data for preventive maintenance management: integrating probabilistic and predictive models. Procedia Manuf. **11**, 1127–1134 (2017)
59. Senderovich, A., Di Francescomarino, C., Ghidini, C., Jorbina, K., Maggi, F.M.: Intra and inter-case features in predictive process monitoring: a tale of two dimensions. In: Carmona, J., Engels, G., Kumar, A. (eds.) BPM 2017. LNCS, vol. 10445, pp. 306–323. Springer, Cham (2017). https://doi.org/10.1007/978-3-319-65000-5_18
60. Senderovich, A., Shleyfman, A., Weidlich, M., Gal, A., Mandelbaum, A.: P^3-folder: optimal model simplification for improving accuracy in process performance prediction. In: La Rosa, M., Loos, P., Pastor, O. (eds.) BPM 2016. LNCS, vol. 9850, pp. 418–436. Springer, Cham (2016). https://doi.org/10.1007/978-3-319-45348-4_24
61. Senderovich, A., Weidlich, M., Gal, A., Mandelbaum, A.: Queue mining – predicting delays in service processes. CAiSE 2014. LNCS, vol. 8484, pp. 42–57. Springer, Cham (2014). https://doi.org/10.1007/978-3-319-07881-6_4
62. Senderovich, A., Weidlich, M., Gal, A., Mandelbaum, A.: Queue mining for delay prediction in multi-class service processes. Inf. Syst. **53**, 278–295 (2015)
63. Si, Y.W., Hoi, K.K., Biuk-Aghai, R.P., Fong, S., Zhang, D.: Run-based exception prediction for workflows. J. Syst. Softw. **113**, 59–75 (2016)
64. Tax, N., Verenich, I., La Rosa, M., Dumas, M.: Predictive business process monitoring with LSTM neural networks. In: Dubois, E., Pohl, K. (eds.) CAiSE 2017. LNCS, vol. 10253, pp. 477–492. Springer, Cham (2017). https://doi.org/10.1007/978-3-319-59536-8_30
65. Teinemaa, I., Dumas, M., Maggi, F.M., Di Francescomarino, C.: Predictive business process monitoring with structured and unstructured data. In: La Rosa, M., Loos, P., Pastor, O. (eds.) BPM 2016. LNCS, vol. 9850, pp. 401–417. Springer, Cham (2016). https://doi.org/10.1007/978-3-319-45348-4_23
66. Tu, T.B.H., Song, M.: Analysis and prediction cost of manufacturing process based on process mining. In: ICIMSA, May 2016
67. Verenich, I., Dumas, M., La Rosa, M., Maggi, F.M., Di Francescomarino, C.: Complex symbolic sequence clustering and multiple classifiers for predictive process monitoring. In: Reichert, M., Reijers, H.A. (eds.) BPM 2015. LNBIP, vol. 256, pp. 218–229. Springer, Cham (2016). https://doi.org/10.1007/978-3-319-42887-1_18
68. Verenich, I., Dumas, M., La Rosa, M., Maggi, F.M., Di Francescomarino, C.: Minimizing overprocessing waste in business processes via predictive activity ordering. In: Nurcan, S., Soffer, P., Bajec, M., Eder, J. (eds.) CAiSE 2016. LNCS, vol. 9694, pp. 186–202. Springer, Cham (2016). https://doi.org/10.1007/978-3-319-39696-5_12

69. Verenich, I., Nguyen, H., La Rosa, M., Dumas, M.: White-box prediction of process performance indicators via flow analysis. In: Proceedings of the 2017 International Conference on Software and System Process, ICSSP 2017 (2017)
70. Wynn, M.T., Low, W.Z., ter Hofstede, A.H.M., Nauta, W.: A framework for cost-aware process management: cost reporting and cost prediction. J. Univ. Comput. Sci. **20**(3), 406–430 (2014)
71. Zeng, L., Lingenfelder, C., Lei, H., Chang, H.: Event-driven quality of service prediction. In: Bouguettaya, A., Krueger, I., Margaria, T. (eds.) ICSOC 2008. LNCS, vol. 5364, pp. 147–161. Springer, Heidelberg (2008). https://doi.org/10.1007/978-3-540-89652-4_14

How Context-Aware Are Extant BPM Methods? - Development of an Assessment Scheme

Marie-Sophie Denner[1], Maximilian Röglinger[1], Theresa Schmiedel[2],
Katharina Stelzl[1(✉)], and Charlotte Wehking[2]

[1] FIM Research Center, University of Bayreuth, 95444 Bayreuth, Germany
{sophie.denner,maximilian.roeglinger,
katharina.stelzl}@fim-rc.de
[2] University of Liechtenstein, 9490 Vaduz, Liechtenstein
{theresa.schmiedel,mauspcharlotte.wehking}@uni.li

Abstract. Context awareness is vital for business process management (BPM) success. Although many academics have called for context-aware BPM, current BPM research and practice do not seem to sufficiently account for various contexts. To examine whether this statement holds true, we developed an assessment scheme that enables determining to which extent existing BPM methods can be applied in various contexts. We identified 25 exemplary BPM methods based on a structured literature review and rated them according to their applicability to different context dimensions, i.e., goal, process, organization and environment dimension. Our results indicate that most BPM methods are rather context-independent, i.e., they are not geared to specific contexts. Accordingly, the investigated BPM methods follow a one-size-fits-all approach and practitioners have no guidance on how to tailor BPM in their organizations. In particular, there is a lack of BPM methods for explorative purposes as well as for knowledge- and creativity-intense business processes. In the digital age, which is characterized by volatility and high pressure for innovation, these domains are very important. Our research is a first step toward context-aware BPM methods and structured guidance for organizations regarding the systematic selection and configuration of BPM methods.

Keywords: BPM methods · Context awareness · Assessment scheme
Literature review

1 Introduction

In the last decades, business process management (BPM) has evolved into an important and mature domain in research and practice alike [49]. Organizations have increasingly adopted BPM in different contexts [19, 26, 52]. In fact, context awareness is one out of ten principles to efficiently and effectively use BPM in organizations [50]. Especially in the digital age, context awareness is one of the key characteristics of successful BPM [25]. As new technologies, customers' expectations, new business models, or additional

© Springer Nature Switzerland AG 2018
M. Weske et al. (Eds.): BPM 2018, LNCS 11080, pp. 480–495, 2018.
https://doi.org/10.1007/978-3-319-98648-7_28

competitors are hurdles that organizations currently need to overcome, organizations need to manage different contexts at the same time [20, 25].

To adequately configure BPM with respect to the requirements of specific contexts, research has started to study context-aware BPM. Context awareness considers specific organizational factors that distinguish one organization from another based on given, situational, and organizational requirements [50]. According to the context framework by vom Brocke et al. [52], context-aware BPM considers the goal of BPM (i.e., exploration or exploitation), certain characteristics of the processes in focus (e.g., repetitiveness or creativity) as well as organizational (e.g., scope or culture) and environmental characteristics (i.e., uncertainty or competitiveness). While the process, organization, and environment dimensions indicate a given context and cannot be modified, the goal dimension can be actively chosen by the organization. An organization consciously decides whether its BPM should strive for exploitation (e.g., improvement), exploration (e.g., innovation), or both simultaneously.

To successfully institutionalize BPM, organizations can choose among a plethora of BPM methods [44, 48, 50]. Even though context awareness is critical for successful BPM, the current body of knowledge does not seem to account for business contexts [45, 52]. One research stream generally addresses context awareness in BPM by investigating various context dimensions [12, 50]. A second research stream focuses on context-aware methods by explicitly stating the application context of a BPM method [1, 13]. Until now, only very few BPM methods seem to consider specific contexts. However, research mainly focuses on BPM methods by following a one-size-fits-all approach, not addressing context specifically [52]. For that reason, several researchers call for context-aware BPM as well as context-aware methods [29, 45, 48, 52]. Against this background, we investigate the following research question: *How context-aware are extant BPM methods?*

To answer this question, we provide an assessment scheme based on the context framework of vom Brocke et al. [52], which enables determining the context awareness of extant BPM methods. We set up the assessment scheme based on the four context dimensions of vom Brocke et al. [52]. Moreover, we identify 25 BPM methods based on a structured literature review. These BPM methods are exemplarily analyzed based on the assessment scheme by determining their applicability with respect to the four context dimensions. Finally, we reason about the context awareness of each dimension, context factor, and characteristic for all examined BPM methods. We also provide further insights into the sample at large. Being aware of the limitations of our literature review, we see our work as a first initial discussion of context-aware BPM methods. The assessment scheme is intended to serve as a starting point, offering guidance for BPM researchers to examine the context awareness of BPM methods. Additionally, the assessment scheme helps practitioners identify suitable BPM methods for specific contexts and goals.

Examining context awareness of existing BPM methods, we proceed as follows. Section 2 provides relevant theoretical background. In Sect. 3, we outline our data collection and analysis method. Section 4 presents and discusses the results of the applicability assessment concerning the identified BPM methods. We conclude in Sect. 5 by summing up the key results, discussing implications and limitations, and pointing to directions for further research.

2   Context-Aware Business Process Management

Organizations need to consider BPM in different contexts to perform efficiently and effectively [50]. The framework offered by vom Brocke et al. [52] helps organizations identify the context in which BPM is applied (Fig. 1). Their framework consolidates a range of the latest research and serves as foundation for context-aware BPM research and practice. It includes four context dimensions, underlying context factors, and various related characteristics [52].

Context factor	Example characteristics		
Goal dimension			
Focus	Exploitation (Improvement, Compliance)		Exploration (Innovation)
Process dimension			
Value contribution	Core process	Management process	Support process
Repetitiveness	Repetitive		Non-repetitive
Knowledge-intensity	Low knowledge-intensity	Medium knowledge-intensity	High knowledge-intensity
Creativity	Low creativity	Medium creativity	High creativity
Interdependence	Low interdependence	Medium interdependence	High interdependence
Variability	Low variability	Medium variability	High variability
Organization dimension			
Industry	Process industry	Product industry	Product & Service industry
Size	Start-up	Small and medium enterprise	Large organization
Culture	Culture supportive for BPM		Culture non-supportive for BPM
Resources	Low organizational resources	Medium organizational resources	High organizational resources
Environment dimension			
Competitiveness	Low competitive environment	Medium competitive environment	Highly competitive environment
Uncertainty	Low environmental uncertainty	Medium environmental uncertainty	High environmental uncertainty

Fig. 1.   Context framework [52].

The goal dimension is crucial for BPM as it directly influences how BPM should be implemented and which methods should be applied [52]. Thereby, several authors recently differentiate between exploitation and exploration which is known as ambidextrous BPM [4, 29, 44]. Exploitative BPM is applied to realize incremental improvements by utilizing known methods [4, 44]. Explorative BPM gears to innovating processes by utilizing creative methods [4, 44]. The process dimension includes various context factors to account for the diversity of processes and their requirements for an appropriate management [52]. Moreover, various context factors of an organization need to be considered, exemplary the size of an organization or the industry in which the organization operates [36, 42]. Finally, the environment dimension includes factors outside the organization, e.g., uncertainty in a rapidly changing environment or competition that influence the selection of BPM methods [43]. Accordingly, a context-aware method considers business process-relevant contextual information which influence the process goal [1, 13, 45]. Therefore, a context-aware BPM method explicitly states the application context [1, 13].

3   Research Method

3.1   Identification of BPM Methods

In this section, we first describe the process of collecting data via a structured literature review to systematically compile relevant BPM methods. Second, we explain our data analysis process (i.e., the structure of our assessment scheme) as well as the process of determining the context awareness of the identified BPM methods.

By conducting a structured literature review, we aim to identify extant BPM methods as a basis for further analysis. According to the nature of this research method, we explain all design decisions regarding suitable outlets, search strings, chosen time frame, and the selection process of relevant articles [51, 53]. We focused on the Business Process Management Journal (BPMJ) and the "Senior Scholars' Basket of Journals" as one of the most recognized outlets of the BPM and Information Systems (IS) discipline. Within the "Senior Scholars' Basket of Journals" [3] highly-rated journals with an eminent influence in the IS community, of which BPM is a part, are included. As for the BPMJ, we assumed that it discusses prevailing and core BPM research problems [9]. Nevertheless, we do not claim for completeness, as many other publication outlets could be included in our literature review. We critically reflect on the limitation in Sect. 5.1, pointing to further ideas for data collection. To cover articles dealing with extant BPM methods, we searched published articles full-text using the search strings summarized in Table 1. To avoid overlooking articles only referring to 'Business Process Management', we also included its abbreviation 'BPM'. Besides 'method', we included 'tool', 'model' and 'framework' as synonyms. As context-aware BPM has gained increasing attention in the past few years, especially with respect to various context dimensions [52] and the goal of ambidextrous BPM [29, 49], we confined the search to a time frame starting from 2014 to the present day.

Table 1. Overview of literature search approach.

Search string I	("Business Process Management" OR "BPM") AND "method"
Search string II	("Business Process Management" OR "BPM") AND "tool"
Search string III	("Business Process Management" OR "BPM") AND "model"
Search string IV	("Business Process Management" OR "BPM") AND "framework"

Having applied the search criteria to all selected journals, the list of search results contained 915 articles. Applying each search strings one after another, the number of search results included several duplicates, which we sorted out, and 255 unique articles remained. To ensure valid results, the final selection was conducted by three researches. We read the articles' titles and abstracts and removed 174 articles that did not match the scope of our research (e.g., article that only cited the BPMJ). The remaining 81 articles were read in full and examined for their relevance to the research topic at hand. We eliminated articles with a descriptive purpose of use that do not develop a BPM method, but focused on case studies, statistical tests, development of capabilities, or a comparison of state of the art methods. Articles were considered relevant if they developed a BPM method. This led to a final removal of 56 articles and thus, the number of relevant articles ended up at 25. As the identified methods did not always have specific names, Table 2 lists the key idea and a short description of each method instead.

Table 2. List of all identified BPM methods.

References	Key Idea (the BPM method helps organizations to…)
[1]	Identify contextual factors which impact processes and their process goals to adapt these
[2]	Assess the social sustainability of processes to diagnose participants resist following modelled process
[5]	Automatically detect potential process weaknesses in semantic process models
[6]	Use modeling and simulation standards to measure process key performance indicators and test improvements
[7]	Derive concrete recommendations for process improvement in a goal-oriented manner
[8]	Fit a probabilistic model according to a data set of past behavior base on predictive modeling
[11]	Assess the maturity of BPM governance practices to identify activities for improvement
[14]	Provide transparency concerning process ownership
[15]	Systematically and automatically analyze and match conceptual legacy process models in different languages
[17]	Create value and improve efficiency based on analyzing strategic operations
[18]	Select the most suitable processes according to organizational objectives during a process
[21]	Compile and structure organizational capabilities to facilitate and implement open innovations
[22]	Facilitate organizational change through BPM
[24]	Decompose BPMN models according to a structured guideline to improve process modelling
[27]	Reduce complexity of an initial BPMN model
[28]	Achieve a process-oriented structure without destroying existing department structures
[30]	Receive a value-oriented and holistic view of open innovation adoption inside the organization
[32]	Build ambidexterity into inter-organizational IT-enabled service processes to meet the needs of their customers
[34]	Investigate their role in the value creation process by identifying potential value creation activities and sources

(*continued*)

Table 2. (*continued*)

References	Key Idea (the BPM method helps organizations to…)
[38]	Provide an overview of process losses and corresponding prioritization steps for the elimination of such losses
[39]	Extract business rules from existing process models
[40]	Capture process knowledge to improve user collaboration and manage ad hoc and semi-structured processes
[41]	Systematize operational processes for managing and improving processes
[47]	Understand the customer needs and integrate the organizations' products and services into customer processes
[54]	Extend context-aware process modeling towards location-awareness to increase organizational objectives

3.2   Development of the Assessment Scheme

To assess the applicability of extant BPM methods, we set up an assessment scheme based on the aforementioned context framework of vom Brocke et al. [52]. Originally, the context framework was developed for classifying the context of an organization in which BPM is applied. As BPM methods can help organizations to overcome the current hurdles of various contexts of an organization and facilitate the innovation or improvement of business processes [50], we classified the context awareness of BPM methods based on the context framework offered by vom Brocke et al. [52].

Structuring our analysis, we split the four context dimensions in the goal dimension, on the one hand, and the process, organization, and environment dimension on the other hand. As the goal dimension can be influenced by organizations, it commands for a conscious decision of the organization [52]. The goal dimension thus needs to be treated separately, as it is orthogonal to the other dimensions. Its characteristics (i.e., exploitation, exploration) build the columns of our assessment scheme. Contrary, the process, organization, and environment dimensions, which represent given context factors, build the lines of our assessment scheme. In case one of these context factors includes three characteristics, one of which reflects a medium level (e.g., medium knowledge-intensity), we decided to exclude the medium level. Besides being impractical for our purposes of assessing the context awareness of BPM methods, the medium level of a context factor is only qualitatively described and thus lacks a clear definition of the term 'medium'. Therefore, the medium level strongly depends on subjective interpretation that might bias the results.

Having set up the assessment scheme, the context awareness of each BPM method can be determined in three consecutive steps. First, the BPM method is analyzed regarding its goal, which means its applicability for exploitation and/or exploration. Thus, the BPM method is allocated to either or both columns of the assessment scheme. Second, regarding the other three context dimensions, the applicability of a BPM method for each specific characteristic is determined. Third, the context awareness of a BPM method is derived based on its applicability. Both, the applicability as well as the context awareness of a BPM method, are rated in terms of a Likert scale. The Likert scale is a scaling technique that can be used to obtain participants' level of agreement with given statements [33]. We used a three-point Likert scale, whereby its assessment criteria are interpreted as ordinal data with odd numbers [23]. Accordingly, the applicability of a BPM method regarding each characteristic in step two and step three, is specified by the following scale: (1) constitutes that the method is not applicable to a specific characteristic meaning the method is context-aware but does not support a specific characteristic, (3) constitutes that the method's applicability is independent of a specific characteristics meaning the method is context-independent, and (5) constitutes that the method is applicable to a specific characteristic meaning that the method is context-aware and supports a specific characteristic. Thus, all BPM methods assessed with either (1) or (5) are context-aware. A method assessed with (3) indicates an independence and neutrality concerning context awareness. This neutrality still implies an application of the method in an organization. Accordingly, a method assessed with either (1) or (5) is superior of (3) since the method is explicitly suitable for a specific context. However, applying a non-suitable method to a specific context is not recommendable. For example, if method A is applicable for a repetitive process, this characteristic is assessed with a value of (5) meaning that the method is context-aware and supports repetitive processes. If method A is not applicable for start-ups, this characteristic is assessed with a value of (1) meaning that the method is context-aware but does not support start-ups. If the applicability of method A is independent of a high knowledge-intensity, this characteristic is assessed with a value of (3) meaning that the method is context-independent and thus unaffected by knowledge-intense processes.

To enable a more detailed discussion, we distinguish three levels of analysis based on the columns and lines of our assessment scheme. Based on the assessment of the applicability of each BPM method in step two, context awareness can be discussed with respect to each characteristic (first level of analysis). Additionally, context awareness is aggregated for all context factors (second level of analysis) as well as for the four context dimensions (third level of analysis). Thereby, the meaning of the assessment criteria (1), (3), and (5) is transferrable from the first to the second and third level of analysis. We further used statistics to calculate the relative frequency of each assessment criteria regarding each characteristic in percent (f_i) and the median for each characteristic, context factor and context dimension (m) (first level of analysis). The relative frequency indicates the number of times an assessment criterion occurred in relation to the number of all assessment criteria [16]. For instance, the relative frequency of (1) for the first level of analysis is calculated by the quantity (1) occurs for core processes divided through the quantity (1), (3), and (5) occurs for core processes.

To assess the extent of context awareness based on the relative frequency, we analyzed their tertile [16]. The tertile splits the distribution into three equal parts, whereas each part contains one third of the distribution. While a relative frequency below 33.3% represents weak context awareness, a relative frequency between 33.3% and 66.6% represents a moderate, and a relative frequency beyond 66.6% represents strong context awareness. The median splits the distribution of assessment criteria into halves and thus calculates the middle score of a ranked set of numbers. The median is calculated instead of the mean for ordinal data as it includes outliers in data [16]. For example, a median of three indicates that there are mostly context-independent BPM methods for one specific characteristic (e.g., support processes), a median of four indicates that BPM methods are moderately context-aware (e.g., management processes), and a median of five indicates that BPM methods are strongly context-aware (e.g., core processes).

To ensure the validity and reliability of assessments between two judges, different metrics are presented [37]. First, the validity of the assessments is verified by calculating hit ratios [37], whose values range from 1 for perfect agreement to 0 for perfect disagreement. Partial agreement is expressed via intermediate values. Considering the frequency of correctly assigned objects, validity is measured through method-specific and overall hit ratios [35]. To measure reliability, we used Cohen's Kappa [10], which mirrors "the proportion of joint judgement in which there is agreement after chance agreement is excluded" [37]. In cases of disagreement, the judges discuss all mismatching assessments and decide on one single statement (i.e., 1, 3, or 5) in the end.

4 Context Awareness of BPM Methods

4.1 Application of the Assessment Scheme

In this section, we first present the results of our assessment of 25 BPM methods and shortly discuss the inter-coder reliability. Second, we discuss the key findings with respect to context awareness of the investigated BPM methods.

Based on the results of our literature review and the developed assessment scheme, we determined the context awareness of 25 BPM methods. Therefore, two authors independently assessed the applicability of these methods regarding the goal dimension (step one) and the other three context dimensions (step two). Finally, the context awareness of the investigated BPM methods was derived (step three). The results are shown in Fig. 2. If a BPM method is context-aware or context-independent regarding

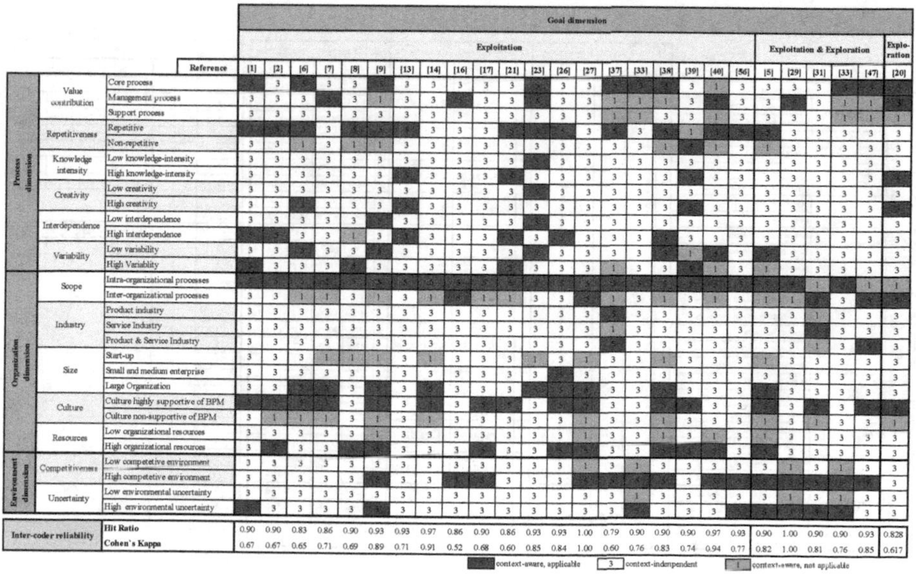

Fig. 2. Context awareness of extant BPM methods. (Color figure online)

one specific characteristic, this is reflected by its assessment criteria (i.e., 1, 3, 5) as well as three corresponding colors.

Moreover, the achieved method-specific hit ratios and the Cohen's Kappa coefficient for assessing inter-coder reliability are included in Fig. 2. The two judging authors achieved method-specific hit ratios between 0.79 and 1.00, yielding an overall average of 91% which was considered to represent a significant agreement [35]. Regarding Cohen's Kappa, the results ranged from 0.52 to 1.00, reflecting the validity and reliability of our assessment [31].

4.2 Discussion of Key Findings

To enable a detailed discussion, we calculated the relative frequency in percent (f_i) as well as the median (m) for all levels of analysis based on our sample size. The results are shown in Fig. 3. In the following, we discuss the results with respect to all levels of analysis. Therefore, we start with some overall findings regarding the relative frequency, before we analyze the four context dimensions (third level of analysis) in detail. Therefore, we present selected highlights for each context factor (second level of analysis) and all underlying characteristics (first level of analysis).

Overall Findings. All in all, our results in Figs. 2 and 3 indicate that the investigated BPM methods are not yet aligned to various context dimensions. About 70% of the investigated BPM methods must be considered as context independent. This is supported by mainly medians of three. Only 30% of the investigated BPM methods are assessed with an assessment criterion of (1) or (5), which indicates context awareness. Moreover, it is interesting to note that 20% of the investigated BPM methods are context-aware and support specific characteristics (5), while only 10% especially address characteristics for which they are not applicable (1). However, characteristics

			Level of analysis 3 - dimension					Level of analysis 2 - context factor					Level of analysis 1 - characteristic					Goal dimension	
			m	f_1	f_3	f_5	$f_{1,5}$	m	f_1	f_3	f_5	$f_{1,5}$	m	f_1	f_3	f_5	$f_{1,5}$	$m_{exploitation}$	$m_{exploration}$
Process dimension	Value contribution	Core process	3	7.4%	76.4%	16.2%	23.6%	3	16.7%	63.3%	20.0%	36.7%	3	3.3%	60.0%	36.7%	40.0%	3	4
		Management process											3	23.3%	53.3%	23.3%	46.7%	3	3
		Support process											3	23.3%	76.7%	0.0%	23.3%	3	2
	Repetitiveness	Repetitive						3	13.3%	60.0%	26.7%	40.0%	4	3.3%	46.7%	50.0%	53.3%	5	3
		Non-repetitive											3	23.3%	73.3%	3.3%	26.7%	3	3
	Knowledge intensity	Low knowledge-intensity						3	0.0%	91.7%	8.3%	8.3%	3	0.0%	96.7%	3.3%	3.3%	3	3
		High knowledge-intensity											3	0.0%	86.7%	13.3%	13.3%	3	3
	Creativity	Low creativity						3	0.0%	91.7%	8.3%	8.3%	3	0.0%	96.7%	3.3%	3.3%	3	3
		High creativity											3	0.0%	86.7%	13.3%	13.3%	3	3
	Interdependence	Low interdependence						3	1.7%	85.0%	13.3%	15.0%	3	0.0%	93.3%	6.7%	6.7%	3	3
		High interdependence											3	3.3%	76.7%	20.0%	23.3%	3	3
	Variability	Low variability						3	8.3%	73.3%	18.3%	26.7%	3	3.3%	73.3%	23.3%	26.7%	3	3
		High Variablity											3	13.3%	73.3%	13.3%	26.7%	3	3
Organization dimension	Scope	Intra-organizational processes	3	13.9%	64.2%	21.9%	35.8%	5	30.0%	16.7%	53.3%	83.3%	5	16.7%	0.0%	83.3%	100%	5	3
		Inter-organizational processes											3	43.3%	33.3%	23.3%	66.7%	3	4
	Industry	Product industry						3	5.6%	87.8%	6.7%	12.2%	3	6.7%	90.0%	3.3%	10.0%	3	3
		Service Industry											3	3.3%	90.0%	6.7%	10.0%	3	3
		Product & Service Industry											3	6.7%	83.3%	10.0%	16.7%	3	3
	Size	Start-up						3	10.0%	77.8%	12.2%	22.2%	3	30.0%	70.0%	0.0%	30.0%	3	3
		Small and medium enterprise											3	0.0%	96.7%	3.3%	3.3%	3	3
		Large Organization											3	0.0%	66.7%	33.3%	33.3%	3	3
	Culture	Culture highly supportive of BPM						3	20.0%	48.3%	31.7%	51.7%	5	0.0%	36.7%	63.3%	63.3%	5	5
		Culture non-supportive of BPM											3	40.0%	60.0%	0.0%	40.0%	3	2
	Resources	Low organizational resources						3	10.0%	71.7%	18.3%	28.3%	3	20.0%	80.0%	0.0%	20.0%	3	3
		High organizational resources											3	0.0%	63.3%	36.7%	36.7%	3	3
Environment dimension	Competitiveness	Low competetive environment	3	7.5%	68.3%	24.2%	31.7%	3	8.3%	61.7%	30.0%	38.3%	3	16.7%	83.3%	0.0%	16.7%	3	3
		High competetive environment											5	0.0%	40.0%	60.0%	60.0%	5	5
	Uncertainty	Low environmental uncertainty						3	6.7%	75.0%	18.3%	25.0%	3	13.3%	86.7%	0.0%	13.3%	3	3
		High environmental uncertainty											3	0.0%	63.3%	36.7%	36.7%	3	5

Exploitation			
f_1	f_3	f_5	$f_{1,5}$
9.3%	70.7%	20.0%	29.3%

Exploration			
f_1	f_3	f_5	$f_{1,5}$
13.2%	68.4%	18.4%	31.6%

Overall findings			
f_1	f_3	f_5	$f_{1,5}$
10.1%	70.2%	19.7%	30%

relative frequency fi median m strong context awareness >66.6% moderate context awareness >33.3%,<66.6% weak context awareness <33.3%

Fig. 3. Relative frequency and median for three levels of analysis.

for which a BPM method is not applicable are important to note as an application would probably lead to undesired results. For example, if a BPM method is developed only for service industries, an application in product industries is not recommendable.

Goal Dimension. The assessment of the identified 25 BPM methods shows that 24 BPM methods are applicable for exploitation, six for exploitation and exploration, and one for exploration only. Thus, the ratio between BPM methods that focus on exploitation and exploration is unbalanced.

Even though the BPM methods which refer to exploration are shorthanded, one third indicate context awareness ($f_{1,5} = 31.6\%$). These six exploration BPM methods are especially applicable for core processes inter-organizational processes meeting the needs of customers in a highly competitive environment. To properly strive for explorative BPM, organizations also need a highly supportive BPM culture. Cultural values such as responsibility, excellence or customer orientation [46] build a solid foundation to successfully apply BPM methods for explorative purpose.

The 24 identified exploitation BPM methods indicate a slightly smaller degree of context awareness than those for exploration ($f_{1,5} = 29.3\%$). Moreover, they are especially applicable for repetitive and intra-organizational processes in a highly competitive environment. As already mentioned for exploration, the support of BPM culture plays a crucial role to properly execute exploitation BPM methods.

Process Dimensions. The third level of analysis shows that a quarter of all investigated BPM methods are context-aware referring to the process dimension ($f_{1,5} = 23.6\%$). In particular, the second level of analysis illustrates that BPM methods are moderately context-aware regarding the context factors of value contribution and repetitiveness, while they are weak context-aware for knowledge-intensity, creativity, interdependence, and variability. The first level of analysis indicates that BPM methods for management and core processes are moderately context- aware, while for support processes the context awareness of BPM methods is weak. Moreover, no context-aware BPM methods are applicable for support processes ($f_5 = 0.0\%$). The same applied for the context factor of repetitiveness. More than half of all BPM methods indicate context awareness for repetitive processes which is also supported by a median of four. Further, the context awareness of BPM methods for non-repetitive processes is weak and most are not applicable for non-repetitive processes ($f_1 = 23.3\%$).

Organization Dimension. Considering the third level of analysis, around one third of the investigated BPM methods are context-aware referring to the organization dimension ($f_{1,5} = 35.8\%$). More precisely, for the context factors of scope and culture the context awareness of BPM methods is strong, while it is weak for industry, size, and resources (second level of analysis). BPM methods are strongly context-aware regarding the context factor of scope as 83.3% are applicable for intra-organizational processes (first level of analysis). This is also supported by a median of five. Only some BPM methods are context-aware and applicable for inter-organizational processes ($f_5 = 23.3\%$), especially if they focus on the goal dimension of exploration. This fact might be explained by an increased customer orientation along with BPM methods for exploration. Moreover, a median of five as well as two thirds of all identified BPM methods show that a highly supportive culture for BPM is required ($f_5 = 63.3\%$) (first level of analysis). Thus, a raising awareness regarding potentially neglected cultural factors is important to supplement existing BPM methods.

Environment Dimension. For the environment dimension (third level of analysis), around one third of all BPM methods are context-aware $(f_{1,5} = 31.7\%)$, whereby the context awareness of BPM methods for the context factor competitiveness is greater than for uncertainty (second level of analysis). As a median of five and the relative frequency of high competitive environments $(f_5 = 60.0\%)$ show, many BPM methods are especially applicable for high competitive environments. In most cases, also high environmental uncertainty requires context-aware BPM methods $(f_5 = 37\%)$ (first level of analysis).

5   Conclusion

5.1   Summary and Implications

In the digital age, organizations face several challenges such as fulfilling customers' changing wishes, facing high uncertainty or surviving in a dynamic competitive environment [20, 26]. To overcome these challenges, organizations need to manage several contexts at the same time [50, 52]. Given the increasing importance of context awareness and appropriate context-aware BPM methods for organizations in the digital age, the purpose of our study was to determine the context awareness of extant BPM methods regarding the four context dimensions (i.e., goal, process, organization and environment dimension) of the BPM context framework [52]. Therefore, we developed an assessment scheme for the context awareness of BPM methods which has been applied for the first time. Based on three levels of analysis, our results show many 'white boxes', which refer to a lack of context-aware BPM methods. In particular, BPM methods that focus on the goal of exploration seem to be rare. Further, our results suggest that BPM methods show room for improvement regarding context-awareness in the process, organization, and environment dimensions. Only very few methods account for these dimensions. Regarding the process dimension, particularly BPM methods geared toward the context factors knowledge-intensity, creativity, interdependence, and variability are seldom. Concerning the organizational dimension, size and industry have not been in focus yet. The context factor culture has been considered in most BPM methods, meaning that a supportive BPM culture is necessary to apply almost any BPM method. With reference to the environment dimension, BPM methods for the context factors of competitiveness and uncertainty need to be in the center of attention. Concluding, BPM research faces several gaps concerning context awareness and corresponding BPM methods. Research should strongly focus on appropriate methods which support different contexts. Emerging BPM methods should always consider and state the specific contexts they refer to.

The results of our research have implications for practice and research. The implications will be structured according to the assessment scheme and final results. Additionally, we point out how our work contributes to the research stream of context-aware BPM and ambidextrous BPM. First, with regard to theoretical implications, we developed an assessment scheme, which is based on the context framework of vom Brocke et al. [52]. The assessment scheme makes it possible to assess the context awareness of extant BPM methods based on three well-structured steps. Additionally, the assessment scheme offers a distinction in the goal dimension (horizontal axis) and the process, organization, and environment dimension (vertical axis). This distinction emphasizes that organizations can actively chose a BPM goal (i.e., exploitation or exploration), while the dimensions on the vertical axis are given and cannot be modified. Thus, an integrated perspective of context-aware and ambidextrous BPM is facilitated and allows for a detailed analysis of context dimensions that are given and influenceable. The assessment scheme also serves as a basis for further conceptual development (e.g., assigning BPM methods to a BPM lifecycle phase for further analysis, or developing a toolbox to assist organizations in systematically selecting context-aware BPM methods). Second, the final results of investigating 25 BPM methods point out a lack of context-aware BPM methods. Many 'white boxes' indicate that the majority of BPM methods are context-independent. Thus, our work serves as a starting point for further discussions and developments. We call for emerging context-aware BPM methods to manage several contexts simultaneously and meet the requirements of the digital age. Therefore, the assessment scheme facilitates the classification and development of a BPM method according to its context awareness in various dimensions.

With regard to the practical implications, the assessment scheme enables practitioners to assess extant BPM methods in a structured way. In particular, practitioners can check the applicability of currently used BPM methods and systematically select new ones. Especially the goal dimension underlines the importance of ambidextrous thinking, which means that an organization should focus on exploitation and exploration simultaneously and needs to consciously decide which goals require which BPM method. Thus, practitioners are able to improve the choice of BPM methods. Second, the final results illustrate that extant BPM methods are rarely context-aware. To be successful in the digital age, practitioners need to specifically select BPM methods from the limited number of context-aware BPM methods. Especially, if an organization strives for exploration, which is of utmost importance in the digital age, appropriate BPM methods need to be identified.

5.2 Limitations and Further Research

Being aware of the limitations of our research, we identified several directions for further research that are outlined in this section.

Our findings build on a few journals and a small sample size of BPM methods. We focused on specific journals such as the BPMJ or the "Senior Scholars' Basket of Journals" since they cover core BPM research. We do not claim for completeness and generalizability of all results. Yet, our results provide first insights about the context awareness of BPM methods. As direction for further research, we propose to broaden the sample size of BPM methods.

Another limitation is that we assessed the applicability of the BPM methods from a researchers' perspective only. We believe, however, that the current assessment is adequate to provide first insights on the context awareness of BPM methods. Further research may assess the applicability of BPM methods with both researchers and practitioners.

In the end, Table 3 presents our "Call for Action" for further research. On the one hand, the table summarizes ideas for further research based on the limitations mentioned above. On the other hand, it includes further research ideas beyond limitations. For instance, how the assessment scheme can be further developed or which potential issues could arise from dealing with context-aware BPM methods. Thus, Table 3 outlines different research areas, their corresponding research problems, and points to some theoretical and practical solution ideas.

Table 3. "Call for Action" for further research.

Area	Research problem	Solution ideas
Context-aware BPM methods	Identify further BPM methods	• Search further literature (e.g., within BPM Handbooks or conferences and journals covering BPM research) • Include methods from other disciplines (e.g., innovation management that may fit explorative purposes)
	Develop context-aware BPM methods	• Identify challenges for each context factor and develop corresponding context-aware BPM methods (e.g., for challenges of knowledge-intense processes)
	Support decision-making for ambidextrous BPM	• Develop a decision model that assists organizations in selecting exploration and exploitation BPM methods
	Meet challenges of realizing ambidextrous BPM	• Identify challenges that occur by simultaneously conducting exploitation and exploration BPM methods • Develop a maturity model that guides an organization to realize ambidextrous BPM
Assessment scheme	Increase the number of judges	• Consult further BPM researchers and practitioners to determine the applicability of extant BPM methods based on our assessment scheme
	Extend the assessment scheme	• Add further context dimensions (i.e., a customer dimension to better account for challenges in the digital age) • Assign each BPM method to a BPM lifecycle phase for further analysis (e.g., process identification or analysis phase)
	Assess the current context based on the four context dimensions	• Operationalize each characteristic to assess the current context of a process and an organization as a foundation to select context-aware BPM methods
	Assist organizations in selecting context-aware BPM methods	• Develop a toolbox that enables organizations to assess their context and decide which BPM methods are required for which goal and lifecycle phases

References

1. Anastassiu, M., Santoro, F.M., Recker, J., Rosemann, M.: The quest for organizational flexibility: driving changes in business processes through the identification of relevant context. Bus. Process Manag. J. **22**(4), 763–790 (2016)
2. Antunes, A.S., Rupino da Cunha, P., Barata, J.: MUVE IT: reduce the friction in business processes. Bus. Process Manag. J. **20**(4), 571–597 (2014)
3. Association for Information Systems: Senior Scholars' Basket of Journals. https://aisnet.org/?SeniorScholarBasket. Accessed 30 May 2018
4. Benner, M.J., Tushman, M.L.: Exploitation, exploration, and process management: the productivity dilemma revisited. Acad. Manag. Rev. **28**(2), 238–256 (2003)
5. Bergener, P., Delfmann, P., Weiss, B., Winkelmann, A.: Detecting potential weaknesses in business processes: an exploration of semantic pattern matching in process models. Bus. Process Manag. J. **21**(1), 25–54 (2015)
6. Bisogno, S., Calabrese, A., Gastaldi, M., Ghiron, N.L.: Combining modelling and simulation approaches: how to measure performance of business processes. Bus. Process Manag. J. **22** (1), 56–74 (2016)
7. Bolsinger, M., Elsäßer, A., Helm, C., Röglinger, M.: Process improvement through economically driven routing of instances. Bus. Process Manag. J. **21**(2), 353–378 (2015)
8. Breuker, D., Matzner, M., Delfmann, P., Becker, J.: Comprehensible predictive models for business processes. MISQ **40**(4), 1009–1034 (2016)
9. Business Process Management Journal. http://www.emeraldgrouppublishing.com/products/journals/journals.htm?id=bpmj. Accessed 30 May 2018
10. Cohen, J.: A coefficient of agreement for nominal scales. Educ. Psychol. Measur. **20**(1), 37–46 (1960)
11. de Boer, F.G., Müller, C.J., Schwengber ten Caten, C.: Assessment model for organizational business process maturity with a focus on BPM governance practices. Bus. Process Manag. J. **21**(4), 908–927 (2015)
12. de Bruin, T., Rosemann, M.: Towards a business process management maturity model. In: Bartmann, D., Rajola, F., Kallinikos, J., Avision, D., Winter, R., Ein Dor, T., et al. (eds.) ECIS 2005 Proceedings of the 13th European Conference on Information Systems, Regensburg, Germany (2005)
13. Denner, M.-S., Püschel, L., Röglinger, M.: How to exploit the digitalization potential of business processes. Bus. Inf. Syst. Eng. **60**(4), 1–19 (2017)
14. do Prado Leite, J.C.S., Santoro, F.M., Cappelli, C., Batista, T.V., Santos, F.J.N.: Ownership relevance in aspect-oriented business process models. Bus. Process Manag. J. **22**(3), 566–593 (2016)
15. Fengel, J.: Semantic technologies for aligning heterogeneous business process models. Bus. Process Manag. J. **20**(4), 549–570 (2014)
16. Field, A.: Discovering Statistics Using SPSS, 3rd edn. Sage Publications Ltd., London (2009)
17. Fiorentino, R.: Operations strategy: a firm boundary-based perspective. Bus. Process Manag. J. **22**(6), 1022–1043 (2016)
18. Hakim, A., Gheitasi, M., Soltani, F.: Fuzzy model on selecting processes in business process reengineering. Bus. Process Manag. J. **22**(6), 1118–1138 (2016)
19. Harmon, P., Wolf, C.: The State of Business Process Management 2014: A BPTrends Report. BPTrends (2014)
20. Harmon, P., Wolf, C.: The State of Business Process Management 2018: A BPTrends Report. BPTrends (2018)

21. Hosseini, S., Kees, A., Manderscheid, J., Röglinger, M., Rosemann, M.: What does it take to implement open innovation? Towards an integrated capability framework. Bus. Process Manag. J. **23**(1), 87–107 (2017)

22. Inês Dallavalle de Pádua, S., Mascarenhas Hornos da Costa, J., Segatto, M., Aparecido de Souza Júnior, M., José Chiappetta Jabbour, C.: BPM for change management: two process diagnosis techniques. Bus. Process Manag. J. **20**(2), 247–271 (2014)

23. Jacoby, J., Matell, M.S.: Three-point likert scales are good enough. J. Mark. Res. **8**(4), 495–500 (1971)

24. Johannsen, F., Leist, S., Tausch, R.: Wand and Weber's good decomposition conditions for BPMN: an interpretation and differences to event-driven process chains. Bus. Process Manag. J. **20**(5), 693–729 (2014)

25. Kerpedzhiev, G., König, U., Röglinger, M., Rosemann, M.: BPM in the Digital Age: BPM Capability Framework. http://digital-bpm.com/bpm-capability-framework/. Accessed 30 May 2018

26. Kerpedzhiev, G., König, U., Röglinger, M., Rosemann, M.: Business Process Management in the Digital Age. http://digital-bpm.com/. Accessed 30 May 2018

27. Khlif, W., Ben-Abdallah, H., Ayed, N.E.B.: A methodology for the semantic and structural restructuring of BPMN models. Bus. Process Manag. J. **23**(1), 16–46 (2017)

28. Khosravi, A.: Business process rearrangement and renaming: a new approach to process orientation and improvement. Bus. Process Manag. J. **22**(1), 116–139 (2016)

29. Kohlborn, T., Mueller, O., Poeppelbuss, J., Roeglinger, M.: Interview with michael rosemann on ambidextrous business process management. Bus. Process Manag. J. **20**(4), 634–638 (2014)

30. Lamberti, E., Michelino, F., Cammarano, A., Caputo, M.: Open innovation scorecard: a managerial tool. Bus. Process Manag. J. **23**(6), 1216–1244 (2017)

31. Landis, J.R., Koch, G.G.: The measurement of observer agreement for categorical data. Biometrics **33**(1), 159–174 (1977)

32. Lavikka, R., Smeds, R., Jaatinen, M.: A process for building inter-organizational contextual ambidexterity. Bus. Process Manag. J. **21**(5), 1140–1161 (2015)

33. Likert, R.: A technique for the measurement of attitudes. Arch. Psychol. **22**(140), 1–55 (1932)

34. Lindman, M., Pennanen, K., Rothenstein, J., Scozzi, B., Vincze, Z.: The value space: how firms facilitate value creation. Bus. Process Manag. J. **22**(4), 736–762 (2016)

35. Moore, G.C., Benbasat, I.: Development of an instrument to measure the perceptions of adopting an information technology innovation. Inf. Syst. Res. **2**(3), 192–222 (1991)

36. Morton, N.A., Hu, Q.: Implications of the fit between organizational structure and ERP: a structural contingency theory perspective. Int. J. Inf. Manag. **28**(5), 391–402 (2008)

37. Nahm, A.Y., Rao, S.S., Solis-Galvan, L.E., Ragu-Nathan, T.S.: The Q-sort method: assessing reliability and construct validity of questionnaire items at a pre-testing stage. J. Modern Appl. Stat. Methods **1**(1), 114–125 (2002)

38. Pereira Librelato, T., Pacheco Lacerda, D., Rodrigues, L.H., Veit, D.R.: A process improvement approach based on the value stream mapping and the theory of constraints thinking process. Bus. Process Manag. J. **20**(6), 922–949 (2014)

39. Polpinij, J., Ghose, A., Dam, H.K.: Mining business rules from business process model repositories. Bus. Process Manag. J. **21**(4), 820–836 (2015)

40. Rangiha, M.E., Comuzzi, M., Karakostas, B.: A framework to capture and reuse process knowledge in business process design and execution using social tagging. Bus. Process Manag. J. **22**(4), 835–859 (2016)

41. Rocha, R.D.S., Fantinato, M., Thom, L.H., Eler, M.M.: Dynamic product line for business process management. Bus. Process Manag. J. **21**(6), 1224–1256 (2015)

42. Roeser, T., Kern, E.-M.: Surveys in business process management – a literature review. Bus. Process Manag. J. **21**(3), 692–718 (2015)
43. Rogers, P.R., Miller, A., Judge, W.Q.: Using information-processing theory to understand planning/performance relationships in the context of strategy. Strateg. Manag. J. **20**(6), 567–577 (1999)
44. Rosemann, M.: Proposals for future BPM research directions. In: Ouyang, C., Jung, J.-Y. (eds.) AP-BPM 2014. LNBIP, vol. 181, pp. 1–15. Springer, Cham (2014). https://doi.org/10.1007/978-3-319-08222-6_1
45. Rosemann, M., Recker, J., Flender, C.: Contextualisation of business processes. Int. J. Bus. Process Integr. Manag. **3**(1), 47–60 (2008)
46. Schmiedel, T., vom Brocke, J., Recker, J.: Which cultural values matter to business process management? Bus. Process Manag. J. **19**(2), 292–317 (2013)
47. Trkman, P., Mertens, W., Viaene, S., Gemmel, P.: From business process management to customer process management. Bus. Process Manag. J. **21**(2), 250–266 (2015)
48. van der Aalst, W.M.P.: Business process management: a comprehensive survey. ISRN Softw. Eng. 1–37 (2013)
49. vom Brocke, J., Mendling, J.: Frameworks for business process management: a taxonomy for business process management cases. In: vom Brocke, J., Mendling, J. (eds.) Business Process Management Cases. MP, pp. 1–17. Springer, Cham (2018). https://doi.org/10.1007/978-3-319-58307-5_1
50. vom Brocke, J., Schmiedel, T., Recker, J., Trkman, P., Mertens, W., Viaene, S.: Ten principles of good business process management. Bus. Process Manag. J. **20**(4), 530–548 (2014)
51. vom Brocke, J., Simons, A., Niehaves, B., Riemer, Kai, Plattfaut, R., Cleven, A.: Reconstructing the giant: on the importance of rigour in documenting the literature search process. In: 17th ECIS, Verona, Italy, pp. 2206–2217 (2009)
52. vom Brocke, J., Zelt, S., Schmiedel, T.: On the role of context in business process management. Int. J. Inf. Manag. **36**(3), 486–495 (2016)
53. Webster, J., Watson, R.T.: Analyzing the past to prepare for the future: writing a literature review. MIS Q. **26**(2), xiii–xxiii (2002)
54. Zhu, X., Recker, J., Zhu, G., Maria Santoro, F.: Exploring location-dependency in process modeling. Bus. Process Manag. J. **20**(6), 794–815 (2014)

Process Forecasting: Towards Proactive Business Process Management

Rouven Poll[1(✉)], Artem Polyvyanyy[2], Michael Rosemann[3],
Maximilian Röglinger[4], and Lea Rupprecht[1]

[1] FIM Research Center, University of Augsburg, Augsburg, Germany
{rouven.poll, lea.rupprecht}@fim-rc.de
[2] The University of Melbourne, Parkville, VIC 3010, Australia
artem.polyvyanyy@unimelb.edu.au
[3] Queensland University of Technology, Brisbane, Australia
m.rosemann@qut.edu.au
[4] FIM Research Center, University of Bayreuth, Bayreuth, Germany
maximilian.roeglinger@fim-rc.de

Abstract. The digital economy is highly volatile and uncertain. Ever-changing customer needs and technical progress increase the pressure on organizations to continuously improve and innovate their business processes. The ability to anticipate incremental and radical process changes required in the future is a critical success factor. However, organizations often fail to forecast future business process designs and process performance. One reason is that Business Process Management (BPM) is dominated by *reactive* methods (e.g., lean management, traditional process monitoring), whereas there are only a few future-oriented approaches (e.g., process simulation, predictive process monitoring). This paper supports the shift towards *proactive* BPM by coining the notion of *process forecasting* – an umbrella concept for future-oriented BPM methods and techniques. We motivate the need for process forecasting by eliciting various types of process forecasting from BPM use cases and create a first understanding of its scope by providing a definition, a reference process, showing the steps to be followed in process forecasting initiatives, and a positioning against related BPM sub-areas. The definition and reference process are based on a structured literature review.

Keywords: Process forecasting · Proactive business process management
Predictive business process management

1 Introduction

Business processes allow organizations to match existing customer demand with the supply of the resources needed to fulfil this demand. In the digital age, ever-changing customer needs and rapid technical progress cause high volatility and uncertainty. Such ongoing changes in market conditions force organizations to continuously adapt their business processes [1], which involves both the adaptation of resources to changes in the quantitative demand (e.g., the number of incoming customer orders) and the provision of business processes in respect to qualitative demand changes (e.g., customers

© Springer Nature Switzerland AG 2018
M. Weske et al. (Eds.): BPM 2018, LNCS 11080, pp. 496–512, 2018.
https://doi.org/10.1007/978-3-319-98648-7_29

seeking new digital channels to interact with organizations). The ability to timely anticipate incremental or radical changes of business processes required in the future is a critical success factor for organizations. However, the BPM state of the art does not provide sufficient tools to manage business processes proactively [2].

We argue that a more prevalent usage of future-oriented methods in BPM will lead to an improved and earlier understanding of future process demands and, thus, enable the timely implementation of required process changes [2]. Concretely, a widespread use of these methods could help shift the predominating focus on reactive BPM (e.g., lean management, process monitoring) towards proactive BPM. Proactive BPM is concerned with sensing process changes required in the future timely and effectively and implementing the identified changes before issues occur or opportunities are missed. For instance, as described in [3], instead of the reactive practice of spotting different types of waste, this would entail proactively identifying *waste-in-the-making* (e.g., emerging re-work) leading to an entire new discipline of *proactive lean management*.

The quality of proactive BPM can be measured by the extent to which it reduces *process latency*, i.e., the time of the occurrence of a process problem and its resolution, reducing the accumulated time during which a process is of unsatisfactory design or execution. The economic benefits of forecasting business processes can be seen in selective practices such as Amazon's predictive shipping where goods are delivered in anticipation that customers will order them. Done successfully, this leads to earlier demand satisfaction and revenue, and positive customer experience [4]. On the cost side, predictive maintenance approaches show how dynamically calculating emerging maintenance actions and embedding them into the production schedule minimizes costs related to significant replacements [5]. Predictive shipping and maintenance, however, are still isolated practices, and only a few future-oriented methods exist in BPM [2].

In this light, our paper aims to sensitize for the need for proactive BPM and to trigger a community-wide discussion on the use of forecasting elements into BPM. Thus, we seek to introduce process forecasting as a concept for gaining early insights into and anticipating future business processes by answering the following research questions: *(RQ1) What are use cases of forecasting in the context of BPM? (RQ2) How can process forecasting be defined? (RQ3) What are main steps of a process forecasting initiative?* Our answers to these questions resulted in three conceptual elements proposed in this paper: process forecasting types, definition, and reference process. The remainder of this paper is structured as follows. In Sect. 2, our methodological approach to develop the three conceptual elements is presented. In Sect. 3, we present the distinct types of process forecasting, demonstrating its wide range of use cases. After having motivated the need for process forecasting, Sect. 4 proposes the definition of process forecasting. In Sect. 5, we propose the reference process for process forecasting initiatives. After discussing the results of the conducted literature review in Sect. 6, Sect. 7 positions process forecasting against other BPM sub-areas. Finally, Sect. 8 concludes the paper by summarizing the findings and pointing to future work.

2 Research Method

To develop the three conceptual elements aiming to coin the notion of process forecasting, we applied the following methodological approach. Firstly, on the basis of BPM literature, we identified different types of process forecasting. Secondly, by means of a structured literature review, we defined process forecasting. Thirdly, we used the knowledge obtained from the literature review to adapt a well-accepted forecasting reference process to the BPM domain, resulting in the proposed process forecasting reference process. Below, we describe our tripartite research method in detail.

Identifying the Types of Process Forecasting. The different types of process forecasting describe use cases of forecasting in BPM. To identify them, we conducted an in-depth analysis of the BPM life cycle and BPM use cases proposed by Van der Aalst [6]. To do so, we mapped the BPM use cases to the phases of the life cycle. Then, each researcher independently analyzed the BPM use cases with regard to whether they can be supported by forecasting. The resulting unstructured collection of forecasting use cases in the BPM domain was then discussed and consolidated in a joint workshop.

Formulating the Definition of Process Forecasting. With the aim to construct a well-founded definition of process forecasting, which comprises all identified types, we performed a structured literature review. Referring to the guidelines of Vom Brocke et al. [7], before conducting a literature search, the research scope needs to be defined. The topic of concern is the concept of process forecasting with its objective to predict process characteristics. To this end, in our literature review, we aimed for a comprehensive coverage of BPM-related research containing a forecasting component. Based on the terminology used in seminal publications [8–10], we formulated the following search phrases: "predict* […] business process*", "forecast* […] business process*", "business process forecast*", "business process prediction", "predictive business process monitoring" and "predictive process monitoring". We applied these to two scholarly databases, i.e., Scopus and Web of Science. Scopus is one of the largest abstract and citation database of peer-reviewed literature and includes scientific journals and books. To account for the fact that in computer science conferences are a significant publication outlet, we also used Web of Science, as this database – besides a large number of journals and books – covers over 180,000 conference proceedings [11]. Thus, a wide coverage of our literature review in information and computer science related topics is ensured. All studies containing at least one of the phrases in the title, keywords, abstract or (for Scopus only) in the full text of the paper were retrieved. Subsequently, we merged and filtered the retrieved papers, i.e., we removed duplicates, manuscripts not written in English, and not published as a journal article, book chapter, proceedings paper or as an article in press. From the resulting 120 papers, 56 were classified as relevant. For these papers, a forward and backward search was conducted, leading to a final set of 65 relevant papers. We selected papers that propose techniques, methods, or approaches supporting the early detection of process issues and opportunities, i.e., the prediction of future values of process characteristics. Examples for exclusions are papers addressing quality and complexity of process models, process discovery, and conformance checking. These were considered irrelevant as they do not contain a forecasting component. As we intended to construct a framework compiling possible input and output parameters of process forecasting, we then classified the methods proposed in the literature by their input and output. After that, we grouped the individual parameters found into categories. The resulting high-level categories of input and output parameters were then conceptualized and defined by means of appropriate literature.

Constructing the Process Forecasting Reference Process. To create a reference process to be followed in process forecasting initiatives, we drew from an accepted forecasting reference process (see [12]). The process we chose as our basis resembles other reference processes in forecasting literature (see e.g., [13]). Making use of forecasting literature is reasonable here, because – in our understanding – process forecasting is a specific type of forecasting and as such, it should inherit its basic, domain-agnostic properties. To instantiate the reference process for BPM, based on the knowledge gained from the literature survey, we carried out domain-specific adaptions to each step.

3   Types of Process Forecasting

The BPM life cycle, as proposed by Van der Aalst [6], includes three phases. In the phase *(re-)design*, a process model is designed. The phase *implement/configure* refers to making a process model executable. Finally, the phase *run and adjust* is concerned with process execution. In the context of the phase *run and adjust*, Van der Aalst [6] emphasizes the need for analysis of expected and past performance, and monitoring of processes (see use cases "analyze performance based on model", "analyze performance using event data", and "monitor" [6]). By combining these use cases and taking a future-oriented perspective, we obtain the first type of process forecasting, viz., "solving the execution problem", which addresses the predictive monitoring and simulation of processes at or shortly before run-time. The second type, "solving the configuration problem", relates to the phase *implement/configure* and adds value when multiple process model variants exist one of which needs to be selected prior to process execution (see use case "configure configurable model" [6]). Thereby, process forecasting enables to anticipate which model is best suited for upcoming process executions, accounting for the future states of the operating environment. The third type, "solving the design problem", relates to the phase *(re-)design* and supports the demand-driven design and improvement of process models (see use cases "design model" and "improve model" [6]). Here, the use case of process forecasting lies in predicting how processes need to be designed to comply with future process demands. The proposed types of process forecasting – visualized in Fig. 1 – are explained below via illustrative examples.

Type 1: Solving the Execution Problem. This type of process forecasting is concerned with anticipating process-related issues of day-to-day operations. This involves two forecasting tasks: forecasting process demand, i.e., the number and type of instances arriving in a future time period (Type 1a) (see e.g., [14]) as well as forecasting the

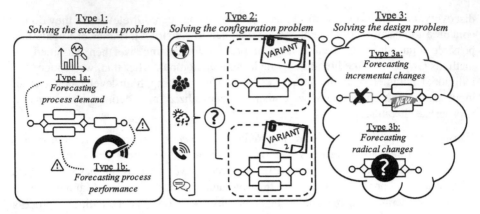

Fig. 1. Visualization of the three types of process forecasting.

expected performance of running and future process instances (Type 1b). The latter involves the prediction of performance indicators (e.g., cycle time) (see e.g., [15]), the process outcome (e.g., the probability of violating business constraints) (see e.g., [8]) as well as the sequence of activities (see e.g., [16]). When focusing on running instances, this may also include forecasting the next activities to be executed. An illustrative example of solving the execution problem can be inferred from [17], where the authors describe a retailer deploying forecasting techniques in order to prevent its stores from running out of stock. In the first step, the retailer may predict how many and which products will be sold in each store to anticipate the most cost-effective point in time when replenishments are needed (Type 1a). This forecast might be derived by taking into account seasonality aspects (e.g., higher demand for certain products shortly before Christmas) or consumer trends. After having ordered the replenishments, the retailer might be interested in the probability of delayed deliveries (Type 1b). As shown in [17], taking contextual information such as future weather conditions and their impact on transport routes into account, the retailer may detect delays before they occur. This enables proactive rescheduling of transport routes to prevent the upcoming delays.

Type 2: Solving the Configuration Problem. This type of process forecasting problem exists when a concrete model from some configurable process model needs to be created (see use case "Configure Configurable Model" in [6]), i.e., when an organization needs to select between alternative process model variants based on contextual variables (e.g., time, location, weather). In this regard, process forecasting can help anticipate the process model variant that is needed at a certain future point in time. This enables organizations to better prepare the execution of a process model variant and, thus, reduces process latency. For example, Rosemann and Recker [18] describe an insurance company that has designed process variants for lodging insurance claims based on different levels of severity of storms during the Australian storm season. As soon as a storm occurs, its severity is evaluated and the execution of the corresponding variant of the process model is triggered. In this case, process latency could be reduced by taking forecasts of the severity of storms into account to predict the process variant that is needed and proactively initiate targeted measures.

Type 3: Solving the Design Problem. The design of a business process is driven by the requirements assigned to the respective process [19]. Here, the use case of process forecasting lies in anticipating the changes in process models that will occur or be demanded in the future. Thereby, in line with the common view in business process improvement literature [20], we distinguish between incremental and radical changes. Whereas incremental changes are adaptions of existing process models (Type 3a), radical changes address the creation of entire new models (Type 3b). An example setting can be derived from [21]. This work describes a bank that aims to decide on the channels that should be offered to customers to conduct their banking activities in the future. As analyzed in [21], the customer use of a certain channel depends on customers' intrinsic attributes (e.g., attitude towards technology or age of customers). Knowing this and building on information about future changes in customer characteristics, process forecasting could discover which channels should or should not be offered in the future. For instance, at a certain future point in time, the fraction of customers using telephone banking might be forecasted to decrease considerably, resulting in the recommendation to shut down the telephone service (Type 3a). This would enable the bank to timely initiate associated actions such as cancelling contracts with external service providers and to plan the re-allocation of resources. Going one step further, process forecasting could also predict entire new process models for conducting banking activities (Type 3b). For instance, accounting for the rapidly expanding usage of virtual voice assistants, forecasting techniques could be able to predict the point in time when money transfers via virtual voice assistants are desired, affordable, and viable. Additionally, by learning from related process model designs (e.g., from other industries having implemented voice-based interactions into their process models), an algorithm could output the bank's future process model supporting money transfers via virtual voice assistants.

4 Definition of Process Forecasting

The use cases outlined in Sect. 3 showed that process forecasting types differ in their objective, input, output, and time horizon. However, all types pursue the same over-arching goal, namely to boost organizational preparedness for future business processes. As a result of the conducted literature review and the identified types of process forecasting, we define process forecasting as an *umbrella concept for BPM methods and techniques that aim to predict future business process demands, performance, and designs*. For the purpose of our research, we deliberately refrain from developing a specific process forecasting method. Rather, we structure the field of action by proposing a framework of relevant input and output parameters. The specific set of input variables used to derive forecasts primarily depends on the applied forecasting method [12]. The method to be used, in turn, might be constrained by data availability [13]. As the literature review showed, process forecasts can generally be achieved by combining or extrapolating data from past and running process executions (see e.g., [8]). More sophisticated are context-aware process forecasts that take the future process environment into account (see e.g., [10]). A schematic view of the input and output parameters for applying process forecasting methods is shown in Fig. 2.

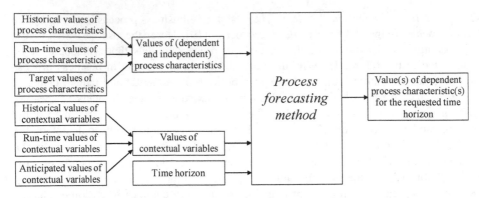

Fig. 2. Schematic view of relevant input and output parameters of process forecasting.

As can be inferred from Fig. 2, a process forecasting method aims to predict the values of one or more dependent process characteristics for a given time horizon based on available historical, run-time, and target values of (independent and dependent) process characteristics as well as historical, run-time, and anticipated values of contextual variables. Thereof, the time horizon is a mandatory input, whereas the other input variables depend on the problem to be solved. Next, we define all the proposed components.

Process characteristics include both *process performance indicators* (PPIs) and the typical *core elements* of business processes. PPIs are measures of the critical success factors of business processes such as cycle time or cost. PPIs can be defined over a single process instance or a group of instances (e.g., instances occurring within a period of time) [22]. The core elements of a business process are essential to its execution and understanding [23]. Referring to de Leoni et al. [24], we differentiate core elements into *data* (e.g., data required to execute a process), *resources* (e.g., the resource performing or supporting a particular activity), *time* (e.g., the duration of an activity), and the *control flow*, i.e., the executed activities and their temporal and logical relationship. The control flow of a process can, for example, manifest itself in the form of an *event log*, i.e., a collection of sequences of observed and recorded events, a *simulation model*, i.e., a conceptual model of a collection of processes with a finite imitation of its operations, or an ordinary *process model*, i.e., a conceptual model of a collection of processes [25]. Process characteristics can be classified as dependent or independent, where the former is the process characteristic to be forecasted and the latter is any other process characteristic taken into account to derive the forecast. The selection of dependent process characteristics primarily hinges on the forecasting problem to be solved and ranges from PPIs, through involved resources, to event logs and process models.

Contextual variables describe the environment in which a business process operates. With regard to the classification presented by Rosemann et al. [23], we distinguish internal, external, and environmental context. Internal context involves the internal environment of an organization having an indirect impact on a business process (e.g., policies, resource capacity, and corporate strategy). External context captures factors beyond an organization's control sphere but within its business network (e.g., characteristics of suppliers and customers, industry-specific factors such as trends driving the demand for an industry's service, and regulations). Environmental context is the environment beyond the business network in which an organization is embedded (e.g., weather, seasonality, and political system).

Time horizon is the period of time for which a forecast is produced. By transferring the classification suggested in general forecasting literature [12], we categorize time horizons into short-, medium- and long-range horizons. In the context of process forecasting, based on the time horizons found in the related literature, we define short-range forecasts such that they cover the prediction of process characteristics of running instances (e.g., forecasting the remaining duration of a running process execution). Medium-range forecasts are based on weekly or monthly time spans from now (e.g., how many employees will be required next Monday to serve arriving customers?). Finally, long-range forecasts cover a (multi-)annual time span (e.g., forecasting the future process model in one year from now).

5 Process Forecasting Reference Process

Below, we present the process forecasting reference process, describing the basic steps to be followed in process forecasting initiatives. Figure 3 provides an overview of the proposed six steps. In the remainder of this section, we describe each step in detail.

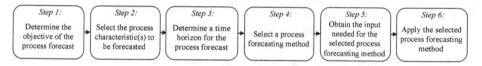

Fig. 3. Process forecasting reference process.

Step 1: Determine the Objective of the Process Forecast. In the first step, the objective of the forecast needs to be defined. This involves questions like "Why is the forecast useful?", "In which way is the forecast used?", and "Who needs the forecasting results?". The overarching goal of process forecasting is to detect and proactively manage process-related issues and opportunities. As can be inferred from the identified types of process forecasting, the instantiations of this objective are manifold. They range from achieving an adequate resource or materials planning and ensuring flawless process executions (solving the execution problem) to reducing process latency (solving the configuration problem), or discovering how future process models will look like (solving the design problem). The defined objective further determines the level of accuracy and the level of aggregation required in the forecast, i.e., whether the forecasting task focuses on predicting single process instances, an aggregation of multiple process instances occurring in a certain time interval, or on predicting complete process models. In general, it is agreed in literature that forecasting for groups (of process instances) is more accurate than for individuals [12]. Moreover, when focusing on a single process instance, one should be aware that this can result in losing sight of the totality of ongoing process executions. For instance, taking actions to prevent a single instance from failing may have negative and unanticipated effects on other running process instances.

Step 2: Select the Process Characteristic(s) to be Forecasted. After determining why to forecast, the next step is to define what to forecast. The selection of the dependent process characteristic(s) depends on the objective of the forecast. For instance, an organization that aims to solve the design problem, e.g., to discover how a process needs to be designed to satisfy future customer needs, should select the future process model as prediction target. In contrast, for an organization interested in an estimation of how a process will perform in the future (solving the execution problem), the prediction of one or several PPIs might be expedient. For instance, in [15], the prediction of cycle time is emphasized, as this enables organizations to provide customers with waiting times guarantees. These were shown to increase customer satisfaction, if they are met [15]. When selecting a dependent process characteristic, analysts should consider that the explanatory power of a forecast can be increased by choosing leading indicators, i.e., the key factors that are known to influence unwanted changes of a certain higher-level process characteristic (lagging indicator), as the prediction target [24]. Leading indicators can be defined on different levels. For instance, human resources involved in a process can be defined as leading indicator for the lagging indicator cycle time (e.g., due to different levels of experience of the employees) [24]. Cycle time, in turn, is a leading indicator for the lagging indicator process cash flow. Based on the executed literature review, we obtained the following relationship: the lower the level of a leading indicator, the higher the forecast's explanatory power. Thus, adhering to the example sketched above, forecasting the employees that will be assigned to work on the process of interest (instead of forecasting the cycle time), may reveal additional information, as it enables analysts to not only detect how but also why a process will perform in a certain manner. More of these cause-effect relationships between process characteristics were discovered as part of the literature review and are shown in Fig. 5.

Step 3: Determine a Time Horizon for the Process Forecast. As a third step, the length of time on which a forecast is based and how far into the future the forecast is generated, needs to be determined. Thereby, it should be taken into account that, in general, the accuracy of forecasts decreases as the time horizon extends [12]. However, short-range forecasts provide decision makers only with little time to act. The earlier an issue is detected, the more can be done to proactively solve it [2]. For example, when solving the execution problem, let us assume that a group of instances of a process is forecasted to be delayed. Detecting this within a short-range forecast, i.e., when the instances are already running, enables decision makers to shift resources working on other processes to the forecasted process. This might eliminate the delay of the forecasted process, but in turn lead to an increased cycle time of the other processes running simultaneously. In contrast, a medium-range forecast might have provided decision-makers with enough time to make resources available without affecting the performance of other processes. When solving the design problem (e.g., implementing a new process), only long-range forecasts may be expedient, as strategical decisions of this sort generally are subject to long lead-times. Further, regarding forecasts that rely on historical data, the decision on a time horizon should be taken in consideration of the time period for which historical data is available. It can be expected that a short observation period leads to less accurate long-range forecasts than a multi-annual one [12].

Step 4: Select a Process Forecasting Method. In the next step, the method for process forecasting needs to be selected. Depending on the characteristics of the problem at hand (e.g., objective of the forecast, type of dependent process characteristic, and data availability), the applicability and suitability of distinct forecasting methods should be evaluated. The chosen forecasting method then determines the set of potential input variables and the way they are processed [12]. As can be inferred from the diversity of methods proposed in the relevant literature, process forecasting is not limited to specific methods, i.e., statistical and judgmental forecasts as well as a combination can be applied equally well. Whereas statistical methods make use of historical data and, thus, underlie the assumption that observed dependencies will continue in the future, judgmental methods such as panel approaches or Delphi studies are based upon opinions of experts [12]. Expert judgments are particularly helpful when historical data is unavailable or unable to "explain" the future properly. This may, for instance, be the case when predicting sales for an entire new product or forecasting radical changes of process models (cf. Type 3b, Sect. 3). Further approaches such as planning algorithms (see e.g., [19]) or cognitive computing (see e.g., [26]) are also conceivable for certain forecasting problems. The methods most commonly used in the literature related to process forecasting – especially in the field of predictive process monitoring – are based on machine learning (increasingly deep learning), constraint satisfaction, and quality-of-service aggregation [27]. Thus, the majority of existing works focuses on statistical forecasts, aiming to learn from past as well as present dependencies between process characteristics and to transpose this knowledge into the future (see e.g., [16, 24]).

Step 5: Obtain the Input Needed for the Selected Process Forecasting Method. After having selected a method, in the next step, the required input needs to be collected. In most cases, forecasts are based on large amounts of data. Against the backdrop of the recent uptake of new methods such as deep learning, analysts are enabled to draw on structured as well as unstructured data (e.g., in the form of images, voice, and videos). Further, the rising availability of micro-grained data about historical and ongoing business process executions, particularly in the form of process logs, pushes the boundaries of data that can be taken into account for a process forecasting task. As pictured in Fig. 2, both *process characteristics* and *contextual variables* should be collected and used as input for a process forecast. The exploitation of the relationship between historical values of the dependent process characteristic and all other *process characteristics* is motivated in [24]. As an example, de Leoni et al. [24] mention the possible dependency between involved resources and customer satisfaction, i.e., certain resources involved in a process execution may lead to a lower customer satisfaction. Besides historical and run-time values, target values of *process characteristics* in the form of future process requirements (e.g., service level agreements that have to be met) may also be necessary to forecast certain prediction targets.

Further, the literature suggests taking *contextual variables* into account. Such data is agreed to have a high explanatory power on process behavior [10]. For example in [24], the consideration of the context variable *weather* as input to forecast the process characteristic *activity duration* is emphasized, assuming that certain resources, which in turn have an impact on the activity duration, are more efficient when the weather is good. Here, besides considering past relationships between context and process executions, it is also conceivable to take future context data (e.g., the weather forecast for next week) into account. Linking this information with the dependency patterns learned in the past may increase the forecast's accuracy. Further, depending on the type of the process forecast, certain data may be mandatory. For instance, the configuration problem can only be solved, if the set of process model variants as well as their extrinsic trigger points are available.

Step 6: Apply the Selected Process Forecasting Method. Having implemented all the previous steps, as a final step, the process forecasting method is applied.

6 Results of the Structured Literature Review

Below, we describe the results of the structured literature review. These built the theoretical backbone of the process forecasting definition and reference process. Following the research approach described in Sect. 2, we classified the methods found in the literature with respect to their input and output. Figure 4 shows the components (dashed borderline) of the proposed definition of process forecasting. These components are divided into sub-components (full borderline) based on their conceptualizations described in Sect. 4. The numbers in brackets indicate the number of methods using the (sub-)components, with the shadings of gray indicating whether a (sub-) component is used in many (dark gray) or few (light gray) methods.[1]

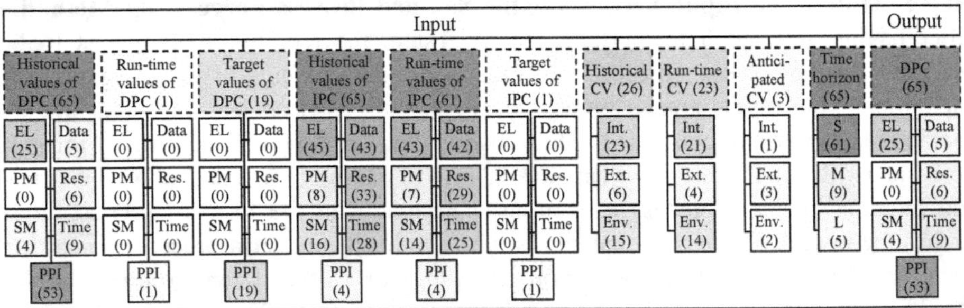

Input										Output
Historical values of DPC (65)	Run-time values of DPC (1)	Target values of DPC (19)	Historical values of IPC (65)	Run-time values of IPC (61)	Target values of IPC (1)	Historical CV (26)	Run-time CV (23)	Antici-pated CV (3)	Time horizon (65)	DPC (65)
EL (25) / Data (5)	EL (0) / Data (0)	EL (0) / Data (0)	EL (45) / Data (43)	EL (43) / Data (42)	EL (0) / Data (0)	Int. (23)	Int. (21)	Int. (1)	S (61)	EL (25) / Data (5)
PM (0) / Res. (6)	PM (0) / Res. (0)	PM (0) / Res. (0)	PM (8) / Res. (33)	PM (7) / Res. (29)	PM (0) / Res. (0)	Ext. (6)	Ext. (4)	Ext. (3)	M (9)	PM (0) / Res. (6)
SM (4) / Time (9)	SM (0) / Time (0)	SM (0) / Time (0)	SM (16) / Time (28)	SM (14) / Time (25)	SM (0) / Time (0)	Env. (15)	Env. (14)	Env. (2)	L (5)	SM (4) / Time (9)
PPI (53)	PPI (1)	PPI (19)	PPI (4)	PPI (4)	PPI (1)					PPI (53)

Abbreviations: dependent process characteristic(s) (DPC), independent process characteristic(s) (IPC), contextual variables (CV), event log (EL), process model (PM), simulation model (SM), resource (Res.), process performance indicator(s) (PPI), internal context (Int.), external context (Ext.), environmental context (Env.), short-range (S), medium-range (M), long-range (L).

Fig. 4. Heat map of the (sub-)components of process forecasting.

---

[1] The individual classification of the methods and further details on the results of the literature review are available at researchgate.net/publication/323691573_Process_Forecasting.

The results show that all of the framework's components are addressed at least once in the literature, whereby none of the existing techniques exploits all components. Regarding the sub-components, the analysis reveals that forecasting a process model has not been addressed and run-time values of the dependent process characteristic(s) as well as target values of independent process characteristic(s) have only been considered by one existing technique each. This is not surprising, as taking run-time values of the dependent process characteristic(s) into account becomes relevant for medium- and long-range time horizons. The majority of the identified and analyzed techniques, however, focus on a short-range time horizon, i.e., on the prediction of running instances. The results disclose that most of the proposed techniques are data-driven. Among the data-driven forecasting methods, a large number uses historical and run-time values from event logs, data attributes assigned to activities or processes, and resources involved in a process execution as independent process characteristics. Concerning the selection of the dependent process characteristic(s), PPIs, particularly the cycle time, have received most attention, whereas the prediction of activity-specific attributes (e.g., involved resources) is fragmentarily addressed. More than a third of the analyzed approaches exploit historical and run-time contextual data to potentially lever the accuracy of the forecasts, among which the lion's share considers the internal context. In contrast, anticipated contextual data has been used rarely.

Most existing techniques focus on the prediction of PPIs. However, the literature emphasizes that the explanatory power of forecasts can be increased by selecting leading instead of lagging indicators as prediction targets. This insight was also integrated into the proposed reference process (cf. Step 2, Sect. 5). In the analyzed literature, numerous causalities between process characteristics are mentioned. Figure 5 schematizes the knowledge scattered across the literature by means of an acyclic directed graph.

The graph illustrates which process characteristics are addressed (this involves either using the process characteristic as a prediction target or highlighting a process characteristic as leading indicator for another process characteristic) and which causal relationships are proposed. In the figure, nodes are process characteristics and edges describe the cause-effect relationships between process characteristics (e.g., "A → B" describes "A contributes to changes in B."). In addition, numbers in the labels of nodes and edges indicate the number of papers selecting a certain process characteristic as prediction target and confirming a certain cause-effect relationship. As in Fig. 4, the shadings of nodes indicate whether a process characteristic is frequently (dark gray) or rarely forecasted (light gray). A number of zero in the label of a node indicates that this process characteristic was not explicitly forecasted in the analyzed techniques. However, it is included in the figure, as it was mentioned in at least one cause-effect relationship. Given a forecasting objective, the graph can guide an analyst through the selection of the dependent process characteristic (cf. Step 2, Sect. 5) or the identification of influential independent process characteristics needed to make an accurate forecast (cf. Step 5, Sect. 5). The further analysts get down the causality chain, the closer they will get to the leading indicators of future process behavior. For instance, let an organization be interested in the cycle time of future process executions. Separately predicting the processing and waiting times of future process executions may further localize the problem. An even higher explanatory power of the forecast can be achieved

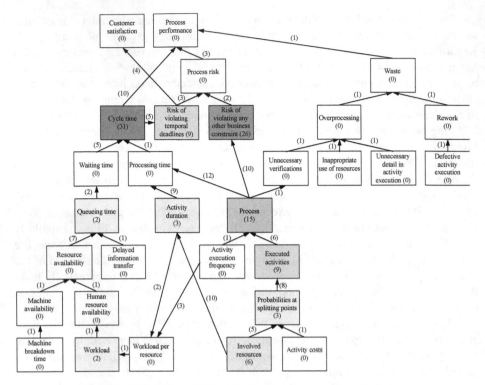

Fig. 5. Cause-effect relationships between process characteristics.

by predicting process characteristics located at the end of a causality chain, e.g., the breakdown time of a machine, as this may reveal that the long waiting times result from the poor condition of a machine involved in future process executions. This can enable the company to proactively take targeted actions and eliminate issues before they occur.

7   Related Work

Due to its interdisciplinary nature, process forecasting is linked to many disciplines beyond BPM (e.g., operations management, demand forecasts, pervasive computing). To position process forecasting as BPM sub-area, we compare it with other BPM sub-areas, namely process flexibility, declarative process modeling, emergent workflow, business process intelligence, process mining, predictive process monitoring, process simulation, and process planning. The positioning is shown in Fig. 6. For the sake of transparency, we deliberately abstract from overlaps between other sub-areas.

Fig. 6. Positioning of process forecasting against BPM-related sub-areas.

Process flexibility enables adapting processes to internal or external triggers without completely replacing them [18]. Flexibility approaches usually do not take a forecasting perspective. Rather, they are concerned with how to quickly react to changes. However, linking extant flexibility types [1] to our process forecasting types reveals that process flexibility is an enabler for process forecasting, because organizations can only benefit from process forecasting, if processes can be adapted flexibly. For instance, the strategy *flexibility by deviation*, i.e., allowing for short-term deviations, enables solving the execution problem. *Flexibility by design*, based on many model variants, builds the foundation for solving the configuration problem. *Flexibility by change* supports solving the design problem in the short-term, as it enables changing process models at run-time. The fourth strategy, *flexibility by under-specification*, allows for the formulation of incomplete process models at design time and the addition of process model fragments at run-time. As such, it helps solve the execution and design problem.

Next, *declarative process modeling* implements flexibility by design and under-specification in a non-procedural way. Thus, it also serves as a tool for realizing process forecasting and enables solving the execution, configuration, and design problem. The basic idea is to model processes via constraints that must be satisfied by every process instance, instead of rigorously defining the control flow [1]. Consequently, more options to proactively manage process-related issues and opportunities are created.

Similarly, *emergent workflow* follows the idea to design or adapt process models at run-time, if needed [28]. This creates the ability to flexibly react to changes or issues that are anticipated by means of process forecasting.

Business process intelligence (BPI) refers to the application of techniques that support the collection, analysis, and presentation of business process information with the aim to enable better decision-making [29]. In contrast to process forecasting, however, most BPI techniques target retrospective analyses, i.e., process instances are analyzed after their termination providing insights into how processes are executed. Among these techniques, an application area that receives a lot of attention is process mining.

Process mining aims to extract knowledge from process logs [30]. As such, it sets the basis for process forecasting methods that rely on process data. As opposed to its predominant focus on understanding past process behavior, in recent years, proactive process mining approaches emerged that support real-time and future-oriented analyses. The majority deals with the performance prediction of running instances and, thus, relates to the execution problem of process forecasting (cf. Type 1b, Sect. 3). This research field is commonly referred to as predictive process monitoring.

Predictive process monitoring is concerned with predicting how running process instances will unfold up to their completion [16]. In essence, existing approaches use past process execution data, partial traces of the monitored process instances, and partly also contextual data to predict the remaining duration (see e.g., [15]), process-related risks, i.e., the likelihood or severity of a process fault [9]), the outcome (e.g., whether or not a running instance will violate a compliance rule [8]), or the future path of a process instance [16]. With the aim to combine the approaches for different but related prediction tasks, a general framework to predict process characteristics of running instances was proposed in [24]. Differently from the above-mentioned works, process forecasting goes beyond the prediction of individual running instances. Predicting process characteristics of a group of running and/or future instances arriving in a distinct time interval is hardly addressed in literature [14].

Further, *process simulation* evaluates the impact of design decision on business processes prior to implementation [31]. Thus, process simulation is a part of process forecasting, as it is concerned with understanding dependencies between process characteristics. Once understood, the discovered dependencies help derive process forecasts.

Finally, *process planning* deals with the automated construction of new process models. It facilitates the design of process models that comply with future process demands [19]. Consequently, process planning is linked to solving the design problem of process forecasting, as it tackles the same issue, i.e., to reduce process latency by enabling a timely preparation of future process models. Thus, process planning can be seen as a specific instantiation of a process forecasting method that uses future process demands as input, namely in the form of target values of independent process characteristics, and forecasts the design of a process model that is conform with these demands.

As summarized above, there are several heterogeneous approaches pursuing the same overarching objective, namely to proactively manage process-related issues and opportunities. In particular, the methods focusing on the prediction of process characteristics differ in terms of the time horizon of the prediction, the input variables taken into account, the process characteristics to be forecasted, their level of aggregation, and the techniques to be used. Process forecasting, as an umbrella concept, consolidates and extends these existing future-oriented BPM methods and reveals that there is a considerable need for future research.

8 Conclusion

In this paper, we proposed the concept of process forecasting. With the aim to stimulate a community-wide discussion about the uptake of proactive BPM, we presented various use cases that motivate the need for process forecasting. We also provided a definition and a reference process as well as a positioning against related BPM sub-areas. Process forecasting is an umbrella concept for future-oriented BPM methods and techniques that tackle process-related issues before they occur, proactively seize process-related opportunities, and reduce process latency. The proposed definition and the reference process were built on the results of a structured literature review.

This paper revealed that the range of existing future-oriented BPM methods and techniques is not sufficient. Whereas the first type of process forecasting, i.e., solving the execution problem, has been addressed by research fields such as predictive process monitoring, forecasting techniques that solve the configuration or design problem hardly exist. This strengthens the need for future research. Further, the limitations of this paper also stimulate future research. As research on process forecasting is still in its early stages, our rationale for this paper was to create a first overall understanding of the need for and scope of process forecasting. With a focus on the interdisciplinary nature of the topic, we acknowledge that the scope of the conducted literature review should be broadened. In future work, we intend to enrich process forecasting by exploiting extant methods from other disciplines. Another avenue for future research is to provide methodological guidance for choosing a suitable forecasting technique within a process forecasting initiative. In addition, we plan to conduct case studies validating the applicability of our concept, to set up a research agenda for process forecasting, and to devise a proactive BPM life cycle. However, we trust that this paper is a solid starting point for discussing and exploring the under-researched potentials of proactive BPM.

References

1. Schonenberg, H., Mans, R., Russell, N., Mulyar, N., van der Aalst, W.: Process flexibility: a survey of contemporary approaches. In: Dietz, J.L.G., Albani, A., Barjis, J. (eds.) CIAO!/ EOMAS -2008. LNBIP, vol. 10, pp. 16–30. Springer, Heidelberg (2008). https://doi.org/10. 1007/978-3-540-68644-6_2
2. Krumeich, J., Werth, D., Loos, P.: Enhancing organizational performance through event-based process predictions. In: Proceedings of the 21st Americas Conference on Information Systems (2015)
3. Verenich, I., Dumas, M., La Rosa, M., Maggi, F.M., Di Francescomarino, C.: Minimizing overprocessing waste in business processes via predictive activity ordering. In: Nurcan, S., Soffer, P., Bajec, M., Eder, J. (eds.) CAiSE 2016. LNCS, vol. 9694, pp. 186–202. Springer, Cham (2016). https://doi.org/10.1007/978-3-319-39696-5_12
4. Leveling, J., Edelbrock, M., Otto, B.: Big data analytics for supply chain management. In: IEEE International Conference on Industrial Engineering and Engineering Management (2014)
5. Selcuk, S.: Predictive maintenance, its implementation and latest trends. J. Eng. Manuf. **231**, 1670–1679 (2017)
6. van der Aalst, W.M.P.: Business process management: a comprehensive survey. ISRN Softw. Eng. **2013**, Article ID 507984, 1–37 (2013). https://doi.org/10.1155/2013/507984
7. Vom Brocke, J., Simons, A., Niehaves, B., Reimer, K., Cleven, A.: Reconstructing the giant: on the importance of rigour in documenting the literature search process. In: ECIS, pp. 2206–2217 (2009)
8. Maggi, F.M., Di Francescomarino, C., Dumas, M., Ghidini, C.: Predictive monitoring of business processes. In: Jarke, M., et al. (eds.) CAiSE 2014. LNCS, vol. 8484, pp. 457–472. Springer, Cham (2014). https://doi.org/10.1007/978-3-319-07881-6_31
9. Conforti, R., De Leoni, M., La Rosa, M., Van Der Aalst, W.M.P., Ter Hofstede, A.H.M.: A recommendation system for predicting risks across multiple business process instances. DSS **69**, 1–19 (2015)
10. Folino, F., Guarascio, M., Pontieri, L.: Discovering context-aware models for predicting business process performances. In: Meersman, R., et al. (eds.) OTM 2012. LNCS, vol. 7565, pp. 287–304. Springer, Heidelberg (2012). https://doi.org/10.1007/978-3-642-33606-5_18
11. Clarivate Analytics: Web of Science Databases. https://clarivate.com/products/web-of-science/databases/. Accessed 17 Mar 2018
12. Heizer, J., Render, B., Munson, C.: Principles of Operations Management: Sustainability and Supply Chain Management. Pearson Education, London (2016)
13. Reid, R.D., Sanders, N.R.: Operations Management: An Integrated Approach. Wiley, Hoboken (2010)

14. Folino, F., Guarascio, M., Pontieri, L.: A prediction framework for proactively monitoring aggregate process-performance indicators. In: IEEE International Enterprise Distributed Object Computing Conference, pp. 128–133 (2015)

15. Rogge-Solti, A., Weske, M.: Prediction of business process durations using non-Markovian stochastic Petri nets. Inf. Syst. **54**, 1–14 (2015)

16. Di Francescomarino, C., Ghidini, C., Maggi, F.M., Petrucci, G., Yeshchenko, A.: An eye into the future: leveraging a-priori knowledge in predictive business process monitoring. In: Carmona, J., Engels, G., Kumar, A. (eds.) BPM 2017. LNCS, vol. 10445, pp. 252–268. Springer, Cham (2017). https://doi.org/10.1007/978-3-319-65000-5_15

17. Metzger, A., Franklin, R., Engel, Y.: Predictive monitoring of heterogeneous service-oriented business networks: the transport and logistics case. In: SRII Global Conference (2012)

18. Rosemann, M., Recker, J.C.: Context-aware process design: exploring the extrinsic drivers for process flexibility. In: CAiSE 2006, pp. 149–158 (2006)

19. Heinrich, B., Klier, M., Zimmermann, S.: Automated planning of process models: design of a novel approach to construct exclusive choices. DSS **78**, 1–14 (2015)

20. Childe, S.J., Maull, R.S., Bennett, J.: Frameworks for understanding business process re-engineering. Int. J. Oper. Prod. Manag. **14**, 22–34 (1994)

21. Sousa, R., Amorim, M., Rabinovich, E., Sodero, A.C.: Customer use of virtual channels in multichannel services: does type of activity matter? Decis. Sci. **46**, 623–657 (2015)

22. del-Río-Ortega, A., Resinas, M., Ruiz-Cortés, A.: Defining process performance indicators: an ontological approach. In: Meersman, R., Dillon, T., Herrero, P. (eds.) OTM 2010. LNCS, vol. 6426, pp. 555–572. Springer, Heidelberg (2010). https://doi.org/10.1007/978-3-642-16934-2_41

23. Rosemann, M., Recker, J.C., Flender, C.: Contextualization of business processes. Int. J. Bus. Process Integr. Manag. **3**, 47–60 (2008)

24. De Leoni, M., Van Der Aalst, W.M.P., Dees, M.: A general process mining framework for correlating, predicting and clustering dynamic behavior based on event logs. Inf. Syst. **56**, 235–257 (2016)

25. Polyvyanyy, A., Ouyang, C., Barros, A., van der Aalst, W.M.P.: Process querying: enabling business intelligence through query-based process analytics. DSS **100**, 41–56 (2017)

26. Roeglinger, M., Seyfried, J., Stelzl, S., Muehlen, M.: Cognitive computing: what's in for business process management? An exploration of use case ideas. In: Teniente, E., Weidlich, M. (eds.) BPM 2017. LNBIP, vol. 308, pp. 419–428. Springer, Cham (2018). https://doi.org/10.1007/978-3-319-74030-0_32

27. Metzger, A., et al.: Comparing and combining predictive business process monitoring techniques. IEEE Trans. Syst. Man Cybern. Syst. **45**, 276–290 (2015)

28. Jorgensen, H., Carlsen, S.: Emergent workflow: planning and performance of process instances. In: Workflow Management (1999)

29. Castellanos, M., Alves De Medeiros, A.K., Mendling, J., Weber, B., Weijters, A.J.M.M.: Business process intelligence. In: Handbook of Research on Business Process Modeling, pp. 456–480 (2009)

30. van der Aalst, W.M.P.: Process Mining: Discovery. Conformance and Enhancement of Business Processes. Springer, Heidelberg (2011). https://doi.org/10.1007/978-3-642-19345-3

31. Jansen-Vullers, M., Netjes, M.: Business process simulation - a tool survey. In: Workshop and Tutorial on Practical Use of Coloured Petri Nets and the CPN Tools (2006)

Correction to: Using Business Process Compliance Approaches for Compliance Management with Regard to Digitization: Evidence from a Systematic Literature Review

Stefan Sackmann, Stephan Kuehnel, and Tobias Seyffarth

Correction to:
Chapter "Using Business Process Compliance Approaches for Compliance Management with Regard to Digitization: Evidence from a Systematic Literature Review"
in: M. Weske et al. (Eds.): *Business Process Management*,
LNCS 11080, https://doi.org/10.1007/978-3-319-98648-7_24

In the originally published version of chapter 24, the total values of table 2 and the text on page 417 were initially published with errors. This has been corrected and the supplementary material has been updated.

The updated version of this chapter can be found at
https://doi.org/10.1007/978-3-319-98648-7_24

Author Index

Armas-Cervantes, Abel   158, 250
Augusto, Adriano   158

Bandara, Wasana   376
Bloemen, Vincent   233
Brocke, Jan vom   3
Burattin, Andrea   250, 322

Carmona, Josep   215, 250
Cecconi, Alessio   121
Combi, Carlo   102
Conforti, Raffaele   158
Corradini, Flavio   83

De Giacomo, Giuseppe   121
De Koninck, Pieter   305
De Weerdt, Jochen   305
Debois, Søren   31
Denisov, Vadim   139
Denner, Marie-Sophie   480
Di Ciccio, Claudio   121
Di Francescomarino, Chiara   462
Dumas, Marlon   158, 176

Ekanayake, Chathura   12
Eshuis, Rik   66

Fahland, Dirk   139, 322
Fischer, Marcus   392
French, Erica   376

Ghidini, Chiara   462

Hildebrandt, Thomas   31

Imgrund, Florian   392

Janiesch, Christian   392

Krogstie, John   426
Kuehnel, Stephan   409

La Rosa, Marcello   158
Leno, Volodymyr   176
Leymann, Frank   12
Lux, Marian   445

Maggi, Fabrizio Maria   176, 462
Marengo, Elisa   48
Mendling, Jan   121, 322
Mikalef, Patrick   426
Milani, Fredrik   462
Montali, Marco   3
Mühlhäuser, Max   271, 288
Muzi, Chiara   83

Nolle, Timo   271, 288
Nutt, Werner   48

Ouyang, Chun   339

Pan, Maolin   339
Perera, Srinath   12
Perktold, Matthias   48
Poll, Rouven   496
Polyvyanyy, Artem   496
Posenato, Roberto   102
Preda, Andrei   445

Re, Barbara   83
Reijers, Hajo A.   322
Reissner, Daniel   158
Rinderle-Ma, Stefanie   445
Röglinger, Maximilian   480, 496
Rosemann, Michael   496
Rossi, Lorenzo   83
Rupprecht, Lea   496

Sackmann, Stefan   409
Schmiedel, Theresa   480
Seeliger, Alexander   271, 288
Seyffarth, Tobias   409
Slaats, Tijs   31
Soffer, Pnina   322

Stelzl, Katharina   480
Syed, Rehan   376

Taymouri, Farbod   215
ter Hofstede, Arthur H. M.   339
Tiezzi, Francesco   83

van de Pol, Jaco   233
van der Aalst, Wil M. P.   139, 233
van Dongen, Boudewijn F.   197, 233, 250
Van Looy, Amy   359
van Zelst, Sebastiaan J.   233, 250
vanden Broucke, Seppe   305
Vanderfeesten, Irene   322

Weber, Barbara   322
Weber, Ingo   3
Weerawarana, Sanjiva   12
Wehking, Charlotte   480
Weidlich, Matthias   322
Weske, Mathias   3
Winkelmann, Axel   392

Yang, Jing   339
Yu, Yang   339

Zerbato, Francesca   102

Printed in the United States
By Bookmasters